A Comprehensive Study of Ecosystems

A Comprehensive Study of Ecosystems

Edited by Molly Ismay

SYRAWOOD
PUBLISHING HOUSE
New York

Published by Syrawood Publishing House,
750 Third Avenue, 9th Floor,
New York, NY 10017, USA
www.syrawoodpublishinghouse.com

A Comprehensive Study of Ecosystems
Edited by Molly Ismay

International Standard Book Number: 978-1-68286-614-6 (Hardback)

Cataloging-in-Publication Data

A comprehensive study of ecosystems / edited by Molly Ismay.
   p. cm.
Includes bibliographical references and index.
ISBN 978-1-68286-614-6
1. Biotic communities. 2. Ecology. 3. Ecosystem management. I. Ismay, Molly.
QH541 .C66 2018
577.82--dc23

TABLE OF CONTENTS

PREFACE

I am honored to present to you this unique book which encompasses the most up-to-date data in the field. I was extremely pleased to get this opportunity of editing the work of experts from across the globe. I have also written papers in this field and researched the various aspects revolving around the progress of the discipline. I have tried to unify my knowledge along with that of stalwarts from every corner of the world, to produce a text which not only benefits the readers but also facilitates the growth of the field.

A group of living organisms along with the biotic and abiotic components of their environment or habitat forms an ecosystem. Biodiversity of flora and fauna affect the processes of an ecosystem. Some external and internal factors such as climate, nutrient recycling, etc. also influence an ecosystem. This book covers in detail some existent theories and innovative concepts revolving around ecosystems. It includes contribution of experts and scientists which will provide innovative insights into this field.

Finally, I would like to thank all the contributing authors for their valuable time and contributions. This book would not have been possible without their efforts. I would also like to thank my friends and family for their constant support.

Editor

Functional Structure of Biological Communities Predicts Ecosystem Multifunctionality

David Mouillot[1]*, Sébastien Villéger[2], Michael Scherer-Lorenzen[3], Norman W. H. Mason[4]

1 Laboratoire Ecologie des Systèmes Marins Côtiers UMR 5119, Université Montpellier 2, Montpellier, France, 2 Laboratoire Evolution et Diversité Biologique UMR 5174, Université Paul Sabatier, Toulouse, France, 3 Faculty of Biology - Geobotany, University of Freiburg, Freiburg, Germany, 4 Landcare Research, Hamilton, New Zealand

Abstract

The accelerating rate of change in biodiversity patterns, mediated by ever increasing human pressures and global warming, demands a better understanding of the relationship between the structure of biological communities and ecosystem functioning (BEF). Recent investigations suggest that the functional structure of communities, i.e. the composition and diversity of functional traits, is the main driver of ecological processes. However, the predictive power of BEF research is still low, the integration of all components of functional community structure as predictors is still lacking, and the multifunctionality of ecosystems (i.e. rates of multiple processes) must be considered. Here, using a multiple-processes framework from grassland biodiversity experiments, we show that functional identity of species and functional divergence among species, rather than species diversity *per se*, together promote the level of ecosystem multifunctionality with a predictive power of 80%. Our results suggest that primary productivity and decomposition rates, two key ecosystem processes upon which the global carbon cycle depends, are primarily sustained by specialist species, i.e. those that hold specialized combinations of traits and perform particular functions. Contrary to studies focusing on single ecosystem functions and considering species richness as the sole measure of biodiversity, we found a linear and non-saturating effect of the functional structure of communities on ecosystem multifunctionality. Thus, sustaining multiple ecological processes would require focusing on trait dominance and on the degree of community specialization, even in species-rich assemblages.

Editor: Tamara Romanuk, Dalhousie University, Canada

Funding: The BIODEPTH project ran from 1996 to 1998 and was funded by the European Commission within the Framework IV Environment and Climate programme (ENV-CT95-0008). The funders had no role in study design, data collection and analysis, decision to publish, or preparation of the manuscript.

Competing Interests: The authors have declared that no competing interests exist.

* E-mail: david.mouillot@univ-montp2.fr

Introduction

Ecosystems are facing ever increasing levels of human pressures which imperil the goods and services they provide to humanity. It is now recognized that both changes in environmental conditions (e.g., global warming) and modifications to biological communities (e.g., biodiversity erosion) affect ecosystem processes [1,2,3], the latter issue having stimulated convincing advances but also controversy [4]. During the last two decades the positive relationship between biodiversity and ecosystem functioning (BEF hereafter) has been demonstrated through experiments manipulating species composition in model assemblages [2,4,5]. These studies helped to place the problems of environmental change and biodiversity loss into the mainstream political agenda [6]. However, there is an urgent need to move beyond the heuristic objective of early biodiversity experiments, and then to disentangle the contributions of the various components of biodiversity on ecosystem processes and, ultimately, to build a predictive framework for BEF research that can forecast the potential effects of biodiversity changes that all ecosystems on earth are experiencing.

To reach this objective, at least two limitations remain. First, biodiversity has been recognized as a multidimensional concept [7,8] but many BEF studies rely solely on species richness for practical reasons and remain silent on the functional structure of communities. Yet, the functional trait composition of biological communities is a key component that most often explains ecosystem functioning better than species richness *per se* whatever the biota [2,9], a functional trait being any morphological, physiological or phenological feature, measurable at the individual level, that determines species effects on ecosystem properties [10]. The limited predictive power of BEF research, even if biodiversity effects were demonstrated to be positive and significant [11,12,13], is certainly due to the clear initial focus on testing diversity effects (mostly on the species richness level) irrespective of other compositional factors, such as species or functional identity, and the resultant lack of an integrative framework where different components of biodiversity were considered altogether as predictor variables. Second, the vast majority of BEF studies have focused on a single ecosystem process (e.g. productivity) while overall ecosystem functioning is sustained by several processes [14]. Recent results suggest that the effect of biodiversity in natural ecosystems may be much larger than currently thought if we consider a multiple-processes framework [15,16,17].

Taxonomic diversity, functional identity and functional diversity of ecological communities are each known to influence ecosystem processes but their relative effects remain largely untested, particularly in predicting rates of multiple ecosystem processes [18,19,20]. Species richness was the first biodiversity component to be related to ecosystem functioning [21] supporting the hypothesis that species complementarity and sampling effects both enhance resource use and productivity. Then, species evenness, or how equitably abundance is distributed among species within a community, was demonstrated to positively influence productivity [22]. The functionally orientated BEF research began as early as 1941 [23] with the study of the effect of particular species functional traits (functional identity) on ecosystem processes (soil formation). Then, other authors have pointed out that ecosystem properties should primarily depend on the identity of dominant species and their functional traits following the 'mass ratio hypothesis' [24,25]. Indeed, functional identity, usually expressed as the biomass-weighted mean trait value for a community, has been demonstrated to be a key driver of ecosystem functioning from local [20] to regional [26] and global [27] scales. Beyond functional identity, functional diversity, defined as the diversity and abundance distribution of traits within a community [28], has been shown to be an accurate predictor of ecosystem functioning [29,30,31], reinforcing the importance of niche complementarity for enhancing ecosystem processes [32].

All these biodiversity components are not mutually exclusive but are unlikely to exert equal influence on ecosystem processes and on the multifunctionality of ecosystems. Thus the question is no longer whether each of the three components of biodiversity (taxonomic diversity, functional identity and functional diversity) matters but whether it still matters after removing the effect of two other components? In other words, we examined the additional effect of each biodiversity component on the prediction of ecosystem processes to determine whether each component has an essential and complementary contribution to the explanation of ecosystem multifunctionality. Further, by including the eight biodiversity indices, embracing all aspects of taxonomic and functional structure of communities, we built a minimum adequate model that reached an unprecedented level of explanatory power with functional identity and functional diversity together as predictor variables of multiple ecosystem processes. Finally, we implemented structural equation models to explore both causal and spurious associations between predictors of ecosystem processes.

We used data on several ecosystem processes including biomass production and decomposition trials within the German BIO-DEPTH experiment (BIODiversity and Ecological Processes in Terrestrial Herbaceous Ecosystems) to predict the effects of biodiversity change on ecosystem functioning. This experiment allows testing all components of biodiversity given that, for each species richness level, different species combinations were constructed.

Materials and Methods

Experiment

Data have been collected at the German site of the pan-European BIODEPTH project [33]. A gradient of plant species richness and number of functional groups (grasses, legumes, non-leguminous herbs) was created by sowing mixtures containing 1, 2, 4, 8 or 16 species, typically found in mid-European hay meadows (total species pool was 31 species). Total seed density was 2,000 viable seeds per m^2, equally divided among all species following a substitutive replacement series design. Each diversity level was replicated with several mixtures differing in species composition. The whole experiment was replicated in a second block with new randomization of plots, yielding a total of 60 plots of 2×2 m in size. Unsown seedlings were continuously weeded, and the plots were not fertilized. Among the 60 plots, we retained only 26 since we had to select only those with species richness ranging from 4 to 16 species to be able to estimate indices of functional community structure.

Indeed, there is no functional volume (functional richness index) with 1 or 2 species. As an alternative, we should use the Functional Diversity (FD) index [34], based on a dendrogram, to cope with species poor communities (less than 3 species) and thus use the entire range of species richness available in BIODEPTH. However, the building of functional dendrograms is contentious [35] and we cannot estimate the other functional diversity components (those including species abundances) with this approach.

Ecosystem processes

Among the ecological processes that were measured at the German BIODEPTH site we selected those that were relatively independent (mean correlation over all selected processes was 0.5) since two highly correlated processes would be trivially ruled by the same biodiversity components. The response variables were cotton decomposition in 1997 and 1998, litter decomposition in 1998, productivity in 1997 and 1998, and nitrogen pool size in aboveground biomass 1998. Cotton decomposition trials are a standard method to test for effects of microenvironmental conditions on decomposition processes. It was measured as dry weight loss $(g.g^{-1}.d^{-1})$ of a standard cotton fabric using strips of 5×12 cm (Shirley Soil Burial Test Fabric, c. 95% cellulose; initial nitrogen concentration of 0.09%) during 10 weeks of field exposure in all experimental communities, with three strips per plot [31,36]. Litter decomposition was the dry weight loss $(g.g^{-1}.d^{-1})$ of plot-specific senescent leaf and stem material, sealed in litter bags of 5×5 cm made of a 0.5 mm nylon mesh, during 10 weeks in autumn 1998.

Macrofauna was excluded with this mesh size. Assuming an equal effect of a small mesh size in all treatments, excluding one decomposer group should not have an effect on our results. This assumption might not be true in case of a diversity effect on macrofauna occurrence. In another experiment, carried out on the same plots and half a year later, we could show, however, that several indices of soil fauna, including different groups of earthworms (litter feeding epigeics and anecics) and nematodes, were not influenced by our plant diversity treatments [37].

The litter bags were placed in a homogeneous patch of an adjacent meadow, thus quantifying the effect of community-specific litter composition and quality on decomposition processes, independent of differences in microenvironmental conditions induced by the experimental communities. Thus, both decomposition trials independently quantified different pathways of potential diversity effects on decomposition processes [31]. Productivity was the sum of two harvests per year (June and September) in 1997 and 1998, as a proxy for annual biomass production (dry weight, $g.m^{-2}$). Standing biomass was cut at a height of 5 cm in two areas of $0.5\ m\times 0.2$ m each within a permanent quadrat placed in the center of each plot [33,36]. Nitrogen pool size in aboveground biomass 1998 was the nitrogen content in dried aboveground biomass of the year 1998, calculated as the product of N concentration and biomass $(gN.m^{-2})$. Nitrogen was measured by dry combustion with an automated C/N analyzer (Carlo Erba NA 1500, Carlo Erba, Mailand, Italy) [33,36]. We also calculated a multifunctionality variable as the mean performance of communities over the four processes after standardizing each community performance (mean of 0 and standard deviation of 1) in order to give them the same weight. When the same process was measured for two years we first calculated a mean value for this process over the two years.

Functional traits

The selected traits were growth form: caespitosa, reptantia, scandentia, semirosulata and rosulata; leaf size: nanophyllous (20–

200 mm^2), microphyllous (2–6 cm^2), submicrophyllous (6–20 cm^2) and mesophyllous (20–100 cm^2); leaf seasonality: summergreen, partly evergreen and evergreen, CN ratio of plant litter [31]; SLA based on measurements in another biodiversity experiment [38]; and leaf angle: predominantly vertical leaf orientation, predominantly inclined leaf orientation and predominantly horizontal leaf orientation [38]. See Table S1 for details by species.

Indices for community structure

We considered two independent variables related to taxonomic composition: species richness and the evenness of abundance distribution among species using the Pielou index [39]. Since we have both quantitative and qualitative traits, we performed a Principal Coordinate Analysis (PCoA) on a Gower distance matrix to provide three independent axes that summarize species distribution within a trait functional space [39]. The functional structure of each community was assessed within this 3-dimensional PCoA space which represents more than 90% of the total inertia. These three independent functional axes from PCoA were used to measure functional identity through biomass-weighted mean trait values for each community.

Three independent variables were related to functional diversity [40] (Figure 1). Functional richness was measured as the amount of functional space filled by the community which is the volume inside the convex hull that contains all trait combinations represented in the community, which basically corresponds to a multivariate functional range [40,41]. Functional evenness was estimated as the regularity of abundance distribution in the multidimensional functional space, i.e. the regularity with which species abundances fill the functional space. Finally, functional divergence quantified whether higher abundances are close to the volume borders, i.e. whether specialist species *sensu* Elton [42] have the highest abundances. See Text S1 for details on functional diversity indices.

Statistical analyses

In order to disentangle the relative effect of each biodiversity component on ecosystem processes, several alternative nested models were tested. We used the generalized likelihood ratio test [43] to determine whether each biodiversity component has a significant additional contribution to the explanation of ecosystem processes. Then the parsimony of each model was assessed using the AICc criteria given the ratio between the number of observations (26) and the number of variables (8) [43].

In order to prioritize the biodiversity indices related to ecosystem processes, and to investigate their effects (coefficients), we followed a multiple regression approach. Starting with a full model including all 8 indices, the relative importance of indices was assessed using a backward selection procedure. The significance of the increase in deviance resulting from the deletion of a variable in the model was estimated using the chi-squared deletion test [44]. The minimal adequate model was selected as the one containing nothing but significant variables. For each response variable (ecosystem process), we performed multiple regressions and we then selected the minimal adequate model. We did not rely on classical AIC, BIC or AICc criteria to select the most parsimonious model, i.e. the one offering an optimal trade-off between increased information (number of explicative variables) and decreased reliability (goodness-of-fit), since the number of potential models with 8 predictors vastly exceeds the number of observations [44]. This may lead to spurious model selection results [43].

To correctly estimate the influence of each biodiversity index on ecosystem processes we need to rely on independent biodiversity predictors, since the inherent collinearity among explanatory variables has blurred many statistical and inferential interpretations in ecology [45]. This potential multicollinearity among predictive variables was tested using the variance inflation factor (VIF) [46].

However, even if VIF values are lower than 10, we may still obtain significant biases in parameter estimates and low statistical power, potentially impairing the identification of significant effects and invalidating approaches assuming no collinearity among predictor variables [45]. To examine the role of co-varying factors, we constructed and applied structural equation models (SEMs) for each ecosystem process. This allows direct and indirect effects of the variables of interest to be teased apart and has already applied in BEF research [20]. On one hand, taxonomic diversity or species composition may have significant effects on ecosystem processes but they should be driven by relationships with functional community structure [4]. On the other hand, taxonomic diversity is not expected to be perfectly correlated with functional structure [37]. SEMs allow us to test simultaneously how well functional structure accounts for any effects of taxonomic diversity on EF, and how strongly taxonomic diversity influences functional structure. This will ultimately provide a causal framework linking taxonomic diversity and EF via functional community structure.

All statistical analyses were carried out using R software and packages 'qpcR', 'car' and 'lavaan'.

Results

Contribution of each biodiversity component

First we ran four linear models for each ecosystem process: the full model including taxonomic diversity (TD), functional identity (FI) and functional diversity (FD) with 8 biodiversity indices (TD+FI+FD), the model without any taxonomic component (FI+FD) where species richness and evenness were removed, the model without any functional identity component (TD+FD) where the biomass-weighted mean trait values were removed, and the model without any functional diversity component (TD+FI) where the 3 functional diversity indices were removed. The most parsimonious model, according to the AICc criteria, was the model without any taxonomic component for all processes but litter decomposition (Table 1). This FI+FD model also provided the highest adjusted R^2 values whatever the process except litter decomposition and productivity in 1997 (Table 1).

Then, we examined whether each of the three biodiversity components added a significant contribution to the explanation of ecosystem processes using generalized likelihood-ratio tests comparing nested models. The taxonomic component (richness and evenness) never made an additional contribution to the explanation of ecosystem processes since the FI+FD model was not significantly outperformed by any full model (TD+FI+FD) with all 8 indices (Table 1). Conversely, functional identity and functional diversity added a significant contribution for, respectively, 3 and 5 processes. We found that all variance inflation factors were lower than the critical heuristic value of 10 suggesting that collinearity among explanatory variables did not strongly affect our results (see Table S2 for values by predictor).

Selection of the minimal adequate model and its explanatory power

For each ecosystem process, we performed a multiple regression including the 8 indices as predictive variables with a backward procedure to select the minimal adequate model (Table 2). Biodiversity indices explained significantly, albeit weakly, cotton decomposition ($R^2 = 0.34$–0.42) but only functional aspects of community structure

Functional Richness

Functional Evenness

Functional Divergence

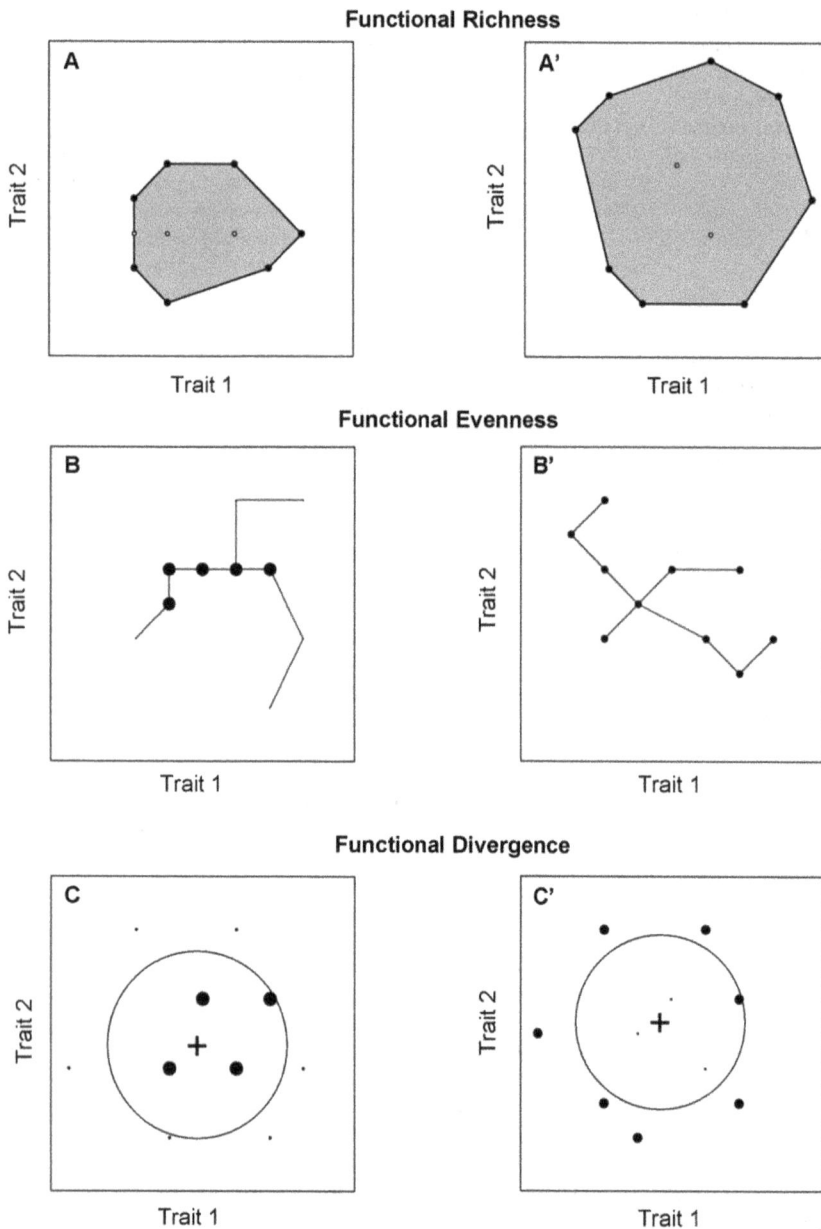

Figure 1. Geometrical presentation of functional diversity indices. For simplicity, only two traits are considered to define a two-dimensional functional space. For the 6 panels, a local community of 10 species (dark disks) is considered among a regional pool of 25 species (grey crosses). Species are plotted in this space according to their respective trait values while the circle areas are proportional to their abundances. Functional diversity of a community is thus the distribution of species and of their abundances in this functional space. Functional richness is the functional space occupied by the community, functional evenness is the regularity in the distribution of species abundances in the functional space and functional divergence quantifies how species abundances diverge from the centre of the functional space. For each component of functional diversity, two contrasting communities are represented, the right column showing an increase of the index value. More details on indices can be found in Text S1.

were retained in the minimal adequate model, with functional divergence having the main effect (positive) over the two years. For litter decomposition, 69% of the variation was explained by community structure with a combination of three indices: species evenness, functional identity on the second axis and functional divergence, the latter having a positive influence. More interestingly, up to 82% of the variation in productivity was explained by community structure with consistent effects of functional divergence and functional identity (first and second axis) over the two years. Similarly, the functional structure of communities explained nitrogen pool size at

84%, with a predominant positive effect of functional divergence, while species richness was not retained in the final model. Finally, 80% of the level of multifunctionality was explained by only three variables: functional identity (first and third PCoA axes) and functional divergence, with functional divergence having the greatest influence (positive). In other words, the aggregated mean position of the community within functional trait space in combination with functional divergence accurately predicts the level of ecosystem multifunctionality. Figure 2a shows the influence of position in functional space for multifunctionality, with communities having

Table 1. Summary of model comparisons for each ecosystem process as well as multifunctionality.

Process	Model	df	AICc	R^2	p	Test	L.Ratio	p
Cottondecom 97	TD+FI+FD	16	−216.9	0.25	0.114			
	FI+FD	18	**−227.0**	**0.313**	0.041	TD+FI+FD vs. FI+FD	0.243	0.787
	TD+FD	19	−226.9	0.225	0.076	TD+FI+FD vs. TD+FD	1.205	0.340
	TD+FI	19	−222.1	0.059	0.306	TD+FI+FD vs. TD+FI	2.612	**0.087**
Cottondecom 98	TD+FI+FD	17	−229.8	0.29	0.073			
	FI+FD	19	**−239.5**	**0.354**	0.022	TD+FI+FD vs. FI+FD	0.149	0.863
	TD+FD	20	−238.8	0.256	0.049	TD+FI+FD vs. TD+FD	1.320	0.301
	TD+FI	20	−228.2	0.119	0.794	TD+FI+FD vs. TD+FI	4.838	**0.013**
Litterdecom	TD+FI+FD	17	−291.9	**0.648**	<0.001			
	FI+FD	19	−295.8	0.599	<0.001	TD+FI+FD vs. FI+FD	2.317	0.129
	TD+FD	20	−297.0	0.572	<0.001	TD+FI+FD vs. TD+FD	2.436	0.100
	TD+FI	20	**−298.1**	0.589	<0.001	TD+FI+FD vs. TD+FI	2.124	0.135
Productivity 97	TD+FI+FD	17	368.1	**0.794**	<0.001			
	FI+FD	19	**361.2**	0.791	<0.001	TD+FI+FD vs. FI+FD	1.139	0.344
	TD+FD	20	367.6	0.701	<0.001	TD+FI+FD vs. TD+FD	4.030	**0.025**
	TD+FI	20	362.9	0.751	<0.001	TD+FI+FD vs. TD+FI	2.418	0.102
Productivity 98	TD+FI+FD	17	367.0	0.713	<0.001			
	FI+FD	19	**358.4**	**0.725**	<0.001	TD+FI+FD vs. FI+FD	0.594	0.563
	TD+FD	20	358.6	0.695	<0.001	TD+FI+FD vs. TD+FD	1.413	0.273
	TD+FI	20	367.5	0.567	<0.001	TD+FI+FD vs. TD+FI	4.381	**0.019**
Npool bm 98	TD+FI+FD	17	131.3	0.823	<0.001			
	FI+FD	19	**122.2**	**0.834**	<0.001	TD+FI+FD vs. FI+FD	0.363	0.701
	TD+FD	20	127.8	0.77	<0.001	TD+FI+FD vs. TD+FD	2.980	**0.061**
	TD+FI	20	134.9	0.698	<0.001	TD+FI+FD vs. TD+FI	5.689	**0.007**
Multifunctionality	TD+FI+FD	17	51.3	0.751	<0.001			
	FI+FD	19	**42.8**	**0.762**	<0.001	TD+FI+FD vs. FI+FD	0.552	0.586
	TD+FD	20	47.7	0.679	<0.001	TD+FI+FD vs. TD+FD	2.916	**0.064**
	TD+FI	20	52.1	0.62	<0.001	TD+FI+FD vs. TD+FI	4.495	**0.017**

The weight of support for the alternative models (TD: taxonomic diversity, FI: functional identity, FD: functional diversity) and estimates of model parameters for each ecosystem process (Cottondecomp: cotton decomposition, Litterdecom 98: litter decomposition in 1998, Productivity: productivity as annual biomass production, Npool bm: nitrogen pool size in aboveground biomass, Multifunctionality: mean performance over all processes). Results of likelihood ratio tests comparing nested models (L.Ratio) and associated p-values. Adjusted R^2s for the ordinary least squares regression models and p-value associated to the multiple regressions are presented. The lowest AICc value for each process, the highest adjusted R^2 and the significant differences between models ($p<0.1$) are in bold.

higher values than −0.1 on the first PCoA axis also have higher levels of multifunctionality than the others while communities with low values on both the first and third PCoA axes have a low average multifunctionality values. In addition, all communities with high functional divergence values (>0.85) show high multifunctionality levels (Figure 2b).

Figure 3 shows two communities containing the same number of species (8) with extreme values along the gradient of multifunctionality level (community a>community b). In the high functioning community a, all the dominant species are specialists (*i.e.* with extreme combinations of traits), which contributes to a high functional divergence value. Community a also has a higher mean value on the first PCoA axis of all communities (indicated by the black triangle in Figure 3a). Conversely, the low functioning community b has a lower functional divergence value with some dominant species being generalists (i.e. close to the center of the functional space occupied by the community) that are functionally redundant (Figure 3b). This community has also a lower mean value on the first PCoA axis.

Structural Equation Model

Using a structural equation model (SEM) for ecosystem multifunctionality (models for other processes are provided in Text S2), we confirm that taxonomic composition of communities had no direct significant influence on ecosystem multifunctionality (Figure 4); only functional identity (through first and third PCoA axes) and functional divergence had a significant direct effect with functional divergence having the greatest influence (positive). Taxonomic diversity did have a significant influence on the functional structure of communities, but the greatest effect was the positive influence of species richness on functional richness, which had no significant effect on multifunctionality. Functional indices were weakly related between each other and only two correlations were significant and positive (functional divergence and first PCoA axis, functional richness and second PCoA axis). The SEM illustrates that despite the co-linearity between the first PCoA axis and functional divergence, both indices had significant independent effects on multifunctionality.

Table 2. Summary of the minimal adequate models.

	S	E	PC1	PC2	PC3	FRic	FEve	FDiv	R^2	p
Cottondecom 97					−2.0*	−2.1**	−1.9*	3.6***	0.34	0.0140
Cottondecom 98					−2.3**		−4.1***	2.5**	0.42	0.0016
Litterdecom homo		−3.1***		2.4**	−1.7*			2.9***	0.69	<0.0001
Productivity 97	2.3**		3.8***		−2.8***			2.7**	0.82	<0.0001
Productivity 98			1.8*		−2.2**	3.0***		2.4**	0.75	<0.0001
Npool bm 98			3.1***		−2.8**	2.4**		3.4***	0.84	<0.0001
Multifunctionality			3.0***		−3.5***			4.2***	0.80	<0.0001

Results of regressions of ecosystem processes (Cottondecomp: cotton decomposition, Litterdecom 98: litter decomposition in 1998, Productivity: productivity as annual biomass production, Npool bm: nitrogen pool size in aboveground biomass, Multifunctionality: mean performance over all processes) against 8 biodiversity indices (S: species richness, E: species evenness, PC1 PC2 and PC3: aggregated mean trait values along three PCoA axes, FRic: functional richness, FEve: functional evenness, FDiv: functional divergence). t-value for each selected variable, adjusted R^2s for the ordinary least squares regression models and p-value associated to the multiple regressions are presented. Explanatory variables (biodiversity indices) were selected using a backward selection procedure starting with a maximal model towards the one containing nothing but significant terms (p<0.1).
p<0.1,
**p<0.05,
***p<0.01.

Discussion

Our results demonstrate that biodiversity components differ greatly in their influence on ecosystem processes. The taxonomic component, after removing the effects of functional identity and diversity, has no additional effect on processes with consistently low and non significant likelihood-ratio values (Table 1). In addition, species richness and evenness were rarely retained in the minimal adequate model (Table 2) or by the SEM (Figure 4) for their direct influence on ecosystem processes (Text S2). This result can be partly explained by the positive relationship between functional richness and species richness (Figure 4) [40] since communities with more species are more likely to hold a higher diversity of traits and thus perform more functions [47]. Therefore, the additional effect of species richness is likely to be weak after removing the effect of functional richness. Similarly, species evenness has no significant influence on ecosystem processes (except litter decomposition) but it influences the

functional structure of communities as revealed by the SEM analysis (Figure 4). We conclude that while the influence of taxonomic structure on ecosystem processes is less important than that of functional identity and diversity, taxonomic composition mediates functional structure. This implies that the taxonomic composition of communities may have indirect effects on ecosystem processes since they are not their proximate, but partly their ultimate, drivers.

While it remains difficult to provide a definite mechanistic explanation for the relationship between functional structure and multifunctionality, existing literature and our own observations may provide some clues. Two of the key functional traits for explaining multifunctionality were leaf phenology (evergreen vs. partly evergreen vs. summergreen) and leaf inclination. There is only very little evidence that increased phenological complementarity can have a positive effect on annual productivity in early successional forb communities, although such an effect might be stronger at low levels of species richness [48]. Evergreen species at

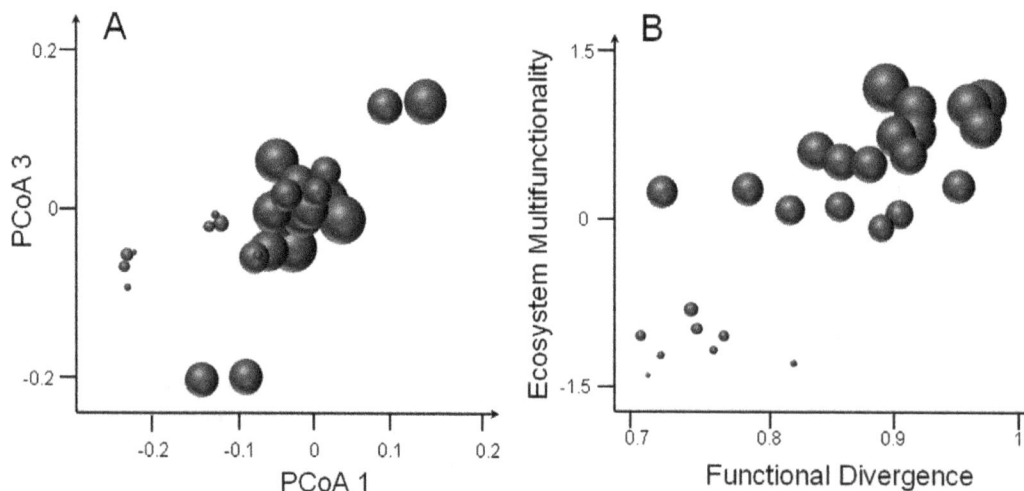

Figure 2. Relationships between community structure and ecosystem multifunctionality. (A) Multifunctionality performance of each community in the functional trait space (first and third axes of the PCoA – PCoA 1 and PCoA 3 respectively). (A) Multifunctionality performance against functional divergence (FDiv). Circle sizes are proportional to performance of communities. See Table 1 for associated statistics.

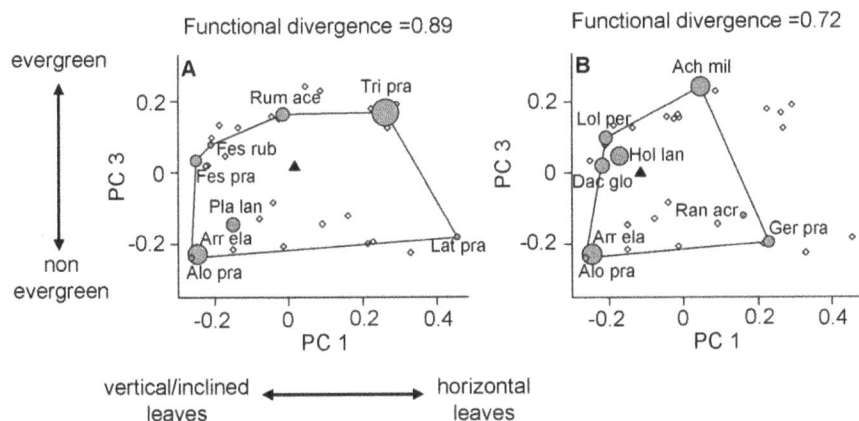

Figure 3. Two species communities represented in functional space with contrasting multifunctionality levels. Two 8-species communities of our experiment with the highest multifunctionality level (a) and the lowest (b). Positions of species are presented in the functional space (first and third PCoA axes). The black triangle labeled "Agg" represents the biomass-weighted mean trait values (aggregated trait) along the two PCoA axes while the lines represent the functional volume occupied by each community. The sizes of grey circles are proportional to species relative abundances. Full species names and trait values can be found in Table S1.

our site might have some photosynthetic activity during mild winter days, but biomass production is very low until the onset of spring. Some of the evergreen or partly evergreen species, however, shown an early onset of growth in spring with an early peak in the season (e.g. *Alopecurus pratensis, Plantago lanceolata*), while the summergreen species have a tendency to peak later in the year (e.g. *Centaurea jacea, Geranium pratense*). Thus, this temporal complementarity of growth might have induced higher productivity with higher functional divergence in leaf phenology. Variability in leaf inclination is known to enhance the photosyn-

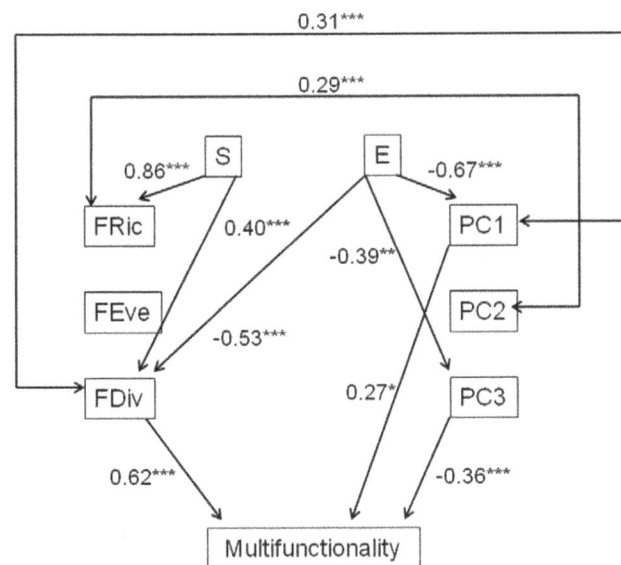

Figure 4. Results of the structural equation model (SEM) linking the multifunctionality of ecosystems to biodiversity indices. (S: species richness, E: evenness in species abundances, PC1 PC2 and PC3: aggregated mean trait values along three PCoA axes, FRic: functional richness, FEve: functional evenness, FDiv: functional divergence.) Numbers next to unidirectional arrows are standardized slopes and those next to bidirectional arrows are correlations. Only significant effects or correlations are shown (* $p<0.1$, ** $p<0.05$, *** $p<0.01$). For detailed statistics and for each process, see Text S2.

thetic light capture of individual tree crowns (e.g. [49]), while in a study of mixed red clover (*Trifolium pratense*) and tall fescue (*Festuca arundinacea*) canopy differences in leaf inclination between the two species increase equality in light partitioning between the taller fescue and shorter clover [50]. In our model plant community, the influence of functional divergence on productivity may be due to temporal and spatial partitioning in light capture via complementarity in phenology and leaf inclination, respectively.

In a previous study on the same site, it has been shown that increasing functional diversity positively influences decomposition rates of plant litter, while species richness had no such effect [31]. These results suggest that this positive effect of functional diversity was due to improved microenvironmental conditions for decomposer fauna, and due to higher litter quality.

With reference to functional identity, increasing dominance of species with more horizontal leaf inclination might enhance productivity by increasing total light capture relative to communities dominated by species with vertical inclination, which might partially explain the influence of PC1 on multifunctionality. Supporting this mechanism, communities dominated by forb species with horizontal leaf inclination also had higher leaf area index than those dominated by grasses with a more vertical inclination. In addition, all nitrogen-fixing legumes planted show a horizontal leaf inclination, partly confounding leaf inclination with N-fixation, the latter being known to positively influence productivity at our site [51]. However, it is unclear how aggregate mean phenology would affect multifunctionality. Perhaps summergreen species are able to grow faster since decreased leaf longevity is associated with increased photosynthetic rates [52]. Short lived leaves also have traits associated with more rapid decomposition rates (e.g. high nutrient content, [27]), which would explain the influence of PC3 on litter decomposition in the minimal adequate regression. The higher nutrient content of summergreen leaves is supported by the negative relationship between PC3 and the amount of nitrogen in biomass in the minimal adequate regression.

The predominance of variables linked to the functional structure of communities over taxonomic variables in predicting ecosystem processes is in accordance with the most recent findings obtained in experiments [20] or with empirical data [9]. Except for decomposition, we show that functional identity and diversity bring independent and additional explanatory power to ecosystem processes with consistently high likelihood-ratio values. Overall, the results suggest that neither functional identity nor functional divergence was more important than the other in explaining ecosystem processes and particularly the multifunctionality. So, contrary to other studies, demonstrating the higher contribution of one component over the others [20,53], we demonstrate that this differential contribution may depend on the process involved, and when considering multiple processes the magnitude of the two component effects is similar. Thus, to reach high levels of predictability in modelling multiple ecosystem processes, functional identity and diversity components have to be taken into account in a common framework [54].

It has been suggested that, since different species often influence different functions, the level of biodiversity needed to sustain multifunctionality in ecosystems is higher than previously thought [15,16]. By integrating across four ecosystem processes in assessing the level of community multifunctionality, we show that both functional divergence and functional identity have a predominant role, while species richness has no direct effects (Table 2) and few indirect effects (Figure 4). We suggest that this absence of a species richness effect is partly explained by the relatively high richness values considered in our study (4 to 16 species) while past evidence for positive effects of species richness on ecosystem processes have often been due to the weak performances of monocultures or very species poor communities [2]. Indeed our results are not in contradiction with previous studies demonstrating positive species diversity effects on ecosystem functioning. Rather, they suggest that, except at the extreme low end of species richness gradients, the taxonomic structure of ecological communities is no longer the main driver of ecosystem processes, with the functional structure being the primary determinant.

Our study reconciles two hypotheses that have been alternatively suggested to primarily underpin ecosystem processes: the complementarity and the mass ratio hypotheses. We suggest that a combined effect of functional identity and functional divergence is the most parsimonious explanation for key ecosystem processes. Taken separately, each biodiversity component has weak explanatory power for ecosystem functioning [20,31,36]. However, the combined effect of biodiversity components related to the functional structure of communities used in our study consistently reached unprecedented levels of predictive accuracy (up to 84%) whatever the process and for all processes together.

Our finding is crucial since recent work has demonstrated that global gradients in decomposition rates, for example, are primarily driven by plant functional traits rather than climate [27], emphasizing the need for better understanding of the interplay between functional structure of communities and ecosystem functioning. The predominance of functional divergence effects on most of ecosystem processes sheds light on the need to preserve specialist species *sensu* Elton (i.e. those that have a particular combination of traits and perform particular functions in the system). However, since under the combined influence of habitat degradation or global change, we are increasingly losing local specialist species [55,56], the level of functional diversity held by communities is declining worldwide [57]. Our results show that modifying the functional structure of communities has a strong impact on ecosystem processes and should receive more attention in assessing and countering the global decline of biodiversity.

Supporting Information

Table S1   Species used in the German BIODEPTH experiment with their traits.

Table S2   Values of the variance inflation factor (VIF) for each biodiversity index (S: species richness, E: species evenness, PC1 PC2 and PC3: aggregated mean trait values along three PCoA axes, FRic: functional richness, FEve: functional evenness, FDiv: functional divergence).

Text S1   Calculation of functional diversity indices.

Text S2   Results from Structural Equation Models (SEM) for each process.

Acknowledgments

We are grateful to those colleagues that commented on earlier versions of this manuscript, including Michel Loreau, Andy Hector and Eric Garnier.

Author Contributions

Conceived and designed the experiments: MSL. Performed the experiments: MSL. Analyzed the data: DM SV NWHM MSL. Contributed reagents/materials/analysis tools: DM SV NWHM MSL. Wrote the manuscript: DM SV NWHM MSL.

References

1. Chapin FS, Walker BH, Hobbs RJ, Hooper DU, Lawton JH, et al. (1997) Biotic control over the functioning of ecosystems. Science 277: 500–504.
2. Hooper DU, Chapin FS, Ewel JJ, Hector A, Inchausti P, et al. (2005) Effects of biodiversity on ecosystem functioning: A consensus of current knowledge. Ecological Monographs 75: 3–35.
3. Smith SD, Huxman TE, Zitzer SF, Charlet TN, Housman DC, et al. (2000) Elevated CO2 increases productivity and invasive species success in an arid ecosystem. Nature 408: 79–82.
4. Loreau M, Naeem S, Inchausti P, Bengtsson J, Grime JP, et al. (2001) Ecology - Biodiversity and ecosystem functioning: Current knowledge and future challenges. Science 294: 804–808.
5. Cardinale BJ, Srivastava DS, Duffy JE, Wright JP, Downing AL, et al. (2006) Effects of biodiversity on the functioning of trophic groups and ecosystems. Nature 443: 989–992.
6. Chapin FS, Sala OE, Burke IC, Grime JP, Hooper DU, et al. (1998) Ecosystem consequences of changing biodiversity - Experimental evidence and a research agenda for the future. Bioscience 48: 45–52.
7. Purvis A, Hector A (2000) Getting the measure of biodiversity. Nature 405: 212–219.
8. Devictor V, Mouillot D, Meynard C, Jiguet F, Thuiller W, et al. (2010) Spatial mismatch and congruence between taxonomic, phylogenetic and functional diversity: the need for integrative conservation strategies in a changing world. Ecology Letters 13: 1030–1040.
9. Danovaro R, Gambi C, Dell'Anno A, Corinaldesi C, Fraschetti S, et al. (2008) Exponential decline of deep-sea ecosystem functioning linked to benthic biodiversity loss. Current Biology 18: 1–8.
10. Violle C, Navas ML, Vile D, Kazakou E, Fortunel C, et al. (2007) Let the concept of trait be functional! Oikos 116: 882–892.
11. Balvanera P, Pfisterer AB, Buchmann N, He JS, Nakashizuka T, et al. (2006) Quantifying the evidence for biodiversity effects on ecosystem functioning and services. Ecology Letters 9: 1146–1156.
12. Tilman D, Reich PB, Knops J, Wedin D, Mielke T, et al. (2001) Diversity and productivity in a long-term grassland experiment. Science 294: 843–845.
13. Hector A, Schmid B, Beierkuhnlein C, Caldeira MC, Diemer M, et al. (1999) Plant diversity and productivity experiments in European grasslands. Science 286: 1123–1127.
14. Reiss J, Bridle JR, Montoya JM, Woodward G (2009) Emerging horizons in biodiversity and ecosystem functioning research. Trends in Ecology & Evolution 24: 505–514.
15. Hector A, Bagchi R (2007) Biodiversity and ecosystem multifunctionality. Nature 448: 188–U186.

16. Gamfeldt L, Hillebrand H, Jonsson PR (2008) Multiple functions increase the importance of biodiversity for overall ecosystem functioning. Ecology 89: 1223–1231.

17. Zavaleta ES, Pasari JR, Hulvey KB, Tilman GD (2010) Sustaining multiple ecosystem functions in grassland communities requires higher biodiversity. Proceedings of the National Academy of Sciences of the United States of America 107: 1443–1446.

18. Hooper DU, Vitousek PM (1997) The effects of plant composition and diversity on ecosystem processes. Science 277: 1302–1305.

19. Diaz S, Lavorel S, de Bello F, Quetier F, Grigulis K, et al. (2007) Incorporating plant functional diversity effects in ecosystem service assessments. Proceedings of the National Academy of Sciences of the United States of America 104: 20684–20689.

20. Mokany K, Ash J, Roxburgh S (2008) Functional identity is more important than diversity in influencing ecosystem processes in a temperate native grassland. Journal of Ecology 96: 884–893.

21. Tilman D, Wedin D, Knops J (1996) Productivity and sustainability influenced by biodiversity in grassland ecosystems. Nature 379: 718–720.

22. Wilsey BJ, Potvin C (2000) Biodiversity and ecosystem functioning: Importance of species evenness in an old field. Ecology 81: 887–892.

23. Jenny H (1941) Factors of Soil Formation. New york: McGraw-Hill.

24. Grime JP (1998) Benefits of plant diversity to ecosystems: immediate, filter and founder effects. Journal of Ecology 86: 902–910.

25. Hillebrand H, Bennett DM, Cadotte MW (2008) Consequences of dominance: A review of evenness effects on local and regional ecosystem processes. Ecology 89: 1510–1520.

26. Garnier E, Cortez J, Billes G, Navas ML, Roumet C, et al. (2004) Plant functional markers capture ecosystem properties during secondary succession. Ecology 85: 2630–2637.

27. Cornwell WK, Cornelissen JHC, Amatangelo K, Dorrepaal E, Eviner VT, et al. (2008) Plant species traits are the predominant control on litter decomposition rates within biomes worldwide. Ecology Letters 11: 1065–1071.

28. Mason NWH, Mouillot D, Lee WG, Wilson JB (2005) Functional richness, functional evenness and functional divergence: the primary components of functional diversity. Oikos 111: 112–118.

29. Tilman D, Knops J, Weldin D, Reich P, Ritchie M, et al. (1997) The influence of functional diversity and composition on ecosystem processes. Science 277: 1300–1302.

30. Hooper DU, Vitousek PM (1998) Effects of plant composition and diversity on nutrient cycling. Ecological Monographs 68: 121–149.

31. Scherer-Lorenzen M (2008) Functional diversity affects decomposition processes in experimental grasslands. Functional Ecology 22: 547–555.

32. Fargione J, Tilman D, Dybzinski R, HilleRisLambers J, Clark C, et al. (2007) From selection to complementarity: shifts in the causes of biodiversity-productivity relationships in a long-term biodiversity experiment. Proceedings of the Royal Society B-Biological Sciences 274: 871–876.

33. Scherer-Lorenzen M, Palmborg C, Prinz A, Schulze ED (2003) The role of plant diversity and composition for nitrate leaching in grasslands. Ecology 84: 1539–1552.

34. Petchey OL, Gaston KJ (2002) Functional diversity (FD), species richness and community composition. Ecology Letters 5: 402–411.

35. Mouchet M, Guilhaumon F, Villeger S, Mason NWH, Tomasini JA, et al. (2008) Towards a consensus for calculating dendrogram-based functional diversity indices. Oikos 117: 794–800.

36. Spehn EM, Hector A, Joshi J, Scherer-Lorenzen M, Schmid B, et al. (2005) Ecosystem effects of biodiversity manipulations in European grasslands. Ecological Monographs 75: 37–63.

37. Gastine A, Scherer-Lorenzen M, Leadley PW (2003) No consistent effects of plant diversity on root biomass, soil biota and soil abiotic conditions in temperate grassland communities. Applied Soil Ecology 24: 101–111.

38. Heisse K, Roscher C, Schumacher J, Schulze ED (2007) Establishment of grassland species in monocultures: different strategies lead to success. Oecologia 152: 435–447.

39. Legendre P, Legendre L (1998) Numerical Ecology. Amsterdam: Elsevier. 853 p.

40. Villeger S, Mason NWH, Mouillot D (2008) New multidimensional functional diversity indices for a multifaceted framework in functional ecology. Ecology 89: 2290–2301.

41. Cornwell WK, Schwilk DW, Ackerly DD (2006) A trait-based test for habitat filtering: Convex hull volume. Ecology 87: 1465–1471.

42. Devictor V, Clavel J, Julliard R, Lavergne S, Mouillot D, et al. (2010) Defining and measuring ecological specialization. Journal of Applied Ecology 47: 15–25.

43. Burnham KP, Anderson DR (1998) Model selection and inference. A practical information-theoretic approach. New York: Springer.

44. Chatterjee S, Hadi AS (2006) Regression analysis by example: John Wiley & Sons. 371 p.

45. Graham MH (2003) Confronting multicollinearity in ecological multiple regression. Ecology 84: 2809–2815.

46. Fox J (2008) Applied Regression Analysis and Generalized Linear Models: Sage.

47. Halpern BS, Floeter SR (2008) Functional diversity responses to changing species richness in reef fish communities. Marine Ecology Progress Series 364: 147–156.

48. Stevens MHH, Carson WP (2001) Phenological complementarity, species diversity, and ecosystem function. Oikos 92: 291–296.

49. Posada JM, Lechowicz MJ, Kitajima K (2009) Optimal photosynthetic use of light by tropical tree crowns achieved by adjustment of individual leaf angles and nitrogen content. Annals of Botany 103: 795–805.

50. Sonohat G, Sinoquet H, Varlet-Grancher C, Rakocevic M, Jacquet A, et al. (2002) Leaf dispersion and light partitioning in three-dimensionally digitized tall fescue-white clover mixtures. Plant Cell and Environment 25: 529–538.

51. Spehn EM, Scherer-Lorenzen M, Schmid B, Hector A, Caldeira MC, et al. (2002) The role of legumes as a component of biodiversity in a cross-European study of grassland biomass nitrogen. Oikos 98: 205–218.

52. Reich PB, Ellsworth DS, Walters MB, Vose JM, Gresham C, et al. (1999) Generality of leaf trait relationships: A test across six biomes. Ecology 80: 1955–1969.

53. Arenas F, Sanchez I, Hawkins SJ, Jenkins SR (2006) The invasibility of marine algal assemblages: Role of functional diversity and identity. Ecology 87: 2851–2861.

54. Schumacher J, Roscher C (2009) Differential effects of functional traits on aboveground biomass in semi-natural grasslands. Oikos 118: 1659–1668.

55. Devictor V, Julliard R, Jiguet F (2008) Distribution of specialist and generalist species along spatial gradients of habitat disturbance and fragmentation. Oikos 117: 507–514.

56. Villeger S, Miranda JR, Hernandez DF, Mouillot D (2010) Contrasting changes in taxonomic vs. functional diversity of tropical fish communities after habitat degradation. Ecological Applications 20: 1512–1522.

57. Flynn DFB, Gogol-Prokurat M, Nogeire T, Molinari N, Richers BT, et al. (2009) Loss of functional diversity under land use intensification across multiple taxa. Ecology Letters 12: 22–33.

2

Enhanced Production of Green Tide Algal Biomass through Additional Carbon Supply

Pedro H. de Paula Silva*, Nicholas A. Paul, Rocky de Nys, Leonardo Mata

School of Marine and Tropical Biology & Centre for Sustainable Tropical Fisheries and Aquaculture, James Cook University, Townsville, Australia

Abstract

Intensive algal cultivation usually requires a high flux of dissolved inorganic carbon (Ci) to support productivity, particularly for high density algal cultures. Carbon dioxide (CO_2) enrichment can be used to overcome Ci limitation and enhance productivity of algae in intensive culture, however, it is unclear whether algal species with the ability to utilise bicarbonate (HCO_3^-) as a carbon source for photosynthesis will benefit from CO_2 enrichment. This study quantified the HCO_3^- affinity of three green tide algal species, *Cladophora coelothrix*, *Cladophora patentiramea* and *Chaetomorpha linum*, targeted for biomass and bioenergy production. Subsequently, we quantified productivity and carbon, nitrogen and ash content in response to CO_2 enrichment. All three species had similar high pH compensation points (9.7–9.9), and grew at similar rates up to pH 9, demonstrating HCO_3^- utilization. Algal cultures enriched with CO_2 as a carbon source had 30% more total Ci available, supplying twenty five times more CO_2 than the control. This higher Ci significantly enhanced the productivity of *Cladophora coelothrix* (26%), *Chaetomorpha linum* (24%) and to a lesser extent for *Cladophora patentiramea* (11%), compared to controls. We demonstrated that supplying carbon as CO_2 can enhance the productivity of targeted green tide algal species under intensive culture, despite their clear ability to utilise HCO_3^-.

Editor: Douglas Andrew Campbell, Mount Allison University, Canada

Funding: This research is part of the MBD Energy Research and Development program for Biological Carbon Capture and Storage. The project is supported by the Advanced Manufacturing Cooperative Research Centre (AMCRC), funded through the Australian Government's Cooperative Research Centre Scheme. This project is also supported by the Australian Government through the Australian Renewable Energy Agency. Pedro de Paula Silva was supported by an AMCRC PhD Scholarship. The funders had no role in study design, data collection and analysis, decision to publish, or preparation of the manuscript.

Competing Interests: The authors have declared that no competing interests exist.

* E-mail: pedro.depaulasilva@my.jcu.edu.au

Introduction

Macroalgal biomass is an emerging resource for sustainable bioenergy [1] and advanced biofuels [2,3]. Bioenergy applications rely on the production of a high volume/low value biomass opening opportunities to develop the culture of new commercial species. Green tide algae have the potential to meet these criteria as they are fast growing species [4] with a tolerance to a broad range of environmental conditions [5]. Furthermore, they are highly suitable as a bioenergy feedstock for ethanol [6], biogas [7,8] and thermo-chemical conversion to biocrude [3,9]. Green tide algae can be cultured extensively in open water culture [10] or harvested from natural blooms [11]. Alternatively, they can be cultured intensively in land-based ponds and tanks integrated into nutrient-rich aquaculture [12–14] and municipal [15] waste streams for bioremediation.

Dissolved inorganic carbon (Ci) is usually the limiting factor for growth in intensive cultivation with nutrient-rich systems, as the rate of Ci assimilation by the algae is greater than the rate of CO_2 diffusion from the air into the water, even when vigorous aeration is used [16]. Total Ci in the water is composed of an equilibrium between carbon dioxide (CO_2), bicarbonate (HCO_3^-) and carbonate (CO_3^{2-}), which are part of a buffered system. The relative amount of each fraction is dependent on pH, and to a lesser extent on salinity and temperature [17]. At pH 6, the molar fraction of the total Ci is divided equally between CO_2 and HCO_3^-, the only usable forms of carbon for most of algae. The concentration of CO_2 at pH 8.5 is negligible, as HCO_3^- is in equilibrium with CO_3^{-2}, which is not a direct source of inorganic carbon for algal photosynthesis [18]. Above pH 9, the relative fraction of HCO_3^- continues to decrease relative to CO_3^{-2} leading to Ci limitation. Ironically, the daily pH fluctuations in a carbon limited system of intensive algal cultivation can cross at least 2 pH units [19], which provide a unique setting to evaluate the benefits of dosing Ci at commercial scales.

Increasing total dissolved CO_2 concentrations in land-based intensive seaweed cultivation can therefore significantly enhance biomass productivity [20–23]. However, CO_2 enrichment can also have no effect or may even be detrimental for some species [24–26]. The lack of widespread positive responses to CO_2 enrichment in algae has been attributed to the presence of carbon concentration mechanisms (CCMs). These mechanisms allow algae to utilize the HCO_3^- pool in seawater, which is the most common form of carbon [27]. However, the efficiency of HCO_3^- use is species specific, with some species relying on HCO_3^- to complement CO_2 as a carbon source, while others can efficiently saturate carbon requirements using HCO_3^- alone [28]. Therefore, enhanced productivity through CO_2 enrichment is affected by the capability and efficiency with which species use HCO_3^-. Quantifying and understanding the response of algae to CO_2 enrichment is a critical first step in optimisation of growth under intensive culture.

The major objective of this study was to quantify the ability of three green tide algal species, *Cladophora coelothrix* Kützing,

Cladophora patentiramea (Montagne) Kützing and *Chaetomorpha linum* (O. F. Müller) Kützing, to utilise alternative carbon sources under intensive culture, and subsequently quantify their growth response to removing carbon limitations. These three species were selected as they are clearly identified targets for biomass production for bioremediation and bioenergy applications in tropical Australia [14,29], and worldwide [8,10,30]. Specifically, we quantified the affinity for HCO_3^- as a carbon source using the pH drift technique to determine the compensation point. We subsequently quantified growth under laboratory conditions at different pH levels with defined carbon sources. Finally, the three species were cultured for four weeks in an outdoor experiment testing the effects of CO_2 enrichment and HCO_3^- affinity on productivity and elemental composition (carbon, nitrogen and ash).

Materials and Methods

Algae collection and stock cultures

Three green tide algal species were collected from private aquaculture facilities in Queensland, Australia. *C. coelothrix* and *C. linum* were collected from the settlement pond and intake channel, respectively, of an intensive fish farm (Latitude: 20.02°S Longitude: 148.22°E, barramundi *Lates calcarifer*). *C. patentiramea* (Montagne) Kützing was collected from the intake dam of an intensive prawn farm (Latitude: 18.26°S Longitude 146.03°E, tiger prawns *Penaeus monodon*). Permission was obtained from owners to collect algae from these sites. Algal samples were hand collected and placed in aerated seawater for transportation to the James Cook University, Marine Aquaculture Research Facility Unit (MARFU). Stock cultures of each algal species were maintained in 70 L tanks within a recirculating system (~27°C, 36‰).

Algal affinity for HCO_3^-

Two approaches were used to quantify the ability of the three algal species to utilise HCO_3^- as a source of Ci; pH drift technique (compensation point), and algal growth response to different pH levels.

pH drift in closed vessel

The pH drift technique is a reliable method to determine HCO_3^- utilization [31]. As the photosynthetic uptake of CO_2 and/or HCO_3^- results in a near stoichiometric production of hydroxyl ions, the pH of the culture media increases in response to photosynthesis. At pH 9, dissolved CO_2 is virtually absent and species without mechanisms of HCO_3^- utilization reach their limit of Ci extraction. Consequently, pH will not increase beyond this level, enabling the ability to utilise HCO_3^- to be evaluated [32].

The pH drift assays were carried out in a culture chamber (Sanyo model MLR-351) with constant temperature (28°C) and irradiance (150 μmol photons m^{-2} s^{-1}). Basal culture media was prepared using filtered sterile seawater (NO_3-N 0.06 mg l^{-1}, PO_4^--P 0.02 mg l^{-1}, Ci 1.9 mM and 32 ‰) enriched with f/2 growth media [33]. Algal samples were collected from the stock cultures, washed clean and pre-incubated for five days in the conditions described above. Approximately 100 mg fresh weight of filaments were incubated in closed airtight 120 ml graduated culture vessels filled with 130 ml of freshly prepared growth media (pH 7.9), leaving a minute air space. Culture vessels were repositioned and stirred hourly during the experiment to minimise any artefacts relating to light source or the formation of a boundary layer.

Thirty-six culture vessels were prepared for each species and three random culture vessels for each species (n = 3) were destructively sampled for pH measurements (YSI 63 pH meter).

The pH drift assays ran for twelve hours. The pH measurements were performed at one and two hours in culture and then repeated every two hours until the maximum pH reached a stable level for at least two consecutive measurements (pH compensation point). This compensation point represents the pH at which the Ci taken up by the algae equals the CO_2 released by respiration and/or photorespiration into the medium.

Effects of pH on algal growth

The HCO_3^- affinity of the algae can be inferred from their growth response at different pH levels because the relative amount of CO_2 and HCO_3^- available for growth is pH dependent. Above pH 8.5, where CO_2 is virtually absent, species with no or little ability to use HCO_3^- experience a steep decrease in growth. In contrast, species with the ability to efficiently use HCO_3^- respond more slowly to the increase in pH as they utilize HCO_3^- for growth.

To test HCO_3^- affinity, algal biomass was transferred from the outdoor stock cultures to the laboratory and pre-cultured in f/2 enriched growth media for five days (in conditions described in the previous section). The growth experiment was carried out in a culture chamber (Sanyo model MLR-351) with constant temperature (28°C) and irradiance (150 μmol photons m^{-2} s^{-1}) with a 12 L:12 D photoperiod. Samples of each algal species (~100 mg fresh weight) were incubated in 100 mL of seawater enriched with f/2 growth media [33] within 120 mL plastic culture vessels with the lid loosely placed on top. The culture media in each treatment was buffered to maintain constant pH (+ 0.1 units), and correspondingly Ci ratios, using biological Tris (Sigma) at a final concentration of 25 mM. The water pH was adjusted to the desired pH levels (7, 7.5, 8, 8.5 and 9) using freshly prepared 1 M NaOH or HCl solutions. The culture media was prepared and replaced every day to renew Ci and to maintain the original CO_2:HCO_3^- ratios for each treatment. Algal filaments were filtered through a mesh screen and resuspended in the new growth media. Samples were again stirred and repositioned daily to a new position in the culture chamber. Treatments were weighted at the beginning and end of a ten day experimental period. Daily growth rates (DGR; % day^{-1}) were then calculated using the following equation:

$$DGR = \left[(Wf/Wi)^{(1/T)} - 1\right] * 100 \qquad (1)$$

where Wi is the initial fresh weight, Wf is the final fresh weight and T is the culture period in days.

Algal productivity under CO_2 enrichment

A CO_2 enrichment experiment was performed outdoors using recirculating cultivation systems at the Marine Research Facility Unit (MARFU) at James Cook University between August and September 2010. Two independent sumps were used, one was supplied directly with CO_2 gas stream (food grade 99.9% – BOC Australia) and regularly adjusted to maintain pH between 6.5 and 7, whereas the other acted as a control sump with no additional CO_2. These systems provided a constant water flow of 2 volumes (vol) h^{-1} to polyethylene white buckets with 5 L capacity, 0.035 m^2 surface area, containing a ring of aeration in the bottom to maintain the algae in tumble culture. The buckets were stocked with 3 g fresh weight L^{-1} (n = 3 for each species*CO_2 treatment).

Cultures were acclimated for two weeks at these conditions and a formal growth experiment conducted over the subsequent four week period. Algal biomass of each tank was harvested weekly to determine productivity and subsequently restocked at the original density of 3 g fresh weight L^{-1}. The algae were collected in mesh bags (0.1 mm mesh) and the biomass drained to a constant fresh weight in a washing machine (spin cycle 1000 rpm). Productivity (g m^{-2} day^{-1} dry weight) was then calculated using equation (2):

$$P = \{[(Bf - Bi)/FW : DW]/A\}/T \qquad (2)$$

where Bi is the initial biomass, Bf is the final biomass, FW:DW is the fresh to dry weight ratio, A is area of culture vessels and T the number of days in culture. The dry weights were acquired individually for each week from excess centrifuged biomass oven dried at 65°C for 48 h. Resulting FW:DW ratios were on average 3.5:1 for *C. coelothrix*, 5:1 for *C. patentiramea* and 5.9:1 *C. linum*.

The water pH, temperature and salinity were measured daily in the inflow and outflow water of seaweed cultures at 08:00, 12:00 and 18:00 using an YSI 63 multi-parameter meter. Throughout the experiments water temperature and salinity averaged 28°C (2±SD) and 35‰ (1±SD), respectively. Ambient surface photosynthetic active radiation (PAR) was measured continuously using a LI-192S (2p) sensor placed near the tanks. Daily average PAR recorded during light hours for the experimental period was 881±152 μmol photons m^{-2} s^{-1}. Water samples were collected twice a week at 12:00 from the inflow and outflow of tumble cultures for alkalinity determination. The samples were fixed with 200 μM of saturated $HgCl_2$ solution, immediately taken to the lab and stored in the fridge until alkalinity analysis. Alkalinity was calculated using potentiometric titration by the Australian Centre for Tropical Freshwater Research (ACTFR) at James Cook University. Ci concentration and sources were calculated using the pH, alkalinity, salinity, and temperature original values of collection time, using the software CO_2sys [34]. Nitrogen and phosphorus were measured from water samples collected from the inflow and immediately analysed by cadmium reduction and ascorbic acid techniques (HACH model DR/890), respectively. Average nitrogen and phosphorus concentrations during the experiment were ~2.4 and 0.16 mg L^{-1}, respectively.

Biomass elemental analysis

The biomass of each tank was harvested at the completion of the four week growth period for elemental analysis (n = 3 for each CO_2*species treatment). Biomass was spun dry and then oven dried at 60°C for 48 h, milled and stored in glass containers prior to analysis. Nitrogen and carbon were quantified for each sample using isotope analysis. Ash was quantified using a Carlo-Erba elemental autoanalyzer (Environmental Biology Group, Australian National University, Canberra).

Statistical analysis

Two-way fixed-effect analyses of variance (ANOVA) were used to compare the growth response of the three green tide algal species to different pH and correspondent CO_2 and HCO_3^- ratios, and the effects of CO_2 enrichment on biomass productivity using the software SYSTAT 12. Post-hoc comparisons were made to assess the differences between treatments in both experiments (Tukey's HSD multiple comparisons). The ANOVA assumptions of homogeneity of variance and normality were assessed by scatter plots and normal curve of the residuals, respectively [35]. To test whether the elemental composition of biomass was influenced by CO_2 enrichment, we used a two-factor permutational multivariate analysis of variance (PERMANOVA, PRIMER v.6). The two fixed-factors were species and CO_2 enrichment, while the dependent variables were the % nitrogen, carbon and ash content

of dried biomass. Biomass composition data was fourth-root transformed for the PERMANOVA.

Results

pH drift in closed vessel

The pH drifted from 7.9 to over 9.7 for all three algal species (Fig. 1). *C. coelothrix* had the highest pH compensation point of 9.9, which was reached after six h in culture. *C. linum* and *C. patentiramea* used the HCO_3^- in the water at a slower rate, taking eight and ten hours to achieve the slightly lower pH compensation points of 9.8 and 9.7, respectively (Fig. 1).The relatively faster rate of pH increase for *C. coelothrix* supports more efficient HCO_3^- use than either *C. linum* and *C. patentiramea*.

Effects of pH on algal growth

All three species had decreasing growth rates with increasing pH above the optimum of pH 7.5. However, there was a significant interaction between the species and the pH levels in which they were cultured (P<0.001, Table 1, Fig. 2), driven by different optimal pH ranges for growth. In other terms, integrating the pH drift results in the previous section, different growth responses were reflective of different HCO_3^- affinities. Both *C. coelothrix* and *C. linum* had higher growth rates at pH levels between 7 and 8.5, whereas the optimum pH range for *C. patentiramea* was between 7 and 8 (Fig. 2). There were no significant differences in growth rates within the optimal pH range for each species (Tukey's HSD, P>0.05). The highest individual growth rates for all three species were measured at pH 7.5, with growth rates of 14.5, 8.8 and 8.2% day^{-1} for *C. linum*, *C. patentiramea* and *C. coelothrix*, respectively (Fig. 2). Growth rates for *C. linum* and *C. coelothrix* decreased above the optimal pH range (from pH 8.5 to pH 9) by 48% and 35% relative to the control, respectively. Growth rates for *C. patentiramea* decreased by 30% relative to the control above the optimal pH range (from pH 8 to pH 8.5). Growth rate further decreased to 47% of the control at pH 9. The highest susceptibility of *C. patentiramea* growth to increasing pH levels (lower $CO_2:HCO_3^-$ ratio) supports a relatively lower capability of using HCO_3^-. In accordance with the pH drift experiment, the lower

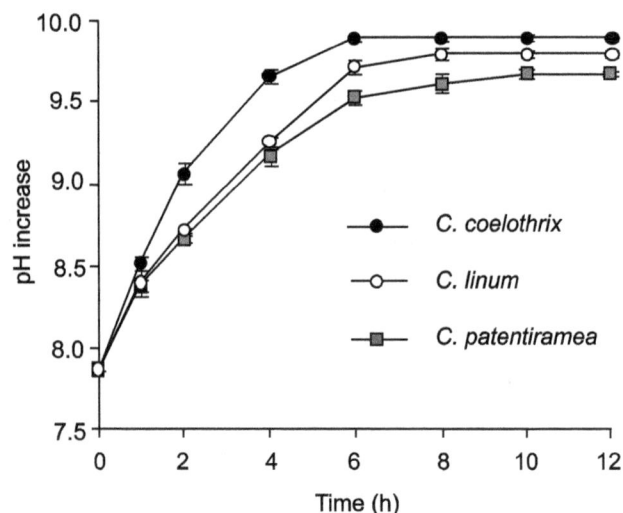

Figure 1. pH drift experiment for *C. coelothrix, C. patentiramea* and *C. linum*. Data show mean pH value (±1 SD) for each sampling time (n = 3).

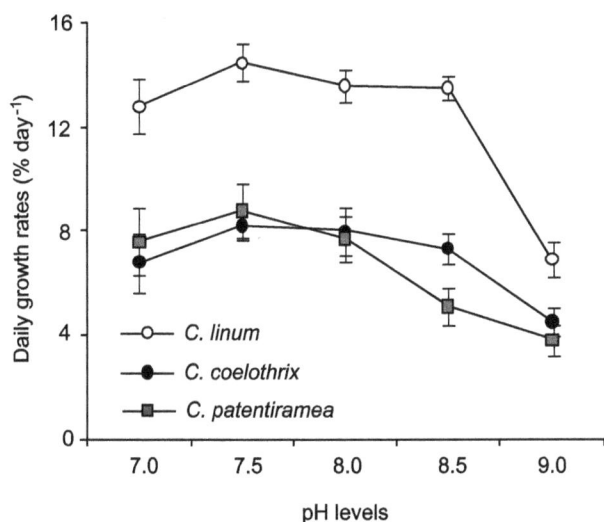

Figure 2. Growth of *C. coelothrix, C. patentiramea* and *C. linum* cultured in different pH levels. Data show mean daily growth rates (± 1 SE) for each pH levels*species (n = 3).

Table 2. Values for pH, dissolved inorganic carbon (Ci), carbon dioxide (CO_2) and bicarbonate (HCO_3^-) for the CO_2 enrichment experiments.

	CO_2 enrichment	pH	Ci (mM)	CO_2 (μM)	HCO_3^- (μM)
Inflow	$+CO_2$	6.73 ± 0.26	2.33 ± 0.20	250 ± 60	1980 ± 160
	Control	7.98 ± 0.16	1.63 ± 0.12	10 ± 05	1400 ± 100
C. coelothrix	$+CO_2$	7.42 ± 0.20	1.60 ± 0.15	40 ± 10	1510 ± 140
	Control	8.52 ± 0.12	1.17 ± 0.13	0	930 ± 130
C. linum	$+CO_2$	7.38 ± 0.19	1.62 ± 0.14	50 ± 10	1520 ± 110
	Control	8.47 ± 0.10	1.20 ± 0.13	0	980 ± 120
C. patentiramea	$+CO_2$	7.35 ± 0.20	1.62 ± 0.11	50 ± 10	1525 ± 100
	Control	8.38 ± 0.13	1.26 ± 0.14	0	1100 ± 140

Data show mean values (± 1 SD) from the inflow and outflow of the green tide algal cultures with additional CO_2 and control (n = 8).

sensitivity of *C. coelothrix* to changes in pH supports its more efficient use of HCO_3^-.

Algal productivity under CO_2 enrichment

Based on the differences in HCO_3^- utilization efficiencies between the three species, the subsequent step was to quantify increases in productivity through the addition of CO_2 in controlled intensive cultures. The addition of CO_2 decreased the pH of the inflowing seawater from ~ pH 8 (control) to pH 6.7. Under these conditions the concentration of CO_2 was twenty five times higher in the CO_2 enriched cultures compared to the control (Table 2). The addition of CO_2 also increased the concentration of HCO_3^- in the CO_2 enriched cultures to 2 mM, compared to 1.4 mM in the control cultures, because the hydration of CO_2 produces carbonic acid, and its subsequent de-protonation leads to the formation of HCO_3^-. After passing through the seaweed tanks, at 2 vol h^{-1}, all CO_2 was depleted from water within the control cultures. In contrast, there was a continual supply of CO_2 in the CO_2 enriched cultures for photosynthesis. HCO_3^- concentration in the control and CO_2 enriched cultures was ~1 mM and 1.5 mM, respectively. *C. coelothrix* cultures had the lowest concentration of all carbon forms (Table 2), and therefore the highest carbon uptake rates of all three species. *C. patentiramea*

Table 1. Summary output for significant interactions of the ANOVA and PERMANOVA analyses.

Source	df	MS	F	P
ANOVA				
Species*pH	12	16.16	7.01	**<0.001**
Species* CO_2	2	5.91	3.73	**0.031**
PERMANOVA				
Species* CO_2	2	7.06	16.33	**<0.001**

ANOVA testing the effects of varying pH on growth and CO_2 enrichment on algal productivity, and PERMANOVA (Species*CO_2) testing effects of CO_2 enrichment on biomass elemental composition.

cultures had the highest Ci concentration in the water, in particular in the control treatment, and therefore carbon uptake rates were lower for *C. patentiramea* when only HCO_3^- was present. These results confirm the laboratory data indicating that *C. patentiramea* has the least effective HCO_3^- utilisation of the three species.

The three algal species had different productivity (growth) responses to CO_2 enrichment, with a significant interaction between CO_2 supply and the species tested (P<0.001, Table 1). The relative productivity of *C. coelothrix* and *C. linum* were significantly enhanced (~26 and 24%, respectively) when supplied with additional CO_2 (Fig. 3) The productivity of *C. coelothrix* increased from 12.5 to 16.8 g DW m^{-2} day^{-1}, and *C. linum* from 9.5 to 12 g DW m^{-2} day^{-1} (Fig. 3). The productivity of *C. patentiramea* (5.2 to 6.2 g DW m^{-2} day^{-1}) to CO_2 enrichment was not significantly different to that of the control (Tukey's HSD, P>0.05).

Biomass elemental analysis

In general, *C. coelothrix* and *C. linum* had higher carbon and nitrogen concentrations and lower ash contents than *C. patentiramea*

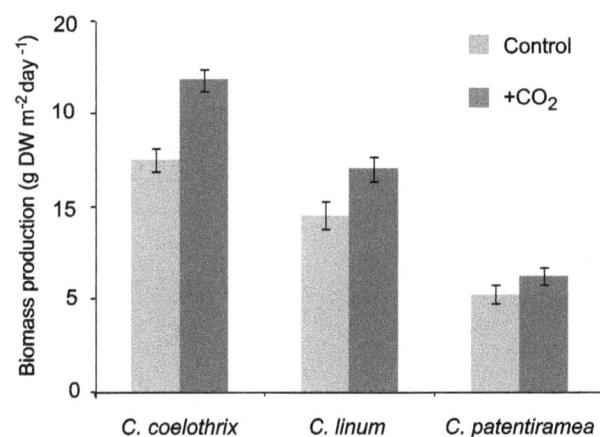

Figure 3. Biomass productivity in response to CO_2 enrichment for *C. coelothrix, C. patentiramea* and *C. linum*. Data show mean biomass productivity (± 1 SE) for each CO_2 level*species (n = 3).

(Table 3). However, CO_2 enrichment influenced the elemental composition of the algal species in different ways (PERMANOVA, Species*CO_2, P<0.001, Table 1). This interaction was mainly driven by positive influence of CO_2 enrichment on carbon and nitrogen content in *C. coelothrix* and *C. linum* compared to the negative influence in *C. patentiramea*, which corresponded with an increase in ash content in the latter (Table 3). *C. coelothrix* biomass increased in carbon and nitrogen content by ~2%, while ash decreased by ~1% relative to the control (Table 3). *C. linum* biomass also increased in carbon and nitrogen content compared to the control, but to different degrees (by ~4% and 2%, respectively). In contrast, the biomass of *C. patentiramea* cultured under CO_2 enrichment had a lower and carbon (4%) and nitrogen (1%) content and high ash content (7%) compared with the control (Table 3).

Discussion

This study demonstrates that three green tide algal species, *C. coelothrix*, *C. linum* and *C. patentiramea*, have the ability to use HCO_3^- as a complementary carbon source to CO_2 for photosynthesis. However, this ability is restricted to a narrower pH range than for many other green algae belonging to the same genera. There is a lower comparative complexity or efficiency of the mechanisms involved in the uptake or conversion of HCO_3^- to CO_2 for these species. Consequently, this corresponds to a relatively higher dependence on CO_2 as a carbon source for photosynthesis and this is reflected in the significant enhancement of productivity of these species when enriched with CO_2 in intensive culture.

Algal affinity for HCO_3^- pH drift in closed vessel

Comparatively, green algae as a taxonomic group photosynthesise at the highest pH levels with compensation points up to pH 10.8 [32]. At these pH levels, CO_2 is absent and HCO_3^- is the only functional form of inorganic carbon, representing less than a quarter of the total Ci. Active photosynthesis at these pH levels is only possible because of diverse and highly efficient mechanisms to overcome CO_2 constraints through the utilization of HCO_3^- [36]. There are at least two mechanisms to utilize HCO_3^- in green macroalgae [37]. The first is the extracellular dehydration of HCO_3^- into CO_2 through the periplasmic carbonic anhydrase (CA) enzyme, followed by diffusion of CO_2 into the cell, and this is the most widely distributed mechanism. The second mechanism is the direct uptake of HCO_3^- through the plasma membrane, mediated by an anion exchange protein [38]. Some species of

Cladophora have a third mechanism with the uptake the Ci through a vanadate-sensitive P-type H^+-ATPase (proton pump) [39]. Species with these mechanisms raise the pH up to 10.5 in a closed vessel. The three species in this study did not raise the pH above 9.9 and therefore have limited HCO_3^- transport. They almost certainly concentrate carbon using the dehydration of HCO_3^- by CA into CO_2, as this is the most common mechanism of HCO_3^- utilization in algae [40]. This mechanism usually operates at ~ pH 8.3, when the proportion of CO_2 in the total Ci pool is below 1% and HCO_3^- is more than 90%. The capacity to utilize HCO_3^- through this mechanism decreases sharply with increased pH, and is ineffective at pH 9.8 [41]. The direct transport of HCO_3^- through an anion exchange protein usually operates at higher pH (~9.3) [39], and is the most probable mechanism for compensation points above 9.5 in the three species. These two mechanisms operate separately in other species of green algae with periplasmic CA activity dominating at lower pH, and direct uptake of HCO_3^- by an anion exchanger at higher pH [39]. The incapacity of the species in this study to raise the pH above 9.7–9.9 suggests that there is no proton pump mechanism involved in HCO_3^- transport. [42] inhibited the proton pump mechanism in *Ulva procera*, an alga capable of a pH compensation point of 10.5, and the pH remained below 9.9, demonstrating a reliance on this third mechanism to elevate the pH compensation to its highest level.

In a comparative context, the two *Cladophora* species in this study are less efficient in the use of HCO_3^- than other species from the same genera with pH compensation points of ~pH 10.5 [32,42]. However, as in this study, some species of *Cladophora* maintain a preference for dissolved CO_2 as a carbon source [43]. These different responses may be related to the environmental niche prior to experiments, as the ability of algae to utilise HCO_3^- is strongly related to habitat [32]. Individuals of the same species can express alternate strategies for carbon acquisition when in different habitats, or the habitat itself might select for survival of genotypes with different carbon acquisition strategies [31]. This relatively limited ability to utilise HCO_3^- is reflected in the growth response of all three species at different pH environments in this study.

Effects of pH on algal growth

As pH increases from 7 to 8, the relative proportion of Ci present as CO_2 is reduced by over 70%, while the relative proportion of HCO_3^- decreases by only 10%. This drastic change in the CO_2:HCO_3^- ratio had no effect on the growth of algae in this study. The comparative ratio of CO_2:HCO_3^- was maintained at each pH throughout the experiment through the addition of a biological buffer. Consequently, CO_2 is always available between pH 7 and 8 at concentrations that meet the carbon requirement for algal photosynthesis and growth. This does not, however, exclude the activation of the CA mechanism at ~pH 8, which supplies additional CO_2 derived from HCO_3^- to compensate for its lower availability at increased pH [19].

From pH 8 to 8.5, CO_2 decreases markedly and photosynthesis and growth depends on the efficiency of HCO_3^- utilization mechanisms. The growth rates of *C. linum* and *C. coelothrix* within this pH range changed little. In contrast, a significant decrease in growth for *C. patentiramea* confirms that it is the least adapted to grow in the absence of CO_2. Above pH 8.5, HCO_3^- is replaced by CO_3^{2-} and growth decreased significantly for all species. Steeper decreases in growth rates for *C. patentiramea* and *C. linum* between pH 8 and 9, compared to *C. coelothrix*, correspond with the slower rate of increasing pH for these two species in the pH drift experiment. These data, together with the highest pH compensation point, confirm that *C. coelothrix* has the most efficient

Table 3. Values for % carbon (C), % nitrogen (N) and % ash from algal biomass cultured with CO_2 enrichment and control.

Species	CO_2 enrichment	% C	% N	% Ash
C. coelothrix	+CO_2	33.29±0.28	6.05±0.03	25.20±0.55
	Control	30.97±0.30	4.07±0.03	26.22±0.56
C. linum	+CO_2	31.67±0.70	5.98±0.10	29.20±0.88
	Control	27.12±0.13	4.12±0.13	33.43±0.47
C. patentiramea	+CO_2	18.05±0.45	3.07±0.08	56.97±1.97
	Control	22.53±1.06	4.27±0.17	50.66±1.53

Data show mean values (±1 SD) of % dried biomass for each species* CO_2 treatment (n = 3).

mechanisms of HCO_3^- utilisation. However, in a broader comparative context, the relatively low pH compensation points and significant decreases in growth rates from pH 8 to 9, again demonstrate that the three species in this study are not as efficient in the use of HCO_3^- as a carbon source compared to many other green tide algal species.

Algal productivity under CO_2 enrichment

The productivity of two of the three species of green tide algae, *C. coelothrix* and *C. linum*, was enhanced through the addition of CO_2. Notably, the enrichment treatment had twenty five times more CO_2 available than the control. This maintained the pH of the enriched water below pH 7.5 (excess CO_2), whereas the pH of the control cultures averaged 8.5 (depleted CO_2). Considering the relatively limited ability of species to utilize the HCO_3^- pool, and that this process has an energetic cost [28], the constant presence of CO_2 at pH 7.5 disproportionately facilitated photosynthetic carbon fixation, and ultimately enhanced biomass productivity. Enhanced growth rates with CO_2 enrichment have also been reported for other species capable of using bicarbonate [20–23]. However, the magnitude of the growth responses to CO_2 enrichment is to some extent dependent on the efficiency of carbon concentrating mechanisms for each species. For example, species depending almost exclusively on CO_2 for photosynthesis can increase their biomass productivity up to three times when cultured in enriched CO_2 culture media [19,44,45]. Any differences in growth relative to enhanced CO_2 can also be due to the effect of CO_2 on the rate of nitrogen assimilation [46,47]. High levels of CO_2 can increase the rate of nitrogen assimilation in some algae by up-regulating nitrate reductase, the main enzyme in the nitrate assimilatory pathway [47,48]. This may be the case for *C. coelothrix* and *C. linum* in this study where they have a higher nitrogen content under CO_2 enrichment. Higher nitrogen and carbon contents on top of increased productivities with CO_2 enrichment represents a clear advantage for integrated systems focused on biomass production for bioremediation of waste streams [14,49].

In contrast, the nitrogen content of *C. patentiramea* decreased under CO_2 enrichment, suggesting no effect on assimilation. The effect of high CO_2 on the assimilation of nitrogen in algae is not consistent with decreases in assimilation for other species of algae

[24,50]. This effect may contribute to the relatively lack of increase in productivity of *C. patentiramea* under CO_2 enrichment. Notably, laboratory experiments suggested that *C. patentiramea* should be the most sensitive species to CO_2 enrichment based on relative capabilities of HCO_3^- utilization, which indicates that controlled, static, laboratory experiments may not be efficient to predict responses in flow environments (e.g. similar to commercial scale), potentially because of boundary layer/water motion effects on Ci distribution [51]. An alternative but related driver to water motion is the morphological differences between the green tide algae. The two rapidly growing species which had enhanced growth under CO_2 enrichment, *C. coelothrix* and *C. linum*, have a fine filamentous morphology suitable for tumble culture. In contrast, *C. patentiramea* has tightly interwoven filaments (e.g. ball-like structure) that restrict light to the inner filaments (auto-shading), thereby potentially limiting photosynthesis and growth. These physical factors may have influenced small density cultures in the laboratory in a different way than the dense cultures in the outdoor experiment, where individual, larger clumps could become limited. Regardless, *C. patentiramea* is not a good option for intensive cultivation because despite the lack of growth response to additional CO_2, high CO_2 affected negatively the nitrogen and carbon content while increasing ash content, and therefore the amount of biomass that can be converted into soil conditioners [52] or biofuels [9] decreases substantially.

In conclusion, intensive cultures of *C. coelothrix* and *C. linum* enriched with CO_2 had significantly enhanced productivity, despite their ability to utilise HCO_3^-. This demonstrates the potential for enhanced production for these species using CO_2 enrichment. This can be integrated with the industrial production of CO_2 and waste water streams from industry [53] to deliver a model where algae provide a bioremediation service of both air (CO_2) and water (nitrogen, phosphorous, metals and trace elements), and an opportunity to utilise this biomass for bio-energy products.

Author Contributions

Conceived and designed the experiments: PPS RdN NP LM. Performed the experiments: PPS. Analyzed the data: PPS NP LM. Wrote the paper: PPS RdN NP LM.

References

1. Ross AB, Jones JM, Kubacki ML, Bridgeman T (2008) Classification of macroalgae as fuel and its thermochemical behaviour. Bioresource Technology 99: 6494–6504.

2. Wargacki AJ, Leonard E, Win MN, Regitsky DD, Santos CNS, et al. (2012) An engineered microbial platform for direct biofuel production from brown macroalgae. Science 335: 308–313.

3. Zhou D, Zhang LA, Zhang SC, Fu HB, Chen JM (2010) Hydrothermal liquefaction of macroalgae *Enteromorpha prolifera* to bio-oil. Energy & Fuels 24: 4054–4061.

4. Raven JA, Taylor R (2003) Macroalgal growth in nutrient-enriched estuaries: A biogeochemical and evolutionary perspective. Water, Air, & Soil Pollution: Focus 3: 7–26.

5. Taylor R, Fletcher RL, Raven JA (2001) Preliminary studies on the growth of selected 'Green tide' algae in laboratory culture: Effects of irradiance, temperature, salinity and nutrients on growth rate. Botanica Marina 44: 327–336.

6. Yanagisawa M, Nakamura K, Ariga O, Nakasaki K (2011) Production of high concentrations of bioethanol from seaweeds that contain easily hydrolyzable polysaccharides. Process Biochemistry 46: 2111–2116.

7. Bruhn A, Dahl J, Nielsen HB, Nikolaisen L, Rasmussen MB, et al. (2011) Bioenergy potential of *Ulva lactuca*: Biomass yield, methane production and combustion. Bioresource Technology 102: 2595–2604.

8. Migliore G, Alisi C, Sprocati AR, Massi E, Ciccoli R, et al. (2012) Anaerobic digestion of macroalgal biomass and sediments sourced from the Orbetello lagoon, Italy. Biomass and Bioenergy 42: 69–77.

9. Zhou D, Zhang SC, Fu HB, Chen JM (2012) Liquefaction of macroalgae *Enteromorpha prolifera* in sub-/supercritical alcohols: direct production of ester compounds. Energy & Fuels 26: 2342–2351.

10. Pierri C, Fanelli G, Giangrande A (2006) Experimental co-culture of low food-chain organisms, *Sabella spallanzanii* (Polychaeta, Sabellidae) and *Cladophora prolifera* (Chlorophyta, Cladophorales), in Porto Cesareo area (Mediterranean Sea). Aquaculture Research 37: 966–974.

11. Bird MI, Wurster CM, de Paula Silva PH, Paul NA, de Nys R (2012) Algal biochar: effects and applications. Global Change Biology Bioenergy 4: 61–69.

12. Neori A (2008) Essential role of seaweed cultivation in integrated multi-trophic aquaculture farms for global expansion of mariculture: an analysis. Journal of Applied Phycology 20: 567–570.

13. Robertson-Andersson D, Potgieter M, Hansen J, Bolton J, Troell M, et al. (2008) Integrated seaweed cultivation on an abalone farm in South Africa. Journal of Applied Phycology 20: 579–595.

14. de Paula Silva PH, McBride S, de Nys R, Paul NA (2008) Integrating filamentous 'green tide' algae into tropical pond-based aquaculture. Aquaculture 284: 74–80.

15. Tsagkamilis P, Danielidis D, Dring MJ, Katsaros C (2010) Removal of phosphate by the green seaweed *Ulva lactuca* in a small-scale sewage treatment plant (Ios Island, Aegean Sea, Greece). Journal of Applied Phycology 22: 331–339.

16. Bidwell RGS, McLachlan J (1985) Carbon nutrition of seaweeds: photosynthesis, photorespiration and respiration. Journal of Experimental Marine Biology and Ecology 86: 15–46.

17. Lobban CS, Harrison PJ (1994) Seaweed ecology and physiology. Cambridge: Cambridge University Press. 366p.

18. Maberly SC (1992) Carbonate ions apper to neither inhibit nor stimulate use of bicarbonate ions in photosynthesis by *Ulva lactuca*. Plant Cell and Environment 15: 255–260.

19. Mata L, Silva J, Schuenhoff A, Santos R (2007) Is the tetrasporophyte of *Asparagopsis armata* (Bonnemaisoniales) limited by inorganic carbon in integrated aquaculture? Journal of Phycology 43: 1252–1258.

20. Gao K, Aruga Y, Asada K, Ishihara T, Akano T, Kiyohara M (1991) Enhanced growth of the red alga *Porphyra yezoensis* Ueda in high CO_2 concentrations. Journal of Applied Phycology 3: 355–362.

21. Gao K, Aruga Y, Asada K, Kiyohara M (1993) Influence of enhanced CO_2 on growth and photosynthesis of the red algae *Gracilaria* sp. and *G. chilensis*. Journal of Applied Phycology 5: 563–571.

22. Demetropoulos CL, Langdon CJ (2004) Enhanced production of Pacific dulse (*Palmaria mollis*) for co-culture with abalone in a land-based system: effects of seawater exchange, pH, and inorganic carbon concentration. Aquaculture 235: 457–470.

23. Zou DH (2005) Effects of elevated atmospheric CO_2 on growth, photosynthesis and nitrogen metabolism in the economic brown seaweed, *Hizikia fusiforme* (Sargassaceae, Phaeophyta). Aquaculture 250: 726–735.

24. Andria JR, Vergara JJ, Pérez-Lloréns JL (1999) Biochemical responses and photosynthetic performance of *Gracilaria* sp (Rhodophyta) from Cadiz, Spain, cultured under different inorganic carbon and nitrogen levels. European Journal of Phycology 34: 497–504.

25. Israel A, Katz S, Dubinsky Z, Merrill JE, Friedlander M (1999) Photosynthetic inorganic carbon utilization and growth of *Porphyra linearis* (Rhodophyta). Journal of Applied Phycology 11: 447–453.

26. Israel A, Hophy M (2002) Growth, photosynthetic properties and Rubisco activities and amounts of marine macroalgae grown under current and elevated seawater CO_2 concentrations. Global Change Biology 8: 831–840.

27. Giordano M, Beardall J, Raven JA (2005) CO_2 concentrating mechanisms in algae: mechanisms, environmental modulation, and evolution. Annual Review in Plant Biology 56: 99–131.

28. Raven JA, Cockell CS, De La Rocha CL (2008) The evolution of inorganic carbon concentrating mechanisms in photosynthesis. Philosophical Transactions of the Royal Society of London, Series B: Biological Sciences 363: 2641–2650.

29. de Paula Silva PH, De Nys R, Paul NA (2012) Seasonal growth dynamics and resilience of the green tide alga *Cladophora coelothrix* in high-nutrient tropical aquaculture. Aquaculture Environment Interactions 2: 253–266.

30. Aresta M, Dibenedetto A, Barberio G (2005) Utilization of macroalgae for enhanced CO_2 fixation and biofuels production: Development of a computing software for an LCA study. Fuel processing technology 86: 1679–1693.

31. Murru M, Sandgren CD (2004) Habitat matters for inorganic carbon acquisition in 38 species of red macroalgae (Rhodophyta) from Puget Sound, Washington, USA. Journal of Phycology 40: 837–845.

32. Maberly SC (1990) Exogenous sources of inorganic carbon for photosynthesis by marine macroalgae. Journal of Phycology 26: 439–449.

33. Ryther JH, Guillard RRL (1962) Studies of marine planktonic diatoms:II. Use of *Cyclotella nana* Hustedt for assays of vitamin B12 in sea water. Canadian Journal of Microbiology 8: 437–445.

34. Lewis E, Wallace DWR (1998) Program developed for CO_2 system calculations. ORNL/CDIAC-105. Carbon dioxide information analysis center, Oak Ridge National Laboratory, U.S. Department of Energy, Oak Ridge, Tennessee, 38pp.

35. Quinn GP, Keough MJ (2002) Experimental design and data analysis for biologists. Cambridge: Cambridge University Press. 527p.

36. Beer S, Bjork M (1994) Photosynthetic properties of protoplasts, as compared with thalli, of *Ulva faciata* (Chlorophyta). Journal of Phycology 30: 633–637.

37. Axelsson L, Larsson C, Ryberg H (1999) Affinity, capacity and oxygen sensitivity of two different mechanisms for bicarbonate utilization in *Ulva lactuca* L.(Chlorophyta). Plant, Cell & Environment 22: 969–978.

38. Larsson C, Axelsson L (1999) Bicarbonate uptake and utilization in marine macroalgae. European Journal of Phycology 34: 79–86.

39. Choo K, Snoeijs P, Pedersén M (2002) Uptake of inorganic carbon by *Cladophora glomerata* (Chlorophyta) from the Baltic sea. Journal of phycology 38: 493–502.

40. Badger MR, Price GD (1994) The role of carbonic anhydrase in photosynthesis. Annual review of plant biology 45: 369–392.

41. Axelsson L, Ryberg H, Beer S (1995) Two modes of bicarbonate utilization in the marine green macroalga *Ulva lactuca*. Plant, Cell & Environment 18: 439–445.

42. Choo KS, Nilsson J, Pedersen M, Snoeijs P (2005) Photosynthesis, carbon uptake and antioxidant defence in two coexisting filamentous green algae under different stress conditions. Marine Ecology Progress Series 292: 127–138.

43. Rivers JS, Peckol P (1995) Interactive effects of nitrogen and dissolved inorganic carbon on photosynthesis, growth, and ammonium uptake of the macroalgae *Cladophora vagabunda* and *Gracilaria tikvahiae*. Marine Biology 121: 747–753.

44. Kubler JE, Johnston AM, Raven JA (1999) The effects of reduced and elevated CO_2 and O_2 on the seaweed *Lomentaria articulata*. Plant Cell and Environment 22: 1303–1310.

45. Mata L, Gaspar H, Santos R (2012) Carbon/nutrient balance in relation to biomass production and halogenated compound content in the red alga *Asparagopsis taxiformis* (Bonnemaisoniaceae). Journal of Phycology 48: 248–253.

46. Rivers JS, Peckol P (1995) Summer decline of *Ulva lactuca* (Chlorophyta) in a eutrophic embayment - Interactive effects of temperature and nitrogen availability. Journal of Phycology 31: 223–228.

47. Gordillo FJL, Niell FX, Figueroa FL (2001) Non-photosynthetic enhancement of growth by high CO_2 level in the nitrophilic seaweed *Ulva rigida* C. Agardh (Chlorophyta). Planta 213: 64–70.

48. Mercado JM, Javier F, Gordillo L, Xavier Niell F, Figueroa FL (1999) Effects of different levels of CO_2 on photosynthesis and cell components of the red alga *Porphyra leucosticta*. Journal of applied phycology 11: 455–461.

49. Israel A, Gavrieli J, Glazer A, Friedlander M (2005) Utilization of flue gas from a power plant for tank cultivation of the red seaweed *Gracilaria cornea*. Aquaculture 249: 311–316.

50. García-Sánchez MJ, Fernández JA, Niell X (1994) Effect of inorganic carbon supply on the photosynthetic physiology of *Gracilaria tenuistipitata*. Planta 194: 55–61.

51. Hurd CL (2000) Water motion, marine macroalgal physiology, and production. Journal of Phycology 36: 453–472.

52. Bird MI, Wurster CM, de Paula Silva PH, Bass AM, de Nys R (2011) Algal biochar - production and properties. Bioresource Technology 102: 1886–1891.

53. Saunders RJ, Paul NA, Hu Y, de Nys R (2012) Sustainable sources of biomass for bioremediation of heavy metals in waste water derived from coal-fired power generation. PloS one 7: e36470.

3

Genome Anchored QTLs for Biomass Productivity in Hybrid *Populus* Grown under Contrasting Environments

Wellington Muchero[1,2]*, **Mitchell M. Sewell**[1¤a], **Priya Ranjan**[1,2], **Lee E. Gunter**[1,2], **Timothy J. Tschaplinski**[1,2], **Tongming Yin**[1¤b], **Gerald A. Tuskan**[1,2]

1 Bioscience Division, Oak Ridge National Laboratory, Oak Ridge, Tennessee, United States of America, **2** BioEnergy Science Center, Oak Ridge National Laboratory, Oak Ridge, Tennessee, United States of America

Abstract

Traits related to biomass production were analyzed for the presence of quantitative trait loci (QTLs) in a *Populus trichocarpa* × *P. deltoides* F$_2$ population. A genetic linkage map composed of 841 SSR, AFLP, and RAPD markers and phenotypic data from 310 progeny were used to identify genomic regions harboring biomass QTLs. Twelve intervals were identified, of which *BM-1*, *BM-2*, and *BM-7* were identified in all three years for both height and diameter. One putative QTL, *BM-7*, and one suggestive QTL exhibited significant evidence of over-dominance in all three years for both traits. Conversely, QTLs *BM-4* and *BM-6* exhibited evidence of under-dominance in both environments for height and diameter. Seven of the nine QTLs were successfully anchored, and QTL peak positions were estimated for each one on the *P. trichocarpa* genome assembly using flanking SSR markers with known physical positions. Of the 3,031 genes located in genome-anchored QTL intervals, 1,892 had PFAM annotations. Of these, 1,313, representing 255 unique annotations, had at least one duplicate copy in a QTL interval identified on a separate scaffold. This observation suggests that some QTLs identified in this study may have shared the same ancestral sequence prior to the salicoid genome duplication in *Populus*.

Editor: Lewis Lukens, University of Guelph, Canada

Funding: This research was supported, in part, by the Biomass Feedstock Development Program, United States Department of Energy; by the Office of Industrial Technology, Energy Efficiency Renewable Energy, United States Department of Energy; by the Office of Science, Biological, and Environmental Research, United States Department of Energy; by the Department of Energy, Office of Science, Biological and Environmental Research, as part of the Plant Microbe Interfaces Scientific Focus Area and by the BioEnergy Science Center. Oak Ridge National Laboratory is managed by UT-Battelle, LLC, for the United States Department of Energy under contract DE-AC05-00OR22725. The funders had no role in study design, data collection and analysis, decision to publish, or preparation of the manuscript.

Competing Interests: The authors have declared that no competing interests exist.

* E-mail: mucherow@ornl.gov

¤a Current address: Mountain Horticultural Crops Research and Extension Center, Mills River, North Carolina, United States of America
¤b Current address: The Key lab of Forest Genetics and Gene Engineering, Nanjin Forestry University, Nanjin, China

Introduction

Hybrid poplars have been intensively cultivated in North America as a short-rotation woody crop species for bioenergy and pulp and paper industries [1,2,3,4]. The recent focus on lignocellulosic biofuels from plant biomass as a complement to fossil fuels has led to increased interest in the genetic characteristics of *Populus* as a rapidly growing biomass feedstock [5]. Several factors account for *Populus* success as a feedstock for biofuels and pulp and paper industries, foremost of which is the interspecific hybridization resulting in hybrids with marked improvement in performance, hybrid vigor (i.e heterosis or over-dominance) compared to parental genotypes [6]. Being a genetically diverse genus which displays considerable variation among its species in such adaptive traits as stem growth, crown architecture, and disease resistance, the genetic diversity can be captured through the ease of hybridization among the approximately 30 species of *Populus*. The most desirable clones from these hybrid combinations can then be easily propagated by the well-developed vegetative systems inherent to *Populus*. Inter-american hybrids [7] generated from crosses between *Populus trichocarpa* Torr. & Gray and *Populus deltoides* Bartr. ex Marsh. (*i.e.*, T×D hybrids) are estimated to produce as much as 35 Mg•ha^{-1}•yr^{-1} of aboveground biomass at age four [8]. Much of the success of the T×D hybrids is thought to result from the complementary combination of desirable traits inherited from each of the parental species in conjunction with the associated hybrid vigor [9]. Although hybrids exhibit superior performance at the overall phenotypic level, out-breeding depression or under-dominance, at the individual locus level resulting from combining alleles that result in poorer performance of the hybrid relative to parental genotypes is also a known genetic phenomenon [10], and may prevent hybrids from achieving maximum possible performance. Therefore, it is of primary interest to identify specific genomic loci contributing toward such hybrid vigor as well as those that contribute toward out-breeding depression for marker-assisted pyramiding of beneficial loci.

Biomass productivity traits are generally quantitative in nature involving numerous genes and genetic pathways whose activity may be modified by the environment leading, sometimes, to environment specific trait expression [11]. Understanding the genotype x environmental interactions of these genetic elements may enable targeted introgression of beneficial loci through ideotype breeding strategies [12]. Numerous studies have identified quantitative trait loci (QTLs) that are involved in biomass accumulation using hybrid *Populus* pedigrees. Results of these studies were generally reproducible among different studies and environments [13–17] suggesting that economically beneficial genes can be isolated for *Populus* improvement.

With the first genome sequence of any tree species [18] in addition to a well developed DNA molecular marker resource, *Populus* has mature genomic resources that should allow for in-depth characterization of loci of interest. Although QTL intervals typically include tens or hundreds of genes, candidate gene mapping has provided some insight into potentially valuable gene targets for improving hybrid *Populus* production. In addition, prioritization of marker saturation can be accurately guided by knowing the physical interval compared to using cM distances whose relationship with physical distance is not always linear due to the heterogeneity of recombination rates across the *Populus* genome [19]. To date, numerous strategies have been applied in mapping efforts to identify potentially viable loci down to the gene level. These strategies include differential gene content in syntenic genomic regions where QTL presence or absence may suggest genetic determinants of economically important traits [20].

The overarching objective of this study was to build on previous work that identified and characterized genetically-driven variation in growth phenotypes within an F_2 *P. trichocarpa* × *P. deltoides* pedigree. In this work, estimates of QTL numbers [12], QTL positions on a genetic linkage map [14], and aspects related to hybrid vigor, genotype-by-environment interaction (G×E), genetic correlations, and broad-sense heritability at the phenotypic level [15] were characterized. However, due to the molecular anonymity of RFLP, STS, and RAPD markers used in these studies and the unavailability of a reference genome at that time, the genomic location and genic features associated with these loci remained anonymous. The completion of the *Populus* genome assembly [18] and the incorporation of SSR markers with known physical positions in the genetic map [21] now provide an opportunity for further characterization of loci described in the precedent work.

Therefore, the goals of this study were to reanalyze phenotypic data for the F_2 pedigree to map QTLs segregating for stem height and diameter using an updated genetic map that incorporated SSR markers with known physical positions. In addition, we sought to provide estimates of the role of G×E interactions and hybrid vigor at the individual QTL level. Finally, we sought to utilize flanking SSR markers to delimit genomic intervals that encompass these QTLs thereby enabling the characterization of genic features associated with these loci.

Results

QTL Analysis and Detection Across Contrasting Environments

Nine putative and three suggestive QTLs were detected for stem height and diameter on eleven linkage groups (LGs) of the family 331 genetic map (Tables 1 and 2). Among the 9 QTLs, 5 were significant in at least one experiment when the more stringent genomewise LOD threshold was used as the cut-off point (Table 1). All QTLs identified in this study were associated with both height and diameter in the same experiment or across different experiments. Of these, six QTLs on LGs I, II, VII, VIII, XIII, and XIV were detected in both Boardman and Clatskanie experimental sites. Three of these on LGs I, II, and XIV were detected in all five datasets analyzed. QTLs *BM-3* and *BM-8* exhibited the highest level of location specificity with QTL x environment interactions showing

significance at Prob (>F) <0.1 (Table 3). All QTLs mapped reproducibly in the same map interval and peak positions were typically associated with no more than three markers in close proximity (Table 1). Figure 1 shows a graphical example of QTL *BM-2* and associated peak on LG II highlighting the close agreements between three different phenotypic datasets used to identify the QTL. Averaged across experiments, the percent phenotypic variance explained ranged from 5.2 to 8.5% for each QTL.

Additive and Dominant Effects

One putative and one suggestive QTL exhibited consistent evidence of over-dominance across experiments and traits, whereas two putative QTLs exhibited consistent under-dominance across traits and environments (Tables 1 and 2). QTL *BM-7* on LG XIV and a suggestive QTL on LG I were detected in all five datasets and exhibited over-dominance in each case (Tables 1 and 2). Putative QTLs *BM-1* and *BM-2* on LGs I and II, respectively, were also detected in all five datasets but each exhibited over-dominance in four of the five instances (Table1). On the other hand, QTLs *BM-4* and *BM-6* on LGs VII and XIII, respectively, exhibited under-dominance across different environments for both height and diameter (Table 1).

Genome Anchoring of QTL Intervals

In this study, seven of the nine putative QTLs *BM-1*, *BM-2*, *BM-3*, *BM-4*, *BM-5*, *BM-6*, and *BM-8* were successfully anchored on the *Populus* genome assembly (Figures 2 and 3 and Table 4). QTL *BM-9* located on LG XIX could not be anchored due to lack of flanking SSR markers with known physical positions. Figure 2 illustrates the use of two SSR markers flanking a QTL interval peaking at marker P_422 on LG II to anchor and estimate the QTL peak position on the genome assembly. The genome assembly position of QTL *BM-7* on LG XIV was reported previously by Ranjan et al. [20]. Interestingly, both markers associated with *BM-7* on LG XIV, CTACG-N1 and AGCGA-14 (Table 1), were associated with a QTL previously identified for root lignin percentage on the same map position [21]. The lignin QTL, *RL-5*, was subsequently anchored on the genome assembly and analyzed for candidate genes by Ranjan et al. [20]. Further, the QTL exhibited the same pattern of over-dominance for the lignin phenotype as it did for height and diameter in the current study. Genomic intervals covered by individual QTLs ranged from 1.3 to 8.8 Mb (Table 4).

Based on the QTL intervals defined here and the results of the SSR marker placement on the genome assembly, we identified 197 previously unmapped SSR markers that occurred within QTL intervals identified in this study (Table S1). The number of additional SSR markers ranged from 9 to 36 within individual QTLs (Tables 4 and S1).

Candidate Gene Identification and Characterization

Intervals spanning the genomic regions summarized in Table 4 were used to identify all genes occurring within the 8 genome-anchored QTLs (Table S2). The number of genes in each interval ranged from 37 for QTL *BM-7* to 721 for QTL *BM-8*. All together, there were 3,031 genes within the 8 genome-anchored QTL intervals. Out of these, 1,892 (62%) had annotations based on PFAM domains and these fell into 290 unique annotations (Table S2). Of the 1,892 annotated gene models, 1,313 (72%) had at least one duplicate in a QTL interval mapping on a different scaffold (Table S3). These represented 255 (88%) of the 290 unique annotations (Table S3).

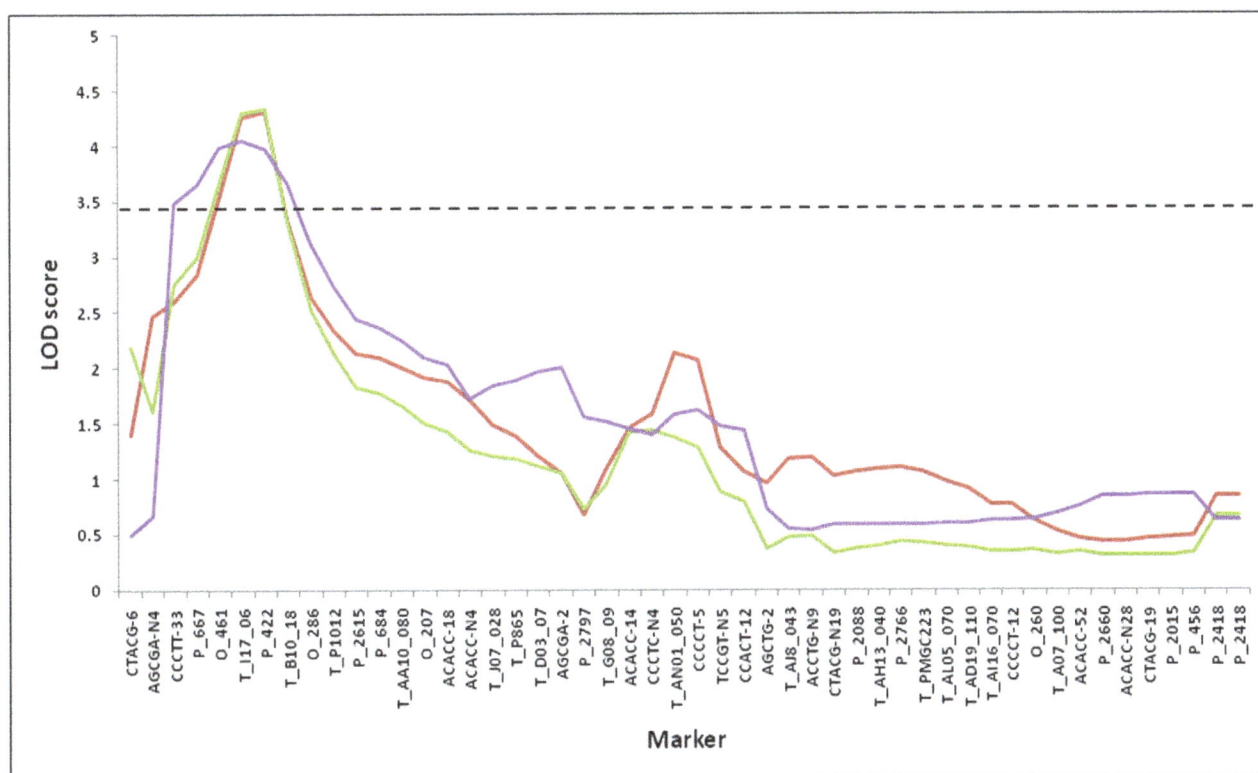

Figure 1. LOD traces for QTL B*M-2* on LG II based on 4-year height (red) and 4-year diameter (green) measured in Clatskanie and 4-year height measured in Boardman (purple). Broken horizontal line represents linkage groupwise LOD significance threshold calculated based on 1,000 permutations at the 0.05 significance level.

Discussion

The genus *Populus* is an economically important tree crop widely grown as feedstock for lignocellulosic biofuels and pulp and paper products, in part, for its rapid growth and ability to thrive in marginal lands that are not suitable for food crop production. Aside from its key importance as an industrial feedstock, *Populus* is also a biological model system for perennial tree crops because of its relatively compact genome, high level of interspecies diversity, and ease of experimental manipulation compared to other genera [4,18]. Increasing biomass productivity using genetic manipulation is of major interest in contemporary efforts to make biofuels production from *Populus* economically viable. Results presented here provide a basis for the isolation of specific genetic determinants that mediate expression of key productivity traits such as height and diameter. Additionally, this work represents a valuable resource in identification and prioritization of genomic intervals that may be targeted for marker-assisted breeding programs in *Populus*. Below, we discuss specific attributes of this work that should facilitate the application of results presented here in genetic improvement of *Populus* feedstocks.

Given the environmental contrast between the Boardman and Clatskanie experimental sites, QTL detection and expression for the stem-growth traits was remarkably consistent across sites. Only three of the twelve QTLs identified in this study exhibited evidence of location specificity. Therefore, these results indicate that the pattern of phenotypic response from each genotypic class was relatively consistent across the two environments. The generally robust detection of QTL regardless of environment is consistent with findings by Rae et al. [11] who evaluated progeny from the same pedigree in France, United Kingdom, and Italy and

identified virtually the same QTL intervals reported in this current study. Given their detection across different geographical environments, QTLs segregating in this pedigree offer valuable targets for further characterization and utilization in improving *Populus* productivity.

Despite the overall hybrid vigor observed for both height and diameter, varying levels of locus-specific QTL mode of action were observed, ranging from under-dominance to over-dominance. The presence of loci exhibiting apparent out-breeding depression (under-dominance) on linkage groups VII and XIII highlight the potential for further improvement of hybrid performance using targeted exclusion of such loci in ideotype breeding approaches. Interestingly, Tschaplinski et al. [10] identified a QTL for osmotic potential exhibiting under-dominance on LG XIII that mapped in the same interval as QTL *BM-6*. Conversely, the QTL peak for *BM-7* on LG XIV, with consistent evidence of over-dominance for both height and diameter, co-located with the same markers which were associated with a root lignin QTL, *RL-5*, which, in a previous study, also exhibited the same over-dominance mode of action [20,21]. Identification of the same QTL interval and the consistency of QTL mode of action between traits suggest that both height and diameter are largely influenced by the same genes. Such pleiotropic effects have been widely described in *Populus* for both related and diverse traits [12,20,21]. Additionally, the association of particular genomic intervals previous associated with phenotypes other than height and diameter suggest that pleiotropy may also exist between apparently non-related traits in *Populus*, although genetic linkage cannot be ruled out based on available data.

Table 1. QTLs associated with height and diameter identified in *Populus* family 331 F$_2$ pedigree based on linkage-group- (LG) and genome-wise LOD significance thresholds (GW).

QTL (LG)	Marker at QTL peak	LOD score	LOD threshold (LG)	LOD Threshold (GW)	Trait	% PVE	mu(ac)†	mu(ad)‡	mu(bc)‡	mu(bd)†	a	d	d/|a|
BM-1 (I)	AGCTG-1	3.26	2.9	4.3	Clatskanie 8 yr height	8.1	122.8	117.9	137.4	123.3	−0.25	4.60	18.4
BM-1 (I)	AGCTG-1	2.52	3.2	4.3	Clatskanie 4 yr height	6.6	8.8	9.8	10.7	8.7	0.05	1.50	30.0
BM-1 (I)	AGCTG-1	2.33	3.0	4.4	Clatskanie 4 yr diameter	6.6	8.4	10.0	11.2	8.6	−0.10	2.10	21.0
BM-1 (I)	AGCTG-1	2.51	3.0	4.4	Boardman 4 yr height	5.4	343.7	338.4	381.3	368.2	−12.25	3.90	0.30
BM-1 (I)	TCCGT-4	2.55	2.9	4.3	Boardman 4 yr diameter	6.8	3.6	3.7	4.3	3.8	−0.10	0.30	3.00
BM-2 (II)	T_D03_07	3.36	2.8	4.3	Boardman 4 yr diameter	6.0	4.1	3.8	4.1	3.6	0.25	0.10	0.40
BM-2 (II)	T_I17_06	4.05	3.1	4.4	Boardman 4 yr height	7.2	331.2	383.2	369.9	341.8	−5.30	40.10	7.56
BM-2 (II)	P_422	4.33	3.0	4.4	Clatskanie 4 yr diameter	7.9	8.4	10.1	11.2	8.2	0.10	2.35	23.50
BM-2 (II)	P_422	4.31	3.4	4.3	Clatskanie 4 yr height	7.8	8.6	9.9	10.6	8.7	−0.05	1.60	32.00
BM-2 (II)	CCCTT-33	2.19	2.8	4.3	Clatskanie 8 yr height	5.9	116.5	133.3	127.5	121.0	−2.25	11.65	5.18
BM-3 (V)	AGCTG-7	7.02	2.7	4.3	Boardman 4 yr diameter	10.5	4.3	3.8	4.0	3.4	0.45	0.05	0.11
BM-3 (V)	AGCTG-7	3.70	2.9	4.3	Boardman 4 yr height	6.4	374.5	345.8	376.4	329.0	22.75	9.35	0.41
BM-4 (VII)	T_AE20_120	3.37	2.6	4.3	Boardman 4 yr height	11.6	389.2	355.0	309.2	365	12.10	−45.00	−3.72
BM-4 (VII)	T_AB14_110	2.76	2.6	4.3	Boardman 4 yr diameter	5.5	4.15	3.8	3.5	3.9	0.12	−0.38	−3.00
BM-4 (VII)	T_AF04_090	2.62	2.6	4.4	Clatskanie 4 yr diameter	5.1	10.9	9.9	7.8	9.4	0.75	−1.30	−1.73
BM-4 (VII)	P_2140	2.52	2.6	4.3	Clatskanie 4 yr height	4.2	10.2	10.0	8.3	9.3	0.45	−0.60	−1.33
BM-5 (VIII)	TCCGT-16	4.94	2.3	4.3	Boardman 4 yr diameter	8.2	4.1	4.1	3.7	3.5	0.30	0.10	0.33
BM-5 (VIII)	TCCGT-16	2.70	2.5	4.4	Boardman 4 yr height	4.5	370.7	373.2	335.4	346.4	12.15	−4.25	−0.35
BM-5 (VIII)	TCCGT-16	2.83	2.4	4.3	Clatskanie 8 yr height	5.2	128.7	130.7	116.2	122.0	3.35	−1.90	−0.57
BM-6 (XIII)	CCCCT-N3	2.83	2.9	4.4	Boardman 4 yr height	6.7	364.6	335.0	347.8	385.8	−10.60	−33.80	−3.19
BM-6 (XIII)	CCCCT-N3	2.45	2.8	4.3	Boardman 4 yr diameter	5.7	4.0	3.6	3.7	4.2	−0.10	−0.45	−4.50
BM-6 (XIII)	AGCTG-27	4.26	2.9	4.3	Clatskanie 8 yr height	10.9	121.5	113.6	127.8	136.3	−7.40	−8.20	−1.11
BM-7 (XIV)	CTACG-N1	4.34	2.7	4.3	Boardman 4 yr diameter	9.0	3.9	4.3	3.9	3.5	0.20	0.40	2.00
BM-7 (XIV)	AGCGA-14	3.94	2.8	4.4	Boardman 4 yr height	6.7	364.8	370.7	365.2	318.5	23.15	26.30	1.14
BM-7 (XIV)	AGCGA-14	4.88	2.8	4.4	Clatskanie 4 yr diameter	8.7	9.4	11.3	9.6	7.4	1.00	2.05	2.05
BM-7 (XIV)	CTACG-N1	3.95	2.8	4.3	Clatskanie 8 yr height	10.5	120.2	132.8	138.8	116.6	1.80	17.4	9.67
BM-7 (XIV)	CTACG-N1	2.26	2.9	4.3	Clatskanie 4 yr height	5.9	9.4	10.3	10.4	8.5	0.45	1.40	3.11
BM-8 (XVII)	T_L09_105	2.50	2.4	4.3	Boardman 4 yr diameter	6.2	4.0	4.0	3.5	3.8	0.10	−0.15	−1.50
BM-8 (XVII)	T_AI06_055	1.94	2.5	4.4	Boardman 4 yr height	4.1	377.3	368.7	342.0	341.8	17.75	−4.2	−0.24
BM-9 (XIX)	ACCTG-24	3.72	2.7	4.4	Clatskanie 4 yr diameter	10.2	7.9	10.7	9.1	11.5	−1.80	0.20	0.11
BM-9 (XIX)	ACCTG-24	3.53	2.5	4.3	Clatskanie 4 yr height	11.2	8.3	11.0	9.0	10.5	−1.10	0.60	0.55
BM-9(XIX)	AGCTG-8	2.51	2.6	4.3	Clatskanie 8 yr height	3.8	120.5	122.1	124.4	133.5	−6.50	−3.75	−0.58

%PVE = percent phenotypic variance explained;

†Mean associated with heterozygous genotypes 'ac' and 'bd' where alleles are derived from the same species;

‡Mean associated with heterozygous genotypes 'ad' and 'bc' where alleles are derived from different species, a = additive; d = dominance; d/\|a\| = QTL mode of action.

Table 2. Suggestive QTLs associated with height and diameter identified in *Populus* family 331 F_2 pedigree.

Linkage group	Marker at QTL peak	LOD score	LOD threshold (LG)	Level of significance LG (GW)†	Trait	% PVE	Mu(ac)	Mu(ad)	Mu(bc)	Mu(bd)	a	d	d/|a|
IA	TCCGT-12	2.14	2.9	0.199 (0.777)	Boardman 4 yr diameter	5.8	3.5	3.9	4.2	3.8	−0.15	0.40	2.67
IA	CCCTT-N6	2.49	3.0	0.103 (0.749)	Boardman 4 yr height	5.1	329.2	360.2	381.3	355.7	−13.25	28.30	2.14
IA	CCCTT-N6	2.15	3.0	0.172 (0.940)	Clatskanie 4 yr diameter	5.6	8.2	10.1	11.1	9.0	−0.40	2.00	5.00
IA	CCCTT-N6	2.18	3.2	0.221 (0.955)	Clatskanie 4 yr height	5.7	8.6	10.1	10.5	9.0	−0.20	1.50	7.50
IA	TCCGT-N11	2.55	2.9	0.078 (0.688)	Clatskanie 8 yr height	3.7	120.3	123.4	134.5	122.1	−0.90	7.75	8.60
VI	O_050	2.95	3.0	0.054 (0.472)	Boardman 4 yr diameter	4.4	4.1	3.7	4.1	3.7	0.20	0.00	0.00
VI	O_050	2.61	3.1	0.098 (0.724)	Boardman 4 yr height	4.0	367.6	340.2	376.1	344.4	11.60	2.15	0.19
VI	O_050	2.49	2.9	0.101 (0.755)	Clatskanie 8 yr height	3.8	123.6	119.4	132.7	122.6	0.50	2.95	5.90
XVIII	ACACC-N21	2.66	2.8	0.061 (0.690)	Clatskanie 4 yr diameter	5.3	8.3	11.0	10.0	9.1	−0.40	1.80	4.50
XVIII	O_480	2.47	2.7	0.080 (0.828)	Clatskanie 4 yr height	4.5	8.5	10.3	9.5	9.7	−0.60	0.80	1.33

†LG = Linkage groupwise significance; GW = Genomewise significance.

Table 3. AMMI analysis results for location specificity in QTL detection between the Clatskanie and Boardman sites.

QTL	Sum of Squares	F	Prob (>F)
BM-1	12.815	4.4956	0.1241
BM-2	13.626	3.0333	0.1799
BM-3	83.833	7.8745	0.0675
BM-4	0.602	0.3765	0.5829
BM-5	62.910	3.5125	0.1576
BM-6	5.471	0.2690	0.6398
BM-7	0.551	0.1091	0.7629
BM-8	16.484	8.7336	0.05977
BM-9	40.208	2.4787	0.2135

Although informative loci have been identified in other studies based on the candidate gene approach in diverse systems such as *Populus* [20], Loblolly pine [22], and cowpeas [23,24], the large genomic intervals encompassed in some of the QTL intervals can make the narrowing down of candidate genes difficult. This limitation was evident in our analyses where only 1 QTL interval (*BM-7*) encompassed less than 100 genes. Despite this limitation, the unique genome duplication event which resulted in synteny between different *Populus* chromosomal segments [18] provides a reasonable starting point at narrowing down the list of candidate genes. The extensive duplication and paralogous relationship between genes found in QTL intervals located on different chromosomes suggests that these QTLs may have been derived from the same ancestral sequences and is consistent with the high levels of inter-chromosomal synteny described previously by Tuskan et al. [18]. This assumption would imply that functional activity for these genes was conserved post-duplication leading to existence of paralogous QTLs on different *Populus* chromosomes. The opposite scenario, where functional divergence or gene loss occurred after the duplication event resulting in deferential QTL presence on otherwise syntenic chromosome intervals was harnessed to identify unique candidate genes for cell-wall chemistry in a previous study [20]. QTL relationships based on ancestral sequences and candidate gene analyses will benefit from further improvements in the genome assembly and annotation of all gene models. At present, 38% of genes present within the 8 QTL intervals did not have adequate annotation information.

The approach described above, however, does not negate the need for additional marker saturation and fine-mapping of these intervals before candidate genes are selected for molecular validation. To that end, we identified 197 unmapped SSR markers that occurred within the 8 QTL intervals. These markers represent a potential source of SSR markers for use in saturating QTL intervals. Since QTLs identified in this study were independently verified in other studies, they potentially harbor key genes mediating biomass productivity as well as the expression of heterosis in *Populus* hybrids. The cumulative results presented in this study provide a basis for further genomic characterization of these high-value QTLs for subsequent use in improving biomass productivity in *Populus*.

Materials and Methods

Experiments were conducted in field sites located in Boardman and Clatskanie, Oregon. These two sites contrast in water

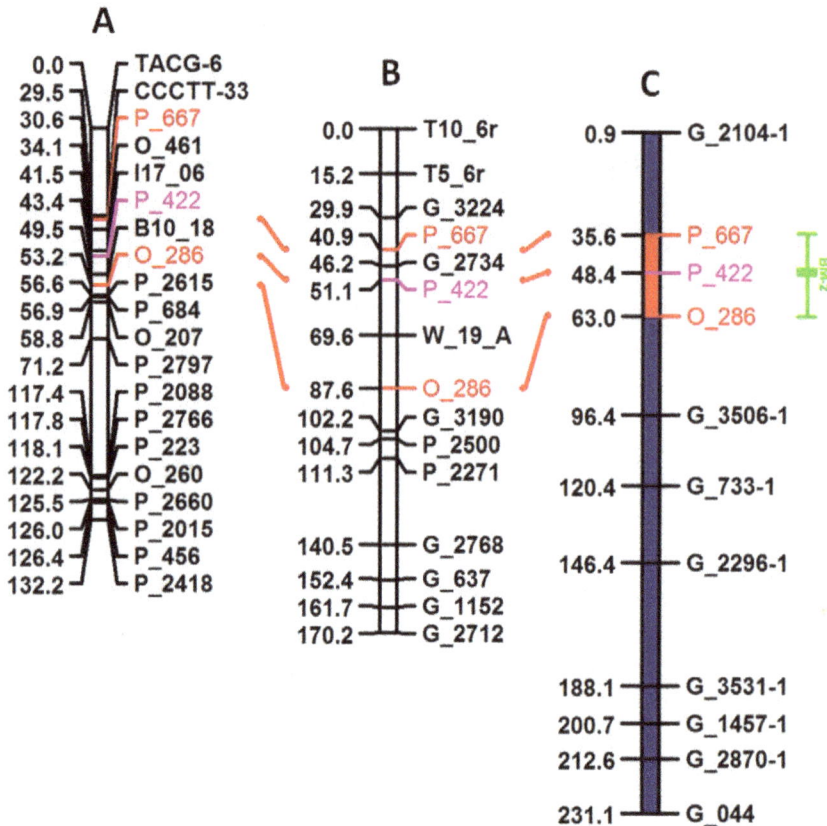

Figure 2. Synteny between (A) Family 331 genetic map LG II, (B) *Populus* **consensus genetic map LG II, and (C) Scaffold 2 of the** *Populus* **genome assembly illustrating the genome anchoring of QTL** *BM-2* **using flanking markers.** Map distance units in A and B represent cM distances and distance units in C represent genomic sequence length (x10Kb).

availability and provide opportunity for characterizing the influence of differential water availability in trait expression [10]. Details related to the mapping population, experimental design, and phenotypic measurements were described previously [12,14,15]. Briefly, an interspecific T×D hybrid population (F₁ Family 53) was generated in 1981 from a cross between a female *P. trichocarpa* (T, black cottonwood clone 93–968) and a male *P. deltoides* (D, eastern cottonwood clone ILL-129). Two resulting hybrids (F₁ clones 53–246 and 53–242, female and male, respectively) were sib-mated in 1988 and again in 1990 to generate an F₂ mapping population, Family 331; of approximately 375 individuals [13,15]. Phenotypic data for stem height and diameter for 310 of these individuals were reanalyzed in this study. Specifically, height and diameter data collected after 4 years of growth were reanalyzed for each site and height data collected after 8 years of growth was reanalyzed for the Clatskanie site.

Genomic Resources

We used the family 331 and *Populus* consensus genetic maps described by Yin et al. [21,25] for QTL identification and genome anchoring. Briefly, the family 331 map was based on 841 AFLP, RAPD, RFLP and SSR markers. Of these, 155 SSR markers were shared with the consensus map and were used to align the family 331 map to the 19 LGs corresponding to the *Populus* haploid chromosome number. Detailed map characteristics, marker nomenclature, and the resulting syntenic relationship between the genetic maps were summarized in Yin et al. [21].

Additionally, we used 2,524 SSR markers with known positions on the *Populus* V2.2 genome assembly (http://www.phytozome.

net/cgi-bin/gbrowse/poplar/) to establish the genome assembly framework used for anchoring QTL. Procedure for assigning SSR markers to physical positions on the genome assembly was previously described by Ranjan et al. [20]. Briefly, the physical position of SSR markers in *Populus* genome sequence was obtained by BLAST search of the corresponding forward and reverse primers. Additional checks were done to ensure that the predicted SSR length based on BLAST result was same as the length of the actual sequenced SSR marker. MapChart 2.2 [26] was used to graphically represent synteny and collinearity between the genetic maps and the genome assembly.

QTL Analysis

The Multiple-QTL Model (MQM) package of MapQTL 6.0 [27] with automatic cofactor selection was used to map putative and suggestive QTL intervals on the family 331 genetic map. The non-parametric Kruskal-Wallis (KW) analysis with a significance threshold of 0.005 was used as secondary confirmation of detected QTL. Mean phenotypic values across ramets and replicates were analyzed separately for each trait and experiment. The criteria for declaring QTL was based on the MQM analysis in which results were subjected to permutation tests [28]. 1000 permutations were conducted separately for each trait and experiment to determine linkage groupwise and genomewise LOD significance thresholds at the 0.05 significance level. A putative QTL was declared when it was detected in at least two experiments or in the same experiment for both traits, with at least one of those instances exceeding the linkage groupwise LOD significance threshold. A suggestive QTL was declared when detected in at least two experiments or in one

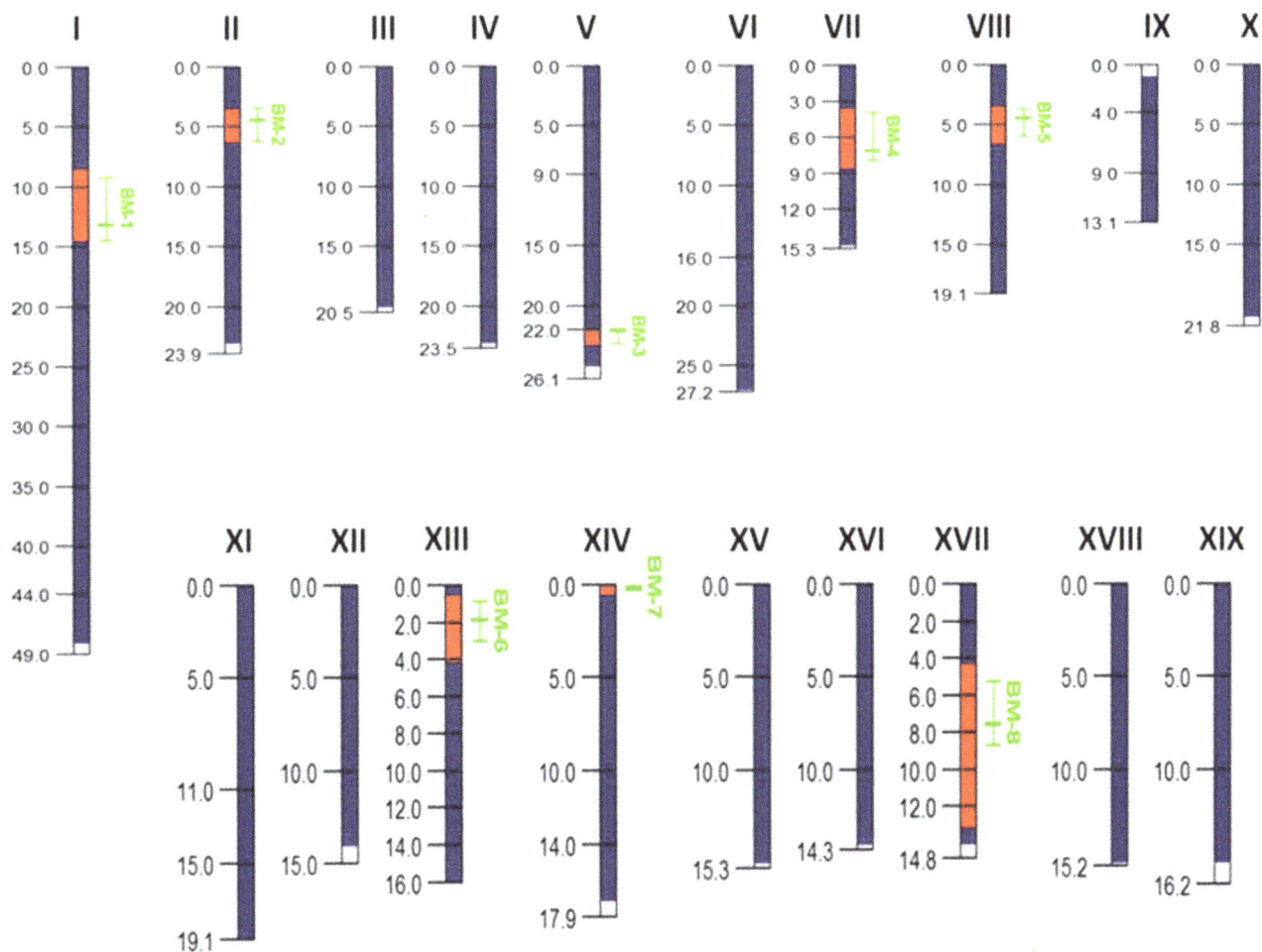

Figure 3. Genome anchored QTL positions on the *Populus* **V2.2 assembly.** Blue bars represent SSR marker coverage for each scaffold, red bars indicate scaffold intervals between flanking SSR markers used for genome anchoring, vertical green lines represent QTL intervals and estimated peak position. Scaffold intervals are represented in Mb.

experiment for both traits, with LOD scores above 2.0 but failing to exceed the LOD significance threshold in any one instance [29]. Results of the KW analysis were used for QTL verification only and were not reported in this manuscript. RAPDS and RFLP markers were excluded in the QTL analysis due to the small population size with available genotypic information for these markers as described by Yin et al. [21]. Nomenclature for naming putative biomass QTL was based on the abbreviation *BM*

Table 4. Genome assembly position for QTLs associated with height and diameter in *Populus* Family F$_2$ pedigree.

QTL	Scaffold	Marker at peak	Genome assembly interval (Mb)	QTL peak Position on assembly (Mb)	Interval size (Mb)	Additional SSRs[†]
BM-1	1	AGCTG-1	9.30–14.50	13.10	5.20	35
BM-2	2	P_422	3.50–6.30	4.80	2.80	12
BM-3	5	AGCTG-7	22.00–23.30	22.20	1.30	12
BM-4	7	T_AE20_120	3.60–8.70	7.20	5.10	32
BM-5	8	TCCGT-16	3.40–6.60	4.60	3.20	35
BM-6	13	CCCCT-N3	0.60–4.10	1.90	3.50	36
BM-7[‡]	14	AGCGA-14/CTACG-N1	0.01–0.50	0.24	0.49	9
BM-8	17	T_L09_105	4.30–13.1	7.50	8.80	32

[†]See Table S1.
[‡]Physical position of the QTL was reported previously by Ranjan et al. (2010).

(biomass) and were numbered according to increasing LG number.

QTL x Environment Interaction

Effects of environment on QTL detection were estimated using the Additive Main Effects and Multiplicative Interaction Model (AMMI) approach [30]. In order to verify QTL-specific changes in response to environment, LOD scores for individual markers within a QTL interval were used in the analysis with environment being the Boardman and Clatskanie sites and each experiment being considered as a replicate for each respective site. Only markers falling within 1-LOD difference on either side of the QTL peak were selected for analysis. The AMMI model equation was:

$$Y_{ger} = \mu + \alpha_g + \beta_e + \sum{}_n \lambda_{gn} \gamma_{gn} \delta_{en} + \rho_{ge} + \varepsilon_{ger}$$

Where Y_{ger} = the observed LOD score of gth genotype (marker) in e^{th} environment for r^{th} replicate;

μ = grand mean; α_g = deviation of mean of gth marker from grand mean; β_e = deviation of mean of the eth environment from the grand mean; λ_{gn} = the singular value for the nth interaction principal component axis (PCA); γ_{gn} = the marker eigenvector for the nth PCA; δ_{en} = the environment eigenvector values for the nth PCA; ρ_{ge} = residual effects; and ε_{ger} = the error term.

Additive and Dominant Effects

Locus specific additive and dominant effects were calculated from mean phenotypic values of the alternate homozygous genotypes and heterozygotes at each QTL. Specifically, locus-specific phenotypic means for heterozygous loci carrying alleles derived from the same species (mu(ac), mu(bd)) and for heterozygous loci carrying alleles derived from both species (mu (ad), mu (bc)) were computed from the MQM mapping procedure using MapQTL 6.0 [27]. Additive (a), dominance (d) effects were calculated as:

$$a = [mu(ac) - mu(bd)]/2;$$
$$d = [mu(ad) + mu(bc)]/2 - [mu(ac) + mu(bd)]/2 \tag{27}$$

QTL mode of action was calculated as the ratio of dominance over additivity, d/|a| [14,31]. d/|a| ratios less than 1 were regarded as reflecting under-dominance, ratios between 0 and 1 reflected partial dominance, and ratios greater than 1 reflected over-dominance as suggested by Hua et al. [31].

Genome Anchoring of QTL Intervals

We used the genome anchoring strategy described by Ranjan et al. [20] with minor variation to establish QTL intervals on the genome assembly. In addition to establishing QTL intervals, we also estimated the physical position of the marker closest to the QTL peak within the QTL interval. In this regard, we used the cM to physical distance ratio determined from SSR markers flanking the QTL to calculate the approximate position of the QTL peak relative to their cM distance from each flanking SSR marker. Where more than one marker was associated with the QTL peak in different experiments, the cM position of the marker with most frequent peak association or the marker with the highest LOD score was used to estimate the QTL peak position on the genome assembly.

Candidate Gene Identification and Characterization

Gene models lying within genome-anchored QTL intervals were identified from the *Populus* genome assembly V2.2 in the Phytozome database (http://www.phytozome.net/cgi-bin/gbrowse/poplar/). Annotations based on PFAM domains were used to exclude gene models with unknown function from the analysis and to establish duplication relationships between genes occurring in different QTL intervals. Paralogous relationships were verified based on information available for each gene model in the GRAMENE database (http://www.gramene.org).

Supporting Information

Table S1   SSR markers mapping in QTL intervals based on physical positions on the *Populus* genome assembly.

Table S2   *Populus* gene models within QTL intervals identified for stem height and diameter in an F$_2$ pedigree.

Table S3   Gene duplication within and between QTL intervals identified from stem height and diameter in an F$_2$ *Populus* pedigree.

Acknowledgments

The authors wish to thank Dr. H.D. Bradshaw for sharing the phenotypic data with us.

Author Contributions

Conceived and designed the experiments: MMS TJT GAT. Performed the experiments: MMS LEG TJT GAT. Analyzed the data: WM MMS PR TY. Wrote the paper: WM MMS.

References

1. Ranney JW, Wright LL, Layton PA (1987) Hardwood energy crops: The technology of intensive culture. Journal of Forestry 85: 17–28.
2. Abelson PH (1991) Improved yields of biomass. Science 252: 1469.
3. Zsuffa L, Giordano E, Pryor LD, Stettler RF (1996) Trends in poplar culture: some global and regional perspectives. In: Stettler RF, Bradshaw HD Jr, Heilman PE, Hinckley TM eds. Biology of Populus and its implications for management and conservation. Ottawa, ON, Canada: NRC Research Press, National Research Council of Canada. Pp. 515–539.
4. Tuskan GA (1998) Short-rotation woody crop supply systems in the United States: what do we know and what do we need to know? Biomass Energy 14: 307–315.
5. Studer MH, DeMartini JD, Davis MF, Sykes RW, Davison B et al. (2011) Lignin content in natural Populus variants affects sugar release. Proceedings of the National Academy of Sciences 108: 6300–6305.
6. Zhuang Y, Adams KL (2007) Extensive allelic variation in gene expression in Populus F₁ hybrids. Genetics 177: 1987–1996.
7. Eckenwalder JE (2001) Descriptions of clonal characteristics In: Dickmann DI, Isebrands JG, Eckenwalder JE, Richardson J eds. (2001) Poplar culture in North America 2001. Ottawa, ON, Canada: NRC Research Press, National Research Council of Canada. 331–382.
8. Scarascia-Mugnozza GE, Ceulemans R, Heilman PE, Isebrands JG, Stettler RF et al. (1997) Production physiology and morphology of Populus species and their hybrids grown under short rotation II Biomass components and harvest index of hybrid and parental species clones. Canadian Journal of Forestry Research 27: 285–294.
9. Stettler RF, Wu R, Zsuffa L (1996) The role of hybridization in the genetic manipulation of Populus In: Stettler RF, Bradshaw HD Jr, Heilman PE, Hinckley TM eds. Biology of Populus and its implications for management and conservation. Ottawa, ON, Canada: NRC Research Press, National Research Council of Canada. 87–112.
10. Tschaplinski TJ, Tuskan GA, Sewell MM, Gebre GM, Todd DE, et al. (2006) Phenotypic variation and QTL identification for osmotic potential in an

interspecific hybrid inbred F$_2$ poplar pedigree growing under contrasting environments. Tree Physiology 26: 595–604.

11. Rae AM, Pinel MPC, Bastien C, Sabatti M, Street NR, et al. (2008) QTL for yield in bioenergy Populus: identifying GxE interactions from growth at three contrasting sites. Tree Genetics and Genomes 4: 97–112.

12. Wu R (1997) Genetic control of macro- and micro-environmental sensitivities in Populus. Theoretical Applied Genetics 94: 104–114.

13. Bradshaw HD Jr, Villar M, Watson BD, Otto KG, Stewart S et al. (1994) Molecular genetics of growth and development in Populus III A genetic linkage map of a hybrid poplar composed of RFLP, STS, and RAPD markers. Theoretical Applied Genetics 89: 167–178.

14. Bradshaw HD Jr, Stettler RF (1995) Molecular genetics of growth and development in Populus IV Mapping QTLs with large effects on growth, form, and phenology traits in a forest tree. Genetics 139: 963–973.

15. Wu R, Stettler RF (1997) Quantitative genetics of growth and development in Populus II The partitioning of genotype × environment interaction in stem growth. Heredity 78: 124–134.

16. Wullschleger SD, Yin TM, DiFazio SP, Tschaplinski TJ, Gunter LE, et al. (2005) Phenotypic variation in growth and biomass distribution for two advanced-generation pedigrees of hybrid poplar. Canadian Journal of Forest Research 35: 1779–1789.

17. Novaes E, Osorio L, Drost DR, Miles BL, Boaventura-Novaes CRD,et al. (2009) Quantitative genetic analysis of biomass and wood chemistry of Populus under different nitrogen levels. New Phytologist 182: 878–890.

18. Tuskan GA, DiFazio S, Jansson S, Bohlmann J, Grigoriev I, et al. (2006) The genome of black cottonwood, Populus trichocarpa (Torr & Gray). Science 313: 1596–1604.

19. Slavov GT, DiFazio SP, Martin J, Schackwitz W, Muchero W, et al. (2012) Genome resequencing reveals multiscale geographic structure and extensive linkage disequilibrium in the forest tree Populus trichocarpa. New Phytologist. doi: 10.1111/j.1469-8137.2012.04258.x.

20. Ranjan P, Yin T, Zhang X, Kalluri UC, Yang X, et al. (2010) Bioinformatics-Based Identification of Candidate Genes from QTLs Associated with Cell Wall Traits in Populus. Bioenergy Research 3: 172–182.

21. Yin T, Zhang X, Gunter L, Ranjan P, Sykes R, et al. (2010) Differential detection of genetic loci underlying stem and root lignin content in Populus. Plos One 5: e14021.

22. Brown GR, Bassoni DL, Gill GP, Fontana JR, Wheeler NC, et al. (2003) Identification of quantitative trait loci influencing wood property traits in loblolly pine (Pinus taeda L) III QTL verification and candidate gene mapping. Genetics 164: 1537–1546.

23. Muchero W, Ehlers JD, Roberts PA (2010) Restriction site polymorphism-based candidate gene mapping for seedling drought tolerance in cowpea [Vigna unguiculata (L.) Walp.]. Theoretical Applied Genetics 120: 509–518.

24. Muchero W, Ehlers JD, Close TJ, Roberts PA (2011) Genic SNP markers and legume synteny reveal candidate genes underlying QTL for Macrophomina phaseolina resistance and maturity in cowpea [Vigna unguiculata (L) Walp]. BMC Genomics 12: 8.

25. Yin T, DiFazio SP, Gunter LE, Zhang X, Sewell MM, et al. (2008) Genome structure and emerging evidence of an incipient sex chromosome in Populus. Genome Research 18: 422–430.

26. Voorrips RE (2002) MapChart: Software for the graphical presentation of linkage maps and QTLs. Heredity 93: 77–78.

27. Van Ooijen JW (2009) MapQTL ® 6, Software for the mapping of quantitative trait loci in experimental populations of diploid species. Kyazma BV, Wageningen, Netherlands.

28. Churchill GA, Doerge RW (1994) Empirical threshold values for quantitative trait mapping. Genetics 138: 963–971.

29. Wu R, Bradshaw HD, Stettler R (1997) Molecular Genetics of Growth and Development in Populus (Salicaceae). V. Mapping Quantitative Trait Loci Affecting Leaf Variation. American Journal of Botany 84: 143–153.

30. Gauch HG (1988) Model selection and validation for yield trials with interactions. Biometrics 44: 705–715.

31. Hua J, Xing Y, Wu W, Xu C, Sun X, et al. (2003) Single-locus heterotic effects and dominance by dominance interactions can adequately explain the genetic basis of heterosis in an elite rice hybrid. Proceedings of the National Academy of Sciences 100: 2574–2579.

4

Salinity Tolerance of *Picochlorum atomus* and the Use of Salinity for Contamination Control by the Freshwater Cyanobacterium *Pseudanabaena limnetica*

Nicolas von Alvensleben[1,2]**, Katherine Stookey**[1,2]**, Marie Magnusson**[1,2]**, Kirsten Heimann**[1,2,3,4]*

1 School of Marine and Tropical Biology, James Cook University, Townsville, Queensland, Australia, **2** Centre for Sustainable Fisheries and Aquaculture, James Cook University, Townsville, Queensland, Australia, **3** Comparative Genomics Centre, James Cook University, Townsville, Queensland, Australia, **4** Centre for Biodiscovery and Molecular Development of Therapeutics, James Cook University, Townsville, Queensland, Australia

Abstract

Microalgae are ideal candidates for waste-gas and –water remediation. However, salinity often varies between different sites. A cosmopolitan microalga with large salinity tolerance and consistent biochemical profiles would be ideal for standardised cultivation across various remediation sites. The aims of this study were to determine the effects of salinity on *Picochlorum atomus* growth, biomass productivity, nutrient uptake and biochemical profiles. To determine if target end-products could be manipulated, the effects of 4-day nutrient limitation were also determined. Culture salinity had no effect on growth, biomass productivity, phosphate, nitrate and total nitrogen uptake at 2, 8, 18, 28 and 36 ppt. 11 ppt, however, initiated a significantly higher total nitrogen uptake. While salinity had only minor effects on biochemical composition, nutrient depletion was a major driver for changes in biomass quality, leading to significant increases in total lipid, fatty acid and carbohydrate quantities. Fatty acid composition was also significantly affected by nutrient depletion, with an increased proportion of saturated and mono-unsaturated fatty acids. Having established that *P. atomus* is a euryhaline microalga, the effects of culture salinity on the development of the freshwater cyanobacterial contaminant *Pseudanabaena limnetica* were determined. Salinity at 28 and 36 ppt significantly inhibited establishment of *P. limnetica* in *P. atomus* cultures. In conclusion, *P. atomus* can be deployed for bioremediation at sites with highly variable salinities without effects on end-product potential. Nutrient status critically affected biochemical profiles – an important consideration for end-product development by microalgal industries. 28 and 36 ppt slow the establishment of the freshwater cyanobacterium *P. limnetica*, allowing for harvest of low contaminant containing biomass.

Editor: Lucas J. Stal, Royal Netherlands Institute of Sea Research (NIOZ), The Netherlands

Funding: This work was funded through the Advanced Manufacturing Cooperative Research Centre, Melbourne, Australia. The funders had no role in study design, data collection and analysis, decision to publish, or preparation of the manuscript.

Competing Interests: The authors have declared that no competing interests exist.

* E-mail: Kirsten.heimann@jcu.edu.au

Introduction

The depletion of fossil energy stores, climate change-associated increasing atmospheric levels of carbon dioxide and freshwater pollution have generated a renewed interest in industrial-scale microalgal biomass production [1]. Industrial algal biomass production can utilize and sequester significant amounts of atmospheric or flue gas carbon dioxide [2] and remove pollutant nutrients such as nitrates, nitrites and phosphates from waste water ponds [3].

To make industrial-scale microalgal cultivation successful, algal strain selection should focus on species with high production of target biochemicals and tolerance to a wide range of environmental conditions, such as salinity, temperature and nutrient or pollutant loads. Such algal 'super-species' should also show high biochemical productivity, which would considerably simplify production regarding standardisation of product quality across a range of production sites.

Industry aims for microalgae cultivation at various power-stations in Australia for CO_2 and NO_x remediation from flue gas with parallel production of value-adding biochemicals. However,

these sites differ in the water quality for cultivation. A cosmopolitan marine microalga, *Nannochloris atomus* Butcher (Chlorophyta, synonym for *Picochlorum atomus* (Butcher) Henley [4]), has a suitable lipid and protein content for aquaculture [5,6], high biomass production and a potentially broad tolerance to variations of salinity [7,8]. However, the influence of salinity on growth patterns, nutrient requirements and biochemical profiles below 36 ppt, which are commonly encountered at potential production sites, have to date not been determined. Establishing species-specific growth parameters will identify optimal inoculation cell numbers and culture durations for achieving highest biomass productivity in the shortest possible timeframe. Understanding species-specific daily nutritional requirements will ensure minimal environmental impact (e.g. eutrophication through discharge of nutrient-rich harvest water effluent [9]), whilst also minimising expenses associated with fertilisation.

Nitrate assimilation involves a two-step reduction reaction from nitrate to nitrite and nitrite to ammonium, ultimately resulting in the production of amino acids [10]. Nitrite reduction is rate-limiting and excessive nitrate provision results in an accumulation of cellular nitrite which is secreted, most likely due to its cytotoxicity at high concentrations [11]. This has implications for the remediation of nitric oxide (NO) flue gas, which can be converted 1:1 to nitrite in water [12] to be then used as a nitrogen source. Similarly, to reduce fertilisation costs, industry aims to remediate nutrient-polluted (waste) waters. Optimal remediation requires correlation of inoculation cell numbers with nutrient loads.

Nitrogen and phosphorus availability also influences cellular protein, carbohydrate, and lipid content, as well as fatty acid profiles [13,14]. Nitrogen limitation reduces the synthesis of chloroplastic proteins and chl a, but increases carotenoid content [15] while the surplus of carbon metabolites are stored as storage lipids and - carbohydrates [13,16]. Higher lipid yields through nitrogen limitation have been obtained for several microalgal species [13,17,18] suggesting that target bio-product yields can be optimised through manipulation of culture nutrient status.

Microalgal culture contamination by rogue organisms is an ever-present risk in aquaculture industries [19]. Common contaminants include bacteria, viruses, fungi, other algae and zooplankton (e.g. ciliates, copepods, rotifers) [20]. Current procedures to minimise culture contamination include pH or salinity manipulations [19,20], the use of ammonium as a nitrogen source, or quinine treatment to reduce amoeba populations [20,21]. Other remedies, such as the addition of antibiotics [22] carry the risk of antibiotic resistance, placing restrictions on the use of the biomass and waste water disposal.

Culture contamination by non-target algae or cyanobacteria generally results in resource competition [23] and/or the release of potentially toxic allelochemicals into the culture medium, inhibiting growth or killing the target species [24]. This often leads to lost productivity associated with disposal of contaminated cultures, sterilisation, re-inoculation and culture re-establishment. Adverse impacts on product quality can further negatively affect industry, even if productivity is unaffected.

Pseudanabaena limnetica (Lemmermann) Komàrek is a filamentous, non-heterocystous [25] and non-toxic [26] freshwater cyanobacterium [27], with a certain degree of halotolerance [28] and is a frequent local nuisance contaminant in outdoor microalgal cultures during the tropical wet season. Consequently, methods must be developed to control levels of contamination, ideally not affecting the target species or influencing final products.

Given the potential importance of *P. atomus* in aquaculture, this study firstly aimed to determine the influence of salinity on growth, nutrient utilisation, biomass and lipid production and effects of nutrient limitation on biochemical profiles to determine end-product choice and industrial-scale cultivation protocols. Additionally, the effectiveness of salinity manipulations for contamination control of the freshwater cyanobacterial contaminant *P. limnetica* was investigated.

It is shown that salinity had no effect on *P. atomus* growth and nutrient utilisation (except at 11 ppt for the latter) and had only a marginal effect on total lipid at 2 ppt and carbohydrate at 8 ppt, respectively, under nutrient-replete conditions. Nutrient status, however, significantly affected total lipid and fatty acid profiles, carbohydrate and protein contents. It is further shown that salinity can be used to control the establishment of *P. limnetica*.

Materials and Methods

Algal culture conditions

Batch cultures of *Picochlorum atomus* (culture accession # NQAIF 284) were maintained (24 °C, with a 12:12 h photoperiod and light intensity of 42 µmol photons m^{-2} s^{-1}) at the North Queensland Algal Identification/Culturing Facility (NQAIF) culture collection (James Cook University, Townsville, Australia), and were individually areated with 0.45 µm filtered air (Durapore; Millipore). Monoclonal cultures with low bacterial numbers (<1 mL^{-1}) were established in a total culture volume of 2 L in modified L1 culture medium [29], with 6 mg instead of 3 mg PO$_4^{3-}$ L^{-1}. Cultures were re-fertilised with nitrate (~55 mg L^{-1}) and phosphate (6 mg L^{-1}) on day 5 after inoculation to generate sufficient biomass for biochemical analyses of nutrient-replete cultures.

The modified L1 culture medium was prepared at six different salinities: 2, 8, 11, 18, 28 and 36 ppt NaCl in filtered seawater (FSW) (pre-filtration Whatman GF/C, followed by 0.45 µm Durapore, Millipore). All materials were sterilised by autoclaving (Tomy, Quantum Scientific) and cultures were handled and inoculated aseptically in a laminar flow (AES Environmental Pty LTD fitted with HEPA filter). Replicate cultures (2 L, n = 3) of *P. atomus* were inoculated at a density of 4×10^9 cells L^{-1} (~100 mg dry weight L^{-1}) for each salinity. Inoculation was carried out from 36 ppt mother-cultures with no acclimation to decreasing salinity. Cultures of *P. atomus* have been maintained at the above salinities for more than 200 generations showing the same growth and nutrient utilisation patterns.

Indirect methods for culture growth determination

Calibration curves were established from triplicate dilution series using *Picochlorum atomus* stock cultures to correlate cells L^{-1} (direct cell counts using a bright-line Neubauer-improved haemocytometer) and dry weights (DW) [g L^{-1}] (gravimetric analysis, modified from Rai et al. [30]) with turbidity (% transmission [% TA at 750 nm, Spectramax Plus; Molecular Devices]). Turbidity and calibration curves were medium blanked for each salinity. Dry weight samples were corrected for salt content using salinity-specific blanks. Results were correlated to generate linear equations ($R^2 > 0.95$) used to determine cell numbers and respective dry weights of cultures of *P. atomus* from turbidity measurements.

Culture growth and nutrient analysis

Growth of *Picochlorum atomus* was determined daily using turbidity, from triplicate 250 µl samples per culture for 20 days and obtained data were transformed to cell numbers and dry weights as described above. Specific growth rates [µ], (eq. 1) were calculated from culture cell numbers [31], as were the derived parameters divisions per day and generation time [days]. Biomass productivies were determined using equation 2 (modified from Su et al. [32]).

$$\mu = \ln(C_2/C_1)/(t_2 - t_1) \qquad \text{(eq.1)}$$

Volumetric biomass productivity $\left[mg \, DW \, L^{-1} day^{-1} \right]$
$= (DW_2 - DW_1)/(t_2 - t_1)$ (eq.2)

where C_1 and C_2 = initial and final cell numbers [cells mL^{-1}], respectively, t_1 and t_2 = initial and final culture timepoints [days] per identified growth period, respectively, DW$_1$ and DW$_2$ = initial and final dry weight [g L^{-1}].

Medium nitrate (NO$_3^-$), nitrite (NO$_2^-$) and phosphate (PO$_4^{3-}$) concentrations were determined every second day and on day 5, following nutrient addition, using the Systea EasyChem (Analytical Solutions Australia (ASA)) auto-analyser following the manufacturer's EPA-approved and certified protocols (Systea User Manual, 2011).

Biochemical analyses

Total lipids, FAME, carbohydrate and protein. Biomass samples for biochemical analyses were harvested from 500 mL samples through centrifugation (20 min at 3000 g (Eppendorf 5810R), followed by 2 min at 16,000 g (Sigma 1–14, John Morris Scientific)) from all cultures when nitrate-replete during late logarithmic growth (day 11) and four days after nitrogen depletion during the initial stationary phase; i.e. days 18 and 22, for cultures at 11 and 2 ppt, respectively, and day 24 for cultures at 8, 18, 28 and 36 ppt. Cultures were classified as nutrient-replete and -deplete based on increasing and decreasing nitrite secretion patterns and the nutrient depletion was assured by harvesting four days after medium nutrient depletion [33]. The biomass pellets were lyophilized (VirTis benchtop 2K, VWR) and stored in air-tight vials under nitrogen at 4 °C until further analysis.

Total lipid determination

Total lipids were determined gravimetrically following a direct extraction and transesterification method adapted from Lewis et al. [34] and modified following Rodriguez-Ruiz et al. [35], and Cohen et al. [36]. Briefly, 2 mL freshly prepared methylation reagent (HPLC-grade methanol : acetyl chloride, 95:5 (v/v)) and 1 mL hexane were added to 30 ± 0.1 mg lyophilized biomass. Following heating (100°C, 60 min), 1 mL MilliQ purified water was added and the samples were centrifuged (1800 g for 5 min at 4°C (Eppendorf 5810R, VWR) to achieve phase separation. The hexane layer was collected and the pellet was extracted twice more with 1 mL hexane, centrifuging as above between washes, to extract all lipids into the organic phase. The hexane extracts were combined (total of 3 mL) in a pre-weighed glass vial and evaporated to dryness under a gentle stream of nitrogen and weighed to determine total lipids.

Fatty acid extraction, transesterification and analysis

Fatty acids in lyophilised samples were simultaneously extracted and transesterified using a method adapted from Rodriguez-Ruiz et al. [35] and Cohen et al. [36] as described in Gosch et al. [37]. Fatty acid analysis was carried out on an Agilent 7890 GC (DB-23 capillary column with a 0.15 μm cyanopropyl stationary phase, 60 m×0.25 mm ID (inner diameter)) equipped with flame ionisation detector (FID) and connected to an Agilent 5975C electron ionisation (EI) turbo mass spectrometer (Agilent technologies), for identification of fatty acid methyl esters (FAMEs) (split injection, 1/50). Injector, FID inlet and column temperatures were programmed following David et al. [38]. Fatty acid quantities were determined by comparison of peak areas of authentic standards (Sigma Aldrich) and were corrected for recovery of internal standard (C19:0) and total fatty acid content (mg g^{-1} DW) was determined as the sum of all FAMEs.

Total lipid and FAME productivities

Total fatty acid productivities were determined using equation 3, where total FAME$_2$ was determined in nutrient-deplete conditions, total FAME$_1$ in nutrient-replete conditions, and t_1 and t_2 = harvest time points for FAME determination.

$$\text{FAME productivity } [mg \, L^{-1} day^{-1}] = \\ (\text{Total FAME}_2 - \text{Total FAME}_1)/(t_2 - t_1) \tag{eq.3}$$

Carbohydrate analysis

Total carbohydrate content was determined using the phenol-sulphuric acid assay [39]. Prior to analysis, lyophilised algal samples were lysed in MilliQ-purified water with a Bullet Blender bead beater (ZrO$_2$ beads, 0.5 mm diameter) (Next Advance, Lomb Scientific) to enable collection of a homogenous sub-sample for extraction.

Ash and protein analysis

Ash-content (mg g^{-1} DW) was determined by combustion in air (500°C, 24 h) (Yokogawa-UP 150, AS1044) while protein content was determined by difference (eq. 4) [40].

$$Protein \, (\% \, wt) = 100 \, \% - (\% \, Ash + \% \, Total \, lipids + \\ \% \, Carbohydrate) \tag{eq.4}$$

Effect of salinity on contamination of *Picochlorum atomus* cultures with *Pseudanabaena limnetica*

To investigate if salinity could be used for contamination control, cultures of *Picochlorum atomus* were raised at 11, 18, 28 and 36 ppt (cultures at 2 and 8 ppt were not established as *P. limnetica* is a freshwater species) and seeded with *Pseudanabaena limnetica* colonies at a ratio of 1:100,000 cells mL^{-1} (*P. limnetica* : *P. atomus*). Cell counts (bright-line Neubauer improved haemocytometer) of both organisms commenced on day 8 after the first visible signs of *P. limnetica* dominance (culture colour change) in the lower salinity cultures (11 and 18 ppt).

Statistical analyses

All statistical analyses were carried out in Statistica 10 (StatSoft Pty Ltd.). Repeated measures ANOVAs were used to determine the effects of salinity on growth rates, nitrite secretion, total nitrogen uptake and contaminant development through culture time. One-way ANOVAs were used to determine the effect of salinity on volumetric biomass productivities. For nutrient uptake analyses, data were divided into pre- and post- nutrient addition (days 0–4 and 5–10, respectively) and the slopes of each uptake period were analysed using one-way ANOVAs. For total lipid, fatty acids, carbohydrate and protein content analyses, factorial ANOVAs were used to determine the effects of salinity, nutrient status and their interaction. Tukey post-hoc tests were used to determine significant differences assigned at $p < 0.05$. Homogeneity of variances and normality assumptions were verified using Cochran-Bartlett tests. Fatty acid and carbohydrate data required log transformation to fulfil normality assumptions.

Results

Effect of salinity on growth and nutrient uptake dynamics of *Picochlorum atomus*

Culture growth of *P. atomus* was divided into three phases (phase I: days 2–5; phase II: days 5–9 and phase III: 9–18) (Fig. 1) for

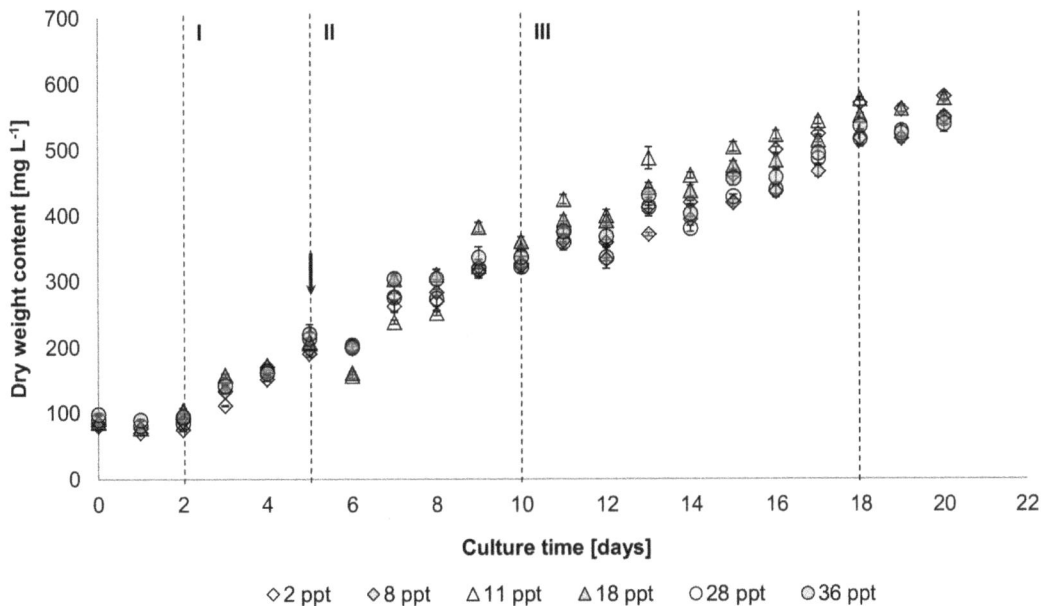

Figure 1. Mean biomass growth [mg DW L^{-1}] of *Picochlorum atomus* at 2, 8, 11, 18, 28 and 36 ppt determined using% transmittance at 750 nm. Arrow: indicates the addition of nutrients. Active growth was divided into 3 phases (I–III) based on log-transformed data for determination of specific growth rates [μ]. n = 3. Standard error is shown. DW: dry weight.

which specific growth rates, divisions per day and generation times were calculated (Table 1). Within each growth phase, salinity had no significant effect ($F_{(5, 12)} = 0.99$, p = 0.46) on growth rates, while the effect of culture phase was significant ($F_{(2, 24)} = 679.67$, p<0.01) as growth rates decreased over culture time. Irrespective of salinity, specific growth rates [μ] were highest for the first two days following a one-day lag phase (μ = 0.21–0.28), then decreased by ~50% during phase II and a further ~50% thereafter during

phase III (Table 1). Nutrient addition on day 5 resulted in culture dilution (Fig. 1).

Biomass productivities during growth phase I were between 34–43 mg L^{-1} day^{-1} and 26–31 mg L^{-1} day^{-1} during phase II, with the exception of cultures at 18 ppt where biomass productivity remained similar at 36 mg L^{-1} day^{-1} (Table 2). Productivities, from the beginning of the logarithmic growth phase to the

Table 1. Effect of salinity on specific growth rates [μ], divisions per day and generation times [days] of *Picochlorum atomus*.

Culture time [days]	2 ppt growth rate [μ]	8 ppt growth rate [μ]	11 ppt growth rate [μ]	18 ppt growth rate [μ]	28 ppt growth rate [μ]	36 ppt growth rate [μ]
Days 2–5	0.28	0.25	0.21	0.27	0.28	0.26
Days 5–9	0.13	0.13	0.11	0.14	0.11	0.11
Days 9–18	0.06	0.05	0.06	0.04	0.05	0.05
Culture time [days]	**2ppt [Div. day^{-1}]**	**8 ppt [Div. day^{-1}]**	**11 ppt [Div. day^{-1}]**	**18 ppt [Div. day^{-1}]**	**28 ppt [Div. day^{-1}]**	**36 ppt [Div. day^{-1}]**
Days 2–5	0.4	0.35	0.3	0.39	0.4	0.37
Days 5–9	0.19	0.18	0.17	0.2	0.15	0.16
Days 9–18	0.09	0.07	0.09	0.06	0.07	0.07
Culture time [days]	**2 ppt generation time [days]**	**8 ppt generation time [days]**	**11 ppt generation time [days]**	**18 ppt generation time [days]**	**28 ppt generation time [days]**	**36 ppt generation time [days]**
Days 2–5	2.47	2.82	3.29	2.58	2.46	2.69
Days 5–9	5.39	5.51	6.04	4.94	6.6	6.35
Days 9–18	11.15	14.08	11.35	18.19	13.85	14.2

Table 2. Effect of salinity on volumetric biomass productivities of *Picochlorum atomus* during growth phases I and II, and overall from days 2–18.

Culture phase	Total dry-weight productivity [mg DW L^{-1} day^{-1}]					
	2 ppt	8 ppt	11 ppt	18 ppt	28 ppt	36 ppt
Days 2–5	38±1	37±3	34±1	35±1	43±2	42±3
Days 5–9	31 ±1	31±1	29±1	36±1	26 ±1	28±1
Days 2–18	30±0.5	27±0.5	29±0.5	28±0.5	27±0.5	27±0.5

n = 3. Average ± standard error.

beginning of the stationary phase were approximately 27–30 mg L^{-1} day^{-1}.

Except for cultures at 11 ppt, salinity had no effect on nitrate uptake of *P. atomus* for the first 4 days of the culture period with ~13–15 mg nitrate L^{-1} day^{-1} being assimilated. Following nutrient replenishment on day 5, a ~50% decrease in nitrate uptake was observed (Fig. 2). Cultures at 11 ppt took up nitrate significantly faster pre- ($F_{(5,\ 12)} = 85.48$, p<0.01) and post- ($F_{(5,\ 12)} = 14.68$, p<0.01) fertilisation, than cultures at the other salinities resulting in an uptake of 60 mg L^{-1} day^{-1} for the first two days and medium nitrate depletion.

In contrast, a significant negative correlation between nitrite release and salinity ($F_{(5,\ 12)} = 6.13$, p<0.05) was observed prior to re-fertilisation, except for cultures at 11 ppt which showed no nitrite release (Fig. 3). Following fertilisation, all cultures released nitrite irrespective of salinity. Nitrite resorption started 4, 6, 10 and 12 days after fertilisation for cultures at 11 ppt, 2 ppt, 18 and 36 ppt, and 8 ppt, respectively, which correlated with medium nitrate depletion in most cultures (compare Fig. 2 and Fig. 3). Total daily nitrogen uptake (Fig. 4) was similar between cultures at 2, 8, 18, 28 and 36 ppt but significantly higher at 11 ppt ($F_{(5,\ 12)} = 34.079$, p<0.01).

Phosphate uptake followed a similar pattern to nitrate with a decrease in uptake rates following fertilisation. Initial phosphate uptake rates were 1.3–2.4 mg L^{-1} day^{-1} (Fig. 5). Following phosphate addition, uptake rates decreased to 0.8–1 mg L^{-1} day^{-1}, except for cultures at 11 ppt. Initially, nitrate to phosphate uptake ratio was 6–9:1 (N:P) and decreased to 4–7:1 (N:P) after nutrient addition.

Effect of salinity and culture nutrient status on the biochemical profile of *Picochlorum atomus*

Post hoc analyses identified marginally significant effects of salinity on total lipid content of *P. atomus* at 2 ppt compared to 28 and 36 ppt under nutrient-replete conditions (Fig. 6A), whereas culture nutrient status had a large effect ($F_{(1,\ 24)} = 229.63$, p<0.01). Nutrient-starved cultures of *P. atomus* had significantly higher lipid content ($F_{(1,\ 24)} = 229.63$, p<0.01) than nutrient-replete cultures (Fig. 6). After 4 days of nutrient starvation, biomass total lipid content increased by 3.5–11% with the lowest increase in cultures at 11 ppt and the highest increase in cultures at 28 and 36 ppt (Fig. 6).

There was no significant effect of salinity on total fatty acid content, but there was a significant effect of culture nutrient status ($F_{(1,\ 24)} = 4.42$, p<0.01) where, as with lipid content, total fatty acid content in nutrient-deplete cultures was significantly higher than in replete biomass.

Fatty acids represented 56–66% of total lipids in nutrient-replete biomass and 66–74% of total lipids in nutrient-deplete cultures, with cultures at 2 ppt showing the highest fatty acid content under both nutrient conditions (Fig. 6). Lowest fatty acid concentrations were recorded in nutrient-replete cultures at 28 ppt and 36 ppt (Fig. 6A). Fatty acid productivities between nutrient-replete and -deplete conditions ranged from 4.7–6.2 mg L^{-1} day^{-1} with cultures at 11 ppt and 2 ppt showing the lowest and highest productivities, respectively (Table 3).

While fatty acid content increased by up to 50% following 4 days of nutrient starvation (Fig. 6), nutrient status also had an influence on fatty acid profiles. A 9 and 11% increase in saturated and mono-unsaturated fatty acids, respectively, and a corresponding decrease in polyunsaturated fatty acids was observed in nutrient-starved *P. atomus* cultures (Table 4). Specifically, C18:1

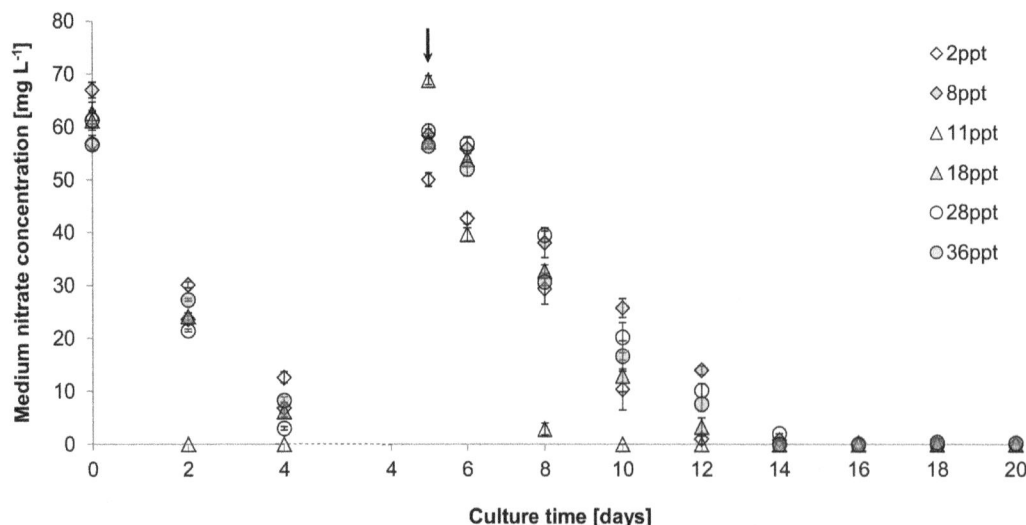

Figure 2. Effect of salinity on nitrate assimilation [mg L^{-1}] of *Picochlorum atomus*. n = 3. Standard error is shown. Arrow: indicates measurements taken after nitrate and phosphate replenishment.

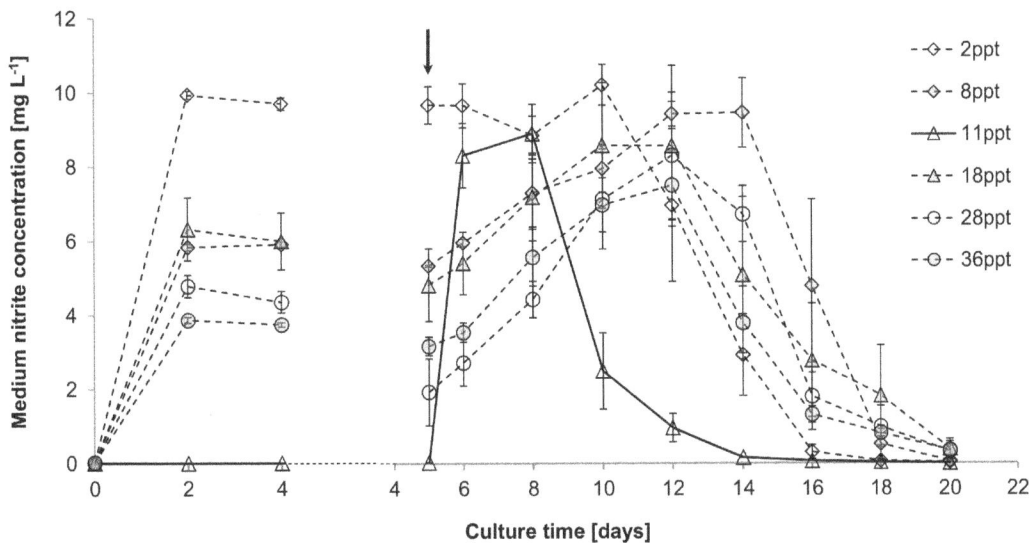

Figure 3. Effect of salinity on media nitrite dynamics [mg L^{-1}] of *Picochlorum atomus.* n = 3. Standard error is shown. Arrow: indicates measurements taken after nitrate and phosphate replenishment.

increased by ~ 13% while C18:3 showed the largest decrease. The most abundant fatty acids were always C18:3 (n-3), C16:0, and C18:2 (n-6), equating to 54–68% of the total fatty acids (Table 4). The observed ~50% decrease in the proportion of omega-3 fatty acids and a small increase of omega-6 fatty acids led to a change in omega-6 to omega-3 ratios (ω6:ω3) from ~0.5:1 to ~1:1 under nutrient-limiting conditions.

Carbohydrate contents were 120–250 mg g^{-1} DW in nutrient-replete cultures, with cultures at 2 ppt and 36 ppt containing the lowest and highest concentrations, respectively. Overall, cellular carbohydrate contents were not affected by salinity, but did increase two to three-fold across all salinities in nutrient-deplete cultures (F$_{(1, 24)}$ = 86.98, p<0.01) (Fig. 7).

Ash content increased with increasing salinity irrespective of nutrient status. Nutrient depletion led to a ~50% decrease in ash content compared to replete cultures. Protein content was significantly higher (F$_{(5, 24)}$ = 5.78, p < 0.01) in cultures at 8 ppt compared to cultures at 28 and 36 ppt in nutrient-replete conditions. Nutrient depletion induced a protein content decrease across all salinities with a significant decrease in cultures at 2 ppt (~40%) and 8 ppt (~30%) (F$_{(1, 24)}$ = 34.34, p<0.01) (Fig. 8). In both nutrient-replete and -deplete conditions, 8 ppt cultures had the highest protein content and cultures at 36 ppt the lowest.

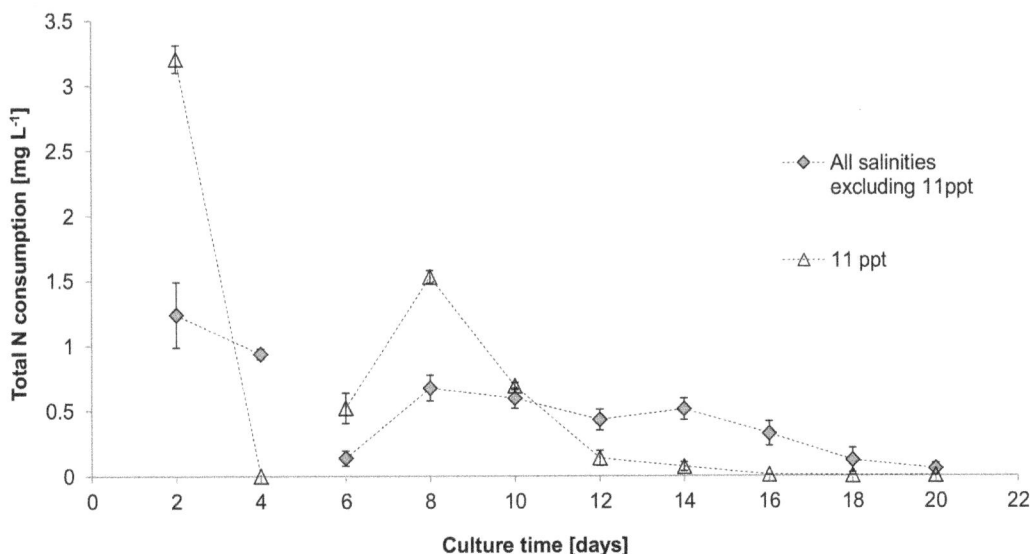

Figure 4. Total daily net nitrogen [N] uptake [mg L^{-1}] of *Picochlorum atomus.* Average total nitrogen consumption is shown for salinities of 2, 8, 18, 28 and 36 ppt, while nitrogen consumption of cultures at 11 ppt is plotted individually to highlight the effect of 11 ppt. n = 3. Standard error is shown.

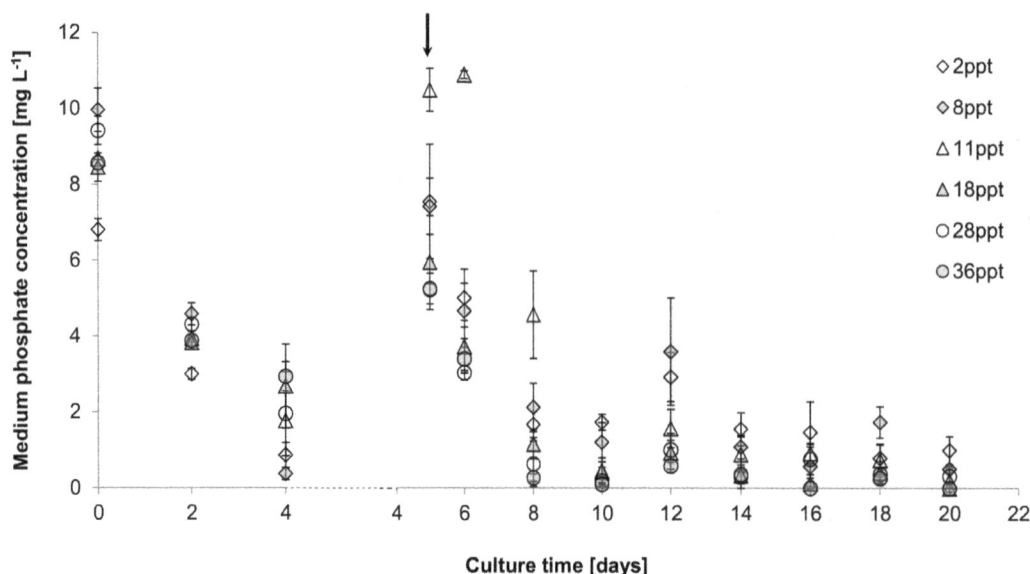

Figure 5. Effect of salinity on phosphate assimilation [mg L^{-1}] of *Picochlorum atomus.* n = 3. Standard error is shown. Arrow: indicates measurements taken after nitrate and phosphate replenishment.

Effect of salinity on contamination of *Picochlorum atomus* cultures with *Pseudanabaena limnetica*

An increase in salinity significantly ($F_{(15, 40)}$ = 5.7, p<0.01) slowed the establishment rate of *P. limnetica* (Fig. 9), resulting in only 10% of contaminant cells in culture at 36 ppt, compared to 60–70% at 11 and 18 ppt on day 8. After 16 days, *P. limnetica* completely dominated cultures at 11 and 18 ppt (90–95%), and reached ~70% dominance at 28 ppt, whereas in cultures at 36 ppt, *P. atomus* was still dominant with 55% (Fig. 9). Specific growth rates [μ] for *P. limnetica* development from day 8 to 10 were ~0.25 in cultures at 11 and 18 ppt and ~0.6 in cultures at 28 and 36 ppt. Overall specific growth rates [μ] from days 8 to 16 were ~0.13 in cultures at 11 and 18 ppt and ~0.25 in cultures at 28 and 36 ppt. This shows that *P. limnetica* at 11 and 18 ppt were in late logarithmic growth around day 8 whereas at 28 and 36 ppt logarithmic growth was just commencing.

Discussion

Effect of salinity on growth and nutrient dynamics of *Picochlorum atomus*

Irrespective of salinity, *Picochlorum atomus* exhibited growth patterns typical of aerated batch cultures [11]. The data established that *P. atomus* is a euryhaline microalga tolerating freshwater to marine salinities without adverse effects on growth

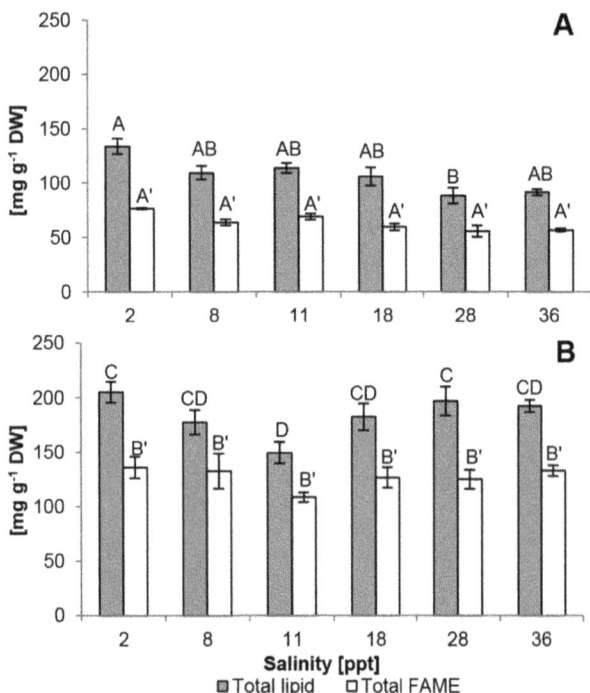

Figure 6. Effect of nutrient availability and salinity on total lipid and fatty acid content. Total lipid and FAME content under nutrient replete conditions (A) and nutrient deplete conditions (B). Grey bars: total lipid, white bars: total FAME. n = 3. Standard error is shown. Statistically significant differences are indicated by different letters. Prime letters are used for total FAME.

Table 3. Total FAME productivities [mg L^{-1} day^{-1}] of *Picochlorum atomus* from nutrient replete to deplete conditions.

Salinity	Total FAME productivity [mg L^{-1} day^{-1}]
2 ppt	6.2±0.25
8 ppt	6.1±0.13
11 ppt	4.7±0.06
18 ppt	6.0±0.09
28 ppt	5.9±0.16
36 ppt	6.2±0.13

n = 3. Average ± standard error.

Table 4. Effect of salinity and culture nutrient status (replete/deplete) on fatty acid profiles (proportion [%] of total FAME) of *Picochlorum atomus*.

	2 ppt		8 ppt		11 ppt		18 ppt		28 ppt		36 ppt	
	Replete	Deplete	Replete	Deplete	Replete	Deplete	Replete	Deplete	Replete	Deplete	Replete	Deplete
Saturated												
C12:0	0.24	0.16	0.48	0.20	0.84	0.45	0.38	0.19	0.50	0.24	0.40	0.21
C14:0	0.38	0.47	0.45	0.53	0.50	0.51	0.46	0.59	0.48	0.56	0.45	0.63
C16:0	14.95	22.50	15.00	22.22	15.68	20.86	15.38	21.69	15.27	21.81	15.82	21.54
C18:0	1.08	4.56	1.20	5.00	1.29	2.75	1.11	3.84	1.32	4.48	1.24	3.74
C20:0	2.21	1.85	2.22	1.94	2.44	1.95	2.29	1.96	2.86	1.94	2.88	2.00
Σ_{SFA}	18.86	29.55	19.35	29.89	20.76	26.51	19.62	28.27	20.42	29.03	20.79	28.11
Monounsaturated												
C16:1 (7)	1.03	1.26	0.93	1.05	1.44	1.26	0.99	1.07	1.09	0.93	0.98	0.85
C16:1 (9)	3.25	1.20	3.17	1.11	3.39	1.59	3.03	1.22	2.89	1.10	2.97	1.24
C18:1 (9)	1.82	14.76	1.99	17.01	4.50	11.63	1.68	14.21	2.07	19.13	1.95	17.96
C18:1 (x)	0.87	0.79	1.03	0.87	1.28	1.12	1.01	0.88	1.17	1.26	1.16	1.43
Σ_{MUFA}	6.99	18.01	7.12	20.04	10.61	15.60	6.71	17.39	7.22	22.43	7.06	21.48
Polyunsaturated												
C16:2 (7,10)	8.80	6.47	7.56	5.64	7.29	6.42	8.15	6.45	7.36	5.16	6.89	5.08
C16:2 (9, 12)	0.37	0.40	0.39	0.39	0.37	0.33	0.37	0.42	0.45	0.40	0.48	0.34
C16:3 (7,10,13)	12.35	6.15	12.81	6.18	9.76	7.47	12.08	6.57	11.69	5.22	11.70	5.82
C16:4 (4,7,10,13)	0.77	0.35	0.83	0.35	0.67	0.41	0.79	0.35	0.79	0.30	0.81	0.30
C18:2	14.26	18.93	12.36	17.20	15.39	17.84	13.23	19.15	13.30	19.31	13.18	19.31
C18:3 (6,9,12)	0.44	0.22	0.40	0.23	0.43	0.30	0.51	0.25	0.48	0.23	0.49	0.23
C18:3 (9,12,15)	34.62	18.48	36.09	18.59	31.90	23.29	35.27	19.60	34.80	16.29	35.19	17.83
Σ_{PUFA}	71.60	50.99	70.44	48.58	65.80	56.06	70.40	52.78	68.87	46.91	68.75	48.90
Sum of ω3	47.74	24.98	49.73	25.12	42.32	31.17	48.14	26.51	47.28	21.81	47.70	23.94
Sum of ω6	23.49	25.62	20.31	23.07	23.11	24.56	21.89	25.85	21.14	24.70	20.56	24.62
ω6:ω3 ratio	0.49	1.03	0.41	0.92	0.55	0.79	0.45	0.97	0.45	1.13	0.43	1.03

n = 3.

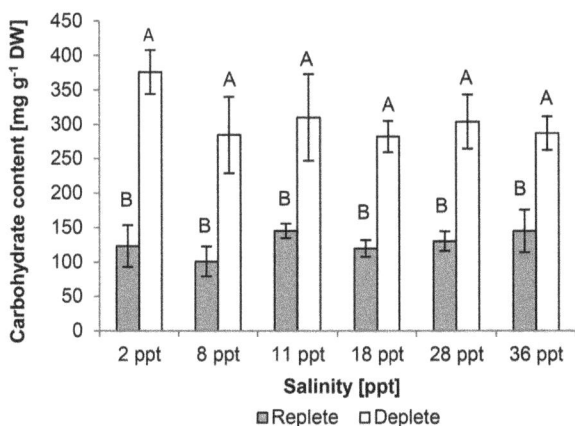

Figure 7. Effect of salinity and culture nutrient status (replete/deplete) on mean carbohydrate content [mg glucose g^{-1} DW] of *Picochlorum atomus*. n = 3. Standard error is shown. Different letters show statistical significance.

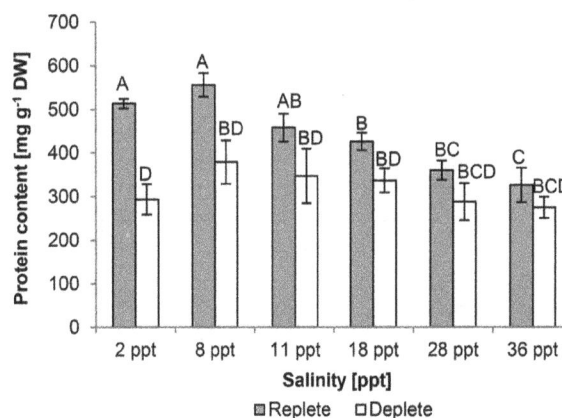

Figure 8. Effect of salinity and culture nutrient status (replete/deplete) on mean protein content [mg protein g^{-1} DW] of *Picochlorum atomus*. n = 3. Standard error is shown. Different letters show statistical significance.

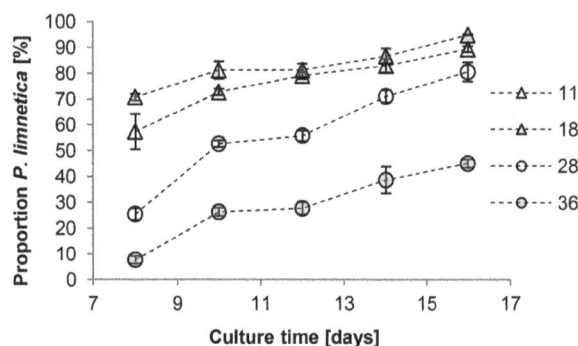

Figure 9. Effect of salinity (11, 18, 28 and 36 ppt) on the proportion [%] of *Pseudanabaena limnetica* in *Picochlorum atomus* cultures. n = 3. Standard error is shown.

and biomass productivities.

Initial specific growth rates [μ] were slightly lower than in previous reports, however maximum biomass [mg DW L^{-1}], maximum cell numbers [cells mL^{-1}] and initial volumetric productivities [mg DW L^{-1} day^{-1}] were comparable to previous reports using similar cultivation procedures for *Picochlorum* spp./ *Nannochloris* spp. (Table 5). Comparisons are however, difficult, as a summary of published biomass at harvest and biomass productivities for *Nannochloris* and *Picochlorum* spp. shows great variability (Table 5). This variation is to be expected [41] and is likely due to a combination of effects, such as species-specific responses and cultivation/environmental parameters, i.e. variable inoculation densities, differing light regimes, cultivation (batch vs. semi-continuous) and productivity calculations (direct vs. indirect) [32,42–44].

The decrease in growth rate during phases II and III (Fig. 1, Table 1) is characteristic of batch cultures [11] and is generally the consequence of individual or combined effects of culture self-shading, nutrient limitation [45] and microalgal/bacterial exudate accumulation [24,46]. Initially, these factors are unlikely to have a considerable effect on culture development, particularly considering the low inoculation densities, adequate nutrient provision and low bacteria cultures used in this study. However, over culture time, the accumulation of algal exudates followed by increased self-shading and bacterial growth-inhibiting exudates (negative allelopathic interactions) are likely to cause the observed decreasing growth rates. Nutrient limitation is unlikely to have affected growth as cultures were maintained nutrient-replete with high

nitrite levels (Fig. 3), indicating cellular nitrogen stores were filled throughout most of the cultivation period [33]. Additionally, culture re-fertilisation on day 5 had no impact on culture growth; also implying cultures were not nutrient limited.

The observed growth patterns for *P. atomus* have direct implications for industrial cultivation, as optimal productivities are achieved in relatively dilute cultures for a brief period. Harvest effort and costs inversely correlate with culture cell densities. Consequently, future studies should investigate whether higher inoculation densities and/or semi-continuous culturing would improve biomass yield and overall productivity. In addition, the accumulation of microalgal/bacterial exudates and their effects on culture development require further investigation, as these may affect water treatment and recycling capacity on industrial-scales.

Nitrogen and phosphorus are essential macronutrients, where the first limiting nutrient reduces microalgal growth rates [45]. Therefore, maximum biomass production requires adequate nutrient availability for each particular species in culture. However, excessive nutrient concentrations in harvest water pose environmental problems and unnecessary costs, unless harvest effluents can be efficiently recycled without compromising culture growth.

Initial nitrate uptake by *P. atomus* was similar at all salinities (except 11 ppt) and comparable to *Nannochloris maculata* [47]. With the exception of cultures at 11 ppt, patterns of nitrite secretion until day 10 can be grouped into high (28 and 36 ppt), intermediate (18 and 8 ppt) and low (2 ppt) salinity patterns, where medium nitrite was highest in low salinity cultures. Medium nitrate depletion resulted in expected nitrite resorption as intracellular nitrogen stores became depleted [33]. Nitrogen fluxes can provide insight into possible osmoregulatory mechanisms, often reflected in changes of biochemical profiles. The production of osmoregulatory solutes, such as proline in response to hyperosmotic stress has been reported for *Nannochloris* sp. [48], which would require higher nitrogen provisions. However, despite the variable nitrite secretion, total nitrogen uptake patterns (except for 11 ppt) were not significantly different. This may indicate that higher nitrite secretion in the lower salinity cultures was potentially due to a slight swelling of cells, increasing cell surface area [49], thereby increasing nitrate uptake. In contrast to nitrate [50], nitrite cannot be stored and is cytotoxic in higher concentrations [11]. Reduction of nitrite to ammonium is limited by nitrite reductase activity (a reaction directly linked to photosynthesis and under circadian control [51]). Thus, when nitrate reduction exceeds the reducing capacity of nitrite reductase, nitrite is secreted.

Table 5. Comparison of growth data and reported ranges in this study with growth data obtained for *Picochlorum* spp./ *Nannochloris* spp. under similar* and deviating cultivation conditions.

Species	Specific growth rate [μ]	Cell numbers [cells mL^{-1}]	Maximum biomass [mg DW L^{-1}]	Volumetric productivities [mg DW L^{-1} day^{-1}]	References
*Picochlorum atomus**	0.21–0.28	~2.2×10^7	~560	~26–43	this study
*Nannochloris atomus**	~0.32–0.38	~3×10^8	–	–	[44,57,68]
Nannochloris spp./ *Picochlorum* spp.*	0.35–0.44	–	~330–410	~40	[6,8,58]
*Nannochloris maculata**	~0.36	~1×10^8	–	–	[47]
*Nannochloris bacillaris**	~0.41	~1×10^7	–	–	[48]
Nannochloris spp./ *Picochlorum* spp.	0.17–2.5	3×10^6–3×10^8	46–1800	7–320	[42,43,47,58,62,69]

The significantly higher nitrogen requirements at 11 ppt are difficult to explain. Typically, higher nitrogen is required mainly for growth [11], which is not the case here (Fig. 1) or hypersaline osmoregulation [4], but no significant differences in protein contents were detected. Although this does not exclude the production of osmolytes such as glycine betaine or proline [49], osmoregulatory responses would be expected to be higher at lower salinities, which should result in greater nitrogen requirements at lower salinities. As this was not observed, we hypothesise that 11 ppt may induce a transitional response where known hypo- or hyper-osmoregulatory responses are not induced.

At 11 ppt the biomass contained twice the amount of C18:1(9) and 2–3% more C18:2 than at other salinities. Fatty acid changes in diacylglycerol (increases in phosphatidyl inositols and hydrolysis of phosphatidyl choline) and an increase in the fatty acid combinations of C16:0/C18:1 and C16:0/C18:2 were observed in *Dunaliella salina* as an immediate but transient response to hyposaline osmotic shock (reducing salinity from 99 to 49 ppt) [52]. This indicates that salinity can affect membrane composition. Hence, 11 ppt could induce changes in membrane lipids, perhaps increasing vacuolar storage capacity for nitrogen, which would explain the rapid uptake and the reduced nitrite secretion at 11 ppt.

Nitrate uptake of *P. atomus* was comparable or higher than reported for other species examined for wastewater treatment, including *Chlorella vulgaris* [53] and *Neochloris oleabundans* [54], suggesting that *P. atomus* could also be used in such applications. Nitrogen uptake potential also has important implications for industrial NO flue gas remediation. *Dunaliella tertiolecta* can remediate 21 mg day^{-1} of nitric oxide (NO) and showed a preference for NO uptake over NO$_3^-$ [55]. Future research should examine *P. atomus*'s nitrogen preferences and NO remediation potential from flue gas emitted by Australian coal-fired power stations.

As for nitrate uptake, initial phosphate uptake across all salinities was comparable to *Nannochloris maculata* [47] and uptake rates were comparable to *Chlorella stigmatophora*, showing potential for urban waste-water remediation [56]. Remediation studies using *Neochloris oleabundans* have shown phosphate uptake to correlate with increasing medium phosphate availability [54]. Consequently, further studies should investigate *P.atomus* phosphate uptake when exposed to higher concentrations.

The N:P ratio of *P. atomus* was similar to *Nannochloris atomus* [57]. The N:P ratio decreased over culture time as nutrient availability per cell decreased and cell numbers increased. Downstream effects of the decreased N availability resulted in reduced total protein contents (Fig. 8).

Effect of salinity and culture nutrient status on the biochemical profile of *Picochlorum atomus*

Culture salinity affected total lipid (at 2 ppt) and protein (at 8 ppt) contents of *Picochlorum atomus* under nutrient-replete conditions. However, nutrient availability was the main driver for significant differences in total lipid, carbohydrate, and protein contents, as well as fatty acid composition. Biochemical profile comparisons between studies are difficult, as species-specificity and environmental factors (nutrient availability, light intensity, photoperiod and cultivation stage) individually and combined affect the proximate chemical composition of microalgae [58–60]. Despite being a marine species, the highest total lipid content was observed when culturing *Picochlorum atomus* at 2 ppt, irrespective of nutrient status. Under nutrient-replete conditions, total lipid content of *P. atomus* was low, whereas nitrogen limitation increased total lipids to ~20%, corresponding to amounts reported for *Nannochloris atomus*

and *Picochlorum* sp. [42,58] and defining it as an oleaginous microorganism with the potential for oil-based biofuel production [61]. In contrast, a higher total lipid content was reported for *Nannochloris* sp. (~ 56%) when CO$_2$ was added [43]. Opportunistic biochemical profiling of very old cultures showed that *P. atomus* can also reach a total lipid content of ~60%. Consequently, studies should investigate high lipid yields in the context of remaining feasible and economically viable on a large-scale.

Total lipid content is not a good indicator for oil-based products, as this fraction contains all other lipid-soluble materials such as pigments. For oil-based products (e.g. biodiesel and bioplastics), the fatty acid content is more important [37,41]. Nutrient-depletion increased fatty acid content by ~10%, suggesting that fertilisation adjustments can improve biomass suitability for such products. Fatty acid proportions of total lipids were comparable to (nutrient-replete) or higher (nutrient-deplete) than those reported for the same genus [42]. Fatty acid profiles were comparable to those described by Volkman et al. [62] but different to others for this genus [42,44,58] (which also differ between each other for many fatty acids). These outcomes highlight the importance to consider culture conditions (e.g. industry location) and species-specificity when considering industrial cultivation. Total fatty acid productivities by *P. atomus* were comparable to other species (e.g. *Nannochloropsis* sp.) (see Lim et al. [41] for summary details).

Nutrient limitation considerably increased amounts of saturated (C16:0) and mono-unsaturated fatty acids (C18:1) but lowered amounts of polyunsaturated fatty acids (C18:3) consistent with responses reported for a wide variety of microalgal species [57]. For nutritional/dietary purposes an ω6:ω3 ratio of approximately 1:1 has been shown to be beneficial for cardio-vascular health [63], suggesting, that under the cultivation conditions reported here, *P. atomus* should be harvested when nutrient-deplete. In contrast, the suggested optimal fatty acid ratio for biofuel of 5:4:1 of C16:1, C18:1 and C14:0, respectively [64] was observed only under nutrient-replete conditions and low concentrations were observed. Identifying species with naturally occurring favourable fatty acid ratios for specific end-products could prove impossible under industrial conditions, therefore blending of fatty acids or oils from different microalgal species [65] and/or fertilisation regimes must be considered to achieve the specifications of a particular end-product. For example, for biofuel production, cultures of *P. atomus* will require nutrient starvation to increase lipid productivity and decrease the PUFA content.

Nutrient status also affected total carbohydrate and protein contents which increased and decreased, respectively, following nutrient limitation. Both carbohydrate and protein contents were similar under nutrient-replete conditions and slightly higher than reported for *Nannochloris atomus* under nutrient limitation [58]. Similar patterns of protein decrease and concurrent carbohydrate increase as a result of nutrient depletion have been observed in a number of microalgal species e.g. *Chlorella vulgaris* and *Scendesmus obliquus* [59], as N-limitation prevents the synthesis of proteins, channelling the photosynthetically acquired carbon into storage. Nutrient-replete *Picochlorum atomus* has been shown to be a promising replacement for *Nannochloropsis oculata* in aquaculture for grouper larval rearing [6], which is rapidly expanding, and already one of the most valuable aquaculture species in Southeast Asia [66].

Contaminant inhibition. In large-scale cultures, contamination by rogue organisms is a serious problem often resulting in significant economic losses [67]. In tropical Australia, the freshwater cyanobacterium *Pseudanabaena limnetica* rapidly outcompetes and dominates other microalgal species in culture. The observed broad salinity tolerance of *P. atomus*, with minimal effects on productivity or biochemical profiles, allows the use of salinity manipulations to inhibit or reduce culture contamination by rogue organisms. Although increased culture salinity does not completely prevent the development of *P. limnetica*, it does delay its establishment and subsequent logarithmic growth at 28 and 36 ppt up to day 8. It is noteworthy however, that while establishment of *P. limnetica* at high salinities is considerably slower, once established, growth rates are high and culture take-over will occur. The extended time for establishment and logarithmic growth of *P. limnetica* provides an extended opportunity to harvest the biomass with low levels of contamination, which is an important aspect for end product quality control.

In conclusion, *Picochlorum atomus* has considerable advantages for large-scale cultivation as it can be cultivated at locations differing in water salinity ranging from 2 – 36 ppt, without adverse effects on biochemical profiles. High carbohydrate and protein content suggests use in aquaculture [8] or as agricultural feed (e.g. for poultry) [5], when harvested under nutrient-replete conditions. In contrast, under nutrient-deplete conditions, fatty acid yields and

the decrease in PUFA content is suitable for lipid-based biofuel production. Similarly, the improved ω6:ω3 ratio under these conditions, would allow cultivation of *P. atomus* as a health food supplement to improve cardio-vascular health. In addition, salinity increase appears to be an effective tool for contamination delay, yielding biomass with guaranteed quality, which allows harvest and minimises economic losses due to culture re-establishment and end-product loss.

Acknowledgments

This research is part of the MBD Energy Research and Development program for Biological Carbon Capture and Storage. Nicolas von Alvensleben was supported by an AMCRC PhD scholarship.

Author Contributions

Conceived and designed the experiments: KH NvA KS MM. Performed the experiments: KH NvA KS MM. Analyzed the data: KH NvA KS MM. Contributed reagents/materials/analysis tools: KH NvA KS MM. Wrote the paper: KH NvA KS MM.

References

1. Stephens E, Ross IL, Mussgnug JH, Wagner LD, Borowitzka MA, et al. (2010) Future prospects of microalgal biofuel production systems. Trends in Plant Science 15: 554–564.
2. de Morais MG, Costa JAV (2007) Isolation and selection of microalgae from coal fired thermoelectric power plant for biofixation of carbon dioxide. Energy Conversion and Management 48: 2169–2173.
3. Grönlund E, Klang A, Falk S, Henænus J (2004) Susatinability of wastewater treatment with microalgae in cold climate, evaluated with emergy and socio-eclogical prinicples. Ecological Engineering 22: 155–174.
4. Henley WJ, Hironaka JL, Guillou L, Buchheim MA, Buchheim JA, et al. (2004) Phylogenetic analysis of the 'Nannochloris-like' algae and diagnoses of *Picochlorum oklahomensis* gen. et sp. nov. (Trebouxiophyceae, Chlorophyta). Phycologia 43: 641–652.
5. Becker EW (2007) Micro-algae as a source of protein. Biotechnology Advances 25: 207–210.
6. Chen TY, Lin HY, Lin CC, Lu CK, Chen YM (2012) *Picochlorum* as an alternative to *Nannochloropsis* for grouper larval rearing. Aquaculture 338: 82–88.
7. Cho SH, Ji SC, Hur SB, Bae J, Park IS, et al. (2007) Optimum temperature and salinity conditions for growth of green algae *Chlorella ellipsoidea* and *Nannochloris oculata*. Fisheries Science 73: 1050–1056.
8. Witt U, Koske PH, Kuhlmann D, Lenz J, Nellen W (1981) Production of *Nannochloris* sp. (Chlorophyceae) in large-scale outdoor tanks and its use as a food organism in marine aquaculture. Aquaculture 23: 171–181.
9. Jarvie HP, Neal C, Withers PJA (2006) Sewage-effluent phosphorus: A greater risk to river eutrophication than agricultural phosphorus? Science of the Total Environment 360: 246– 253.
10. Barea JL, Cardenas J (1975) Nitrate reducing enzyme system of *Chlamydomonas reinhardtii*. Archives of Microbiology 105: 21–25.
11. Becker EW (1994) Microalgae: biotechnology and microbiology. Cambridge University Press, New York.
12. Ignarro LJ, Fukuto JM, Griscavage JM, Rogers NE, Byrns RE (1993) Oxidation of nitric oxide in aqueous solution to nitrite but not nitrate: comparison with enzymatically formed nitric oxide from L-arginine. Proceedings of the National Academy of Sciences 90: 8103–8107.
13. Huerlimann R, de Nys R, Heimann K (2010) Growth, lipid content, productivity and fatty acid composition of tropical microalgae for scale-up production. Biotechnology and Bioengineering 107: 245–257.
14. Ahlgren G, Hyenstrand P (2003) Nitrogen limitation: Effects of different nitrogen sources on nutritional quality of two freshwater organisms *Scenedesmus quadricauda* (Chlorophyceae) and *Synechococcus* sp. (Cyanophyceae). Journal of Phycology 39: 906–917.
15. Geider RJ, Macintyre HL, Graziano LM, McKay RML (1998) Responses of the photosynthetic apparatus of *Dunaliella tertiolecta* (Chlorophyceae) to nitrogen and phosphorus limitation. European Journal of Phycology 33: 315–332.
16. Roessler PG (1990) Environmental control of glycerolipid metabolism in microalgae: commercial implications and future research directions. Journal of Phycology 26: 393–399.
17. Li Y, Horsman M, Wang B, Wu N (2008) Effects of nitrogen sources on cell growth and lipid accumulation of the green alga *Neochloris oleoabundans*. Applied Microbiology and Biotechnology 81: 629–636.
18. Sharma KK, Schuhmann H, Schenk PM (2012) High lipid induction in microalgae for biodiesel production. Energies 5: 1532–1553.
19. Meseck SL (2007) Controlling the growth of a cyanobacterial contaminant, *Synechoccus* sp., in a culture of *Tetraselmis chui* (PLY429) by varying pH: Implications for outdoor aquaculture production. Aquaculture 273: 566–572.
20. Borowitzka MA (2005) Culturing microalgae in outdoor ponds. In: Andersen RA, editor. Algal Culturing Techniques: Elsevier Academic Press. pp. 205–220.
21. Lincoln EP, Hall TW, Koopman B (1983) Zooplankton control in mass algal cultures. Aquaculture 32: 331–337.
22. Churro C, Fernandes AS, Alverca E, Sam-Bento F, Paulino S, et al. (2010) Effects of tryptamine on growth, ultrastructure, and oxidative stress of cyanobacteria and microalgae cultures. Hydrobiologia 649: 195–206.
23. Joint I, Henriksen P, Fonnes GA, Bourne D, Thingstad TF, et al. (2002) Competition for inorganic nutrients between phytoplankton and bacterioplankton in nutrient manipulated mesocosms. Aquatic Microbial Ecology 29: 145–159.
24. Hay ME (2009) Marine chemical ecology: Chemical signals and cues structure marine populations, communities and ecosystems. Annual Review of Marine Science 1: 193–212.
25. Komarek J (2003) Planktic oscillatorialean cyanoprokaryotes (short review according to combined phenotype and molecular aspects). Hydrobiologia 502: 367–382.
26. Mischke U (2003) Cyanobacteria associations in shallow polytrophic lakes: influence of environmental factors. Acta Oecologica-International Journal of Ecology 24: S11–S23.
27. Willame R, Boutte C, Grubisic S, Wilmotte A, Komarek J, et al. (2006) Morphological and molecular characterization of planktonic cyanobacteria from Belgium and Luxembourg. Journal of Phycology 42: 1312–1332.
28. Acinas SG, Haverkamp THA, Huisman J, Stal LJ (2009) Phenotypic and genetic diversification of *Pseudanabaena* spp. (cyanobacteria). ISME Journal 3: 31–46.
29. Andersen RA, Berges JA, Harrison PJ, Watanabe MM (2005) Recipes for freshwater and saltwater media. In: Andersen RA, editor. Algal Culturing Techniques: Elsevier Academic Press. pp. 429–538.
30. Rai LC, Mallick N, Singh JB, Kumar HD (1991) Physiological and biochemical characteristics of a copper tolerant and a wild-type strain of *Anabaena doliolum* under copper stress. Journal of Plant Physiology 138: 68–74.
31. Levasseur M, Thompson PA, Harrison PJ (1993) Physiological acclimation of marine-phytoplankton to different nitrogen sources. Journal of Phycology 29: 587–595.
32. Su CH, Chien LJ, Gomes J, Lin YS, Yu YK, et al. (2011) Factors affecting lipid accumulation by *Nannochloropsis oculata* in a two-stage cultivation process. Journal of Applied Phycology 23: 903–908.
33. Malerba ME, Connolly SR, Heimann K (2012) Nitrate-nitrite dynamics and phytoplankton growth: Formulation and experimental evaluation of a dynamic model. Limnology and Oceanography 57.
34. Lewis T, Nichols PD, McMeekin TA (2000) Evaluation of extraction methods for recovery of fatty acids from lipid-producing microheterotrophs. Journal of Microbiological Methods 43: 107–116.
35. Rodriguez-Ruiz J, Belarbi EH, Sanchez JLG, Alonso DL (1998) Rapid simultaneous lipid extraction and transesterification for fatty acid analyses. Biotechnology Techniques 12: 689–691.
36. Cohen Z, Vonshak A, Richmond A (1988) Effect of environmental conditions on fatty acid composition of the red alga *Porphyridium cruentum*: correlation to growth rate. Journal of Phycology 24: 328–332.

37. Gosch BJ, Magnusson M, Paul NA, de Nys R (2012) Total lipid and fatty acid composition of seaweeds for the selection of species for oil-based biofuel and bioproducts. Global Change Biology Bioenergy 4: 919–930.

38. David F, Sandra P, Wylie PL (2002) Improving the analysis of fatty acid methyl esters using retention time locked methods and retention time databases. Agilent Technologies application note 5988–5871EN.

39. Dubois M, Gilles KA, Hamilton JK, Rebers PA, Smith F (1956) Colorimetric method for determination of sugars and related substances. Analytical Chemistry 28: 350–356.

40. Sims GG (1978) Rapid estimation of carbohydrate in formulated fish products - protein by difference. Journal of the Science of Food and Agriculture 29: 281–284.

41. Lim DKY, Garg S, Timmins M, Zhang ESB, Thomas-Hall SR, et al. (2012) Isolation and evaluation of oil-producing microalgae from subtropical coastal and brackish waters. PLoS ONE 7.

42. de la Vega M, Diaz E, Vila M, Leon R (2011) Isolation of a new strain of *Picochlorum* sp. and characterization of its potential biotechnological applications. Biotechnology Progress 27: 1535–1543.

43. Negoro M, Shiogi N, Miyamoto K, Miura Y (1991) Growth of microalgae in high CO_2 gas and effect of SO_x and NO_x. Applied Biochemistry and Biotechnology 28/29: 877–886.

44. Roncarati A, Meluzzi A, Acciarri S, Tallarico N, Melotti P (2004) Fatty acid composition of different microalgae strains (*Nannochloropsis* sp., *Nannochloropsis oculata* (Droop) Hibberd, *Nannochloris atomus* Butcher and *Isochrysis* sp.) according to the culture phase and the carbon dioxide concentration. Journal of the World Aquaculture Society 35: 401–411.

45. MacIntyre HL, Cullen JJ (2005) Using cultures to investigate the physiological ecology of microalgae. In: Andersen RA, editor. Algal Culturing Techniques Elsevier Academic Press. pp. 287–326.

46. Chiang IZ, Huang WY, Wu JT (2004) Allelochemicals of *Botryococcus braunii* (Chlorophyceae). Journal of Phycology 40: 474–480.

47. Huertas E, Montero O, Lubian LM (2000) Effects of dissolved inorganic carbon availability on growth, nutrient uptake and chlorophyll fluorescence of two species of marine microalgae. Aquacultural Engineering 22: 181–197.

48. Brown LM (1982) Photosynthetic and growth responses to salinity in a marine isolate of *Nannochloris bacillaris* (Chlorophyceae). Journal of Phycology 18: 483–488.

49. Kirst GO (1990) Salinity tolerance of eukaryotic marine algae. Annual Review of Plant Physiology and Plant Molecular Biology 41: 21–53.

50. Dortch Q, Clayton JR Jr, Thoresen SS, Ahmed SI (1984) Species differences in accumulation of nitrogen pools in phytoplankton. Marine Biology 81: 237–250.

51. Rajasekhar VK, Oelmuller R (1987) Regulation of induction of nitrate reductase and nitrite reductase in higher plants Physiologia Plantarum 71: 517–521.

52. Ha KS, Thompson GA (1991) Diacylglycreol metabolism in the green alga *Dunaliella salina* under osmotic stress: Possible roles of dicaylglycerols in phospholipase C-mediated signal transduction. Plant Physiology 97: 921–927.

53. Sydney EB, da Silva TE, Tokarski A, Novak AC, de Carvalho JC, et al. (2011) Screening of microalgae with potential for biodiesel production and nutrient removal from treated domestic sewage. Applied Energy 88: 3291–3294.

54. Wang B, Lan CQ (2011) Biomass production and nitrogen and phosphorus removal by the green alga *Neochloris oleoabundans* in simulated wastewater and secondary municipal wastewater effluent. Bioresource Technology 102: 5639–5644.

55. Nagase H, Yoshihara K, Eguchi K, Okamoto Y, Murasaki S, et al. (2001) Uptake pathway and continuous removal of nitric oxide from flue gas using microalgae. Biochemical Engineering Journal 7: 241–246.

56. Arbib Z, Ruiz J, Alvarez P, Garrido C, Barragan J, et al. (2012) *Chlorella stigmatophora* for urban wastewater nutrient removal and CO_2 abatement. International Journal of Phytoremediation 14: 714–725.

57. Reitan KI, Rainuzzo JR, Olsen Y (1994) Effect of nutrient limitation on fatty-acid and lipid content of marine microalgae. Journal of Phycology 30: 972–979.

58. Benamotz A, Tornabene TG, Thomas WH (1985) Chemical profile of selected species of microalgae with emphasis on lipids. Journal of Phycology 21: 72–81.

59. Piorreck M, Baasch K-H, Pohl P (1984) Biomass production, total protein, chlorophylls, lipids and fatty acids of freshwater green and blue-green algae under different nitrogen regimes. Phytochemistry 23: 207–216.

60. Shifrin NS, Chisholm SW (1981) Phytoplankton lipids - Interspecific differences and effects of nitrate, silicate and light-dark cycles. Journal of Phycology 17: 374–384.

61. Hu Q, Sommerfeld M, Jarvis E, Ghirardi M, Posewitz M, et al. (2008) Microalgal triacylglycerols as feedstocks for biofuel production: perspectives and advances. Plant Journal 54: 621–639.

62. Volkman JK, Jeffrey SW, Nichols PD, Rogers GI, Garland CD (1989) Fatty-acid and lipid composition of 10 species of microalgae used in mariculture. Journal of Experimental Marine Biology and Ecology 128: 219–240.

63. Simopoulos AP (2002) The importance of the ratio of omega-6/omega-3 essential fatty acids. Biomedicine & Pharmacotherapy 56: 365–379.

64. Schenk PM, Thomas-Hall SR, Stephens E, Marx UC, Mussgnug JH, et al. (2008) Second generation biofuels: High-efficiency microalgae for biodiesel production. Bioenergy Research 1: 20–43.

65. Cha TS, Chen JW, Goh EG, Aziz A, Loh SH (2011) Differential regulation of fatty acid biosynthesis in two *Chlorella* species in response to nitrate treatments and the potential of binary blending microalgae oils for biodiesel application. Bioresource Technology 102: 10633–10640.

66. Harikrishnan R, Balasundaram C, Heo MS (2010) Molecular studies, disease status and prophylactic measures in grouper aquaculture: Economic importance, diseases and immunology. Aquaculture 309: 1–14.

67. Meseck SL, Wikfors GH, Alix JH, Smith BC, Dixon MS (2007) Impacts of a cyanobacterium contaminating large-scale aquaculture feed cultures of *Tetraselmis chui* on survival and growth of bay scallops, *Argopecten irradians*. Journal of Shellfish Research 26: 1071–1074.

68. Sunda WG, Hardison DR (2007) Ammonium uptake and growth limitation in marine phytoplankton. Limnology and Oceanography 52: 2496–2506.

69. Shifrin NS, Chisholm SW (1981) Phytoplankton lipids: interspecific differences and effects of nitrate, silicate and light dark cycles. Journal of Phycology 17: 374–384.

5

Productivity, Disturbance and Ecosystem Size Have No Influence on Food Chain Length in Seasonally Connected Rivers

Danielle M. Warfe[1]*[¤], Timothy D. Jardine[2,3], Neil E. Pettit[4], Stephen K. Hamilton[5], Bradley J. Pusey[3,4], Stuart E. Bunn[3], Peter M. Davies[4], Michael M. Douglas[1]

1 Research Institute for the Environment and Livelihoods, Charles Darwin University, Darwin, Northern Territory, Australia, 2 Toxicology Centre, University of Saskatchewan, Saskatoon, Saskatchewan, Canada, 3 Australian Rivers Institute, Griffith University, Nathan, Queensland, Australia, 4 Centre of Excellence in Natural Resource Management, The University of Western Australia, Albany, Western Australia, Australia, 5 Kellogg Biological Station and Department of Zoology, Michigan State University, Hickory Corners, Michigan, United States of America

Abstract

The food web is one of the oldest and most central organising concepts in ecology and for decades, food chain length has been hypothesised to be controlled by productivity, disturbance, and/or ecosystem size; each of which may be mediated by the functional trophic role of the top predator. We characterised aquatic food webs using carbon and nitrogen stable isotopes from 66 river and floodplain sites across the wet-dry tropics of northern Australia to determine the relative importance of productivity (indicated by nutrient concentrations), disturbance (indicated by hydrological isolation) and ecosystem size, and how they may be affected by food web architecture. We show that variation in food chain length was unrelated to these classic environmental determinants, and unrelated to the trophic role of the top predator. This finding is a striking exception to the literature and is the first published example of food chain length being unaffected by any of these determinants. We suggest the distinctive seasonal hydrology of northern Australia allows the movement of fish predators, linking isolated food webs and potentially creating a regional food web that overrides local effects of productivity, disturbance and ecosystem size. This finding supports ecological theory suggesting that mobile consumers promote more stable food webs. It also illustrates how food webs, and energy transfer, may function in the absence of the human modifications to landscape hydrological connectivity that are ubiquitous in more populated regions.

Editor: Nicolas Mouquet, CNRS, University of Montpellier II, France

Funding: This project forms part of the Tropical Rivers and Coastal Knowledge (TRaCK) Program. TRaCK received major funding for its research through the Australian Government's Commonwealth Environment Research Facilities initiative (http://www.environment.gov.au/about/programs/cerf/); the Australian Government's Raising National Water Standards Program (http://archive.nwc.gov.au/rnws); Land and Water Australia (http://lwa.gov.au); the Fisheries Research and Development Corporation (http://www.frdc.com.au) and the Queensland Government's Smart State Innovation Fund (http://www.qld.gov.au). The funders had no role in study design, data collection and analysis, decision to publish, or preparation of the manuscript.

Competing Interests: The authors have declared that no competing interests exist.

* E-mail: danielle.warfe@uwa.edu.au

¤ Current address: Centre of Excellence in Natural Resource Management, The University of Western Australia, Albany, Western Australia, Australia

Introduction

The food web is a central organizing theme in ecology, depicting the feeding relationships between species in a community [1,2] and providing a framework for understanding energy transfer and biogeochemical processes [3], biodiversity and trophic interactions [4], consumer behaviour and movement [5,6], and community stability and persistence in the face of perturbation [2,7,8]. Food web structure is often summarised by emergent properties such as food chain length (FCL), which measures the number of energy transfers between the base and the top of a food web, and is considered a central attribute of ecological communities [9]. Food chain length influences structural attributes of communities such as species diversity, trophic interactions and predator abundance [10,11], as well as functional attributes such as population stability, primary and secondary production, material cycling, and contaminant bioaccumulation [1,12–14].

Variation in FCL has long been observed in natural communities [1] and is hypothesised to be controlled by basal productivity, disturbance and/or ecosystem size [13,15]. The productivity or resource availability hypothesis states that because energy is lost through each successive transfer up the food chain, FCL is limited by available energy resources [16]. The disturbance hypothesis predicts shorter food chains in more disturbed ecosystems due to either longer food chains being less resilient to perturbations than shorter food chains [17], or species at higher trophic levels being rarer and more likely to be lost during disturbance events [18,19]. The ecosystem size hypothesis [20] predicts that larger ecosystems will have longer food chains because they support greater species richness [21], support more basal resources [22], promote coexistence of predators and prey [15,23], promote population persistence through enhanced colonisation opportunity [11,23], and/or support greater functional trophic diversity and less omnivory [20].

Despite having been proposed decades ago, the empirical support for any one of these environmental determinants being a dominant influence on FCL remains equivocal; rather, it is more likely that multiple factors control FCL [11,13,24]. Productivity has been shown to have either neutral or positive effects on FCL, disturbance tends to limit FCL, and ecosystem size generally lengthens food chains (Table 1). A recent meta-analysis of the 13 field studies that tested one or more determinants (using the correlation coefficient as an index of effect size) found that productivity and ecosystem size both positively influenced FCL, whereas disturbance did not significantly shorten food chains [25]. Intriguingly, this meta-analysis also showed that although productivity generally increased FCL, the magnitude of ecosystem size and disturbance effects were highly variable and could include positive, neutral and negative effects on FCL [25]. Only two studies, both in temperate riverine ecosystems, have tested all three environmental determinants concurrently: both found FCL was not affected by productivity, but increased with ecosystem size and decreased with disturbance [26,27] (Table 1). These studies showed that either larger ecosystems attenuate the effects of disturbance, thereby enhancing environmental stability and supporting longer food chains [26], or concluded that effects of

disturbance on productivity are exacerbated in smaller systems leading to increased omnivory and shorter food chains [27].

Such variable findings, and conclusions, are likely due to the fact that FCL is an aggregate property of food webs, reflecting changes in food web structure that can be generated by multiple mechanisms [4,28]. Food chain length can be altered by the addition or removal of a top consumer (additive mechanism), the addition or removal of an intermediate consumer (insertion), or a change in the degree of trophic omnivory shown by a top consumer (omnivory) [28,29]. In particular, the degree of omnivory or the strength of intraguild predation displayed by a top predator has been theoretically shown to mediate the influence of the above-mentioned environmental determinants, limiting FCL under increasing productivity or reduced disturbance but increasing FCL in larger ecosystems [15]. Therefore, examining the trophic role of top predators concurrently with FCL responses to environmental determinants is likely to be instructive in understanding the mechanisms by which these determinants control FCL.

We used carbon and nitrogen stable isotopes to assess the influence of productivity, disturbance and ecosystem size on food chain length, as well as the trophic role of the top predator, in river-floodplain ecosystems of the wet-dry tropics in northern

Table 1. Summary of findings from studies which have concurrently tested one or more environmental determinants of food chain length.

Study	Ecosystem type	Sample size	Environmental determinant			Determinants independent?
			productivity	ecosystem size	disturbance	
Pimm and Kitching 1987 [19]	Artificial treeholes	3	0		−	yes
Jenkins et al. 1992 [18]	Artificial treeholes	15	+		−	yes
Warren and Spencer 1996 [78]	Pond mesocosms	4	0		0	yes
Spencer and Warren 1996 [79]	Laboratory microcosms	12	0	+		yes
Schneider 1997 [80]	Temperate ponds	7			−	
Kaunzinger and Morin 1998 [81]	Laboratory microcosms	12	+			
Townsend et al. 1998 [68]	Temperate streams	10	+		0	yes
Vander Zanden et al. 1999 [29]	Temperate lakes	14	0	+		no
Post et al. 2000 [20]	Temperate lakes	25	0	+		yes
Jennings and Warr 2003 [82]	Marine	74	0		−	
Thompson and Townsend 2005 [24]	Temperate streams	18	+	0		yes
Williams and Trexler 2006 [83]	Tropical wetlands	20	0		−	yes
Hoeinghaus et al. 2008 [44]	Tropical rivers & reservoirs	10	+			
Stenroth et al. 2008 [84]	Temperate lakes	18	+	0		yes
Takimoto et al. 2008 [85]	Tropical islands	36		+	0	yes
Walters and Post 2008 [67]	Temperate streams	6			0	
Doi et al. 2009 [66]	Temperate ponds	15	+	+		yes
McHugh et al. 2010 [27]	Temperate streams	16	0	+	−	no
Sabo et al. 2010 [26]	Temperate rivers	36	0	+	−	no
Reid et al. 2012 [86]	Temperate billabongs	10		+		

+ indicates significant positive effect on FCL.
− indicates significant negative effect on FCL.
0 indicates non-significant effect.
Absence of symbol indicates the determinant was not tested.

Australia. The strongly seasonal, wet-dry climate and the relatively unimpeded flow regimes of this region [30] give rise to spatiotemporal gradients in hydrological connection and isolation [31,32]. This regime of hydrological connectivity can influence patterns in biotic assemblage composition [33], in the strength of coupling between consumers and their local resources [34], in local environmental conditions affecting ecosystem structure [35,36], and in the movement of top predators [6,37]. Together, these patterns suggest that food web structure, and hence food chain length, should vary according to local environmental conditions and provide an opportunity to investigate the mechanisms underpinning such variation. Accordingly, we predicted that 1) more productive sites, as indicated by nutrient concentrations, would have longer food chains, 2) more hydrologically isolated sites, which serve as an analog for more disturbed sites in this landscape setting, would have shorter food chains, and 3) larger ecosystems would have longer food chains. We also predicted that the strength of these relationships would be related to degree of trophic omnivory in the top predator, where food webs with omnivores (i.e. intraguild predators) rather than piscivores as the top predator would have shorter food chains but would still show a positive relationship between FCL and ecosystem size [15].

We show that in fact, none of these classic determinants have any influence on FCL in our seasonally-connected rivers, nor is FCL related to the trophic role of the top predator. Our finding is a striking exception to the literature and well-established patterns in food web ecology [25] (Table 1), and illustrates how food webs, and thus energy transfer, may be structured in the absence of human modifications that disrupt hydrological connectivity across landscapes.

Materials and Methods

Ethics Statement

All field sampling and collection of tissue samples was conducted under animal ethics permits from Charles Darwin University (A08008), Griffith University (ENV/08/08/AEC) and The University of Western Australia (RA/3/100/765), faunal sampling permits from the Northern Territory (DPIF S17/2666), Queensland (DAFF 89212) and Western Australian (DEC SF0063279, DOF 2008-46) Governments, and research permits from the Northern and Kimberley Land Councils to work on Aboriginal land. The giant freshwater whipray (*Himantura chaophraya*) and the freshwater sawfish (*Pristis microdon*), both threatened under the Australian Government's EPBC Act and on the IUCN Red List, were very occasionally sampled during electrofishing but returned to the water unharmed.

Study area

The wet-dry tropics of northern Australia cover approximately one fifth of the continent's land-area (about 1.3 million km^2; Fig. 1). The region is generally of low topographical relief (under 550 m altitude) and dominated by grassy woodland savanna that supports a large cattle grazing industry. Population density is very low (1 person per 2.5 km^2), with approximately 90,000 people in the largest urban centre, Darwin. Consequently, infrastructure is minimal and many of these river systems are remote and inaccessible, largely ungauged, and among the least impacted in the country [30] and the world [38]. Annual rainfall varies from 300–600 mm along the southern boundary of the region, increasing to up to 1000–2000 mm along the coast (Bureau of Meteorology, www.bom.gov.au), with most falling predictably during the summer monsoon season from October to April. Peak

discharges in rivers during the wet season can be large but show high inter-annual variation [32]. Lowland floodplains can represent up to a third of the catchment area [39] and while vast areas can be inundated, the duration of inundation can be highly variable, lasting from days to weeks and, in a few catchments, months [34]. As rainfall ceases, many rivers across the region recede to a series of disconnected waterholes during the winter dry season. Hydrological classifications of rivers across northern Australia characterise them as either perennial, which are groundwater-fed and relatively uncommon, seasonally intermittent with flow ceasing for the dry season (the most common river type), or extremely intermittent, which only flow for short periods during the wet season [32,40].

We collected carbon and nitrogen stable isotope data on food webs from 66 sites across three catchments during the 2008 dry season: 26 sites in the Daly River catchment (Northern Territory), 22 sites in the Mitchell River catchment (Queensland), and 18 sites in the Fitzroy River catchment (Western Australia; Fig. 1). Sites ranged from 13–18° S latitude and 124–145°E longitude, and were stratified according to whether they occurred on main river channels, floodplain waterholes or tributaries (Fig. S1), the latter being more common across this landscape. Sites were selected to cover gradients in productivity, disturbance and ecosystem size and on the basis of accessibility, so were representative rather than random. Post-hoc power analysis showed that sampling 18 sites (the minimum number of sites within a catchment) was sufficient to detect an effect size of $r = 0.60$ at $\alpha = 0.05$ significance, providing 0.86 power of not making a Type II error. This effect size was based on the largest average effect size in a meta-analysis on the effects of productivity ($r = 0.50$), disturbance ($r = -0.28$) and ecosystem size ($r = 0.60$) on FCL [25].

Food web sampling

Potential sources and consumers were sampled from multiple locations across each site to encompass the range of habitats present and obtain as representative a food web as possible (full sampling details are provided in Jardine et al. [34]). Primary sources included plant material from within and outside the water. Whole samples of conditioned leaf litter (cleaned of biofilm), grasses and emergent and floating-leaved macrophytes were collected to represent terrestrial production because they obtain CO_2 from the air, while aquatic sources included submerged macrophytes, charophytes, filamentous algae and biofilm. Biofilm was scrubbed from submerged surfaces (rocks, wood and/or macrophytes), and left undisturbed in a 1 L measuring cylinder for 20 min to allow sediments and detritus to settle out, leaving the top greenish fraction that we extracted and filtered [34]. Consumers included zooplankton, macroinvertebrates, crustaceans and fish. Zooplankton were collected from main channel and waterhole sites by conducting sub-surface tows with a 150 µm net, and benthic and epiphytic macroinvertebrates were sampled using a combination of dip-netting, kick-sampling and baited traps. Macroinvertebrates were live-picked and identified to Family, and enough material collected from across the site to obtain multiple samples for each Family present. Fish and larger crustaceans were collected using both backpack and boat-mounted electrofishing units, for at least 50 min fishing time and intentionally targeting the full range of habitats present. At least three individuals of each fish species, covering the range of body sizes sampled, were kept for white muscle tissue samples from the dorsal muscle, although occasionally non-lethal clips of the anal fins (with isotope ratios that are strongly correlated with muscle tissue [41]) were collected from large individuals. All samples were kept on ice or frozen for transport back to the laboratory and were

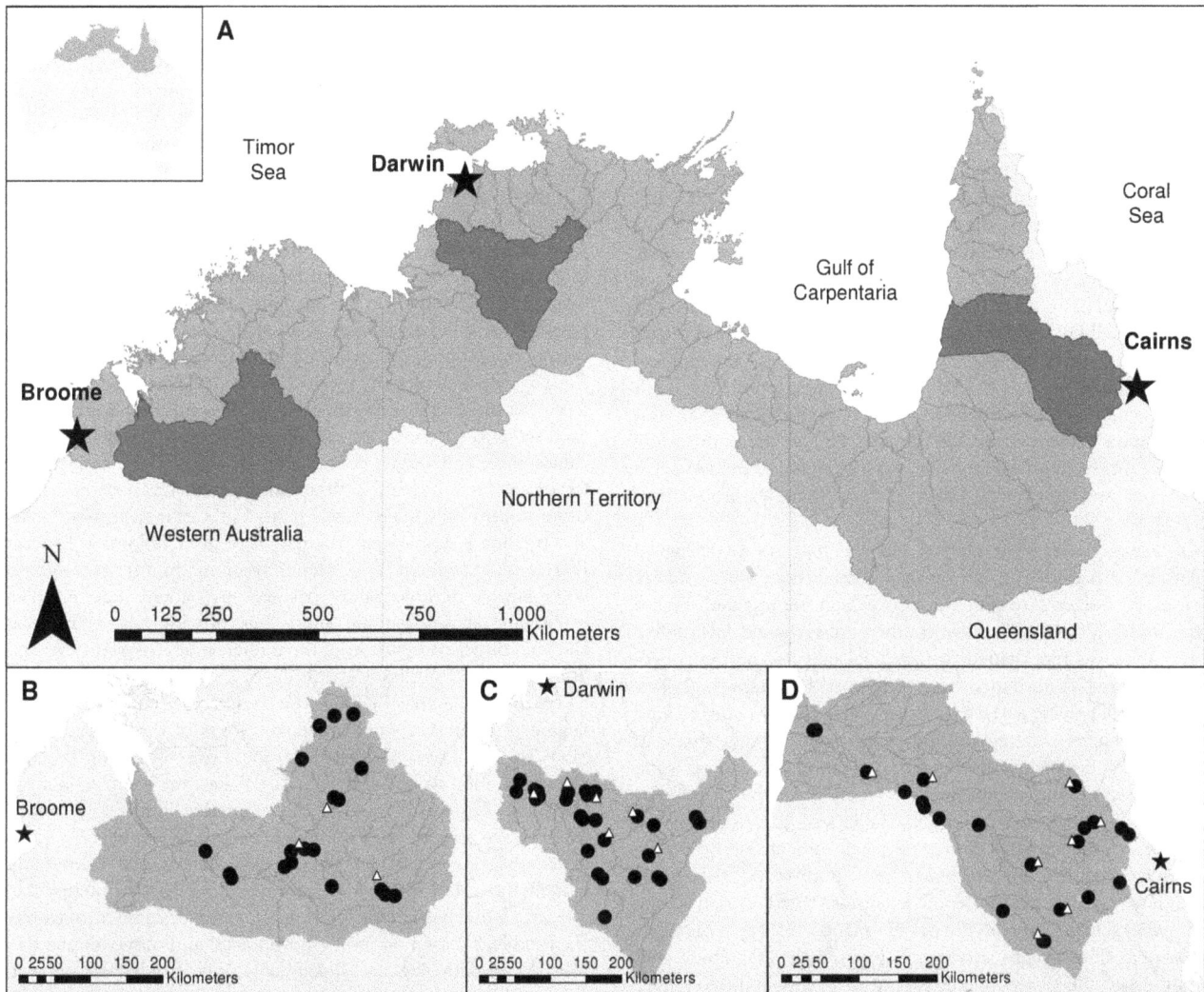

Figure 1. Map of the study area and sample sites. A) The region of the wet-dry tropics of northern Australia (mid-grey), with study catchments highlighted (dark grey). States and territories are labelled, as are major towns (stars) and gauging stations (white triangles). B) The Fitzroy River catchment (96,000 km^2, n = 18 sites), C) the Daly River catchment (55,000 km^2, n = 26 sites), and D) the Mitchell River catchment (72,000 km^2, n = 22 sites) showing sampling locations (black circles).

prepared and analysed for their carbon and nitrogen stable isotopes as described in Jardine et al. [34]. All stable isotope data and environmental data collected from each site are available from the publicly-accessible Tropical Rivers and Coastal Knowledge (TRaCK) Digital Atlas website (http://atlas.track.org.au/).

Food chain length was defined as maximum trophic position (TP) = $\lambda + (\delta^{15}N_{consumer} - \delta^{15}N_{base})/2.54$ [12,20], where λ is the trophic position of the organism used as $\delta^{15}N_{base}$ (in this case $\lambda = 2$ for mayflies, considered to be primary consumers), $\delta^{15}N_{consumer}$ is measured directly, and 2.54 is the average enrichment in $\delta^{15}N$ per trophic level, appropriate for Australian [42] and tropical [43,44] consumers, and resembling the average enrichment we observed between our primary and secondary consumers (2.52‰ $\delta^{15}N$). We used mayflies from the families Baetidae and Leptophlebiidae as our $\delta^{15}N$ baseline (with a separate baseline per site) because they are longer-lived than the periphyton that is their main food source [34] and which supports most of the fish biomass in our northern Australian rivers, particularly large-bodied fish [6,45]. There is evidence that some fish obtain their energy from floodplain periphyton during the wet season to subsidise their river

periphyton sources during the dry season, however, these fish still obtain approximately a third of their biomass from river periphyton sources [6]. Further, $\delta^{13}C$ values of our fish ranged from −36‰ to −13‰, corresponding to the range observed for mayflies across our sites (−41‰ to 14‰), and we found no significant differences in mayfly $\delta^{15}N$ between the wet and the dry seasons ($t_9 = -1.99$, $P = 0.08$), nor between early and late dry seasons ($t_{16} = 0.45$, $P = 0.66$) at subsets of our sites, indicating there was little variation in our choice of baseline across the spatial and temporal range of our consumers that may have confounded calculations of FCL.

Measurement of environmental determinants

We used dissolved nutrient concentrations as a proxy for resource availability or productivity (e.g. [20]), which we combined into a single index of productivity. While gross primary productivity would have been a more accurate measure, the large scale of the survey, the distance between sites and the remoteness of terrain meant that only a limited time was able to spent at each location, precluding our ability to obtain such data [34]. Many northern Australian rivers are oligotrophic [46] and their primary production has been shown to be nutrient- rather than light- or carbon-limited [47–52]. In the Australian wet-dry tropics and adjacent dryland ecosystems, nutrient limitation of benthic algal production can consequently limit fish production [45,53]. Algal species are likely to be limited by either N or P, therefore, both N and P are likely to limit production in algal communities and should be considered together [54]. Following Death and Winterbourn [55] and McHugh et al. [27], Principal Components Analysis (PCA) was used to combine measures of total dissolved nitrogen (TDN; <detection limits to 0.1 mg/L) and total dissolved phosphorus (TDP; <detection limits to 0.064 mg/L), obtained from filtered water samples collected from sites at the time of sampling. Both TDN and TDP loaded strongly onto PC1 (both $r > 0.7$), which had an eigenvalue over 1.0, and explained 52% of the variation between sites. We retained PC1 as the multivariate index for productivity, adding 10 to each score to ensure they were all positive and that higher values indicated greater productivity [27,54].

Disturbance was characterised on the basis of hydrological isolation, or period of hydrological disconnection, where longer periods of isolation during the dry season represented higher disturbance [26,56]. We defined three disturbance levels: low disturbance was represented by perennial sites (i.e. no disconnection), moderate disturbance was represented by intermittent sites that were still flowing at the time of sampling (mid-dry season), and high disturbance was represented by intermittent sites that had already ceased flowing at the time of sampling, and so were disconnected for the longest period. Northern Australia's strongly seasonal wet-dry climate means that the lack of rainfall during the dry season results in most rivers being intermittent and becoming disconnected during the dry season [32]. Both biotic and abiotic conditions in these disconnected waterholes tend to deteriorate with increasing period of hydrological isolation [35,36]. Therefore, while peak-flow events vary annually in their magnitude and duration and contribute to hydrological variability in these systems [32,34,37], we focussed instead on the low-flow events and used the period of hydrological isolation as our measure of disturbance. Because many rivers across northern Australia are ungauged or have limited flow data [32] (only 17 of our 66 sites were gauged), we were unable to use hydrological time series to quantify the period of hydrological isolation at all of our sites. Catchment characteristic such as topography, drainage density and vegetation cover and type can be successfully used as a proxy to classify flow regimes [40], which we matched with the existing ecohydrological classification of gauged rivers in northern Australia [32], and supplemented with local Aboriginal knowledge. We also took into account the flow conditions at the time of sampling, as sites that were already disconnected at the time of sampling during the mid-dry season (May-August) were already disconnected for a longer period than those still flowing (because sites predictably start flowing again in the early wet season, around November). Accordingly, sites were designated along a gradient of increasing disturbance as perennial (n = 25, flowing all the time and representing low disturbance), intermittent flowing (n = 23, intermittent but flowing at the time of sampling, representing moderate disturbance), or intermittent not-flowing (n = 18, these sites had stopped flowing so were hydrologically isolated for the longest period and represented sites of high disturbance). Given we have only defined three levels in our disturbance variable, we also provide a supplementary analysis of the relationship between FCL and the number of zero-flow days, obtained from 20-yr

hydrological records from the 17 gauged sites within our total 66 sites.

Ecosystem size, like productivity, was represented by a multivariate index that combined catchment area (0 to 62,000 km^2), distance from the estuary via watercourse (1 to 695 km), elevation (7 to 521 m.a.s.l.), and active channel width estimated at the time of sampling (<10–1,000 m). This enabled us to include all our sites, including waterholes that received flow inputs as local runoff, sheet flow, or as a variable proportion of overbank flooding from main and distributary channels [35] so their catchment area could not be accurately measured. Principal Components Analysis on the normalised variables resulted in PC1 having the only eigenvalue over 1.0, explaining 62% of the between-site variation, with all variables loading onto it (all $R > 0.6$). As we did for the productivity index, we retained the PC1 scores as the multivariate index of ecosystems size, adding 10 to each score to ensure they were all positive [27,54] and that higher values indicated larger ecosystems, having wider channels and larger catchment areas, together with lower elevations and being closer to the estuary (e.g. main channel sites near river mouth). We also provide a supplementary analysis of the relationship between FCL and catchment area alone, resulting in the exclusion of floodplain waterhole sites (because we could not calculate catchment area for these sites), but providing a relationship allowing direct comparison of the influence of ecosystem size with other published studies (e.g. [26]).

To assess the role of food-web architecture and whether the degree of omnivory would mediate effects of the above mechanisms on FCL, we classified the trophic role of the top predator from each food web (Table S1). At 64 sites, the top predator was one of 23 fish species, and at the remaining four sites it was an invertebrate species. Following Jepsen and Winemiller [57], the trophic class of each fish species was determined from habit, morphology and published summaries of gut contents data [58,59], and invertebrate consumers were designated a trophic class similarly based on habit, morphology and observational data [60] (M.M. Douglas, *unpublished data*). Nine trophic classes were defined and numerically ranked according to increasing trophic level (Table 2).

Data analysis

Relationships among productivity, ecosystem size and trophic role were explored using ordinary least squares (OLS) linear regression. Relationships between these determinants and disturbance (being a categorical variable) were explored using non-parametric analysis of variance and permutation tests of significance on Euclidean distance matrices [61] in PERMANOVA+, the software addition to PRIMER 6 (Primer-E, Plymouth, UK). The data were normally distributed and did not require transformation, nor did they display any spatial autocorrelation. We applied a false discovery rate (FDR) correction to control for the possibility of increased Type I errors associated with multiple tests [62].

We assessed the relative support for each of the determinants (productivity, disturbance, ecosystem size and trophic role of the top predator) hypothesised to control FCL using an information-theoretic model-selection approach [63]. Distance-based linear modelling [61] was performed (using PERMANOVA+), which accommodated using correlated predictor variables and both continuous and categorical variables. Each environmental determinant (normalised) was regressed against the FCL resemblance matrix (Euclidean dissimilarity). The model with the strongest support was identified using values derived from Akaike's Information Criterion corrected for small sample size (AIC$_c$),

Table 2. Trophic roles of the top predator in each food web from our 66 sampled sites, along with example species in each group, classified according to increasing trophic level.

Class	Trophic role	Example taxon	Major dietary items					
			Algae & aquatic plants	Detritus	Micro-crustaceans	Macro-invertebrates	Crustaceans	Fish
1	Filtering macroinvertebrates	Philopotamidae		>67%				
2	Predatory macroinvertebrates	Nepidae, Coenagrionidae				>67%		
3	Herbivorous fishes	*Scortum ogilby* (Gulf grunter)	>67%					
4	Benthivorous fishes	*Neosilurus hyrtlii* (Hyrtls tandan)		>33%		>33%		
5	Omnivorous fishes	*Hephaestus fuliginosus* (sooty grunter)	>25%			>33%	>33%	
6	Invertivorous fishes	*Craterocephalus stramineus* (strawman)			>33%	>33%		
7	Insectivorous fishes	*Glossogobius giurus* (flathead goby)				>67%		
8	Generalist predator fishes	*Leiopotherapon unicolor* (spangled perch)				~30%	~30%	~30%
9	Piscivorous fishes	*Strongylura krefftii* (longtom)						>67%

Taxon (macroinvertebrate (n = 4) or fish (n = 23)), habit, morphology, and observational data contributed to defining trophic classes, but designation was largely based on published summaries of gut contents [58,59] according to relative proportions of major dietary items.

specifically Δ_i, (i.e. = AIC_{ci} − min[AIC_c]), Akaike weights w_i (i.e. $w_i = e^{(−0.5\Delta i)}/\Sigma e^{(−0.5\Delta i)}$), and the evidence ratio (i.e. w_{top}/w_i). Using distance-based linear modelling in PERMANOVA+ also provided a permutation test of significance of the proportion of variation explained by each model.

Results

At each site we sampled an average of 22±6.6 (SD) consumers and 8.6±2.8 sources. Consumers were represented by 48 fish species and 32 macroinvertebrate taxa. Consumer $\delta^{13}C$ averaged −27.3±5.0‰ and consumer $\delta^{15}N$ averaged 6.7±2.7‰ (ranges are presented in Table S1). Food chain length averaged 4.5±0.6, ranging across three trophic levels from 3.2 in a floodplain waterhole to 6.1 in a tributary of the Mitchell River.

Twenty-seven different species represented the top predator across the 66 food webs, and no species was the top predator in more than 12 food webs (Table S1). The generalist predator *Leiopotherapon unicolor* (spangled perch) was the top predator in 12 food webs, the piscivore *Strongylura krefftii* (longtom) was the top predator in 9 food webs, and no other species was the top predator in more than 5 food webs. The most common trophic class of top predator was generalist predators (n = 22 sites), such as *L. unicolor* and *Glossamia aprion* (mouth almighty) that consumed equal proportions of fishes, crustaceans and macroinvertebrates (Table 2). Omnivores (n = 20 sites) were the next most common top predator and included *Hephaestus fuliginosus* (sooty grunter) and *Melanotaenia australis* (rainbowfish) and consumed at least 25% plant material along with crustaceans and macroinvertebrates (Table 2). Piscivores such as *S. krefftii* and *Lates calcarifer* (barramundi) that had a diet dominated by fishes (>67%, Table 2) represented the next most common top predator at 12 sites. Top predators at the remaining 12 sites spanned the remaining six trophic classes (Table S1).

Relationships among environmental determinants

The hypothesised determinants of food web structure were not independent of each other in our study, but not in the manner observed in previous studies (e.g. [26,27]). There was a significantly positive, albeit weak, relationship between productivity (dependent variable) and ecosystem size across our 66 sites ($R^2 = 0.135$, $P = 0.003$), where larger ecosystems were more productive (Fig. 2A), but there was no significant relationship between productivity and disturbance ($R^2 = 0.070$, $F_{2,63} = 2.455$, $P = 0.088$; Fig. 2B). We found a significant U-shaped relationship between ecosystem size (dependent variable) and disturbance ($R^2 = 0.378$, $F_{2,63} = 4.706$, $P < 0.014$), where ecosystems experiencing low disturbance or high disturbance were larger than those experiencing moderate disturbance (Fig. 2C). Trophic class was not predictable from productivity ($R^2 = 0.052$, $P = 0.080$), ecosystem size ($R^2 = 0.029$, $P = 0.170$), or disturbance ($R^2 = 0.268$, $F_{2,63} = 2.542$, $P = 0.089$).

Relationships between environmental determinants and food chain length

Variation in FCL was best explained by disturbance, which had the lowest AIC_c and represented 65% of model weight (Table 3). However, it only explained 7% of variation in FCL among sites: none of the environmental determinants, including disturbance, explained a significant proportion of variation in FCL (Table 3). This was reflected by the lack of a relationship between FCL and productivity ($R^2 = 0.000$, $P = 0.914$; Fig. 3A), ecosystem size ($R^2 = 0.000$, $P = 0.927$; Fig. 3B), disturbance ($R^2 = 0.019$, $F_{2,63} = 2.429$, $P = 0.098$; Fig. 3C) and the trophic class of the top predator ($R^2 = 0.003$, $P = 0.681$; Fig. 3D). Our supplementary analyses also showed no significant relationship between catchment area and FCL (n = 54, $R^2 = 0.001$, $P = 0.785$; Fig. 3E), and no significant relationship between the number of zero-flow days and FCL (n = 17, $R^2 = 0.023$, $P = 0.561$; Fig. 3F).

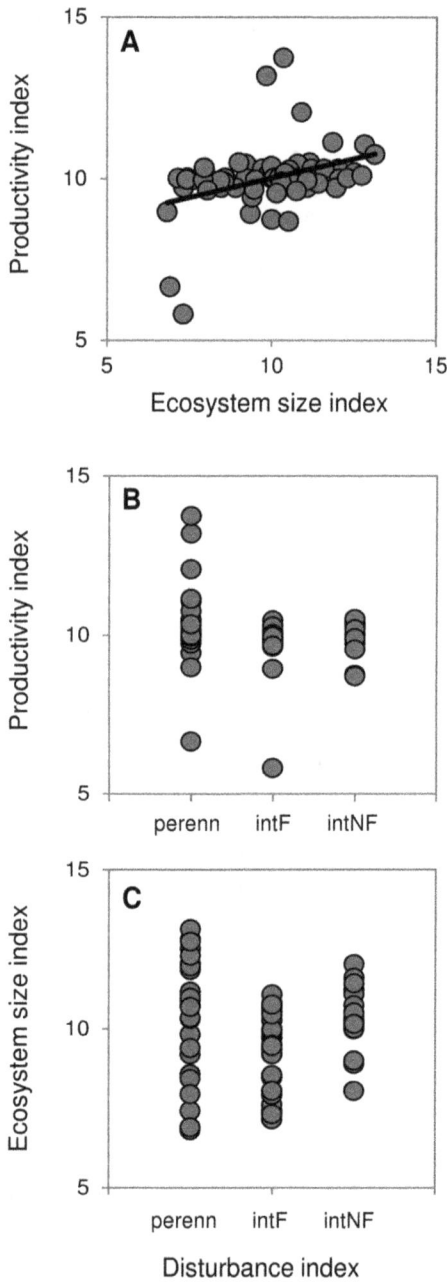

Figure 2. Relationships among environmental determinants. Relationships between A) productivity and ecosystem size ($R^2 = 0.135$, $P < 0.003$), B) productivity and disturbance ($R^2 = 0.070$, $F_{2,63} = 2.455$, $P = 0.088$), and C) ecosystem size and disturbance ($R^2 = 0.378$, $F_{2,63} = 4.706$, $P < 0.014$). For all relationships n = 66 sites. For the disturbance index, "perenn" indicates perennially-flowing sites, "intF" are sites that are intermittent but flowing at the time of sampling, and "intNF" are intermittent non-flowing sites.

Table 3. Model selection results for evaluating the hypothesised determinants of food chain length.

Hypothesised determinant	AIC_c	Δ_i	w_i	Evidence ratio	R^2	p
Productivity	−76.32	2.69	0.17	3.82	<0.01	0.913
Disturbance	−79.01	0.00	0.65	1.00	0.07	0.105
Ecosystem size	−76.31	2.70	0.17	3.85	<0.01	0.925
Trophic role of top predator	−69.50	9.51	0.01	166.67	0.14	0.326

AIC_c is Akaike's Information Criterion corrected for small sample size, Δ_i is the AIC_c difference between a given model and that with the lowest AIC_c value, and w_i is the Akaike weight. The evidence ratio is the relative weight compared to the top model. R^2 is the coefficient of determination, and the p-value is the significance of the proportion of variation explained by each determinant as assessed by marginal permutation tests.

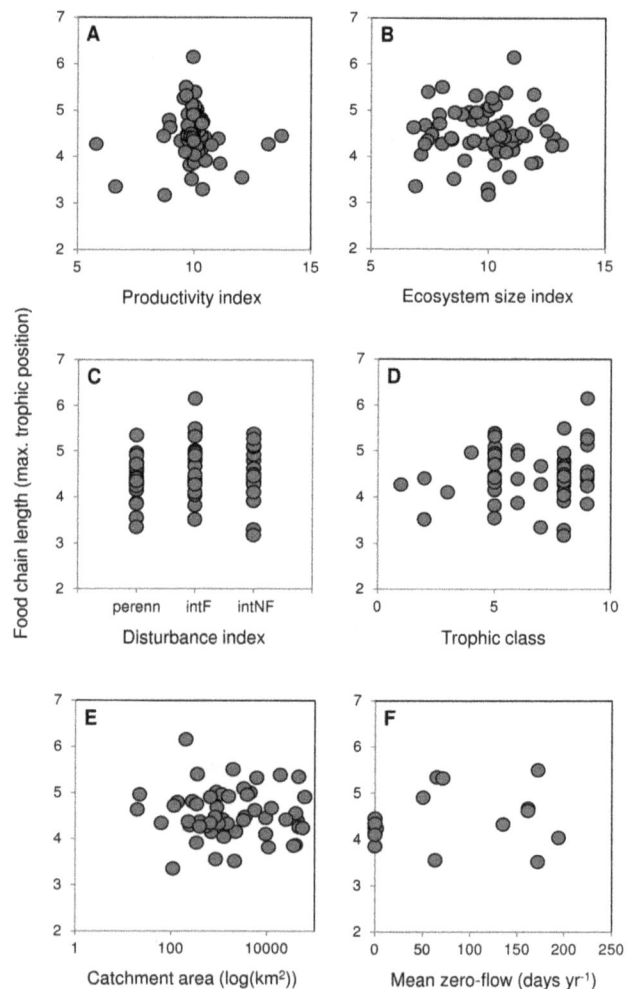

Figure 3. Relationships between environmental determinants and food chain length. Relationships between FCL and A) productivity ($R^2 = 0.000$, $P = 0.914$), B) ecosystem size ($R^2 = 0.000$, $P = 0.927$), C) disturbance ($R^2 = 0.019$, $F_{2,63} = 2.429$, $P = 0.098$), and D) trophic class of the top predator ($R^2 = 0.003$, $P = 0.681$). These relationships all had n = 66, and the disturbance categories are labelled as in Fig. 2. Also presented are supplementary relationships between FCL and E) catchment area (n = 54, $R^2 = 0.001$, $P = 0.785$), and F) the mean annual number of zero-flow days (n = 17, $R^2 = 0.023$, $P = 0.561$).

Discussion

Contrary to our expectations and the well-established patterns in the literature [25] (Table 1), we found none of the classic environmental determinants had any relationship with FCL, nor was FCL related to the trophic role of the top predator. This was despite our large sample size and the considerable variation in FCL among our food webs. Food chain length in our food webs averaged 4.5±0.6, which does not support previous predictions of short food chains in these wet-dry tropical rivers [31], but is still within the range observed in aquatic systems elsewhere [44,64]. The effects of environmental determinants tend to appear at a local scale (e.g. [64]) and we surmise that larger, regional-scale processes might be driving our observed variation in FCL, as theoretically demonstrated by previous authors [23,65]. We hypothesise that the hydrological reconnection via seasonal inundation across large tracts of the landscape, even if brief (e.g. [6]), effectively "opens" these food webs and buffers the effects of local environmental determinants on FCL.

Productivity was not related to FCL, suggesting that the availability of resources is not an important driver of food web structure in these wet-dry tropical ecosystems. Other studies that found no effect of productivity simultaneously found a positive effect of ecosystem size on FCL (Table 1), suggesting that spatial considerations are a more important influence on food web structure [20,26,29]. Our results did not support this suggestion. It is possible that our use of nutrient concentrations was too insensitive a measure of productivity, as they indicate total production rather than the proportion of production that is actually available to consumers [66]. However, we also found that FCL was also unrelated to benthic chlorophyll a at a subset of sites ($n = 26$, $R^2 = 0.01$, $P = 0.515$; T.D. Jardine, *unpublished data*). Benthic algae are known to be an important, if not dominant, energy source supporting fish biomass in Australia's northern and dryland rivers [31,45,53], and their biomass is positively related to gross primary productivity [53] (N.E. Pettit, *unpublished data*). Furthermore, nutrient availability limits primary producer biomass in these rivers, more so than light or carbon [47,51,52], so we considered nutrient concentrations to be an appropriate proxy for productivity, consistent with other studies on FCL (e.g. [20]). However, we also note that our nutrient concentrations were low, typical of northern Australian rivers [46], so either we did not have a large enough gradient to show a relationship with FCL (despite the considerable variation in FCL), or these low concentrations indicate rapid nutrient turnover and sufficient nutrient supply [49], such that there is no limitation of FCL.

We found no influence of disturbance, as measured by the degree of hydrological isolation, on FCL. Longer food chains are hypothesised to be less resilient to disturbance than short food chains [17], hence systems experiencing larger or more frequent disturbances are predicted to have shorter food chains. Research that has tested the influence of disturbance on FCL has found either neutral or negative effects (Table 1), and on average, no effect [25]. In the only study to have experimentally manipulated disturbance, Walters and Post [67] found no effect of low-flow disturbance on FCL in stream food webs; the authors suggested that local refugia mitigate disturbance effects, a conclusion also reached by Townsend et al. [68]. This is entirely possible in Australia's northern rivers, where isolated waterholes, both in-channel and floodplain, can represent the only aquatic habitat in the landscape and are thus a critical refuge during the dry season for a range of biota [37,69]. However, both abiotic and biotic conditions in these waterholes tend to deteriorate over the course of the dry season as available habitat contracts, such that their "refuge quality" is markedly reduced [35,36], our rationale for considering hydrological isolation a disturbance.

One of the mechanisms proposed for the disturbance hypothesis is that disturbance results in the loss of top predators, shortening food chains [18,19,26]. In a related analysis of biotic assemblage structure across our sites, we found that fish diversity was lower at intermittent than at perennial sites [33], potentially supporting this mechanism. However, we found no relationship between distur-

bance and the trophic role of the top predator, indicating that although species may be lost from the food web in more hydrologically isolated sites, they are not necessarily top piscivorous predators, and the trophic levels represented by the remaining species are equivalent to those represented at less disturbed, perennial sites. Research on macroinvertebrate assemblages from other wet-dry rivers in northern Australia indicates that although biodiversity is influenced by the degree of intermittency, generalist feeding strategies result in food web structure being buffered from hydrological disturbance [70]. The modelling that led to the hypothesis that longer food chains are less resilient [17] was based on the assumption that only basal species show self-regulation, i.e. intraspecific interactions that negatively affect population size. By extending the assumption of self-regulation to higher trophic levels, a more realistic assumption accommodating density-dependent feedback and intraspecific competition etc., Sterner et al. [71] showed that longer food chains are actually *more* resilient. This suggests that longer food chains are not less stable and, theoretically, not limited by disturbance, potentially explaining the lack of a clear effect on FCL in the broader literature [11,25] and supported by our findings here.

We also found no effect of ecosystem size on FCL, an unexpected result given that most studies which have tested ecosystem size have found a positive effect on FCL [25]. The ecosystem size hypothesis has more support in the literature than either the productivity or disturbance hypotheses (Table 1), although variability in effect magnitude has led to predictions that field tests of ecosystem size may find non-positive effects on FCL [25], a prediction our findings confirm. As outlined earlier, there are numerous mechanisms proposed to explain the influence of ecosystem size. The productive space hypothesis predicts larger ecosystems have more resources and therefore support longer food chains [22]. Although we had a positive relationship between productivity and ecosystem size, FCL did not show a positive relationship with either determinant, so our findings do not support this mechanism. Larger ecosystems are hypothesised to support greater functional trophic diversity and less omnivory [20], a mechanism also not supported by our findings as there was no relationship between ecosystem size and the degree of omnivory shown by the top predator.

Larger ecosystems can support more species, suggested to result in longer food chains [21]. Assemblage composition data collected during this research indicated that fish assemblages (but not macroinvertebrate or vegetation assemblages) in northern Australia can be more species-rich at perennial than at intermittent sites [33], and perennial sites were more likely to be larger on average (Fig. 2C). However, there was more variability in ecosystem size among perennial sites, and non-flowing intermittent sites also represented larger ecosystems (Fig. 2C) but did not show related increases in species richness [33]. Further, there was no relationship between FCL and the number of consumers in each food web ($n = 66$, $R^2 = 0.013$, $P = 0.369$), indicating species richness, via ecosystem size, did not contribute to our observed variation in FCL.

Related to species richness are the mechanisms of enhanced colonisation opportunity and the promotion of predator-prey coexistence that may explain the influence of ecosystem size on FCL [15,23]. These mechanisms suggest that larger ecosystems are better able to support intraguild predation and longer food chains so long as the intraguild prey are not limited in their dispersal and are good colonisers [15]. Our complementary analysis of community assembly at a subset of the study sites ($n = 46$) has found that dispersal limitation is not a strong factor structuring biotic assemblages [33], supporting this prediction, but the absence of a relationship between ecosystem size and FCL here does not support this mechanism of the ecosystem size hypothesis.

Underpinning many of the above hypotheses explaining effects on FCL are references to food web architecture, i.e. proximate structural mechanisms. These mechanisms suggest that the degree of dietary specialisation (e.g. piscivory) versus omnivory in the top predator is likely to modify FCL itself, or mediate the effects of environmental determinants on FCL [15,28]. We found that generalist predators and omnivores were equally among the most common trophic roles displayed by top predators in our food webs (n = 22 and 20, respectively), but this did not alter observed FCL, nor did it modify the effects of any environmental determinant on FCL. Previous research suggests that ecosystem size increases FCL because larger-bodied top predators tend to be absent from smaller ecosystems [20,27,28], or that the insertion of new species at lower trophic levels increases the trophic position of the top predator [28,29]. These mechanisms depend on food webs having a strong size-structure, where top predators are notably larger than prey from lower trophic levels [44]. The food webs in our wet-dry tropical rivers are not strongly size-structured: for example, piscivorous longtom (*Strongylura krefftii*) are long, slender with elongated jaws and generally reach 500 mm SL whereas the widespread generalist predator spangled perch (*Leiopotherapon unicolor*) is a robust species often only reaching 150 mm SL [58]. Omnivorous fish also show a range of sizes and include sooty grunter (*Hephaestus fuliginosus*), a moderately deep-bodied fish commonly up to 350 mm SL, and rainbowfish (*Melanotaenia* spp.) that grow to about 100 mm SL. A common prey fish is the largely herbivorous bony bream or gizzard shad (*Nematolosa erebi*), a deep-bodied fish commonly 150–300 mm SL [58]. The weak size-structure of our food webs mirrors that observed in food webs from the Neotropics [43,44] and suggests that if larger-bodied fish are absent from smaller ecosystems, these fish are not necessarily predators and could be from a number of trophic levels, and so are unlikely to show relationships with FCL [43]. Another feature of such reticulate but weakly size-structured food webs is widespread omnivory [72], where predators can consume from multiple trophic levels so increases in FCL are likely to occur through the insertion of intermediate trophic levels rather than the addition of new top predators [15,44]. This may be what is occurring in our food webs, where those with longer food chains have more intermediate trophic levels, however, as noted above, we found no relationship with FCL and the number of consumers.

It is possible that the fish we sampled are actually not apical predators in our food webs. Large-bodied predators such as elasmobranchs, crocodiles and piscivorous waterbirds can also be present in this landscape, potentially increasing FCL. While quantitative sampling of these consumers was beyond the scope of the present study, opportunistic sampling of freshwater crocodiles (*Crocodylus johnstoni*) from four sites and bull sharks (*Carcharhinus leucas*) from two sites indicated these predators occupied an equivalent trophic position to piscivorous fishes. But waterbirds, sampled opportunistically from 23 sites, often had more enriched $\delta^{15}N$ than piscivorous fishes (up to 5‰ more enriched, D.M. Warfe, *unpublished data*). Piscivorous waterbirds may therefore occupy a trophic level higher than piscivorous fish, but are not restricted to aquatic habitats so have the capacity to link food webs across larger spatial scales than fish, a possibility which supports the proposed scale-invariance of food web architecture [2]. However, food webs that include waterbirds are effectively open and less likely still to respond to local environmental determinants.

We conclude that our inability to identify environmental factors explaining the observed variation in FCL among our food webs is

due to regional processes [23,65] and a degree of plasticity in trophic dynamics. Both fish and invertebrate consumers from northern Australia can show considerable variation in diet, potentially allowing them to take advantage of scarce resources during the dry season when aquatic habitats are greatly contracted [31,58,59,70], as well as abundant resources during the wet season [6]. While limited dispersal has been theoretically shown to limit FCL at a metacommunity scale [65], associated research in this landscape has shown that dispersal limitation plays only a minor role in species assembly [33] and that floodplain carbon contributes to the biomass of predatory fish caught in permanent waterbodies [6], suggesting that fishes are not restricted in their capacity to move across the landscape.

The seasonal hydrological connection of rivers and floodplains across the landscape, even if relatively brief, can facilitate the movement of fishes onto the floodplain during the wet season where they feed and grow, thereby subsidising stream and river food webs during the dry season and temporarily linking spatially disparate food webs [2,6,73]. We propose that such seasonal linkage creates a "meta-foodweb" during the wet season, which, like a metacommunity [74], could be considered as a set of local food webs that are connected by the landscape-scale movements of high-order consumers. This meta-foodweb then splits into sub-foodwebs as sites become hydrologically disconnected during the dry season, preventing the movement of consumers. This can lead to stochasticity in assemblage structure among sites, similar to that observed in Neotropical river-floodplain systems [75], such that the number (and type) of trophic groups represented is variable, leading to variability in FCL. This hypothesis supports theoretical predictions that mobile consumers that are able to respond to, and exploit, spatial variability in resources can counteract the destabilising effects of local perturbations and thereby confer stability and persistence to food-web dynamics [2,7,8,76,77]. We suggest that the seasonal hydrological reconnection is a predominant influence on food web structure in these wet-dry tropical systems [31,37], overriding local effects of productivity, disturbance and ecosystem size, and potentially conferring resilience to the structure of biotic assemblages and food webs [36].

Wet-dry tropical regions cover extensive areas across South America, Africa, India and southeast Asia, representing a large fraction of the earth's land area, so the occurrence of meta-foodwebs linked by seasonal hydrological connectivity and fish movement could potentially be relatively widespread [73]. The corollary to this is that structures (e.g., dams and levees) and processes (e.g., flow regime alteration and saltwater intrusion) that disrupt the timing, duration and frequency of hydrological connectivity across the landscape, and thereby reduce the capacity of fish to reconnect food webs, may lead to food web structure becoming less resilient to anthropogenic perturbations.

Supporting Information

Figure S1 Photos of selected sampling sites.

Table S1 Ranges (and consumer identity) of $\delta^{13}C$ and $\delta^{15}N$ values, and the trophic class of top consumers, from each food web.

Acknowledgments

We acknowledge the traditional Aboriginal owners of the country from which samples were collected in the Daly (Wardaman, Wagiman, Malak Malak, Jawoyn and Nauiyu), Fitzroy (Bayulu, Muludja, Yiyili, Noonkanbah, Nykina/Mangala, Bunuba, Gooniyandi and Ngaringin) and Mitchell

(Kokominjena, Kokoberra, Kunjen, Western Gugu Yalanji, Mulliridgee, Barbarum, Kuku Djunkan and Gugu Mini) River catchments. We are also grateful to the Kowanyama Aboriginal Land and Natural Resource Management Office, and Fossil Downs, Mornington, Elizabeth Downs and Tipperary Stations. The Ngaringin, Bayulu, Wagiman and Mt. Pierre Station ranger groups, along with P. Close, D. Tunbridge, Q. Allsop, P. Kunroth, I. Dixon, D. Loong, M. & J. Street, P. Palmer, H. Malo, X. Pettit, D. Valdez, R. Hunt, C. Mills, K. Masci and S. Faggotter are thanked for assistance with sampling. We thank R. Diocares, A. Wood, V. Fry and L. Jardine for help with isotope sample processing and analysis, and R. Naiman for helpful suggestions on the manuscript.

Author Contributions

Conceived and designed the experiments: DMW TDJ NEP SKH BJP SEB PMD MMD. Performed the experiments: DMW TDJ NEP SKH BJP. Analyzed the data: DMW TDJ NEP. Wrote the paper: DMW TDJ NEP SKH BJP SEB PMD MMD.

References

1. Elton C (1927) Animal Ecology. London: Sidgwick and Jackson.
2. McCann KS, Rooney N (2009) The more food webs change, the more they stay the same. Philos T R Soc B 364: 1789–1801.
3. Schindler DE, Carpenter SR, Cole JJ, Kitchell JF, Pace ML (1997) Influence of food web structure on carbon exchange between lakes and the atmosphere. Science 277: 248–251.
4. Hairston NGJ, Hairston NGS (1993) Cause-effect relationships in energy flow, trophic structure, and interspecific interactions. Am Nat 142: 379–411.
5. Polis GA, Anderson WB, Holt RD (1997) Toward an integration of landscape and food web ecology: the dynamics of spatially subsidised food webs. Annu Rev Ecol Syst 28: 289–316.
6. Jardine TD, Pusey BJ, Hamilton SK, Pettit NE, Davies PM, et al. (2012) Fish mediate high food web connectivity in the lower reaches of a tropical floodplain river. Oecologia 168: 829–838.
7. Levin SA (1998) Ecosystems and the biosphere as complex adaptive systems. Ecosystems 1: 431–436.
8. Polis GA (1998) Stability is woven by complex webs. Nature 395: 744–745.
9. Briand F, Cohen JE (1987) Environmental correlates of food chain length. Science 238: 956–960.
10. Carpenter SR, Kitchell JF, Hodgson JR, Cochran PA, Elser JJ, et al. (1987) Regulation of lake primary productivity by food web structure. Ecology 68: 1863–1876.
11. Post DM (2002) The long and short of food-chain length. Trends Ecol Evol 17: 269–277.
12. Cabana G, Rasmussen JB (1994) Modelling food chain structure and contaminant bioaccumulation using stable nitrogen isotopes. Nature 372: 255–257.
13. Sabo JL, Finlay JC, Post DM (2009) Food chains in freshwaters. The Year in Ecology and Conservation Biology, 2009: Ann NY Acad Sci 1162: 187–220.
14. DeAngelis DL, Bartell SM, Brenkert AL (1989) Effects of nutrient recycling and food-chain length on resilience. Am Nat 134: 778–805.
15. Takimoto G, Post DM, Spiller DA, Holt RD (2012) Effects of productivity, disturbance and ecosystem size on food-chain length: insights from a metacommunity model of intraguild predation. Ecol Res 27: 481–493.
16. Hutchinson GE (1959) Homage to Santa Rosalia, or, why are there so many kinds of animals? Am Nat 93: 145–159.
17. Pimm SL, Lawton JH (1977) Number of trophic levels in ecological communities. Nature 268: 329–331.
18. Jenkins B, Kitching RL, Pimm SL (1992) Productivity, disturbance and food web structure at a local spatial scale in experimental container habitats. Oikos 65: 249–255.
19. Pimm SL, Kitching RL (1987) The determinants of food chain lengths. Oikos 50: 302–307.
20. Post DM, Pace ML, Hairston Jnr NG (2000) Ecosystem size determines food-chain length in lakes. Nature 405: 1047–1049.
21. Cohen JE, Newman CM (1991) Community area and food-chain length: theoretical predictions. Am Nat 138: 1542–1554.
22. Schoener TW (1989) Food webs from the small to the large. Ecology 70: 1559–1589.
23. Holt RD (2002) Food webs in space: in the interplay of dynamic instability and spatial processes. Ecol Res 17: 261–273.
24. Thompson RM, Townsend CR (2005) Energy availability, spatial heterogeneity and ecosystem size predict food-web structure in streams. Oikos 108: 137–148.
25. Takimoto G, Post DM Environmental determinants of food-chain length: a meta-analysis. Ecol Res In press.
26. Sabo JL, Finlay JC, Kennedy T, Post DM (2010) The role of discharge variation in scaling of drainage area and food chain length in rivers. Science 330: 965–967.
27. McHugh PA, McIntosh AR, Jellyman PG (2010) Dual influences of ecosystem size and disturbance on food chain length in streams. Ecol Lett 13: 881–890.
28. Post DM, Takimoto G (2007) Proximate structural mechanisms for variation in food-chain length. Oikos 116: 775–782.
29. Vander Zanden MJ, Shuter BJ, Lester N, Rasmussen JB (1999) Patterns of food chain length in lakes: a stable isotope study. Am Nat 154: 406–416.
30. Stein JL, Stein JA, Nix HA (2002) Spatial analysis of anthropogenic river disturbance at regional and continental scales: identifying the wild rivers of Australia. Landscape Urban Plan 60: 1–25.
31. Douglas MM, Bunn SE, Davies PM (2005) River and wetland food webs in Australia's wet-dry tropics: general principles and implications for management. Mar Freshwater Res 56: 329–342.
32. Kennard MJ, Pusey BJ, Olden JD, Mackay S, Stein J, et al. (2010) Classification of natural flow regimes in Australia to support environmental flow management. Freshwater Biol 55: 171–193.
33. Warfe DM, Pettit NE, Magierowski R, Pusey BJ, Davies PM, et al. (2013) Hydrological connectivity structures concordant plant and animal assemblages according to niche rather than dispersal processes. Freshwater Biol 58: 292–305.
34. Jardine TD, Pettit NE, Warfe DM, Pusey BJ, Ward DP, et al. (2012) Consumer-resource coupling in wet-dry tropical rivers. J Anim Ecol 81: 310–322.
35. Ward DP, Hamilton SK, Jardine TD, Pettit NE, Tews EK, et al. (2013) Assessing the seasonal dynamics of inundation, turbidity and aquatic vegetation in the Australian wet-dry tropics using optical remote sensing. Ecohydrology 6: 312–323.
36. Pettit NE, Jardine TD, Hamilton SK, Sinnamon V, Valdez D, et al. (2012) Seasonal changes in water quality and macrophytes and the impact of cattle on tropical floodplain waterholes. Mar Freshwater Res 63: 788–800.
37. Warfe DM, Pettit NE, Davies PM, Pusey BJ, Hamilton SK, et al. (2011) The "wet-dry" in the wet-dry tropics drives river ecosystem structure and processes in northern Australia. Freshwater Biol 56: 2169–2195.
38. Vörösmarty CJ, McIntyre PB, Gessner MO, Dudgeon D, Prusevich A, et al. (2010) Global threats to human water security and river biodiversity. Nature 467: 555–561.
39. Stein JL, Hutchinson MF, Pusey BJ, Kennard MJ (2009) Appendix 8. Ecohydrological classification based on landscape and climate data. Ecohydrological regionalisation of Australia: a tool for management and science. Canberra: Land & Water Australia.
40. Moliere DR, Lowry JBC, Humphrey CL (2009) Classifying the flow regime of data-limited streams in the wet-dry tropical region of Australia. J Hydrol 367: 1–13.
41. Jardine TD, Hunt RJ, Pusey BJ, Bunn SE (2011) A non-lethal sampling method for stable carbon and nitrogen isotope studies of tropical fishes. Mar Freshwater Res 62: 83–90.
42. Vanderklift MA, Ponsard S (2003) Sources of variation in consumer-diet $\delta^{15}N$ enrichment: a meta-analysis. Oecologia 136: 169–182.
43. Layman CA, Winemiller KO, Arrington DA, Jepsen DB (2005) Body size and trophic position in a diverse tropical food web. Ecology 86: 2530–2535.
44. Hoeinghaus DJ, Winemiller KO, Agostinho AA (2008) Hydrogeomorphology and river impoundment affect food-chain length of diverse Neotropical food webs. Oikos 117: 984–995.
45. Jardine TD, Hunt RJ, Faggotter SJ, Valdez D, Burford MA, et al. Carbon from periphyton supports fish biomass in waterholes of a wet-dry tropical river. River Res Appl In press.
46. Brodie JE, Mitchell AW (2005) Nutrients in Australian tropical rivers: changes with agricultural development and implications for receiving environments. Mar Freshwater Res 56: 279–302.
47. Webster IT, Rea N, Padovan AV, Dostine P, Townsend SA, et al. (2005) An analysis of primary production in the Daly River, a relatively unimpacted tropical river in northern Australia. Mar Freshwater Res 56: 303–316.
48. Ganf GG, Rea N (2007) Potential for algal blooms in tropical rivers of the Northern Territory, Australia. Mar Freshwater Res 58: 315–326.
49. Townsend SA, Schult JA, Douglas MM, Skinner S (2008) Does the Redfield ratio infer nutrient limitation in the macroalga Spirogyra fluviatilis? Freshwater Biol 53: 509–520.
50. Burford MA, Revill AT, Palmer DW, Clementson L, Robson BJ, et al. (2011) River regulation alters drivers of primary productivity along a tropical river-estuary system. Mar Freshwater Res 62: 141–151.
51. Hunt RJ, Jardine TD, Hamilton SK, Bunn SE (2012) Temporal and spatial variation in ecosystem metabolism and food web carbon transfer in a wet-dry tropical river. Freshwater Biol 57: 435–450.
52. Townsend SA, Webster IT, Schult JH (2011) Metabolism in a groundwater-fed river system in the Australian wet-/dry tropics: tight coupling of photosynthesis and respiration. J N Am Benthol Soc 30: 603–620.
53. Bunn SE, Davies PM, Winning M (2003) Sources of organic carbon supporting the food web in an arid zone floodplain river. Freshwater Biol 48: 619–635.
54. Francoeur SN (2001) Meta-analysis of lotic nutrient emendment experiments: detecting and quantifying subtle responses. J N Am Benthol Soc 20: 358–368.
55. Death RG, Winterbourn MJ (1994) Environmental stability and community persistence: a multivariate perspective. J N Am Benthol Soc 13: 125–139.
56. Lake PS (2003) Ecological effects of perturbation by drought in flowing waters. Freshwater Biol 48: 1161–1172.

57. Jepsen DB, Winemiller KO (2002) Structure of tropical river food webs revealed by stable isotope ratios. Oikos 96: 46–55.

58. Pusey BJ, Kennard MJ, Arthington AH (2004) Freshwater Fishes of North-Eastern Australia. Collingwood, Australia: CSIRO.

59. Davis AM, Pusey BJ, Thorburn DC, Dowe JL, Morgan DL, et al. (2010) Riparian contributions to the diet of terapontid grunters (Pisces: Terapontidae) in wet-dry tropical rivers. J Fish Biol 76: 862–879.

60. Gooderham J, Tsyrlin E (2002) The Waterbug Book. Collingwood, Australia: CSIRO.

61. McArdle BH, Anderson MJ (2001) Fitting multivariate models to community data: a comment on distance-based redundancy analysis. Ecology 82: 290–297.

62. Benjamini Y, Hochberg Y (1995) Controlling the false discovery rate: a practical and powerful approach to multiple testing. J Roy Stat Soc B 57:289–300.

63. Burnham KP, Anderson DR (2002) Model Selection and Multimodal Inference: a practical information-theoretic approach. New York, USA: Springer-Verlag.

64. Vander Zanden MJ, Fetzer WW (2007) Global patterns of aquatic food chain length. Oikos 116: 1378–1388.

65. Calcagno V, Massol F, Mouquet N, Jarne P, David P (2011) Constraints on food chain length arising from regional metacommunity dynamics. P Roy Soc B 278: 3042–3049.

66. Doi H, Chan K-H, Ando T, Ninomiya I, Imai H, et al. (2009) Resource availability and ecosystem size predict food-chain length in pond ecosystems. Oikos 118: 138–144.

67. Walters AW, Post DM (2008) An experimental disturbance alters fish size structure but not food chain length in streams. Ecology 89: 3261–3267.

68. Townsend CRT, Thompson RM., McIntosh AR, Kilroy C, Edwards ED, et al. (1998) Disturbance, resource supply, and food-web architecture in streams. Ecol Lett 1: 200–209.

69. Bunn SE, Thoms MC, Hamilton SK, Capon SJ (2006) Flow variability in dryland rivers: boom, bust and the bits in between. River Res Appl 22: 179–186.

70. Leigh C, Burford MA, Sheldon F, Bunn SE (2010) Dynamic stability in dry season food webs within tropical floodplain rivers. Mar Freshwater Res 61: 357–368.

71. Sterner RW, Bajpal A, Adams T (1997) The enigma of food chain length: absence of theoretical evidence for dynamic constraints. Ecology 78: 2258–2262.

72. Winemiller KO (1990) Spatial and temporal variation in tropical fish networks. Ecol Monogr 60: 331–367.

73. Winemiller KO, Jepsen DB (1998) Effects of seasonality and fish movement on tropical river food webs. J Fish Biol 53 (Supplement A): 267–296.

74. Leibold MA, Holyoak M, Mouquet N, Amarasekare P, Chase JM, et al. (2004) The metacommunity concept: a framework for multi-scale community ecology. Ecol Lett 7: 601–613.

75. Arrington DA, Winemiller KO (2006) Habitat affinity, the seasonal flood pulse, and community assembly in the littoral zone of a Neotropical floodplain river. J N Am Benthol Soc 25: 126–141.

76. McCann KS, Rasmussen JB, Umbanhowar J (2005) The dynamics of spatially coupled food webs. Ecol Lett 8: 513–523.

77. Van de Koppel J, Bardgett RD, Bengtsson J, Rodriguez-Barrueco C, Rietkerk M, et al. (2005) The effects of spatial scale on trophic interactions. Ecosystems 8: 801–807.

78. Warren PH, Spencer M (1996) Community and food-web responses to the manipulation of energy input and disturbance in small ponds. Oikos 75: 407–418.

79. Spencer M, Warren PH (1996) The effects of habitat size and productivity on food web structure in small aquatic microcosms. Oikos 75: 419–430.

80. Schneider DW (1997) Predation and food web structure along a habitat duration gradient. Oecologia 110: 567–575.

81. Kaunzinger CMK, Morin PJ (1998) Productivity controls food-chain properties in microbial communities. Nature 395: 495–497.

82. Jennings S, Warr KJ (2003) Smaller predator-prey body size ratios in longer food chains. P Roy Soc B 270: 1413–1417.

83. Williams AJ, Trexler JC (2006) A preliminary analysis of the correlation of food-web characteristics with hydrology and nutrient gradients in the southern Everglades. Hydrobiologia 569: 493–504.

84. Stenroth P, Holmqvist N, Nystrom P, Berglund O, Larsson P, et al. (2008) The influence of productivity and width of littoral zone on the trophic position of a large-bodied omnivore. Oecologia 156: 681–690.

85. Takimoto G, Spiller DA, Post DM (2008) Ecosystem size, but not disturbance, determines food-chain length on islands of the Bahamas. Ecology 89: 3001–3007.

86. Reid MA, Delong MD, Thoms MC (2012) The influence of hydrological connectivity on food web structure in floodplain lakes. Riv Res Appl 28: 827–844.

6

Water- and Plant-Mediated Responses of Ecosystem Carbon Fluxes to Warming and Nitrogen Addition on the Songnen Grassland in Northeast China

Li Jiang[1], Rui Guo[2], Tingcheng Zhu[1], Xuedun Niu[3], Jixun Guo[1]*, Wei Sun[1]*

1 Key Laboratory for Vegetation Ecology, Ministry of Education, Institute of Grassland Science, Northeast Normal University, Changchun, Jilin Province, P. R. China, **2** Institute of Environment and Sustainable Development in Agriculture, Chinese Academy of Agricultural Sciences, Key Laboratory of Dryland Agriculture, Ministry of Agriculture, Beijing, P. R. China, **3** Key Laboratory of Molecular Enzymology and Engineering of Ministry of Education, College of Life Sciences, Jilin University, Changchun, Jilin Province, P. R. China

Abstract

Background: Understanding how grasslands are affected by a long-term increase in temperature is crucial to predict the future impact of global climate change on terrestrial ecosystems. Additionally, it is not clear how the effects of global warming on grassland productivity are going to be altered by increased N deposition and N addition.

Methodology/Principal Findings: In-situ canopy CO_2 exchange rates were measured in a meadow steppe subjected to 4-year warming and nitrogen addition treatments. Warming treatment reduced net ecosystem CO_2 exchange (NEE) and increased ecosystem respiration (ER); but had no significant impacts on gross ecosystem productivity (GEP). N addition increased NEE, ER and GEP. However, there were no significant interactions between N addition and warming. The variation of NEE during the four experimental years was correlated with soil water content, particularly during early spring, suggesting that water availability is a primary driver of carbon fluxes in the studied semi-arid grassland.

Conclusion/Significance: Ecosystem carbon fluxes in grassland ecosystems are sensitive to warming and N addition. In the studied water-limited grassland, both warming and N addition influence ecosystem carbon fluxes by affecting water availability, which is the primary driver in many arid and semiarid ecosystems. It remains unknown to what extent the long-term N addition would affect the turn-over of soil organic matter and the C sink size of this grassland.

Editor: Ben Bond-Lamberty, DOE Pacific Northwest National Laboratory, United States of America

Funding: This research was supported by the National Natural Science Foundation of China (31170303), the Strategic Priority Research Program of the Chinese Academy of Sciences (XDA 05050601), the Research Fund for the Doctoral Program of Higher Education of China (20090043110007), the State Key Laboratory of Vegetation and Environmental Change (LVEC2012kf01) and the Fundamental Research Funds for the Central Universities (11NQJJ028). The funders had no role in study design, data collection and analysis, decision to publish, or preparation of the manuscript.

Competing Interests: The authors have declared that no competing interests exist.

* E-mail: gjixun@nenu.edu.cn (JXG); sunwei@nenu.edu.cn (WS)

Introduction

Due to rising atmospheric concentrations of CO_2 and other greenhouse gases, global mean air temperatures increased by $\approx 0.2°C$ per decade in the past 30 years, and are projected to increase continually by 1.0 to more than $4.0°C$ by the end of the 21st century [1,2]. Such temperature changes, unprecedented in modern times, are predicted to substantially influence ecosystem processes and global carbon cycling [3]. Grasslands are considered to be highly relevant for future projections of the global carbon budget, as they cover approximately one-third of the earth's terrestrial surface and store 10–30% of the world's soil carbon [4].

Effects of warming on carbon fluxes of grassland ecosystems have been investigated to a certain extent [5–11]. In most of these studies, warming stimulated both gross ecosystem productivity (GEP) and ecosystem respiration (ER) with the net effects on the carbon balance depending on the temperature sensitivity of carbon release by respiration relative to carbon uptake through photosynthesis [12]. Lack of consistent pattern both across ecosystems [3] and within a grassland among growing seasons indicates that a general trend of warming on carbon balance cannot be easily predicted, as biomes exhibit a system specific and temporarily dynamic response to warming [13]. This dynamic response is due to temporal shifts in species composition and co-limitation by other abiotic resources. Dominant drivers of productivity in many grassland systems are water and nitrogen availability, both of which could strongly interact with changes in temperature [14,15].

Many temperate ecosystems are predicted to experience rates of atmospheric N deposition as high as $2–5\ g\ m^{-2}\ yr^{-1}$ above the preindustrial rates over this century [16]. In addition to the projected changes in temperature and water availability, increased N input by wet deposition [17] and increased grassland land-use intensity in combination with N fertilizer application are determining factors of carbon balance of grassland ecosystems. Higher N availability in grassland systems is expected to increase aboveground productivity because of enhancement in leaf area index, and improvement in ecosystem water use efficiency [18,19].

Thus, increased N availability may, by improving water use efficiency, relieve warming-induced water deficits in semiarid grassland systems. Moreover, tightly coupled C and N cycles are strongly regulated by water availability in arid and semiarid grasslands with the combined effects of the availability of these resources ultimately determining the net impact of global change on the carbon balance of grassland systems [10,20].

To understand the effects of global warming and changes in nitrogen availability on the carbon fluxes of grassland ecosystems, we conducted a 4-year artificial warming and N addition experiment in a meadow steppe in Northeast China. The specific questions we addressed in this study included: (1) to what extent do warming and N addition affect GEP and ER and (2) what are the interactive effects between warming and N addition on ecosystem carbon fluxes in the Songnen grassland?

Materials and Methods

Ethics Statement

No specific permissions were required for the described field studies, because the Songnen Grassland Ecological Research Station is a department of Northeast Normal University. No specific permissions were required for this study, because the performance of this study follows the guidelines set by Northeast Normal University. No specific permissions were required for these locations/activities. No location is privately-owned or protected in any way and the field studies did not involve endangered or protected species.

Study Site

The experiment was conducted at the Grassland Ecosystem Experimental Station ($44°30'-44°45'N$, $123°31'-123°56'E$) of the Northeast Normal University in Jilin Province, China. The Songnen grassland is 30500 km^2 in size and located at the Eastern end of the Eurasian steppe belt [21,22]. This area has a typical meso-thermal monsoon climate with a mean annual temperature of $6.4°C$, and a frost-free period of 141 days. Mean annual rainfall is 471 mm, and occurs mainly from June to August. Annual potential evapotranspiration is 2–3 times greater than the annual rainfall. The growing season was limited to late April to early October. Chernozem is the main soil type with 2.0% of soil organic carbon content, 1.4% of soil humus, 0.15% of total N, and pH >9.0 [22]. In the Songnen grassland, *Leymus chinensis* is the dominant species; *Phragmites australis*, *Chloris virgata*, *Kalimeris integrifolia*, *Carex duriuscula*, and *Artemisia mongolica* are abundance. C$_3$ grasses represent approximately 90% of the total plant biomass [22,23]. Mean annual net aboveground primary productivity at the experimental site is 300–400 g m^{-2} yr^{-1} with peak leaf area index (LAI) of up to 4 [24,25].

Experimental Design

The experiment was carried out in a complete randomized block factorial experimental design with warming and N addition as fixed factors and each factor has two levels. The treatment combinations were un-heated and unfertilized treatment (C); unheated and fertilized treatment (N); heated and unfertilized treatment (W); and both heated and fertilized treatment (WN). Each treatment combination had 6 replications. The twelve subplots (each had an area of 3 m × 4 m) were arranged in three rows with 3 m distance between adjacent rows and 1.5 m distance between adjacent plots within each row. Warming treatment was randomly assigned to 6 of the 12 subplots; the other 6 subplots were treated as control. For each of the warming and control subplots, half of the area was treated with N addition. The

airborne N deposition was estimated to be as high as 80–90 g·m^{-2}·yr^{-1} and much higher N deposition is expected in the future owing to increasing anthropogenic activities and land-use change [26,27]. For the studied temperate grassland ecosystem, the atmospheric N deposition rate was up to 2.7 g·m^{-2}·yr^{-1} during the past ten years; however the saturation rate of N deposition was approximately 10.0 g·m^{-2}·yr^{-1} [27,28]. Thus, we applied 10 g·m^{-2} N in form of NH$_4$NO$_3$ in early June of each year in the N addition plots. Warming plots were heated continuously from June 2006 to October 2009 using MSR-2420 infrared radiators (165 cm × 15 cm; Kalglo Electronics, Bethlehem, PA, USA), suspended 2.25 m above the ground. Infrared radiators supplied 1600 W·m^{-2} of thermal radiation, resulting in a soil-surface warming of $1.7±0.2°C$. The warming effects of infrared radiators were tested by measuring soil surface temperature at 20 points diagonally arranged across the heated plots in June of 2007 and 2008. Warmed plots had significantly ($P = 0.003$) higher temperature during the experimental period. In unheated control plots, 'dummy' heaters with the same shape and size as the infrared radiator were suspended 2.25 m above soil surface in order to simulate the shading effects of the heater.

Climate Data, Soil Temperature and Soil Water Content

Climate data in 2006 were obtained from the Pasture Ecology Research Station of the Northeast Normal University, Jilin Province, China ($123°44'$ E, $44°40'$ N), approximately 3 km distant from the study site. Data in 2007, 2008 and 2009 were obtained from an eddy tower roughly 200 m distant from the experimental plots. Soil moisture sensors and soil temperature probes were installed in the middle of each of the 24 plots to measure soil temperature at 10 cm depth and volumetric soil water content (SWC) at 0–10 cm soil depth, respectively. Soil temperature and SWC were continuously recorded at 1-h interval from June 2006 to October 2009 and data were recorded with an ECH$_2$O dielectric aquameter (EM50/R Decagon Ltd, Pullman, WA, USA) from June 2006 to October 2009.

Leaf Area Index

Leaf area index (LAI) was measured monthly with a plant canopy analyzer (LAI-2000 Plant Canopy Analyzer, Li-Cor, Lincoln, NE, USA).

Measurements of Ecosystem Gas Exchange

Ecosystem gas exchange was measured once a month over the growing seasons (from May to October) from 2006 to 2009. Measurements were performed under cloud-free conditions from 9 a.m. to 1 p.m. To avoid potential effects of temporal variation in CO$_2$ exchange between ecosystem and the atmosphere during the 4-hour measuring period, we measured the 6 treatment combinations sequentially. In May 2006, a square aluminum frames (0.5 m × 0.5 m) were inserted into the soil to a depth of approximately 3 cm on each plot with a distance of 30 cm to the plot's border. Care was taken to minimize soil disturbance during installation. Frames provided a plane base for the mounting of the mobile canopy chamber (0.5 m × 0.5 m × 0.9 m, Polymethyl Methacrylate).

Ecosystem gas exchange (CO$_2$ and water fluxes) was measured with a closed-flow infrared gas analyzer (LI-6400, Li-Cor, Lincoln, NE, USA) attached to the transparent canopy chamber. During measurements, the chamber was sealed to the soil surface by attaching it to the permanently fixed aluminum base in each plot. Four small fans ran continuously to mix the air inside the chamber during the measurement period. Nine consecutive recordings of CO$_2$ concentrations were taken at 10-

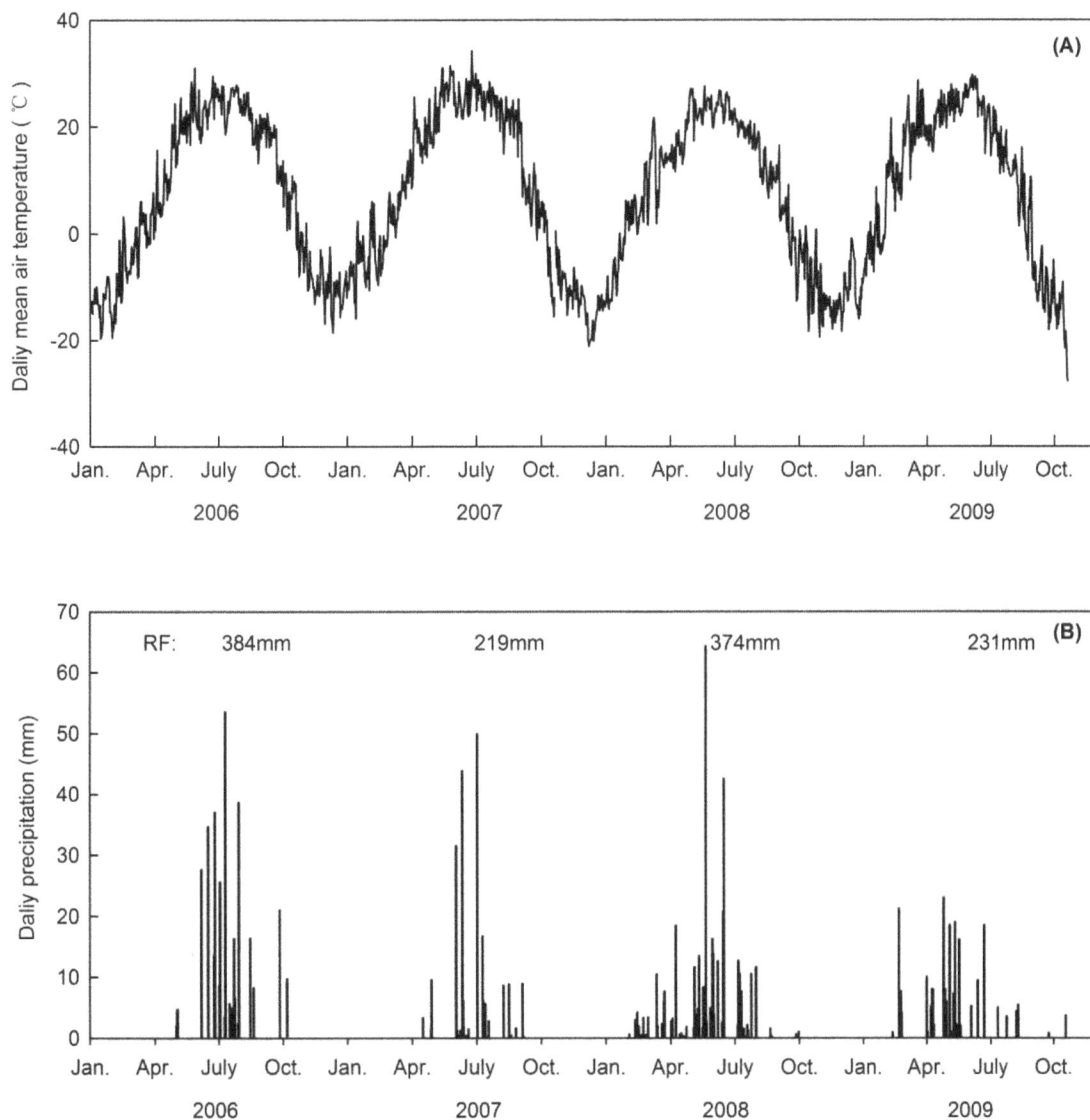

Figure 1. Daily mean air temperature, annual rainfall (RF), and daily precipitation in 2006, 2007, 2008 and 2009. Data for 2006 were obtained from a weather station located in the Chang Ling Horse Breeding Farm in Jilin Province, 3 km in distance from the study site. Data in 2007, 2008 and 2009 were obtained from an eddy tower roughly 200 m distant from the experimental plots.

s intervals during a 90-s period after steady-state conditions were achieved within the chamber (approximately 30 seconds after the placement of chamber). Net ecosystem CO_2 exchange (NEE) was determined from the time-courses of chamber CO_2 concentration change. We followed the approach of Steduto et al. [29] for the conversion from concentration change to flux per unit soil surface area. The NEE data are presented in the notation commonly used by the terrestrial sciences community whereby positive values represent net fluxes from the atmosphere to the ecosystem (carbon uptake) and negative values represent from the terrestrial ecosystem to the atmosphere (carbon release). Details on these static-chamber flux calculations can be found in the soil-flux calculation procedure on Page 6-2 in the LI-6400 manual (Li-Cor Inc., 2004). Following the NEE measurements, the chamber was vented, replaced on the aluminum frame, and covered with an opaque cloth to measure ecosystem respiration (ER). The ER values were obtained by monitoring steadily increasing in chamber CO_2

concentration (usually after 1 min). Gross ecosystem productivity (GEP) was calculated as the difference between NEE and ER.

Leaf Gas-exchange Measurements

Leaf net photosynthetic rate and stomatal conductance in *L. chinensis* and *P. australis* were measured with a Li-6400 gas analyzer (Li-Cor, Lincoln, NE, USA) between 9–11 a.m. on a monthly basis during 2006–2009. The upper most fully expanded leaves were used for leaf gas exchange measurements. Gas exchange parameters were recorded at light intensities of 1400–1600 μmol m^{-2} s^{-1} after readings stabilized (approximately 5 min). In order to compare chamber measurements of GEP with leaf gas exchange data, results from the June measurements are reported in this study.

Statistical Analysis

Seasonal mean values used in this study were calculated from the monthly mean values, which were first averaged from all

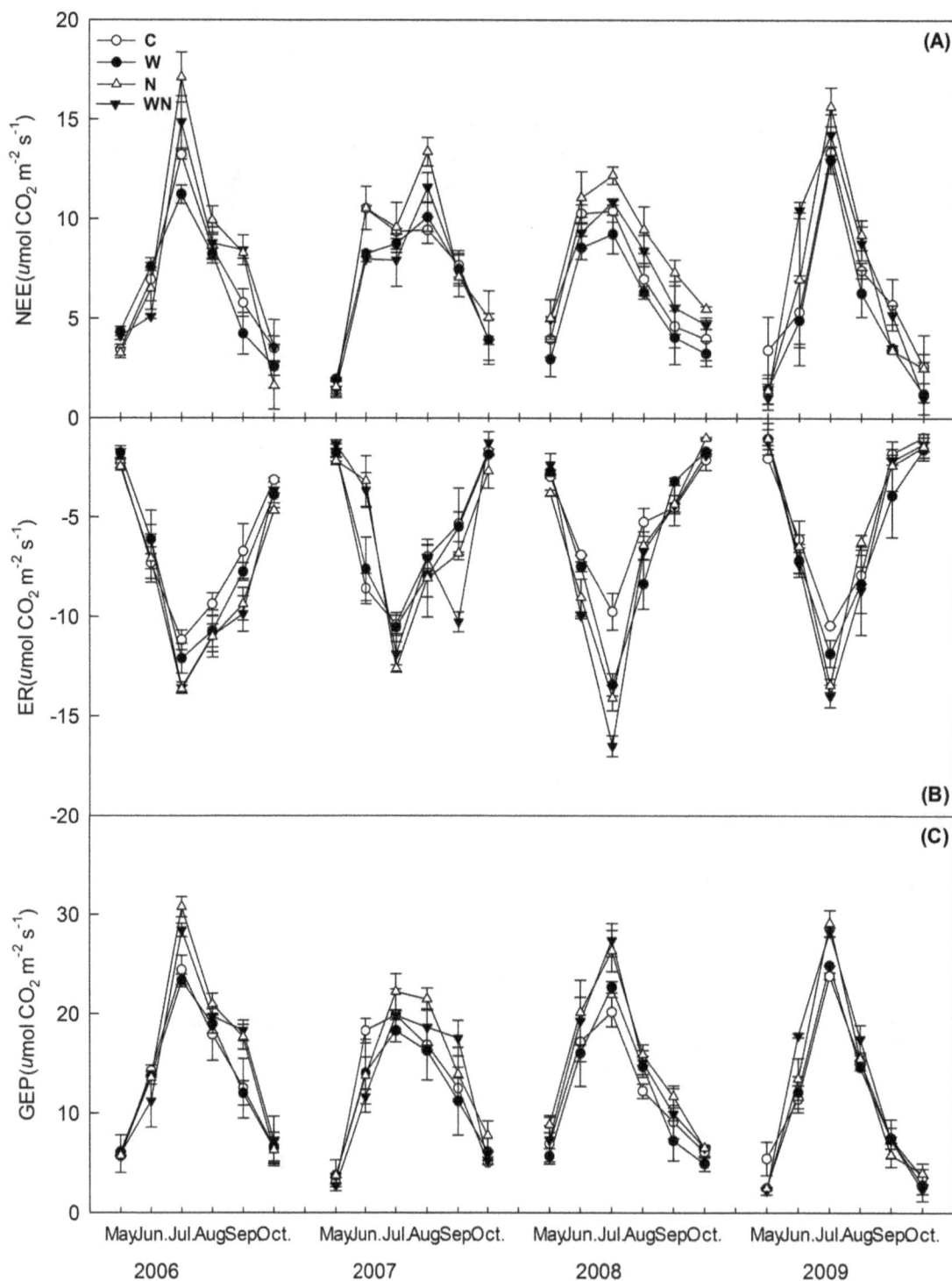

Figure 2. Seasonal dynamics (mean ± SE; n = 6) of net ecosystem CO$_2$ exchange (NEE), ecosystem respiration (ER), gross ecosystem productivity (GEP) in response to warming (1.8°C) and N addition (10 g m^{-2} yr^{-1}) in the Songnen grassland. C = control, W = warming, N = N addition, WN = combined warming and N addition.

measurements in the same month. Three-way ANOVA was used to examine the effects of year, warming, N addition, and their possible interactions on ecosystem C fluxes. If significant inter-annual variation was detected (year effect $P<0.05$), Repeated Measures ANOVA (RMANOVA) were used to examine the temporal (inter- or intra-annual) variation and the effects of

warming and N addition effects on ecosystem C fluxes over the growing seasons in 2006, 2007, 2008 and 2009, respectively. Between-subject effects were evaluated as warming, N addition and their interactions, and within-subject effects were year (or measuring time within season) variation and the effects of warming and N addition on ecosystem C fluxes. Bonferroni correction was

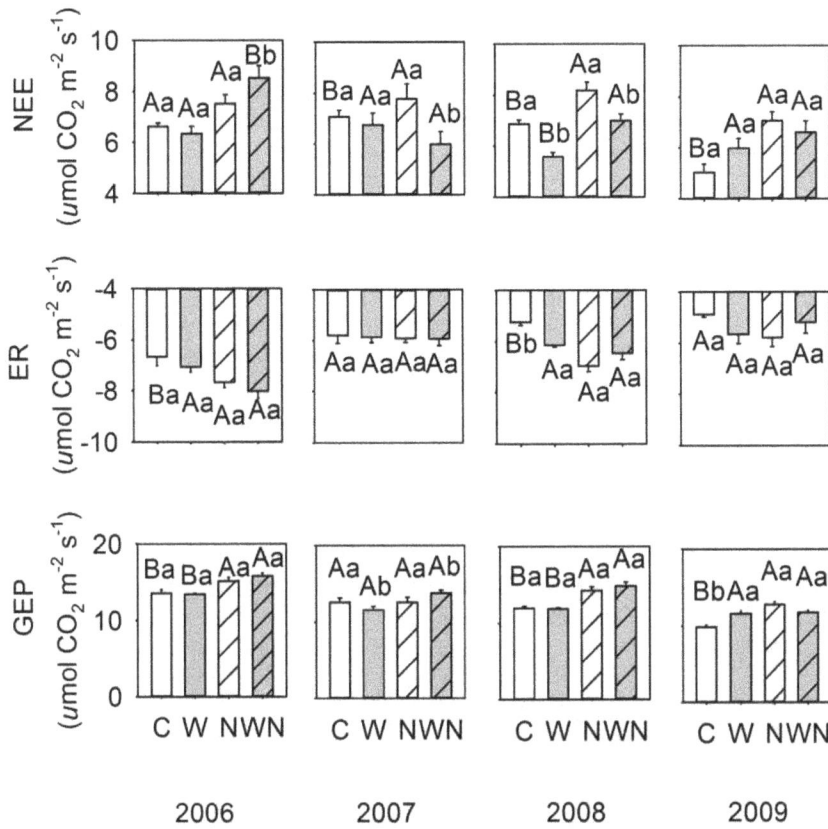

Figure 3. Growing season mean (n = 6) net ecosystem CO₂ exchange (NEE), ecosystem respiration (ER), and gross ecosystem productivity (GEP). Data are presented as mean ± SE. Different letters indicate significant differences (P<0.05) among treatments with capital and small letters indicating differences between N and warming treatments, respectively. C = control, W = warming, N = N addition, WN = combined warming and N addition.

used to adjust for multiple comparisons. Regression with correction for autocorrelation and stepwise multiple linear analyses were used to examine the relationships of ecosystem C fluxes with soil moisture, and leaf area index. All statistical analyses were conducted with SAS software (SAS Institute Inc., Cary, NC, USA).

Table 1. DF, F values and P values from three-way ANOVA on the effects of warming (W), nitrogen (N), and year (Y), and their interactions on net ecosystem CO₂ exchange (NEE), ecosystem respiration (ER) and gross ecosystem productivity (GEP).

Source	DF	NEE		ER		GEP	
		F Value	P	F Value	P	F Value	P
W	1	5.21	0.0151	5.56	0.0251	0.29	0.5935
N	1	42.53	<0.0001	14.74	0.0006	77.27	<0.0001
Y	3	8.03	0.0004	47.28	<0.0001	53.73	<0.0001
W×N	1	1.36	0.2522	0.93	0.3425	2.90	0.0990
Y×W	3	6.67	0.0014	4.22	0.0133	8.87	0.0002
Y×N	3	4.96	0.0065	1.38	0.2682	5.94	0.0026
Y×W×N	3	4.15	0.0143	2.89	0.0518	0.70	0.5615

Results

Soil Microclimate

For the four experimental years, annual mean air temperatures were higher and annual precipitation amounts were lower compared to 30-year (1980–2010) averages of 6.4°C and 471 mm (Fig. 1A, B). The manipulated warming significantly elevated soil temperature (P=0.045). Throughout the whole experimental period, mean soil temperature at 10-cm depth in the heated plots was 2.34°C higher compared to that of the control plots. However, the warming effects were less pronounced in the N addition plots, with only 1.31°C increase in 10-cm soil temperature.

Seasonal Variation in Ecosystem C Fluxes

Temporal dynamics of net ecosystem CO₂ exchange (NEE), ecosystem respiration (ER), and gross ecosystem productivity (GEP) during the vegetation growth period followed the seasonal patterns of air temperature in the four experimental years with higher values of CO₂ exchange being observed in summer and lower values in both spring and autumn. Comparing the seasonal dynamics of NEE of the four years, peak values were found in July of 2006 and 2009, whereas no such clear peaks were observed in the other two years. ER and GEP were always highest in July (Fig. 1 and 2).

Table 2. DF, F values and P values from repeated measures ANOVA on the effects of warming (W), nitrogen (N), and measuring date (D), and their interactions on net ecosystem CO_2 exchange (NEE), ecosystem respiration (ER) and gross ecosystem productivity (GEP) for 2006, 2007, 2008 and 2009.

Source	DF	NEE		ER		GEP	
		F Value	P	F Value	P	F Value	P
2006							
W	1	1.111	0.323	0.662	0.44	2.124	0.183
N	1	20.964	0.002	4.712	0.062	29.271	0.001
D	5	21.606	0.002	293.997	<0.001	657.640	<0.001
W×N	1	3.680	0.091	5.572	0.046	0.001	0.979
D×W	5	4.454	0.068	1.577	0.189	3.657	0.008
D×N	5	7.644	0.024	6.321	<0.001	22.535	<0.001
D×W×N	5	114.223	<0.001	0.26	0.932	11.161	<0.001
2007							
W	1	4.843	0.059	2.804	0.133	7.765	0.024
N	1	0.001	0.973	2.289	0.163	0.462	0.516
D	5	294.033	<0.001	254.932	<0.001	469.165	<0.001
W×N	1	2.336	0.165	3.140	0.144	0.309	0.594
D×W	5	20.692	<0.001	3.783	0.007	18.103	<0.001
D×N	5	17.465	<0.001	24.125	<0.001	24.722	<0.001
D×W×N	5	8.259	<0.001	10.506	<0.001	13.290	0.005
2008							
W	1	28.119	0.001	22.054	0.002	1.078	0.329
N	1	34.156	<0.001	26.159	0.001	32.291	<0.001
D	5	153.749	<0.001	799.759	<0.001	681.520	<0.001
W×N	1	0.024	0.88	2.280	0.169	0.332	0.58
D×W	5	3.073	0.019	26.653	<0.001	7.977	<0.001
D×N	5	1.370	0.255	31.553	<0.001	10.691	<0.001
D×W×N	5	1.583	0.187	8.699	<0.001	3.126	0.018
2009							
W	1	0.433	0.529	3.106	0.116	11.058	0.01
N	1	12.162	0.008	0.250	0.63	32.985	<0.001
D	5	512.464	<0.001	688.011	<0.001	1032.304	<0.001
W×N	1	3.430	0.101	0.028	0.871	6.839	0.031
D×W	5	6.154	<0.001	2.241	0.045	6.544	<0.001
D×N	5	37.015	<0.001	13.425	0.002	35.417	<0.001
D×W×N	5	21.219	<0.001	5.396	0.014	8.739	<0.001

Warming Effects on Ecosystem C Fluxes

Pooling the data from the four experimental years together, we observed that warming had significant effects on NEE and ER, but not on GEP (Table 1). As indicated by the significant interaction between years and treatment factors (Table 1; $P=0.0143$), warming effects on growing season NEE were not consistent over the 4 experimental years (Table 2). On unfertilized plots, warming decreased NEE in three of the four experimental years, but not in 2009 (Fig. 2A and 3; Table 2). Average growing season ER and GEP were interactively affected by warming and year (ER, $P=0.0133$; GEP, $P=0.0002$; Table 1). Warming increased ER in 2008 and 2009 whereas had no apparent effects in 2006 and 2007 (Fig. 2B

and 3; Table 2). Warming decreased GEP in both 2007 and 2009; whereas it had no effects on GEP in 2006 and 2008 (Fig. 2C and 3; Table 2).

No effects of warming on *L. chinensis* relative coverage were observed in years of either 2006 or 2007 (Table 3). Warming marginally reduced *L. chinensis* relative coverage in years of 2008 and 2009 (11.25%, $P=0.053$; 6.25%, $P=0.069$; Table 3).

Nitrogen Effects on Ecosystem C Fluxes

Compared to the control plots, NEE in the N addition and un-heated plots increased slightly in 2006, but to a large extent in 2007, 2008 and 2009 (Fig. 2A). The enhancement of N addition on ER was consistent throughout the growing seasons. N addition increased GEP in all experimental years except in 2007 with the most pronounced enhancement, up to 21%, observed in 2008. N addition marginally increased *L. chinensis* relative coverage by 8.4% ($P=0.063$), 11.25% ($P=0.048$), and 6.25% ($P=0.054$) in 2007, 2008 and 2009, but had no effects in 2006 (Table 3).

Interactive Effects of Warming and N Addition on Ecosystem C Fluxes

No interactive effects of warming plus N addition on NEE, ER, and GEP were detected when we pooled the data from the four experimental years together (Table 1). In 2006, warming plus N addition significantly increased ER, and warming plus N addition significantly increased GEP in 2009 (Table 2).

Leaf-level Gas Exchange

For *L. chinensis*, the significant warming effects on leaf photosynthetic rate (P_n) were observed in 2006, but there were no significant effects in 2007, 2008 and 2009 (Table 3). In contrast, significant N addition stimulation on stomatal conductance (G_s) of *L. chinensis* was detected only in 2009 (Table 3). The P_n of *P. australis* was 20.8% lower in the N addition than in the control treatment in 2008, but had no significant effects on leaf-level gas exchange in 2006, 2007 and 2009. The G_s values of *P. australis* were 6.1 and 28.4% lower in the warming than in the control treatment in 2007 and 2008, but there are no significant effects in 2006 and 2009 (Table 3). In contrast, N addition significantly decreased G_s of *P. australis* in 2008, but increased G_s of *P. australis* in 2009 (Table 3).

Relationships between C Fluxes, Leaf Area Index and Soil Water Content

SWC at 0–10 cm depth fluctuated greatly over the growing season. The warming treatment decreased seasonal mean SWC by 12.9% in 2006, 7.4% in 2007, 11.4% in 2008, and 25.0% in 2009 ($P<0.05$). Conversely, the nitrogen addition increased seasonal mean SWC by 10.3, 8.7, 0.5 and 10.4% in 2006, 2007, 2008 and 2009, respectively. There were no significant relationship between NEE and soil temperature (data not shown); whereas variation in NEE during the growing seasons was strongly correlated with difference in the monthly mean values of SWC across the treatments (Fig. 4; Table 4).

We observed strong positive correlations between GEP and LAI during May and June from 2006 to 2009 for the four treatments (Fig. 3 and 5; Table 4). The slopes of these linear regression functions were of similar magnitude among all treatments, ranging from 7.1 to 9.9 μmol CO_2 $(m_{Leaf})^{-2}$ s^{-1}.

Table 3. Net photosynthetic rate (Pn), stomatal conductance (Gs) and relative coverage (mean ± SE) of *Leymus chinensis* and *Phragmites australis* measured in June of 2006, 2007, 2008 and 2009.

		Leymus chinensis				Phragmites australis			
		2006	2007	2008	2009	2006	2007	2008	2009
Pn (μmol $CO_2 m^{-2} s^{-1}$)	C	10.5±0.44 Aa	7.97±0.63 Aa	15.9±0.96 Aa	17.13±0.37 Aa	13.39±1.41 Aa	16.86±1.48 Aa	22.25±1.64 Aa	24.63±1.57 Aa
	W	9.58±0.60 Ab	8.53±1.31 Aa	16.4±0.96 Aa	17.72±2.21 Aa	15.59±0.74 Aa	14.96±1.62 Aa	19.36±0.98 Aa	27.43±0.8 Aa
	N	10.9±0.43 Aa	9.22±1.02 Aa	17.37±0.83 Aa	18.15±0.75 Aa	17.03±1.3 Aa	14.71±1.44 Aa	17.63±0.44 Ba	28.12±0.34 Aa
	WN	7.97±0.31Bb	8.79±1.08 Aa	17.42±1.15 Aa	16.72±0.8 Aa	16.61±0.77 Aa	20.23±2.56 Ba	19.78±0.66 Aa	24.03±1.72 Ab
Gs ($mmol$ $m^{-2} s^{-1}$)	C	0.45±0.04 Aa	0.15±0.02 Aa	0.38±0.05 Aa	0.11±0.01 Aa	0.47±0.03 Aa	0.33±0.04 Aa	0.67±0.12 Aa	0.41±0.08 Aa
	W	0.38±0.02 Aa	0.14±0.01 Aa	0.40±0.04 Aa	0.15±0.0 Aa	0.53±0.04 Aa	0.31±0.03 Bb	0.48±0.07 Ab	0.40±0.02 Aa
	N	0.46±0.04 Aa	0.15±0.20 Aa	0.35±0.04 Aa	0.16±0.01Ba	0.55±0.06 Aa	0.303±0.03 Aa	0.55±0.03 Ba	0.54±0.03 Ba
	WN	0.35±0.03 Aa	0.17±0.16 Bb	0.46±0.04 Ab	0.16±0.02 Aa	055±0.03 Aa	0.40±0.04 Bb	0.58±0.05 Ba	0.34±0.04 Bb
Relative coverage (%)	C	92±2 Aa	83±3 Aa	80±3 Aa	80±2 Aa	3±2 Aa	10±1 Aa	4±1 Aa	13±1 Aa
	W	91±2 Aa	82±3 Aa	71±2 Aa	75±2 Ab	5±2 Aa	13±1 Aa	3±1 Aa	10±2 Ab
	N	93±1 Aa	90±4 Ba	89±3 Ba	85±3 Ba	2±1 Aa	5±1 Ba	4±2 Aa	11±2 Ba
	WN	93±1 Aa	82±2 Ab	90±1 Bb	82±1 Bb	3±1 Aa	13±1 Aa	5±1 Aa	13±2 Bb

Measurements were performed on the upper most fully expanded leaves from 9.00 to 11.00 a.m. Different letters indicate significant differences ($P<0.05$) among treatments with capital and small letters indicating differences between N and warming treatments, respectively. C = control; W = warming; N = N addition; WN = combined warming and N addition.

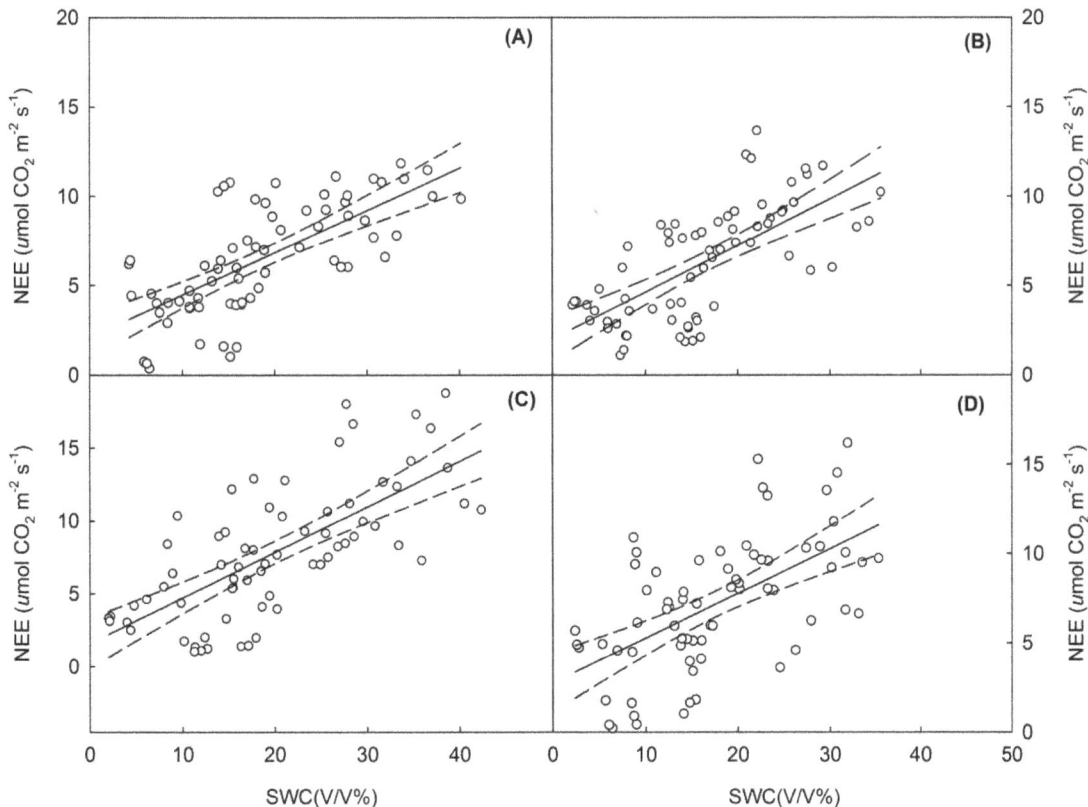

Figure 4. Temporal dependence of net ecosystem CO_2 exchange (NEE) on soil water content (SWC) across the four growing seasons. Regression lines (solid) and 95% confidence level (dotted lines) are showed. All the linear regressions are statistically significant ($P<0.05$). Regression equations and significance levels are provided in Table 4. (A), control; (B), warming; (C), N addition; (D), combined warming and N addition.

Table 4. Summary of linear regressions describing the relationship between carbon flux component and soil water content (SWC), leaf area index (LAI) and the relationship between ecosystem respiration (ER) and gross ecosystem productivity (GEP).

		Treatment			
		Control	Warming	Nitrogen	Warming×Nitrogen
NEE($umolCO_2m^{-2}s^{-1}$) vs.SWC(V/V%)	y-intercept	2.066	2.0044	1.594	2.7729
	slope	0.2365	0.261	0.3119	0.2496
	r^2	0.4886	0.4659	0.4972	0.3402
	MS	314.90	316.75	689.63	313.74
	P	<0.0001	<0.0001	<0.0001	<0.0001
ER($umolCO_2m^{-2}s^{-1}$) vs.SWC(V/V%)	y-intercept	0.9146	0.6552	0.4292	−0.2314
	slope	0.2589	0.3416	0.3001	0.3913
	r^2	0.5177	0.5662	0.5326	0.5317
	MS	377.32	542.41	638.37	771.08
	P	<0.0001	<0.0001	<0.0001	<0.0001
GEP($umolCO_2m^{-2}s^{-1}$) vs.SWC(V/V%)	y-intercept	3.0095	2.6596	2.0193	2.6494
	slope	0.4954	0.6026	0.6120	0.6364
	r^2	0.5426	0.5815	0.5781	0.5000
	MS	1381.62	1688.15	2654.78	2039.59
	P	<0.0001	<0.0001	<0.0001	<0.0001
NEE($umolCO_2m^{-2}s^{-1}$) vs.LAI	y-intercept	0.5412	−0.2915	−4.2295	−1.3917
	slope	4.2316	4.6648	6.3741	4.4301
	r^2	0.2792	0.5243	0.8201	0.7253
	MS	61.19	80.9279	264.93	42.2482
	P	0.0242	0.0007	<0.0001	<0.0001
ER($umolCO_2m^{-2}s^{-1}$) vs.LAI	y-intercept	−0.4930	0.4373	−1.6812	−1.4304
	slope	4.2868	3.3815	3.4772	3.7369
	r^2	0.5325	0.3872	0.6059	0.5844
	MS	62.7906	42.5253	78.84	115.25
	P	0.0006	0.0058	0.0001	0.0002
GEP($umolCO_2m^{-2}s^{-1}$) vs.LAI	y-intercept	0.0482	0.1458	−5.9107	−2.8220
	slope	8.5184	8.0463	9.8513	8.1670
	r^2	0.4024	0.5054	0.8241	0.6932
	MS	247.94	240.78	632.82	550.50
	P	0.0047	0.0009	0.0001	<0.0001
ER($umolCO_2m^{-2}s^{-1}$) vs.GEP($umolCO_2m^{-2}s^{-1}$)	y-intercept	−0.5914	−0.6659	−0.4017	−1.2563
	slope	0.5170	0.5503	0.4793	0.5700
	r^2	0.9351	0.9119	0.8852	0.9108
	MS	713.26	902.39	1101.40	1383.80
	P	<0.0001	<0.0001	<0.0001	<0.0001

Discussion

Effects of Warming on Ecosystem C Fluxes

Warming can increase NEE by increasing GEP relative to ER. This, by definition, should be expected in temperature-limited environments and/or growth phases. Conversely, warming could decrease NEE if it caused a greater increase in ER than GEP. Warming induced reduction in NEE occurs when air temperature is beyond the ecosystem's optimum temperature for carbon assimilation and/or ER responds more strongly to temperature increase than GEP. Moreover, warming could indirectively influence NEE through altering community composition [30,31], and the availability of both water and nutrients [5,8,32]. Previous studies suggest that steppe is subject to a highly variable control of NEE by temperature (during the transition from winter to spring and during the carbon and nitrogen re-allocation phase at the end of the growing season), water (with transient or permanent water deficits most strongly impacting on carbon gain during the main growth period, May to August), and a co-limitation of productivity by N availability in wet years [19,33,34].

In this experiment on the Songnen grassland, which is subject to intensive land-use change [35], warming significantly decreased NEE and increased ER, but had no apparent effects on GEP (Table 1) when we pooled 4 years data. The effects of warming on ecosystem carbon fluxes were not consistent through the experimental period (Fig. 3 and Table 2). For instance, in 2008

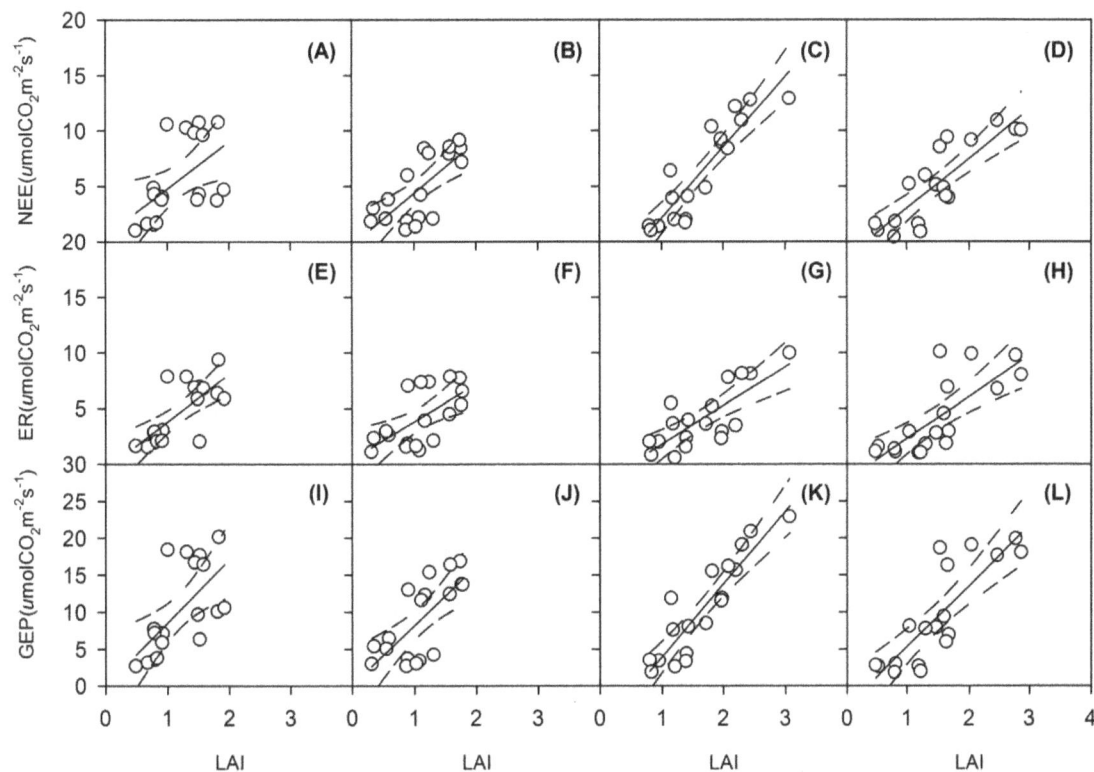

Figure 5. Temporal dependence of seasonal mean net ecosystem CO_2 exchange (NEE), ecosystem respiration (ER) and gross ecosystem productivity (GEP) on leaf area index (LAI) across the four treatments in May and June of 2006–2009. Regression lines (solid) and 95% confidence level (dotted lines) are showed. All the linear regressions are statistically significant ($P<0.05$). Regression equations and significance levels are provided in Table 4. Control plots (A,E,I); Warming plots (B,F,J); N addition plots (C,G,K); Warming and N addition plots (D,H,L).

the negative effects of warming on NEE could be explained by a warming-induced increase of ER with GEP remaining un-affected; whereas in 2009, warming increased NEE due to a proportionally greater increase of GEP than ER. Interpretation of these results is further complicated by interactions between warming and N addition with partially opposite effects of warming on fertilized and un-fertilized plots.

In the studies of Niu et al. [10] and Wan et al. [36] negative warming effects on NEE were predominantly of indirect nature, as changes in NEE were largely attributed to lower soil moisture. Directly supporting this finding, variation in NEE in the four experimental years coincided with changes in average monthly SWC (Fig. 4). Low SWC constrained plant growth and ecosystem productivity particularly during early spring (May). The observed reduction of CO_2 fluxes on the same grassland in 2007 by Dong et al. [37] was also related to early spring drought; these findings together illustrate that warming alters ecosystem carbon fluxes through affecting water availability, which is one the primary drivers of carbon fluxes in this semi-arid grassland.

In the two of the 4 experimental years, we observed significant enhancement in ER in warming plots (Fig. 2 and 3; Table 2). This finding is contrary to findings of Niu et al. [10] and Xia et al. [38] who found no or only slight effects of warming on ER. Similarly, our results don't support the finding of Xia et al. [38] that GEP was more sensitive to inter-annual climatic variation than ER. This finding implies that both carbon fluxes in our study were functionally closely linked and that both exhibited apparently similar sensitivity to intra- and inter-annual variation in climatic conditions. The results of this study, along with those performed in

other biomes [39], suggest ER apparently is not independent of the carbon gain and functionality of grassland in general.

Response of C Fluxes to N Addition

Our study focused on the analysis of interactive effects of N fertilizer application and warming on the carbon budget of the Songnen grassland during the vegetation growth periods. N addition stimulated NEE in three of the four years (Fig. 2A; Table 1), largely due to N-induced increases in GEP. N stimulation of ecosystem C fluxes in the studied grassland is in accordance with positive N responses of grassland productivity in other ecosystems [38,40–42].

N stimulation of ecosystem NEE could have resulted from higher canopy photosynthesis rates. However, the slope of the linear regression between GEP and LAI indicates that there were no significant differences in canopy photosynthesis rates among the four treatments (Fig. 5). Leaf net assimilation rates were, as well, not affected by treatments (Table 3). The overall higher biomass accumulation and NEE in N addition plots were reflected in an increases in ER (Fig. 3). Enhanced aboveground activity apparently resulted in stimulation of belowground C input, and subsequent root and microbial activities. It remains unknown to what extent the long-term application of N would affect the turn-over of soil organic matter and the C sink size in the studied grassland.

Although there were no significant interactive effects of warming and N addition on ecosystem C-exchange over the experimental period (Table 1), the observed greater rates of NEE, ER and GEP in the WN plots, relative to the W plots (Fig. 5B and

C), suggests that warming produces, to some extent, positive effects on ecosystem C-exchange when combined with N addition. We therefore speculate that, given enough time, there may be a significant interactive effects between warming and N addition on ecosystem carbon fluxes because of potential impacts of warming plus N addition on the dynamics of litter and soil organic C.

Conclusions

NEE in the Songnen grassland in Northeast China was significantly affected by warming and N addition treatments, however there were no interactive effects between warming and N addition. The observed inter-annual variation in the effects of warming and N addition treatments suggests that water availability is a key driver of ecosystem carbon fluxes in the studied grassland. Contrary to common expectations of stimulated ecosystem C-fluxes in warming plots, we found reductions in all the measured C-fluxes on the Songnen grassland in northeastern China. The observed decrease in ecosystem C-fluxes could be attributed to offsetting of the direct and positive effects of elevated temperature by the indirect and negative effects via exacerbating water stress. These results highlight the uniqueness of the grassland in the semi-arid and salinate region of northeastern China in response to the

enhanced temperature and nutrition levels. Increased N not only enhanced ecosystem C-fluxes, but also ameliorated the negative C balance caused by experimental warming on ecosystem C-fluxes.

Supporting Information

Figure S1   Seasonal development of leaf area index (LAI, mean ± SE) of the four treatments during the vegetation growth period (May to October). C = control, W = warming, N = N addition, WN = combined warming and N addition.

Acknowledgments

We thank two anonymous reviewers for their constructive comments, which helped in improving the manuscript; Gang Dong and Song Gao provided meteorological data. Luegeng Ma, Wenjuan Hou and Chao Wang assisted field work.

Author Contributions

Conceived and designed the experiments: LJ TCZ JXG. Performed the experiments: LJ. Analyzed the data: LJ WS RG XDN TCZ. Wrote the paper: LJ RG XDN JXG WS. Contributed materials and analysis tools: LJ.

References

1. IPCC (2007) IPCC WGI Fourth assessment report. Climatic change 2007: the physical science basis. Geneva: Intergovernmental Panel on Climate Change.
2. Hansen J, Sato M, Ruedy R, Lo K, Lea DW (2006) Global temperature change. PNAS 103: 14288–14293.
3. Luo Y (2007) Terrestrial carbon–cycle feedback to climate warming. Annu. Rev. Ecol. Evol. Syst. 38: 683–712.
4. Eswaran H, Van Den Berg E, Reich P (1993) Organic carbon in soils of the world. Soil Sci. Soc. Am. J. 57: 192–194.
5. Rustad LE, Campbell JL, Marion GM, Norby RJ, Mitchell MJ, et al. (2001) A meta-analysis of the response of soil respiration, net nitrogen mineralization, and aboveground plant growth to experimental ecosystem warming. Oecologia 126: 543–562.
6. Shaw MR, Zavaleta ES, Chiariello NR, Cleland EE, Mooney HA, et al. (2002) Grassland responses to global environmental changes suppressed by elevated CO$_2$. Science 298: 1987–1990.
7. Suyker AE, Verma SB, Burba GG (2003) Interannual variability in net CO$_2$ exchange of a native tall grass prairie. Global Change Biol. 9: 255–265.
8. Wan S, Hui D, Wallace LL, Luo Y (2005) Direct and indirect warming effects on ecosystem carbon processes in a tallgrass prairie. Global Biogeochem. Cycles 19: 1–13.
9. Zhou X, Sherry RA, An Y, Wallace LL, Luo Y (2006) Main and interactive effects of warming, clipping, and doubled precipitation on soil CO$_2$ efflux in a grassland ecosystem. Global Biogeochem. Cycles 20: GB1003.
10. Niu S, Wu M, Han Y, Xia J, Li L, et al. (2008) Water-mediated responses of ecosystem carbon fluxes to climatic change in a temperate steppe. New Phytol. 177: 209–219.
11. Wu Z, Dijktra P, Koch G, Penuela J, Hungate B (2011) Responses of terrestrial ecosystems to temperature and precipitation change: a meta-analysis of experimental manipulation. Global Change Biol. 17: 927–942.
12. Kirschbaum MUF (1995) The temperature dependence of soil organic matter decomposition, and the effect of global warming on soil organic storage. Soil Biol. Biochem. 27: 753–760.
13. Smith TM, Shugart HH (1993) The potential response of global terrestrial carbon storage to a climatic change. Water, Air Soil Poll. 70: 629–642.
14. Ineson P, Benham DG, Poskitt J, Harrison AF, Taylor K, et al. (1998) Effects of climate change on nitrogen dynamics in upland soils. 2. A soil warming study. Global Change Biol. 4: 153–161.
15. Li SG, Asanuma J, Eugster W, Kotani A, Liu JJ, et al. (2005) Net ecosystem carbon dioxide exchange over grazed steppe in central Mongolia. Global Change Biol. 11: 1941–1955.
16. Galloway JN, Dentener FJ, Capone DG, Boyer EW, Horwarth RW, et al. (2004) Nitrogen cycles, past, present, and future. Biogeochemistry 70: 152–226.
17. Galloway JN, Townsend AR, Erisman JW, Bekunda M, Cai ZC, et al. (2008) Transformation of the nitrogen cycle: recent trends, questions, and potential solutions. Science 320: 889–892.
18. Baldocchi D, Xu L, Kiang N (2004) How plant functional-type, weather, seasonal drought, and soil physical properties alter water and energy fluxes of an oak–grass savanna and an annual grassland. Agric. For. Meteorol. 123: 13–39.

19. Brueck H, Erdle K, Gao Y, Giese M, Zhao Y, et al. (2010) Effects of N and water supply on water use-efficiency of a semiarid grassland in Inner Mongolia. Plant Soil 328: 495–505.
20. Saleska SR, Harte J, Torn MS (1999) The effect of experimental ecosystem warming on CO$_2$ fluxes in a montane meadow. Global Change Biol. 5, 125–141.
21. Zhu T, Li J, Yang D (1983) A study of the ecology of yang-cao (*Leymus chinensis*) grassland in northern china. Proc. 14th international Grassland congress Lexington 1981: 429–431.
22. Qu G, Guo J (2003) The relationship between different plant communities and soil characteristics in Songnen grassland. Acta Prataculturae Sinica 12: 18–22.
23. Han M, Yang L, Zhang Y, Zhou G (2006) The biomass of C3 and C4 plant function groups in *Leymus chinensis* communities and theirs response to environmental change along Northeast China transect. Acta Ecologica Sinica 6: 1825–1832.
24. Ripley EA, Wang R, Zhu T (1996) The climate of the Songnen plain, northeast china. Intern. J. Ecol. Environm. Sci. 22: 1–21.
25. Wang R, Zhu T (2003) Responses of interspecific relationships among main herbaceous plants to flooding disturbance in Songnen Plain, Northeastern China Chin. J. Appl. Ecol. 14: 892–896.
26. He CE, Liu XJ, Andreas F, Zhang FS (2007) Quantifying the total airborne nitrogen input into agro ecosystems in the North China Plain. Agr Ecosyst and Environ 121: 395–400.
27. Zhang Y, Zheng LX, Liu XJ, Jickells T, Cape JN, et al. (2008) Evidence for organic N deposition and its anthropogenic sources in China. Atmospheric Environment 42: 1035–1041.
28. Bai YF, Wu JG, Clark CM, Naeem S, Pan QM, et al. (2010) Tradeoffs and thresholds in the effects of nitrogen addition on biodiversity and ecosystem functioning: evidence from Inner Mongolia grasslands. Global Change Biol 16: 358–372.
29. Steduto P, Çetinköкü Ö, Albrizio R, Kanber R (2002) Automated closed-system canopy-chamber for continuous field-crop monitoring of CO$_2$ and H$_2$O fluxes. Agric. For. Meteorol. 111: 171–186.
30. Shaver GR, Canadell J, Chapin FS III, Gurevitch J, Harte J, et al. (2000) Global warming and terrestrial ecosystems: a conceptual framework for analysis. Bioscience 50: 871–882.
31. Welker JM, Fahnestock JT, Henry GR, O'Dea KW, Chimners RA (2004) CO$_2$ exchange in three Canadian High Arctic ecosystems: response to long-term experimental warming. Global Change Biol. 10: 1981–1995.
32. Schimel J, Bilbrough CB, Welker JM (2004) The effect of changing snow cover on year-round soil nitrogen dynamics in Arctic tundra ecosystems. Soil Biol. Biochem. 36: 217–227.
33. Hooper D, Johnson L (1999) Nitrogen limitation in dry land ecosystems: Responses to geographical and temporal variation in precipitation. Biogeochem. 46: 247–293.
34. Gao Y, Giese M, Lin S, Sattelmacher B, Zhao Y, et al. (2008) Belowground net primary productivity and biomass allocation of a grassland in Inner Mongolia is affected by grazing intensity. Plant Soil 307: 41–50.

35. Wang Z, Song K, Zhang B, Liu D, Ren C, et al. (2009) Shrinkage and fragmentation of grasslands in the West Songnen Plain, China. Agric. Ecosys. Environm. 129: 315–324.

36. Wan S, Norby RJ, Ledford J, Weltzin JF (2007) Responses of soil respiration to elevated CO_2, air warming, and changing soil water availability in a model old-field grassland. Global Change Biol. 13: 2411–2424.

37. Dong G, Guo J, Chen J, Sun G, Gao S, et al. (2011) Effects of spring drought on carbon sequestration, evapotranspiration and water use efficiency in the Songnen meadow steppe in northeast China. Ecohydrology 4: 211–224.

38. Xia J, Niu S, Wan S (2009) Response of ecosystem carbon exchange to warming and nitrogen addition during two hydrologically contrasting growing seasons in a temperate steppe. Global Change Biol., 15: 1544–1556.

39. Raich JW, Nadelhoffer KJ (1989) Belowground carbon allocation in forest ecosystems: global trends. Ecology 70: 1346–1354.

40. Elser JJ, Bracken MES, Cleland EE, Gruner DS W, Harpole S, et al. (2007) Global analysis of nitrogen and phosphorus limitation of primary producers in freshwater, marine and terrestrial ecosystems. Ecology Letters 10: 1135–1142.

41. Harpole WS, Potts DL, Suding KN (2007) Ecosystem responses to water and nitrogen amendment in a California grassland. Global Change Biol. 13: 2341–2348.

42. LeBauer DS, Treseder KK (2008) Nitrogen limitation of net primary productivity in terrestrial ecosystems is globally distributed. Ecology 89: 371–379.

7

Food Sources for *Ruditapes philippinarum* in a Coastal Lagoon Determined by Mass Balance and Stable Isotope Approaches

Tomohiro Komorita[1,2]*, Rumiko Kajihara[1], Hiroaki Tsutsumi[2], Seiichiro Shibanuma[1], Toshiro Yamada[3], Shigeru Montani[1]

1 Graduate School of Environmental Science, Hokkaido University, Sapporo, Japan, 2 Faculty of Environmental and Symbiotic Science, Prefectural University of Kumamoto, Tsukide, Kumamoto, Japan, 3 Nishimuragumi Co. Ltd., Hokkaido, Japan

Abstract

The relationship between the food demand of a clam population (*Ruditapes philippinarum* (Adams & Reeve 1850)) and the isotopic contributions of potential food sources (phytoplankton, benthic diatoms, and organic matter derived from the sediment surface, seagrass, and seaweeds) to the clam diet were investigated. In particular, we investigated the manner in which dense patches of clams with high secondary productivity are sustained in a coastal lagoon ecosystem (Hichirippu Lagoon) in Hokkaido, Japan. Clam feeding behavior should affect material circulation in this lagoon owing to their high secondary productivity (ca. 130 g C m^{-2} yr^{-1}). Phytoplankton were initially found to constitute 14–77% of the clam diet, although phytoplankton nitrogen content (1.79–4.48 kmol N) and the food demand of the clam (16.2 kmol N d^{-1}) suggest that phytoplankton can constitute only up to 28% of clam dietary demands. However, use of isotopic signatures alone may be misleading. For example, the contribution of microphytobenthos (MPB) were estimated to be 0–68% on the basis of isotopic signatures but was subsequently shown to be 35±13% (mean ± S.D.) and 64±4% (mean ± S.D.) on the basis of phytoplankton biomass and clam food demand respectively, suggesting that MPB are the primary food source for clams. Thus, in the present study, the abundant MPB in the subtidal area appear to be a key food source for clams, suggesting that these MPB may sustain the high secondary production of the clam.

Editor: Arga Chandrashekar Anil, CSIR- National institute of oceanography, India

Funding: This study was funded by MEXT Grants-in-Aid for Scientific Research, Nos. 22380102 (to SM) and 15201001 (to SM). The funders had no role in study design, data collection and analysis, decision to publish, or preparation of the manuscript.

Competing Interests: Toshiro Yamada is employee of Nishimuragumi Co. Ltd.

* E-mail: komorita@pu-kumamoto.ac.jp

Introduction

Suspension-feeding bivalves occurring in coastal waters and on tidal flats often establish high-density populations with extremely large standing stocks of biomass (i.e. over several kilograms in wet weight per square meter) and exhibit much higher productivity than other common members of macrobenthic communities. For example, the annual secondary productivity of dense patches of mussels, *Mytilus edulis* (L. 1758), reached 11 MJ m^{-2} yr^{-1} (275 g C m^{-2} yr^{-1}, using 0.025 g C KJ^{-1} based on Brey [1]) in the sublittoral zone at Bellevue, Newfoundland, Canada [2], whereas that of surf clams (*Donax serra* (Röding 1798)) inhabiting the highly exposed sandy beaches of Namibia was reported to be 167–637 g ash free dry mass (AFDM) m^{-2} yr^{-1} (81.2–310 g C m^{-2} yr^{-1},using 0.486 g C gAFDM^{-1} based on Brey [1]) [3]. Moreover, the high secondary productivity of suspension-feeding bivalves corresponds to the range of primary productivity (165–320 g C m^{-2} yr^{-1}[4–6]) of microalgae that are suspended in the water in which the bivalves live, including phytoplankton and microphytobenthos (MPB); these microalgae are a primary source of food for the bivalves.

Based on the rate of conversion of food to somatic growth in invertebrates, which is typically assumed to be 20% [7], the amount of food consumed by suspension-feeding bivalves should be at least five times greater than the secondary productivity for the same population. Therefore, such bivalves require efficient mechanisms to allow the collection of sufficient food from the surrounding environment to sustain their high secondary productivities. However, the relationship between the secondary productivity of suspension-feeding bivalves and the availability of primarily produced organic matter such as phytoplankton and MPB remains poorly understood [8,9].

Stable carbon and nitrogen isotope ratios indicate that suspension-feeding bivalves typically feed on primarily produced organic matter (such as phytoplankton, detritus, and resuspended MPB) suspended in water [10–12]. The resuspension of MPB and particulate matter derived from sediments have been observed to consistently occur shortly after maximum current velocity has been reached in muddy tidal flats, witch typically exhibit a relatively low threshold current velocity (ca. 15 cm s^{-1}) [13]. The short-necked clam, *Ruditapes philippinarum* (Adams & Reeve 1850) as a junior synonym of *Tapes philippinarum*, is one of the most abundant suspension-feeding bivalves on sandy tidal flats along the Japanese coast [14]. The diet of this clam exhibit a spatial gradient that is primarily driven by tidal hydrodynamics within bays and by

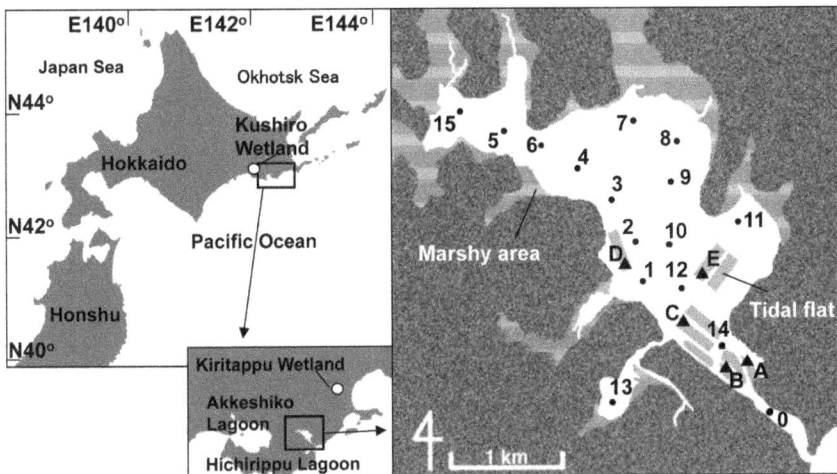

Figure 1. Study area and locations of sampling stations. The oval and rectangular grey areas represent naturally occurring (stations A and B) and artificial tidal flats (the other stations), respectively.

land-use characteristics within catchments [15], both of which can effect changes in the food supply [16]. Thus, the relative contributions of phytoplankton and MPB to the diet of *R. philippinarum* may simply be dependent on their relative availability [16–18].

The proportional contribution of different food sources to the diet of organisms is typically determined using only isotopic mass balance, which assesses the quality of diet components. Quantitative estimation of the food demand of dense patches of suspension-feeding bivalves has rarely been attempted with reference to the contributions of different food sources (i.e. the quantity of diet components). However, adopting such a quantitative approach is important for the precise estimation of material circulation via suspension-feeding bivalves, and this can allow a more realistic determination of the contribution of each food source.

The present study was conducted as part of a long-term research project that aimed to quantify population dynamics and associated energy and/or material flows at various trophic levels in the coastal lagoon ecosystem of Hichirippu Lagoon, located in the eastern part of Hokkaido, northern Japan. On the tidal flats of this lagoon, *R. philippinarum* constitutes one of the dominant macrobenthic communities, and its biomass often exceeds 265 g dry weight (DW) m^{-2} (approximately 5 kg wet weight (WW) m^{-2} using 0.053 gDW gWW^{-1}) [19]. Abundant primarily produced organic matter is required to sustain the secondary productivity of dense patches of the clam; such matter is available for suspension feeding in the water overlying the sediment. Our recent study revealed that the areal biomass of MPB occurring in the surface layer of the sediment (i.e. within 0.5 cm of the surface) is about 100 times greater than the biomass of phytoplankton contained in the water column up to a depth of about 1 m during high tide in summer [19]. Therefore, both phytoplankton and MPB are likely available as primary food resources for clams; however, their availability varies seasonally. Moreover, through their suspension-feeding activities, clams exert a considerable influence on the flow of energy and materials between the primary (i.e. the microalgae) and dominant secondary (i.e. the clams themselves) producers in the tidal flat ecosystem.

In the present study, field surveys to monitor the physicochemical environmental conditions of water and sediment and performed quantitative sampling of macrobenthic animals includ-

ing dense patches of the clam, *R. philippinarum*, were conducted between February 2005 and April 2006. The observed seasonal fluctuations in the abundance and biomass of the macrobenthic organisms and estimate the secondary productivity of the clam population for the entire tidal flat area in the lagoon were described. Furthermore, we compare the food demand of the clams to the amount of potential food sources (phytoplankton, benthic diatoms, and organic matter derived from the sediment surface, seagrass, and seaweeds) and the stable carbon and nitrogen isotope ratios of the clams with those of potential food sources both quantitatively and qualitatively. Finally, how the dense patches and high secondary productivity of the clam population are sustained in the tidal flat ecosystem of Hichirippu Lagoon is discussed.

Materials and Methods

Study Area

Hichirippu Lagoon bordering the Pacific Ocean in Hokkaido, Japan (44°03' N, 145°03' E, Figure 1) covers an area of approximately 3.56 km^2 and is shallow (mean water depth: ~1 m) and brackish. The maximum tidal flow at the entrance to the lagoon reaches approximately 40 cm s^{-1} during spring tide [19], and the flow direction is primarily from northwest to southeast. Freshwater has a relatively small effect on the water budget, ranging from 0.9 to 8.0% of total volume, and salinity ranges between 27.5–34.1 throughout the year [20]. Temperature of the surface sediment (approximately 5 cm depth) ranges from −9.5°C to 19.7°C at a tidal flat station [21]. The sediment in the subtidal area of the lagoon is muddy sand with the mud (less than 63 μm in particle diameter) content ranging between 8.7 and 28.1% [22]. At extreme low water spring tides, the area of the tidal flat exposed reaches approximately 0.19 km^2; this represents around 5.3% of the total lagoon area. These tidal flats have typically been used as fishery sites to harvest the clam, *R philippinarum*, with 30–92 metric tons in wet weight of clam harvested per year during 1998–2004.

Sampling Procedures

Five sampling stations on the tidal flats (stations A – E) to represent clam habitats and 16 stations (stations 0–15) in the subtidal area to represent a reservoir of food supply were selected

(Figure 1). Quantitative sampling of macrobenthic animals were conducted and the concentration of chlorophyll-a (Chl-a) and organic matter in the surface sediment were measured at the tidal flat stations and Chl-a concentration in the water column and at the sediment surface were also monitored at the subtidal stations. This monitoring was conducted between February 2005 and April 2006, with monthly sampling from April to October and bimonthly sampling from December to April. Surface water samples were collected during flood tide. At each tidal flat station, sediment samples for geochemical analysis were randomly collected at 10 different sites within a 1 m radius of the station using an acrylic core tube (3 cm in diameter). The samples were extruded carefully and their surface layers (i.e. up to 0.5 cm depth) were retained to determine Chl-a and organic matter content of the sediment (SOM). At subtidal stations, sediment samples were carefully collected using an Ekman–Birge grab sampler, which sampled a 20×20 cm area to a depth of 20 cm. The topmost 0.5 cm of the sediment was also collected using an acrylic core tube.

For the quantitative survey of macrobenthic animals, 3–5 replicates of each sediment sample were collected using a stainless-steel core sampler ($10 \times 10 \times 10$ cm) and sieved through a 1 mm mesh screen. The residues of each sediment sample were stored in a plastic bag. Furthermore, 18–30 additional sediment samples were collected using the steel core sampler to improve the accuracy of the quantitative data describing the occurrence (in terms of density and biomass) of adult clams that were present at low densities. These additional samples were sieved through a 5 mm mesh screen and treated in the same manner as the other quantitative samples for macrobenthic animals.

From June to October 2005, 4 L of samples of surface water were collected in a sampling bucket at stations 0, 6, and 10 during flood tides to determine the stable carbon and nitrogen isotopic signatures of the suspended particulate organic matter (POM) in the water; these samples were stored in plastic bottles. The tidal flats near station 14 was selected as MPB sampling site, where locates on the center portion of the tidal flats. MPB samples were collected from the sediment surface by following a modified version of the method developed by Couch [23] for the determination of the stable isotopic signatures of carbon and nitrogen [22].

For quantitative analysis, macrophytobenthos (seagrass and macroalgae) were collected with bottom sediment using a handy core sampler ($25 \times 25 \times 10$ cm); sampling was performed five times for each station during August 2004 and October 2004. The samples were sieved through a mesh bag with a 1-mm opening and the residues of each sample in the mesh bag were stored in plastic bags. Seagrass (*Zostera japonica* L. 1753) and benthic macroalgae (*Ulva pertusa* L. 1753) were collected with the grab sampler in August 2005 and October 2005 for stable isotope analysis.

Sample Treatment and Analysis

To determine the Chl-a content of the water, the water samples were filtered through glass fiber filters (Whatman GF/F) and Chl-a was extracted from residues on the filters in test tubes using 90% acetone. The test tubes were stored in darkness for 24 h at $-20°C$ and then sonicated for 5 min. The concentrations of Chl-a in the supernatants were analyzed to estimate Chl-a standing stocks using a spectrophotometer (Turner 10-AU-5, Turner Designs) according to the method of Lorenzen [24]. To determine the Chl-a content of the surface sediment, Chl-a was extracted from duplicate subsamples of the wet surface sediment ($\sim$0.5 g) in test tubes using

90% acetone. The concentrations of the supernatants were also determined by spectrophotometry [25].

Macrobenthic animals were sorted from the residues of the quantitative samples, identified, counted, and weighed by species. For the population study of *R. philippinarum*, the shell lengths of all specimens were measured using a digital caliper to produce a size–frequency distribution for the population. To determine the relationship between the shell length and dry weight of clam body tissues, 361 individuals were randomly collected in different seasons and their shell lengths were measured using the digital caliper. All of these clams were stored in filtered seawater for evacuation for 6–8 h. The body tissues were then removed from the shells, freeze-dried, and weighed. The relationship between the shell length and dry weight of body tissues for individual clams were obtained as follows:

$$DW = 0.007e0.1063\_SL, \qquad (1)$$

where $r^2 = 0.945$, n $= 361$, DW is dry weight, and SL is shell length.

The shell lengths of all specimens of the clams were measured for stable isotope analysis of carbon and nitrogen. The clams were then opened and their posterior adductor muscles were dissected, frozen at $-20°C$ for storage, freeze-dried, and ground to powder with a mortar. Prior to analysis, the samples were treated with a chloroform–methanol mixture solution (2:1, v/v) for 24 h to remove lipids, filtered with a precombusted GF/F filter (450°C, 5 h), rinsed with ethanol, and freeze-dried. The macroalgae and seagrass tissue used for stable isotope analysis were separated from the sample, rinsed with distilled water, freeze-dried, and ground into fine powder with a mortar. The organic matter derived from the surface sediment (SOM) was also ground into a powder with a mortar. Prior to analysis, the samples were treated with 1 N HCl to remove inorganic carbon, rinsed with deionized and distilled water to remove the acid, and freeze-dried. For the analysis of POM, 4 L of the water sample were filtered with a precombusted GF/F filter (450°C, 5 h), and the residues on the filter were used for stable isotope analysis. For the analysis of MPB, the MPB adhering to glass beads was rinsed with filtered seawater and sieved with a 125 μm mesh screen to remove the glass beads, which were then filtered using a precombusted GF/F filter (450°C, 5 h). Prior to analysis, the POM or MPB filters were treated with 1 N HCl to remove inorganic carbonates, rinsed with deionized and distilled water to remove the acid, and freeze-dried.

Analysis of Stable Isotope Ratios of Carbon and Nitrogen

The stable isotope ratios of carbon and nitrogen were determined using a mass spectrometer (DELTA V Plus, Thermo Electron) connected directly to an elemental analyzer (Flash Elemental Analyzer 1112 Series, Thermo Electron). All isotopic data are reported in conventional delta notation (in ‰) as follows:

$$\_X = (Rsample/RStandar - 1)\_1000, \qquad (2)$$

where X is ^{13}C or ^{15}N and R is $^{13}C/^{12}C$ or $^{15}N/^{14}N$ for carbon and nitrogen, respectively. Pee Dee Belemnite and air N_2 were used as the standards for carbon and nitrogen, respectively. The overall analytical error was within ±0.2 ‰.

Estimation of the Secondary Production of the Clam Population

The shell lengths of all clams in the quantitative samples were measured using a digital caliper to obtain the size–frequency

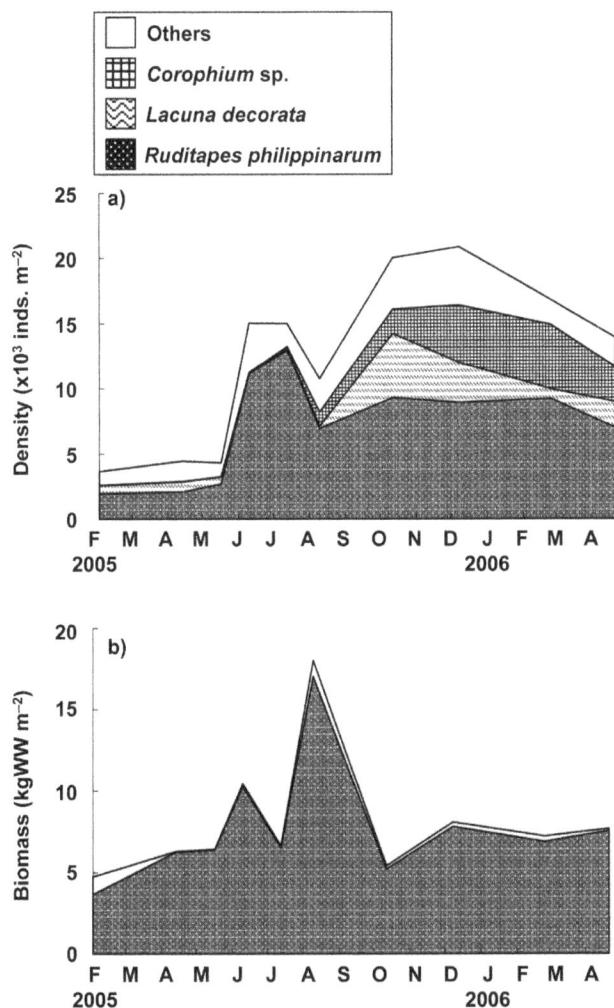

Figure 2. Seasonal variability of the macrobenthic animals at station B. (a) density and (b) biomass.

distribution of the clam population. Version 4.0 of the PROGEAN software [26] was used for generation analysis of the clam population by a graphic method. Mean shell length and the density of each cohort were determined from the size–frequency distribution of the population. Secondary productivity of the clam population was calculated in dry weight according to an incremental summation method [27] using density and mean shell length data for each cohort (obtained by generation analysis of the population) and the relationship between the shell length and biomass (in dry weight) of the clams (Equation 1).

P(t) represents the productivity of the clam population between sampling times t and ($t-1$) and is expressed as follows:

$$P(t) = \sum_{i=1}^{i=n} \frac{1}{2} \times (D(t)_i + D(t-1)_i) \times (B(t)_i - B(t-1)_i) \quad (3)$$

where i is the cohort number, $D(t)i$ is the density of cohort i at time $z\ t$, and $B(t)i$ is the biomass of the individual with the modal shell size in cohort i at time t.

The estimated productivity of the clam population is represented by the amount of carbon and nitrogen based on the carbon and nitrogen contents of body tissues (40.1% and 12.3%, respectively),

which allows evaluation of the importance of this productivity in material flow in the tidal-flat area.

Data Analysis

POM from the lagoon, MPB, SOM, *U. pertusa*, and *Z. japonica* leaves were considered as potential trophic resources for the short-necked clam. The IsoSource software [28,29] was used to determine the relative contribution of each source to the mixed signature of the clam adductor muscle at the tidal flats stations. The trophic enrichments of *R. philippinarum* (0.6 ‰ for $\delta^{13}C$ and 3.4 ‰ for $\delta^{15}N$; [30]) were subtracted from the adductor muscle values before IsoSource analysis.

The carbon content (C/Chl) of phytoplankton and MPB were assumed to be 20–50 [31] and 10–150 [32], respectively, and the elemental compositions (C/N) of phytoplankton and MPB were assumed to be 6.6 [33] and 7.5 [25], respectively. The nitrogen content of *Z. japonica* and *U. pertusa* determined by the elemental analyzer were 2.8% and 1.9%, respectively. The standing stock of nitrogen in the microalgae was obtained by multiplying these elemental ratios with the microalgal standing stock.

Ethics statement. The field survey of this study is approved by the Chirippu Fishery Cooperative Union.

Results

Macrobenthic Community at the Representative Station

Station B was defined as a representative station for the macrobenthic animals, because it was suitable to analyze the cohort of the clam population at this station with the high density and biomass of the clam. Figure 2 illustrates the seasonal fluctuations in the density and biomass of macrobenthic animals at station B. The short-necked clam (*Ruditapes philippinarum*) was found to dominate the benthic community by density, accounting for 57.9% of all macrobenthic animals collected in the present study. The density of the clam peaked at 12960 ind. m^{-2} on July 19, 2005 and decreased gradually to 7000 ind. m^{-2} by April 27, 2006. A small gastropod, *Lacuna decorata* (Adams 1861), and an amphipod, *Corophium* sp., were the second and third most dominant species in the macrobenthic community. The density of *L. decorata* tended to increase in late summer to autumn and reached a peak density of 4930 ind. m^{-2} in October 2005 (Figure 2a). *Corophium* sp. also increased from 1,870 ind. m^{-2} to 4940 ind. m^{-2} between October 2005 and March 2006. In terms of biomass, *R. philippinarum* dominated the macrobenthic community exclusively (Figure 2b). The biomass of the clam was found to be 3.65–17.0 kgWW m^{-2}, with a mean biomass of 7.77 ± 3.68 kgWW m^{-2} (all values are given as mean ± SD), throughout the duration of the present study; this represents 95.9% of the total biomass in wet weight of the specimens collected.

Biomass, Secondary Productivity, and P/B of
R. philippinarum

The size–frequency distributions of the clam population between February 2005 and April 2006 are presented in Figure 3; these distributions were obtained by combining data collected from quantitative samples covering two different size ranges for macrobenthic organisms. The presence of five different cohorts (C1–C5) can be recognized in these size–frequency distributions, and the shell growth curves of these five cohorts are plotted in Figure 4a. In the study area, the clam breeds in September and planktonic larvae settle on tidal flat sediments in October [21]. In the size–frequency distributions, a new cohort (C5) initiated recruitment in April 2005, reaching a peak in recruitment in July 2005 with a density of 11250 ind. m^{-2}.

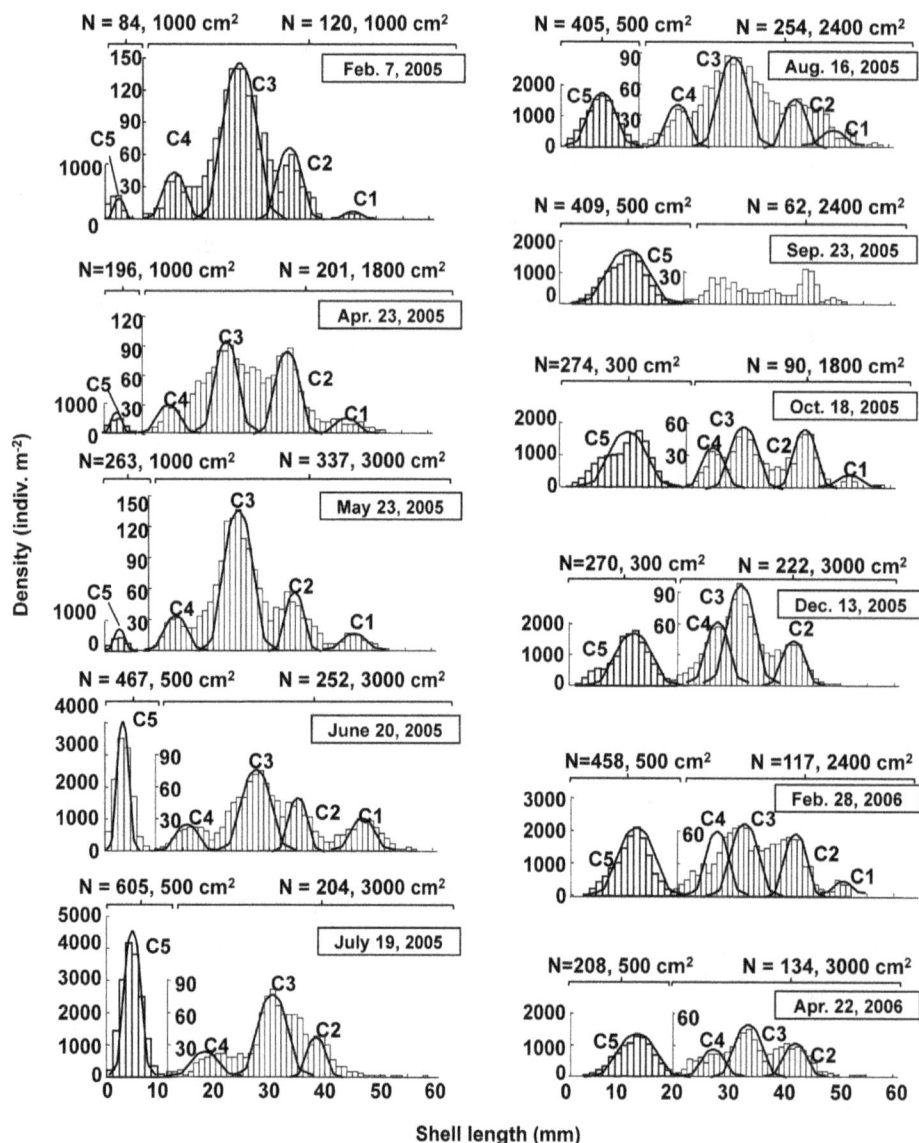

Figure 3. Size–frequency distribution of *Ruditapes philippinarum* population at station B. The left and right scales correspond to samples obtained using mesh sizes of 1 and 5 mm, respectively (C = cohorts; n = number of individuals).

Moreover, all five cohorts recognized here exhibited increases in shell length of 5–12 mm in the warm season between May 2005 and October 2005. In contrast, their growth was extremely depressed or ceased altogether in the cold season between November 2005 and April 2006. From the growth curves produced, the clam was considered as approximately 1 year and 5 years to reach shell lengths of 10 mm and 50 mm, respectively, in the area investigated.

Figure 4b illustrates seasonal fluctuations in the secondary productivity of the clam population estimated from changes in the size–frequency distributions of the five cohorts. The annual secondary productivity of the clam population was 130 g C m^{-2} yr^{-1} between April 2005 and April 2006, peaking at 994 mg C m^{-2} d^{-1} in August 2005 and decreasing to 0–64.0 mg C m^{-2} d^{-1} between December 2005 and April 2006 owing to the depression or cessation of growth. Seasonal variation in the ratio of the daily secondary productivity of the clam to its biomass in kg wet weight (P/B_{kgWW}) at each sampling instance exhibited a

pattern similar to that of secondary productivity and varied between 0 and 0.16 (Figure 4b). The period between June and October was defined as the productive period, i.e. the period during which secondary production was greater than 500 mg C m^{-2} d^{-1}.

Figure 5 illustrates seasonal variation in biomass and secondary productivity for the tidal-flat area. The biomass of clams was found to be heterogeneously distributed among tidal-flat stations (Figure S1); although, the clam accounted for at least 70% of the biomass of the benthic community. The mean values were found to be relatively stable, ranging from 1.98±1.77 kgWW m^{-2} to 5.59±7.51 kgWW m^{-2} (Figure 5a). The secondary productivity at the tidal-flat stations, calculated based on the mean biomass (Figure 5a) and P/B_{kgWW} (Figure 4b) at the representative station, exhibited a clear seasonal pattern: productivity peaked at 840±634 mg C m^{-2} d^{-1} in October 2005 and decreased to 9.6±5.9 and 28.0±33.8 mg C m^{-2} d^{-1} by December 2005 and April 2006, respectively. Thus, the annual secondary productivity

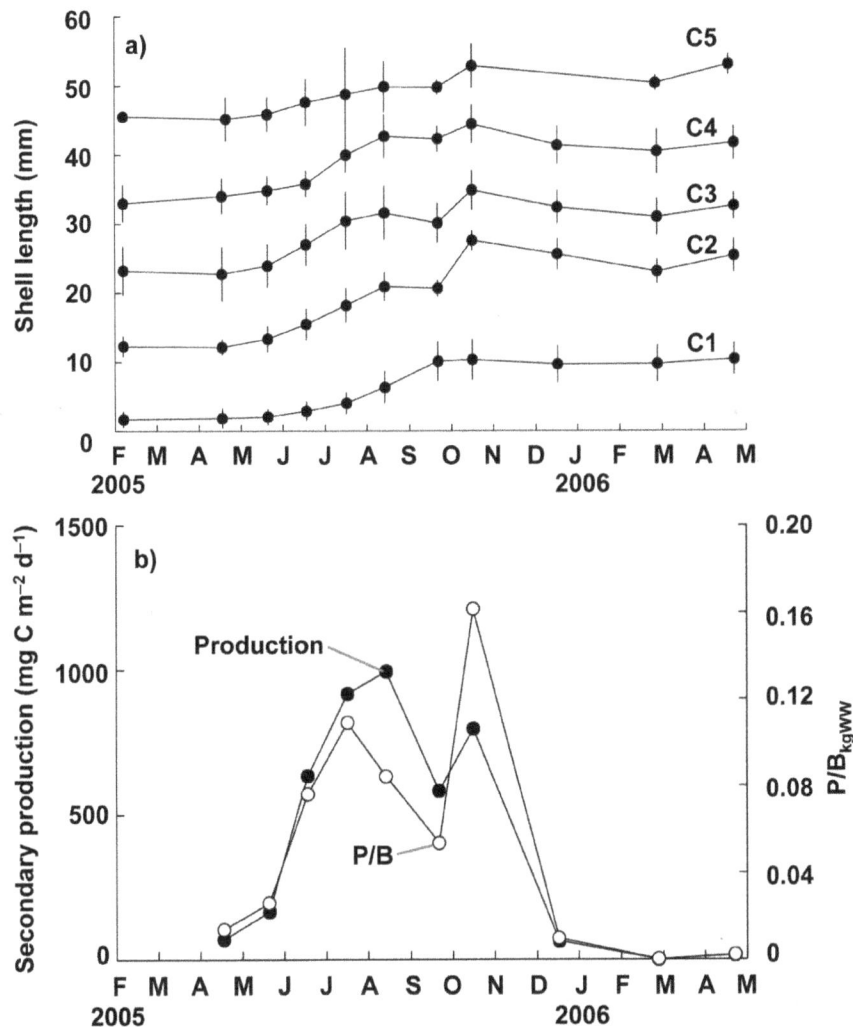

Figure 4. Seasonal variation of growth of the clam (*Ruditapes philippinarum*) at station B. Changes in (a) mean shell length of each cohort of the clam *Ruditapes philippinarum* and (b) daily secondary productivity and P/B$_{kgWW}$ at station B.

was estimated to be 96.5 g C m^{-2} yr^{-1} throughout the tidal flats. During the productive period (i.e. for 120 days during June–October 2005), the integrated secondary productivity accounted for 90.6% (87.5 g C m^{-2}) of the annual productivity.

Food Demand of the Clam and Quantification of Potential Food Sources

Table 1 presents the net growth efficiency (i.e. growth/consumption) of six different species of bivalves including *R. philippinarum*. The food demand of the clam population (mg C m^{-2} d^{-1}) was determined using an assimilation number of 18.8%, which was the mean value obtained from nine different energy budgets for the bivalve species (Table 1).

During the productive period, nitrogen-based secondary productivity and the food demand of the clam was 1.92 mol N m^{-2} and 10.2 mol N m^{-2}, respectively, and the total secondary productivity and food demand of the clam throughout the tidal flats (i.e. for 0.19 km^2) were 364 kmol N and 1939 kmol N, respectively. The daily secondary productivity and food demand of the clam during the productive season were 3.04 kmol N d^{-1} and 16.2 kmol N d^{-1}, respectively, for the entire tidal flat area (0.19 km^2) (Table 2).

During the productive season, the mean biomass per unit area of MPB in the surface sediment (depth: 0.5 cm; 140.6±79.8 mg Chl-*a* m^{-2} in the subtidal area, n = 56; 140.6±52.5 mg Chl-*a* m^{-2} on the tidal flats, n = 39) was approximately 70 times greater than that of phytoplankton in the water column (depth: 1 m; 2.0±1.3 mg Chl-*a* m^{-2}) (n = 123). The mean standing stocks of *Zostera japonica* and *Ulva pertusa* were 98.8±102.4 g DW m^{-2} and 19.9±129.2 g DW m^{-2}, respectively, in the subtidal area (n = 30). The nitrogen-based biomasses of MPB, phytoplankton, *Z. japonica*, and *U. pertusa* were 213–3200 mmol N m^{-2} (for both the tidal flats and the subtidal area), 6.04–15.1, 203, and 9.22 mmol N m^{-2}, respectively.

The total nitrogen content of phytoplankton (1.79–4.48 kmol N) was lower than that for all other organisms in the lagoon, and phytoplankton accounted for only 12–28% of the daily food demand of the clam population (Table 2). MPB in tidal flats also imposed limits on the daily food demand of the clams, accounting for 42–104% of the daily food demand. However, none of the other food sources studied imposed limits on the daily food demand of the clams.

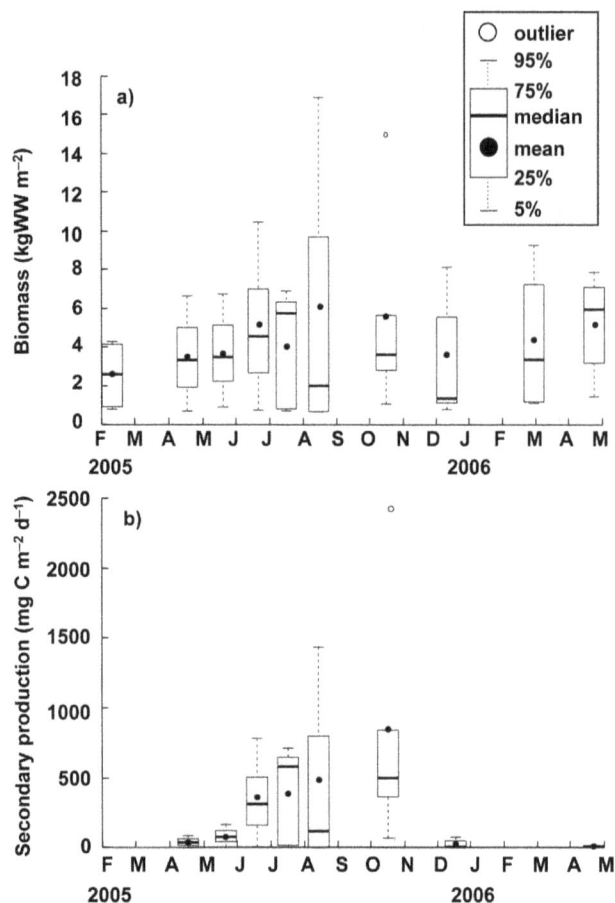

Figure 5. Seasonal variation of the clam (*Ruditapes philippinarum*) population at tidal flats stations. (a) biomass (kgWW m^{-2}) and (b) secondary productivity (g C m^{-2} d^{-1}) of the clam at five tidal flat stations, expressed as box plots.

Carbon and Nitrogen Isotopic Signatures and Contribution to the Diet of the Clam

The carbon stable isotope ratios (δ^{13}C) of the clam diet and potential food sources in the lagoon ranged from -18.5 ± 1.2 ‰ for POM to -9.7 ± 0.6 ‰ for *Z. japonica*, whereas the nitrogen isotope ratios (δ^{15}N) ranged from 4.8 ± 0.9 ‰ for SOM to 6.8 ± 0.5

‰ for *U. pertusa* (Figure 6). Thus, the diet of the clam was found to exhibit -16.4 ± 0.9 ‰ δ^{13}C and 5.9 ± 0.9 ‰ δ^{15}N, which corresponds to the isotopic signature of MPB (-17.1 ± 1.1 ‰ δ^{13}C and 5.5 ± 0.6 ‰ δ^{15}N). At our study site, the δ^{13}C value of POM fell within the ranges of previously reported for temperate marine phytoplankton (-24 ‰ to -18 ‰) from other estuarine and coastal waters [17,34–36]. Thus, the isotope signature of the POM samples was regarded in this study to primarily reflect the presence of phytoplankton.

According to the IsoSource mixing model, POM was the primary contributor to the clam diet ($43\pm12\%$) (all values are given as mean $\pm$ SD) followed by MPB ($20\pm15\%$) has a wider range (min.–max.: 0–69%) and *U. pertusa* ($20\pm6\%$) has a narrower range (min.–max.: 3–34%); SOM, and *Z. japonica* contributed, on average, $11\pm8\%$, and $6\pm4\%$, respectively (Figure 7).

Comparison of the total amount of phytoplankton present to the daily food demand of the clam indicates that phytoplankton account for 12–28% of the daily food demand of the clam (Table 2). A curve was constructed to a represent the mean contribution (and associated standard deviations) of potential food sources at each contribution (i.e. for every 1% increase) of POM, thus illustrating the entire range of solutions using only isotopic constraints (for all regions). Then, a subset of the output (left side of the dotted line) was extracted; this subset contains only solutions that satisfy the food demands of the clam population (Figure 8). For this subset, MPB was the primary contributor ($35\pm13\%$ to $64\pm4\%$), followed by *U. pertusa* ($19\pm1\%$ to $23\pm4\%$). The contribution of SOM and *Z. japonica* to the clam diet were $3\pm3\%$ to $13\pm9\%$ and $0\pm0\%$ to $2\pm1\%$, respectively.

Discussion

The short-necked clam (*R. philippinarum*) was found to be the dominant macrobenthic species in the present study area (Figure 2a, b). Comparison of the secondary productivity of the clam with that of the other suspension-feeding bivalve species present demonstrates that higher secondary productivity (i.e. 228–996 gC m^{-2} yr^{-1}) was found in the following mytilid species: *Choromytilus meridionalis* (Krauss 1848) [37], *Perna picta* (Born 1778) [38,39], and *M. edulis* [2]. These species typically form extremely dense populations, which accounts (at least primarily) for their high productivity. During the productive season, secondary productivity of the clam population in the lagoon was estimated to be 364 kmol N (per 120 days) based on both the mean secondary productivity of the clam and the area of the tidal flats

Table 1. Net growth efficiency (%) of bivalves.

Species	Country	Net growth efficiency	Method*	Reference
Cardium edule	Sweden	15.4	in situ+lab	[7]
Mya arenaria	Sweden	15.5	in situ+lab	[7]
Macoma balthica	Netherland	4.4	in situ+lab	[47]
Ruditapes philippinarum	USA	41.7	lab	[48]
Scrobicularia plana	Wales	14.8	in situ+lab	[49]
Tellina tenuis	Scotland	13.6	core	[50]
Tellina tenuis	Scotland	16.2	core	[50]
Tellina tenuis	Scotland	28.7	core	[50]
	Mean	18.8		

*lab: laboratory experiments, core: field core incubation experiments.

Table 2. Food demand of the clam (*Ruditapes philippinarumm*) and total amount of potential food sources during the productive season.

Food demand of the clam	Region*	kmol N d^{-1}	
	Tidal flats	**16.2**	
Amount of potential food sources		kmol N (min.–max.)	Amount/Demand (min.–max.)
Phytoplankton	Whole	1.79–4.48	0.11–0.28
Microphytobenthos	Tidal flats	6.8–16.9	0.42–1.04
	Subtidal	120–300	7.41–18.5
Zostera japonica	Subtidal	685	42.3
Ulva pertusa	Subtidal	31.1	1.92

Area: entire (3.56 km^2), subtidal (3.37 km^2), and tidal-flat (0.19 km^2).

(0.19 km^2). A recent study showed that ingress of nitrate from the Pacific Ocean through the tidal inlet is necessary to maintain the lagoon ecosystem, and the dissolved inorganic nitrogen budget of this lagoon has been estimated to be 258 kmol N yr^{-1} [20]; this is comparable to the secondary productivity of the clam. Therefore, the feeding of the clam should have a considerable impact on the material circulation of the lagoon.

Yokoyama et al. [17] suggested that the relatively large biomass of phytoplankton and the scarcity of benthic microalgae could have produced the phytoplankton-based trophic structure (including *R. philippinarum*) found in Ariake Bay, southern Japan. In contrast, several authors have suggested that the clam depends more on MPB than on phytoplankton based on the biomass of microalgae [16,18,35]. In the present study, the relatively low contribution of the phytoplankton (at most 28%) as a food source for the clam should correspond to the occurrence of low phytoplankton biomass (Table 2); however, this is not the case. Although the contribution of the phytoplankton was estimated initially to be 14–77% based on isotopic data alone (Figure 7a), the stable isotope signatures obtained indicate that the phytoplankton

contribute (at most) 28% of the food resources required for the clam population in the study area (Figure 8). Thus, the contributions derived using only isotopic signatures may be somewhat misleading.

Assuming a POM contribution of 28% based on the phytoplankton biomass, the clam population within the tidal flats (i.e. an area of 0.19 km^2) must consume all of the phytoplankton present in the entire lagoon (i.e. an area of 3.56 km^2) each day. Filtration experiments conducted on the clam [40] have demonstrated its weight-specific filtration rate for Chl-*a* is 2.3 L gDW^{-1} h^{-1} under temperatures of 24°C and mean Chl-*a* concentrations of 3.1 μg L^{-1}. However, the temperature and Chl-*a* concentration for the water column in the present study were somewhat lower than those for the filtration experiment (Table 3) [20,22]. Assuming that the weight-specific filtration rate of the clam (which has a mean biomass of 246–276 gDW m^{-2} in the study area) remains constant throughout the day, the filtration rate for clams on the tidal flats of the present study can be estimated to be 2.89–7.17 kg Chl-*a* d^{-1}. The total amount of Chl-*a* in the lagoon was 3.56–9.90 kg; thus, the dense clam population appears to filtrate

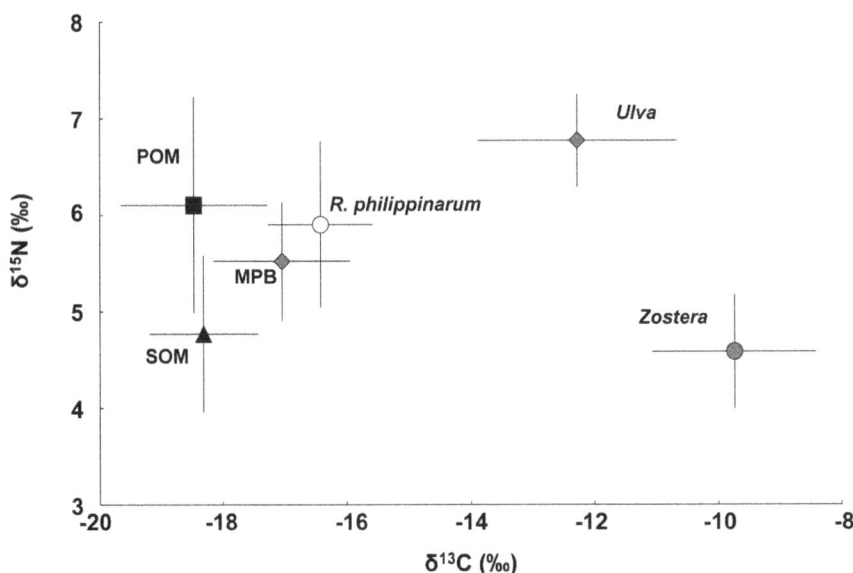

Figure 6. A dual isotope plot for carbon and nitrogen for the expected diet of the clam (*Ruditapes philippinarum*) and their potential food sources in Hichirippu Lagoon. See text in Materials and methods for diet-tissue fractionation for the clam to determine expected diet (*R. philippinarum*). Error bars are standard deviations.

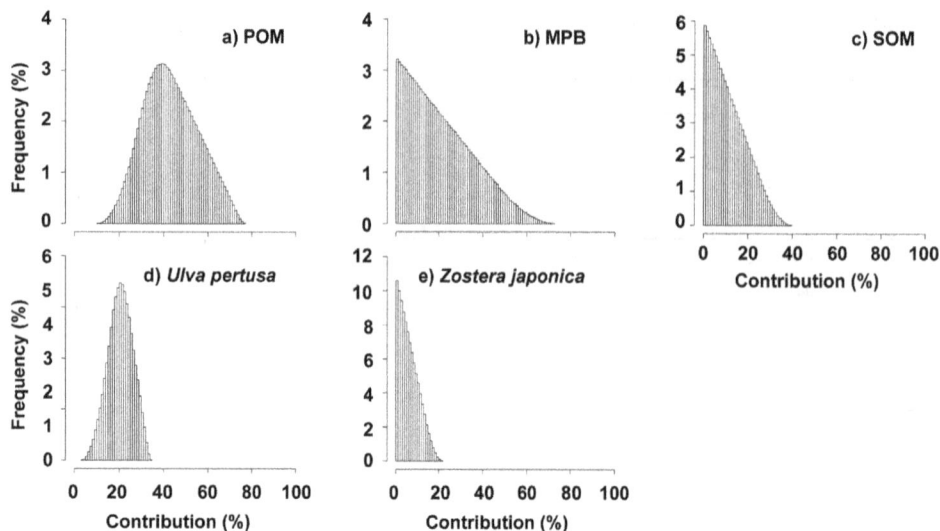

Figure 7. Histograms of the proportional contribution of each food source to the diet of the clam (*Ruditapes philippinarum*) based on the IsoSource mixing model.

72.5–80% of the Chl-*a* of water in the lagoon each day. The feeding activity of the clam population should have a considerable effect on the water column phytoplankton; however, it is highly unlikely that the clams present could consume all of the phytoplankton in the lagoon water owing to the circadian nature of suspension-feeding bivalves [41,42]. Thus, it is reasonable to assume that the contribution of the phytoplankton to the clam diet is less than 28%.

Assuming the lower limit of the phytoplankton contribution to be accurate, the clam population would tend toward consuming

almost all of (or more than) the total MPB available in the tidal-flat area each day (Table 2). Resuspended primarily produced organic particles (i.e. those originating from MPB) in tidal flats are considered to be important food sources for benthic animals [43], and their lateral transport to the habitat of suspension-feeding benthic bivalves (*R. philippinarum* and *Mactra veneriformis* Reeve 1854) is thought to be essential to sustain the high secondary productivity of these bivalves [9]. In the inner part of Ariake Bay, resuspended MPB was supplied to the benthic community on tidal flats and provided an additional source of food in addition to that

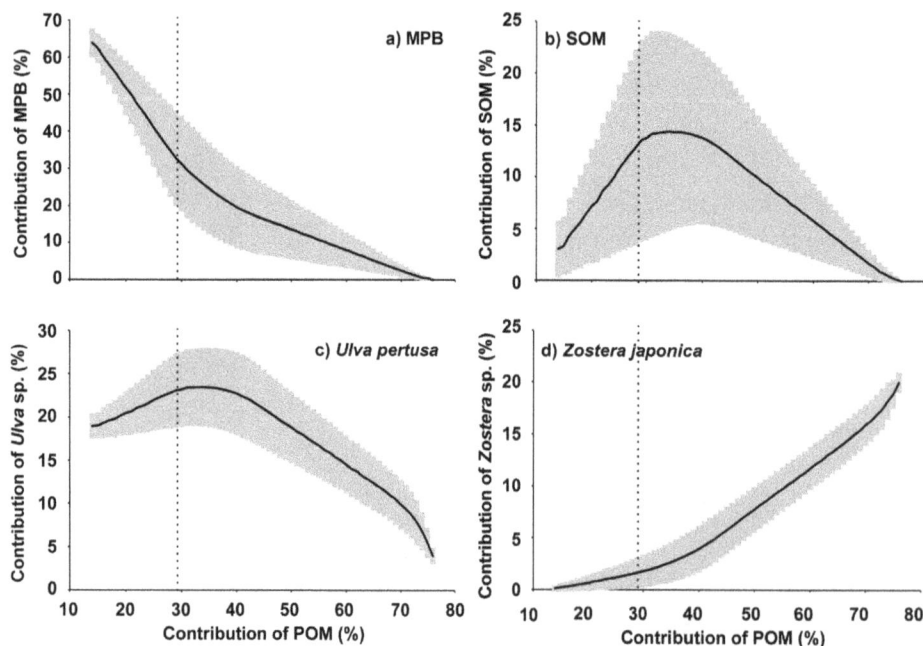

Figure 8. Scatter plot illustrating the relationship between the contributions of phytoplankton and other food sources. (a) MPB, (b) SOM, (c) *U. pertusa*, and (d) *Z. japonica*. Solid lines and grey region indicate mean and standard deviations for every 1% increase of POM contribution, respectively. Vertical dotted line indicates the upper limit to the phytoplankton contribution based on their abundance and the food demand of the clam.

Table 3. Estimation of the rate of filtration by the clam (*Ruditapes philippinarum*) in Hichirippu Lagoon during productive season.

	June	August	October
Clam biomass (gDW m^{-2})	246 (211)	296 (398)	276 (304)
Chl-*a* conc. (µg L^{-1})	2.78 (1.51)	1.49 (1.16)	1.00 (0.67)
Filteration rate (L gDW^{-1} h^{-1})	2.3		
Capacity of filteration (kg Chl-a d^{-1})	7.17	4.63	2.89
Amount of Chl-*a* (kg)	9.90	5.30	3.56
Filteration/Amount (%)	72.5	87.2	81.3

Standard deviations for biomass Chl-*a* are indicated inparentheses.

found in the subtidal region, thus increasing the food availability by 50% [44]. In the present study, the abundant MPB in the subtidal area (where MPB are approximately 18 times more abundant than in the tidal flat area) appear to have been a key clam food source, particularly in its resuspended form; in this manner, MPB could have sustained the high secondary productivity of the clam.

In the muddy tidal flats of Ariake Bay, southern Japan, a certain portion of the MPB typically becomes resuspended during high tide, thus contribution to the total biomass in the water column [13]. In the central part of the area investigated in the present study, resuspension of MPB due to the tidal cycle has been estimated to contribute up to 74% of water column Chl-*a* [45]. Therefore, the biomass of phytoplankton has likely been somewhat overestimated in the present study owing to the lack of consideration of resusupended MPB.

Interestingly, the contribution of *U. pertus* for the extracted subset to the clam diet (based on the amount of phytoplankton) was found to be 12–28% (Figure 8c). Previously, clam feeding experiments have demonstrated that the clam can filtrate a

maximum particle size of 200 µm [40]. Furthermore, the hard clam *Mercenaria mercenaria* L. 1758, which is in the same family (Veneridae) as *R. philippinarum*, can filtrate aggregates up to approximately 1 mm in diameter [46]. These previous studies suggest that seaweeds must be broken down into fine particles (i.e. less than 1 mm diameter) to be filtrated. During this process, the numerous bacteria that are attached to the debris form a community and act as a further food source for the clam.

The results of the present study suggest a clear relationship between the isotopic signature and food demand of the clam population. The contribution of MPB was unclear before the extraction; however, the final results, which are revised with respect to the combination of food available and the total abundance of primary producers in the lagoon, indicate that MPB are likely the main food source for the clam population in this environment. The dense patches of clams with high productivity (130 g C m^{-2} yr^{-1}) are likely sustained by the supply of particulate organic matter derived from MPB (35–64%) in the subtidal area, where the standing stock of MPB is approximately 18 times that for the tidal flats.

Supporting Information

Figure S1 Seasonal variations of macro-benthic biomass (in kgWW m^{-2}) at each tidal flat station.

Acknowledgments

We would like to express our gratitude to Ms. Y Inaba, Ms. A. Hamada and Mr. N. Kawanishi for their assistance in the field survey, and to Dr. I. Kudo, and 2 anonymous reviewers for their constructive comments of this manuscript.

Author Contributions

Conceived and designed the experiments: TK RK SM SS TY HT. Performed the experiments: TK RK SS TY. Analyzed the data: TK RK SM. Contributed reagents/materials/analysis tools: TK RK. Wrote the paper: TK RK SM HT.

References

1. Brey T (2001) Population dynamics in benthic invertebrates. A virtual handbook. Version 01.2. Available: http://www.thomas-brey.de. Accessed 2013 Sept 14.

2. Thompson R (1984) Production, reproductive effort, reproductive value and reproductive cost in a population of the blue mussel *Mytilus edulis* from a subarctic environment. Mar Ecol Prog Ser 16: 249–257. Available: http://www.int-res.com/articles/meps/16/m016p249.pdf. Accessed 2013 April 22.

3. Laudien J, Brey T, Arntz W (2003) Population structure, growth and production of the surf clam *Donax serra* (Bivalvia, Donacidae) on two Namibian sandy beaches. Estuar Coast Shelf Sci 58S: 105–115. Available: http://www.sciencedirect.com/science/article/pii/S0272771403000441. Accessed 2013 April 22.

4. MacIntyre H, Geider R, Miller D (1996) Microphytobenthos: The ecological role of the "secret garden" of unvegetated, shallow-water marine habitats. I. Distribution, abundance and primary production. Estuaries 19: 186–201. Available: http://link.springer.com/article/10.2307/1352224. Accessed 2013 April 22.

5. Cahoon LB (1999) The role of benthic microalgae in neritic ecosystems. Oceanogr Mar Biol Annu Rev 37: 47–86.

6. Underwood G, Kromkamp J (1999) Primary production by phytoplankton and microphytobenthos in estuaries. Adv Ecol Res 29: 93–153. Available: http://www.sciencedirect.com/science/article/pii/S0065250408601920. Accessed 2013 April 22.

7. Loo L, Rosenberg R (1996) Production and energy budget in marine suspension feeding populations: *Mytilus edulis, Cerastoderma edule, Mya arenaria* and *Amphiura filiformis*. J Sea Res 35: 199–207. Available: http://www.sciencedirect.com/science/article/pii/S1385110196907479. Accessed 2013 April 15.

8. Nakaoka M (1992) Spatial and seasonal variation in growth rate and secondary production of *Yoldia notabilis* in Otsuchi Bay, Japan, with reference to the influence of food supply from the water column. Mar Ecol Prog Ser 88: 215–223. Available: http://www.int-res.com/articles/meps/88/m088p215.pdf.

9. Yamaguchi H, Tsutsumi H, Tsukuda M, Nagata S, Kimura C, et al. (2004) Utilization of photosynthetically produced organic particles by dense patches of suspension feeding bivalves on the sand flat of Midori River Estuary, Kyusyu, Japan. Benthos Res 59: 67–77. Available: https://www.jstage.jst.go.jp/article/benthos1996/59/2/59_2_67/_pdf. Accessed 2013 April 22.

10. Kang C, Sauriau P (1999) the infaunal suspension-feeding bivalve *Cerastoderma edule* in a muddy sandflat of Marennes-Oleron Bay, as determined by analyses of carbon and nitrogen. Mar Ecol Prog Ser 187: 147–158. Available: http://www.int-res.com/articles/meps/187/m187p147.pdf. Accessed 2013 Sept 14.

11. Kanaya G, Nobata E, Toya T, Kikuchi E (2005) Effects of different feeding habits of three bivalve species on sediment characteristics and benthic diatom abundance. Mar Ecol Prog Ser 299: 67–78. Available: www.int-res.com/articles/meps2005/299/m299p067.pdf.

12. Kasai A, Horie H, Sakamoto W (2004) Selection of food sources by *Ruditapes philippinarum* and *Mactra veneriformis* (Bivalva: Mollusca) determined from stable isotope analysis. Fish Sci 70: 11–20. Available: http://onlinelibrary.wiley.com/doi/10.1111/j.1444-2906.2003.00764.x/full. Accessed 2012 Feb 16.

13. Koh C, Khim J, Araki H (2006) Tidal resuspension of microphytobenthic chlorophyll a in a Nanaura mudflat, Saga, Ariake Sea, Japan: flood-ebb and spring-neap variations. Mar Ecol Prog Ser 312: 85–100. Available: http://www.int-res.com/abstracts/meps/v312/p85-100/. Accessed 2013 Sept 12.

14. Ponurovsky S, Yakovlev Y (1992) The reproductive biology of the Japanese littleneck, *Tapes philippinarum* (A. Adams and Reeve, 1850) (Bivalvia: Veneridae). J Shellfish Res 22: 814–823.

15. Dang C, Sauriau P, Savoye N (2009) Determination of diet in Manila clams by spatial analysis of stable isotopes. Mar Ecol Prog Ser 387: 167–177. Available: http://www.int-res.com/abstracts/meps/v387/p167-177/. Accessed 2012 Aug 31.

16. Kanaya G, Takagi S, Nobata E, Kikuchi E (2007) Spatial dietary shift of macrozoobenthos in a brackish lagoon revealed by carbon and nitrogen stable

isotope ratios. Mar Ecol Prog Ser 345: 117–127. Available: http://www.int-res.com/abstracts/meps/v345/p117-127/. Accessed 2012 Aug 14.

17. Yokoyama H, Tamaki A, Koyama K, Ishii Y, Shimoda K, et al. (2005) Isotopic evidence for phytoplankton as a major food source for macrobenthos on an intertidal sandflat in Ariake Sound, Japan. Mar Ecol Prog Ser 304: 101–116. Available: http://www.int-res.com/abstracts/meps/v304/p101-116/. Accessed 2013 April 13.

18. Watanabe S, Katayama S, Kodama M, Cho N, Nakata K, et al. (2009) Small-scale variation in feeding environments for the Manila clam *Ruditapes philippinarum* in a tidal flat in Tokyo Bay. Fish Sci 75: 937–945. Available: http://link.springer.com/10.1007/s12562-009-0113-1. Accessed 2013 Sept 12.

19. Komorita T, Kajihara R, Tsutsumi H, Shibanuma S, Yamada T, et al. (2010) Reevaluation of the nutrient mineralization process by infaunal bivalves (*Ruditapes philippinarum*) in a shallow lagoon in Hokkaido, Japan. J Exp Mar Bio Ecol 383: 8–16. Available: http://linkinghub.elsevier.com/retrieve/pii/S0022098109004870. Accessed 2013 Sept 18.

20. Komorita T, Tsutsumi H, Kajihara R, Suga N, Shibanuma S, et al. (2012) Oceanic nutrient supply and uptake by microphytobenthos of the Hichirippu Lagoon, Hokkaido, Japan. Mar Ecol Prog Ser 446: 161–171. Available: http://www.int-res.com/abstracts/meps/v446/p161-171/. Accessed 2013 Feb 13.

21. Komorita T, Shibanuma S, Yamada T, Kajihara R, Tsukuda M, et al. (2009) Impact of low temperature during the winter on the mortality in the post-settlement period of the juvenile of short-neck clam, *Ruditapes philippinarum*, on the tidal flats in Hichirippu Lagoon, Hokkaido, Japan. Plankt Benthos Res 4: 31–37. Available: https://www.jstage.jst.go.jp/article/pbr/4/1/4_1_31/_pdf. Accessed 2013 Feb 16.

22. Kajihara R, Komorita T, Hamada A, Shibanuma S, Yamada T, et al. (2010) Possibility of direct utilization of seagrass and algae as main food resources by small gastropod, *Lacuna decorata*, in a subarctic lagoon, Hichirippu, eastern Hokkaido, Japan. Plankt Benthos Res 5: 90–97. Available: https://www.jstage.jst.go.jp/article/pbr/5/3/5_3_90/_pdf.

23. Couch C (1989) Carbon and nitrogen stable isotopes of meiobenthos and their food resources. Estuar Coast Shelf Sci 28: 433–441. Available: http://www.sciencedirect.com/science/article/pii/0272771489900905. Accessed 2013 May 14.

24. Lorenzen C (1967) Determination of cholorophyll and pheopigments: spectrophotometric equations. Limnol Oceanogr 22: 1096–1109. Available: http://aslo.org/lo/toc/vol_12/issue_2/0343.pdf. Accessed 2013 April 22.

25. Montani S, Magni P, Abe N (2003) Seasonal and interannual patterns of intertidal microphytobenthos in combination with laboratory and areal production estimates. Mar Ecol Prog Ser 249: 79–91. Available: www.int-res.com/articles/meps2003/249/m249p079.pdf.

26. Tsutsumi H, Tanaka M (1994) Cohort analysis of size frequency distribution with computer programs based on a graphic method and simplex's method. Benthos Res 46: 1–10. doi:10.1521/jaap.2011.39.4.753.

27. Crisp D (1984) Energy flow measurements. In: Holm NA, McIntyre AD, editors. Methods for the study of marine benthos, IBP Handbook No. 16. Blackwell Scientific publication, Vol. 16. pp. 416–426.

28. Phillips DL, Gregg JW (2003) Source partitioning using stable isotopes: coping with too many sources. Oecologia 136: 261–269. Available: http://www.epa.gov/naaujydh/pages/models/stableIsotopes/publications/Phillips %26 Gregg 2003.pdf. Accessed 2013 Feb 4.

29. Phillips DL, Newsome SD, Gregg JW (2005) Combining sources in stable isotope mixing models: alternative methods. Oecologia 144: 520–527.Available: http://www.epa.gov/naaujydh/pages/models/stableIsotopes/publications/Phillips et al 2005.pdf. Accessed 2013 Feb 4.

30. Yokoyama H, Tamaki A, Harada K (2005) Variability of diet-tissue isotopic fractionation in estuarine macrobenthos. Mar Ecol Prog Ser 296: 115–128. Available: http://www.int-res.com/abstracts/meps/v296/p115-128/. Accessed 2013 Feb 12.

31. Eppley R (1968) An incubation method for estimating the carbon content of phytoplankton in natural samples. Limnol Ocean 13: 574–582. Available: http://tube.aslo.net/lo/toc/vol_13/issue_4/0574.pdf. Accessed 2013 Jan 30.

32. Jonge V De (1980) Fluctuations in the organic carbon to chlorophyll a ratios for estuarine benthic diatom populations. Mar Ecol Prog Ser 2: 343–353. Available: http://www.int-res.com/articles/meps/2/m002p345.pdf. Accessed 2013 Sept 13.

33. Redfield A, Ketchum B, Richards F (1963) The influence of organisms on the composition of sea-water. In: Hill M, editor. The Sea. New York: Wiley, Vol. 22. pp. 26–77.

34. France R (1995) Carbon-13 enrichment in benthic compared to planktonic algae: foodweb implications. Mar Ecol Prog Ser 124: 307–312. Available: http://www.int-res.com/articles/meps/124/m124p307.pdf. Accessed 2013 Sept 14.

35. Kanaya G, Nakamura Y, Koizumi T, Yamada K, Koshikawa H, et al. (2013) Temporal changes in carbon and nitrogen stable isotope ratios of macro-zoobenthos on an artificial tidal flat facing a hypertrophic canal, inner Tokyo Bay. Mar Pollut Bull 71: 179–189. Available: http://www.ncbi.nlm.nih.gov/pubmed/23602262. Accessed 2013 Sept 12.

36. Yokoyama H, Sakami T, Ishihi Y (2009) Food sources of benthic animals on intertidal and subtidal bottoms in inner Ariake Sound, southern Japan, determined by stable isotopes. Estuar Coast Shelf Sci 82: 243–253. Available: http://dx.doi.org/10.1016/j.ecss.2009.01.010. Accessed 2013 Feb 16.

37. Griffiths R (1981) Production and energy flow in relation to age and shore level in the bivalve *Choromytilus meridionalis* (Kr.). Estuar Coast Shelf Sci 13: 477–493. Available: http://www.sciencedirect.com/science/article/pii/S0302352481800539. Accessed 2013 April 22.

38. Shafee M (1992) Production estimate of a mussel population *Perna picta* (Born) on the Atlantic coast of Morocco. J Exp Mar Bio Ecol 163: 183–197. Available: http://www.sciencedirect.com/science/article/pii/002209819290048F. Accessed 2013 April 22.

39. Hicks DW, Jr JWT, Mcmahon RF (2001) Population dynamics of the nonindigenous brown mussel *Perna perna* in the Gulf of Mexico compared to other world-wide populations. Mar Ecol Prog Ser 211: 181–192. Available: http://www.int-res.com/articles/meps/211/m211p181.pdf.

40. Nakamura Y (2001) Filtration rates of the Manila clam, *Ruditapes philippinarum*: dependence on prey items including bacteria and picocyanobacteria. J Exp Mar Bio Ecol 266: 181–192. Available: http://www.sciencedirect.com/science/article/pii/S0022098101003549. Accessed 2013 May 27.

41. García-March JR, Sanchís Solsona MÁ, García-Carrascosa M (2008) Shell gaping behaviour of *Pinna nobilis* L., 1758: circadian and circalunar rhythms revealed by in situ monitoring. Mar Biol 153: 689–698. Available: http://www.springerlink.com/index/10.1007/s00227-007-0842-6. Accessed 2012 April 3.

42. Rodland DL, Schöne BR, Helama S, Nielsen JK, Baier S (2006) A clockwork mollusc: Ultradian rhythms in bivalve activity revealed by digital photography. J Exp Mar Bio Ecol 334: 316–323. Available: http://linkinghub.elsevier.com/retrieve/pii/S0022098106001110. Accessed 2013 April 16.

43. Jonge V De, Beuselom J Van (1992) Contribution of resuspended micro-phytobenthos to total phytoplankton in the EMS estuary and its possible role for grazers. Netherlands J Sea Res 22: 445–459. Available: http://www.sciencedirect.com/science/article/pii/007775799290049K. Accessed 2013 Sept 13.

44. Yoshino K, Tsugeki NK, Amano Y, Hayami Y, Hamaoka H, et al. (2012) Intertidal bare mudflats subsidize subtidal production through outwelling of benthic microalgae. Estuar Coast Shelf Sci 109: 138–143. Available: http://linkinghub.elsevier.com/retrieve/pii/S0272771412001874. Accessed 2013 Sept 13.

45. Suga N, Kajihara R, Shibanuma S, Yamada T, Montani S (2011) Estimation of microphytobenthic resuspension fluxes in a shallow lagoon in Hokkaido, Japan. Plankt Benthos Res 6: 115–123. Available: https://www.jstage.jst.go.jp/article/pbr/6/2/6_2_115/_pdf. Accessed 2013 Feb 13.

46. Kach DJ, Ward JE (2007) The role of marine aggregates in the ingestion of picoplankton-size particles by suspension-feeding molluscs. Mar Biol 153: 797–805. Available: http://www.springerlink.com/index/10.1007/s00227-007-0852-4. Accessed 2012 March 12.

47. Hummel H (1985) An energy budget for a *Macoma balthica* (mollusca) population living on a tidal flat in the Dutch Wadden Sea. Netherlands J sea Res 19: 84–92. Available: http://www.sciencedirect.com/science/article/pii/0077757985900456. Accessed 2013 Sept 13.

48. Langton R, Winter J, Roels O (1977) The effect of ration size on the growth and growth efficiency of the bivalve mollusc *Tapes japonica*. Aquaculture 12: 283–292. Available: http://www.sciencedirect.com/science/article/pii/0044848677902071. Accessed 2013 Sept 13.

49. Hughes R (1970) An energy budget for a tidal-flat population of the bivalve *Scrobicularia plana* (Da costa). J Anim Ecol 39: 357–179. Available: http://www.jstor.org/stable/10.2307/2976. Accessed 2013 Sept 13.

50. Trevallion A (1971) Studies on *Tellina tenuis* Da Costa. III Aspects of general biology and energy flow. J Exp Mar Bio Ecol 7: 95–122. Available: http://www.sciencedirect.com/science/article/pii/0022098171900062. Accessed 2013 Sept 13.

8

Selecting Reliable and Robust Freshwater Macroalgae for Biomass Applications

Rebecca J. Lawton*, Rocky de Nys, Nicholas A. Paul

School of Marine and Tropical Biology, James Cook University, Townsville, Queensland, Australia

Abstract

Intensive cultivation of freshwater macroalgae is likely to increase with the development of an algal biofuels industry and algal bioremediation. However, target freshwater macroalgae species suitable for large-scale intensive cultivation have not yet been identified. Therefore, as a first step to identifying target species, we compared the productivity, growth and biochemical composition of three species representative of key freshwater macroalgae genera across a range of cultivation conditions. We then selected a primary target species and assessed its competitive ability against other species over a range of stocking densities. *Oedogonium* had the highest productivity (8.0 g ash free dry weight m^{-2} day^{-1}), lowest ash content (3–8%), lowest water content (fresh weigh: dry weight ratio of 3.4), highest carbon content (45%) and highest bioenergy potential (higher heating value 20 MJ/kg) compared to *Cladophora* and *Spirogyra*. The higher productivity of *Oedogonium* relative to *Cladophora* and *Spirogyra* was consistent when algae were cultured with and without the addition of CO_2 across three aeration treatments. Therefore, *Oedogonium* was selected as our primary target species. The competitive ability of *Oedogonium* was assessed by growing it in bi-cultures and polycultures with *Cladophora* and *Spirogyra* over a range of stocking densities. Cultures were initially stocked with equal proportions of each species, but after three weeks of growth the proportion of *Oedogonium* had increased to at least 96% ($\pm$7 S.E.) in *Oedogonium-Spirogyra* bi-cultures, 86% ($\pm$16 S.E.) in *Oedogonium-Cladophora* bi-cultures and 82% ($\pm$18 S.E.) in polycultures. The high productivity, bioenergy potential and competitive dominance of *Oedogonium* make this species an ideal freshwater macroalgal target for large-scale production and a valuable biomass source for bioenergy applications. These results demonstrate that freshwater macroalgae are thus far an under-utilised feedstock with much potential for biomass applications.

Editor: Vishal Shah, Dowling College, United States of America

Funding: This research is part of the MBD Energy Research and Development program for Biological Carbon Capture and Storage. The project is supported by the Advanced Manufacturing Cooperative Research Centre (AMCRC), funded through the Australian Government's Cooperative Research Centre Scheme. The funders had no role in study design, data collection and analysis, decision to publish, or preparation of the manuscript.

Competing Interests: The authors have declared that no competing interests exist.

* E-mail: rebecca.lawton@jcu.edu.au

Introduction

Macroalgae have diverse biomass applications as a source of food and hydrocolloids [1], as fertiliser and soil conditioners [2], and more recently as a targets for a broad range of biofuels [3–6]. The majority of these applications utilise marine macroalgae (seaweed) and no significant production of freshwater macroalgae exists. However, this is likely to change. Demand for biofuels is increasing and there is widespread recognition that a viable biofuels industry must be based around feedstocks that use minimal amounts of freshwater and commercial fertilisers and do not directly compete with food production [7–9]. Macroalgae satisfy all three requirements when cultivated in industrial waste water and their bioenergy potential is favourable (e.g. [6]). Concurrently, as freshwater ecosystems become threatened by industrial pollution and excessive nutrient loading [10], the use of live algae to remove pollutants and excess nutrients from water – algal bioremediation – is receiving increased attention due to the low costs of implementation compared to alternative physico-chemical treatment methods [11] and the ability to directly grow algae in waste waters [12–14]. As most major industries and waste water streams are based around freshwater rather than saltwater (e.g. agriculture, mineral processing, energy production, municipal waste), increasing development of both an algal biofuels industry and algal bioremediation is likely to result in increased cultivation of freshwater macroalgae, supported by concepts derived from a mature seaweed industry.

In contrast to seaweed, target species of freshwater macroalgae for intensive mono-culture are yet to be identified. Several key characteristics are desirable in a target species, irrespective of the biomass application. As most industrial applications and potential end-product uses of macroalgae require large amounts of biomass, it is essential for target species to have high "areal" biomass productivity, expressed as grams of dry weight per unit area (m^2) per time (day) [15,16]. Additionally, species should be able to grow across a wide range of conditions with the aim of year round production in open culture systems and controlled water motion to maximise photosynthetic yields [16,17]. Target species should therefore be competitively dominant to prevent cultures becoming overgrown by nuisance species, a problem that has plagued long-term production of algal monocultures [17]. Finally, low variation in biochemical composition over a range of cultivation conditions is also desirable to ensure a consistent source of biomass for end-product applications. This is particularly the case for biofuel applications, where the productivity of the organic component of the biomass is paramount to bioenergy potential which is typically expressed as the higher heating value in MJ/kg.

Therefore, as a first step to identifying target species of freshwater macroalgae for biomass applications, we compared the productivity, growth and biochemical composition of three species representative of key freshwater macroalgae genera across a range of cultivation conditions representative of intensive culture systems. We then selected a primary target species and assessed its competitive ability against other species over a range of stocking densities. Our overall objective was to identify a freshwater macroalga suitable for large scale cultivation in industrial waste water streams to provide biomass for a range of end-product applications. To do this we focus on filamentous species of freshwater macroalgae from the genera *Cladophora*, *Spirogyra* and *Oedogonium*. These genera were chosen as they all have broad geographic distributions, are representative of the macroalgae available in many freshwater environments, have rapid growth and can become pest species when nutrient levels are high [18,19].

Methods

Study Species

This study compared three types of freshwater macroalgae from the genera *Cladophora*, *Spirogyra* and *Oedogonium* (Fig. 1). *Cladophora* species are branching algae with reasonably large filaments (cell diameter 66–133 µm) that commonly form thick mats and turfs. *Spirogyra* species have intermediate sized unbranched filaments (cell diameter 65–88 µm) and typically form dense floating mats. *Oedogonium* species have very fine unbranched filaments (cell diameter 18–32 µm) and commonly grow attached to aquatic vegetation, but can also form floating mats. Both *Cladophora* and *Spirogyra* are late successional species that are commonly found in established macroalgal communities [20]. Species were identified using taxonomic keys [21,22] and subsequently with DNA sequencing analysis (Supporting information, Text S1). However, identification was only possible to genus level using taxonomic keys as algae lacked species-specific defining characteristics, and DNA sequencing failed to identify unique species (hereafter we refer to genera only: *Cladophora*, *Spirogyra* and *Oedogonium*). For *Oedogonium*, 3 of the 4 most closely related species from DNA sequencing analysis are located in a clade formed by the monoecious taxa (Clade B [23]), suggesting that our *Oedogonium* species also falls within this clade (Table S1). All new genetic sequences were deposited in GenBank (Accession numbers: KC701472, KC701473, KC701474).

Culture Methods

Stock cultures of the three species were collected from outdoor ponds at the Baramundi Fishing Farm Townsville and Good Fortune Bay Fisheries Ltd Kelso. Permission was obtained from owners to collect algae from these sites. Stock cultures were grown in a greenhouse in 60 L plastic buckets with ambient natural light at the Marine and Aquaculture Research Facility Unit, James Cook University. Cultures were provided with aeration by a continuous stream of air entering the cultures through multiple inlets around the base of the buckets. Additional dissolved inorganic carbon was provided to some cultures in the form of CO_2 intermittently pulsed directly into the culture water though an airstone between the hours of 8 am and 4 pm. Culture water was enriched (0.1 g L^{-1}) with MAF growth medium (Manutech Pty Ltd, 13.4% N, 1.4% P), which was non-limiting in nitrogen and phosphorus for our culture system (Text S2, Table S2). Stock cultures were maintained for a period of at least four weeks prior to the start of each experiment to allow acclimation to the culture system and ensure that all algae were pre-exposed to identical conditions. All experimental replicates were maintained in 20 L plastic buckets under the same conditions and ambient light. Water temperature and pH were measured daily in each culture. To simulate environments with low water flow that the algae would likely be grown in if cultured in industrial waste water (e.g. settlement ponds, ash dams), the water in each culture was partially exchanged twice a week at a rate equating to a 10% replacement of the total water volume per day. The species selection and competition experiments were run two months apart.

Species Selection Experiment

To determine which species had the highest growth and productivity under a range of different culture conditions, four replicate cultures of each species were grown with and without CO_2 under each of three aeration treatments (no aeration, low aeration and high aeration). Supplying CO_2 has been shown to significantly increase algal productivity [16,24] as it provides additional dissolved inorganic carbon (DIC), which can become limiting under intensive culture conditions [25,26]. Cultures had an average pH of 8.2 (± 2.0 S.D.) for the CO_2 treatment and 10.5 (± 1.5 S.D.) for the treatment without CO_2. Bottom aeration of

Figure 1. Study species. The three study species - *Cladophora* (A), *Spirogyra* (B) and *Oedogonium* (C).

macroalgae cultures is proposed to increase areal productivity by generating vertical movement and water turbulence within the culture, exposing stock to optimal light and increasing the flow of nutrients around the algal surface [27–29]. Air flow for the low aeration treatment was set as the minimum amount required to keep algae in constant motion (2 L min^{-1}). This flow rate was quadrupled for the high aeration treatment (8 L min^{-1}). To provide a proxy for the relative level of water movement these different aeration rates provided, dissolution rates of gypsum balls in each aeration treatment were measured. Dissolution rates in the high aeration treatment were approximately double those of the low aeration treatment (high aeration: 0.40 g $hour^{-1}$ (± 0.03 S.E), low aeration: 0.21 g $hour^{-1}$ (± 0.05 S.E)), indicating that four times as much airflow is required to double water movement in our system. We used a low and high aeration treatment to generate two levels of water movement as increasing water flow and turbulence can enhance productivity and growth [30,31]. Average water temperature was $27.7°C$ (± 1.6 S.D.) and cultures received an average of 30.9 mol photons m^{-2} day^{-1} (± 3.0 S.D.). Cultures were stocked at a rate of 0.5 g fresh weight (FW) L^{-1} and harvested and weighed after 7 days. Biomass samples were taken from each replicate upon harvesting and dried in an oven at $65°C$ for at least 24 hours to determine fresh weight : dry weight (FW:DW) ratios for each individual replicate for each week of growth. The ash content of each replicate was quantified by combusting a 500 mg subsample of dried biomass at $550°C$ in a muffle furnace until constant weight was reached. Following harvesting, stocking density was reset back to 0.5 g FW L^{-1} by removing excess biomass in each culture. The experiment was run for a total of three weeks, providing for three harvests.

Both ash free dry weight (AFDW) productivity (g AFDW m^{-2} day^{-1}) and specific growth rate (SGR) were calculated for each replicate for each week as each provide different metrics. AFDW productivity is a measure of the amount of organic biomass produced per unit area, whereas SGR provides information on the relative growth rates of individuals within the culture. AFDW was calculated using the equation $P = \{ [(B_f - B_i)/FW{:}DW\,]*(1\text{-}ash)\ \} / A/T$, where B_f and B_I are the final and initial algal biomasses (g), FW:DW is the fresh weight to dry weight ratio, ash is the proportional ash content of the dried biomass, A is the area (m^2) of our culture tanks and T is the number of days in culture. Specific growth rate was calculated using the equation SGR (% day^{-1}) = $Ln(B_f/B_i)/T*100$, where B_f and B_I are the final and initial algal biomasses (g) and T is the number of days in culture. Permutational analyses of variance (PERMANOVAs) were used to analyse the effect of week, species, CO_2 and aeration on AFDW productivity, specific growth rate, FW:DW ratios and ash content (Table S3).

Biomass samples from replicates of each species cultured with and without CO_2 at the high aeration level from week 3 were analysed for carbon, hydrogen, oxygen, nitrogen and sulphur (ultimate analysis) (OEA Laboratories UK). To quantify the suitability of biomass as a potential biofuel the higher heating value (HHV) was calculated for each sample. The HHV is based on the elemental composition of the biomass and is a measure of the amount of energy stored within. The HHV was calculated using the equation $HHV\ (MJ/kg) = 0.3491*C +1.1783*H +0.1005*S -0.1034*O -0.0151*N -0.0211*ash$, where C, H, S, O, N and ash are the carbon, hydrogen, sulphur, oxygen, nitrogen and ash mass percentages of the algae on a dry basis [32].

Competition Experiment

Oedogonium was selected as our target species as it had the highest AFDW culture productivity in five of the six aeration and CO_2 treatment combinations and the most favourable biochemical composition for end-product applications (see Results and Discussion). To investigate the competitive ability of this species, *Oedogonium–Cladophora* and *Oedogonium-Spirogyra* bi-cultures and a polyculture of all three species were grown at each of three different stocking densities (total densities of 0.25 g FW L^{-1}, 0.5 g FW L^{-1}, 1 g FW L^{-1}). Three replicate cultures of each treatment were established with equal quantities of FW biomass of each species summed to each stocking density. Cultures were grown under high aeration with CO_2 as *Oedogonium* AFDW productivity was highest under these conditions in the first experiment (see Results and Discussion). Three replicate *Oedogonium* monocultures were also established at each of the three stocking densities as controls. Cultures had an average pH of 9.7 (± 0.2 S.D.), average water temperature was $30.1°C$ (± 1.8 S.D.), and cultures received an average of 35.5 mol photons m^{-2} day^{-1} (± 3.7 S.D.). Cultures were harvested and weighed after 7 days and a biomass sample was taken from each replicate. Individual FW:DW ratios and ash contents were calculated for each replicate as described above. To estimate the proportional composition of species in all bi-culture and polyculture treatments a biomass sample of 0.4 g FW was sub-sampled from each replicate and suspended in 200 mL dechlorinated water prior to being fixed in Lugols solution (1%). Subsequently, ten replicate sub-samples of each biomass sample were photographed under a dissecting microscope and the proportional species composition calculated by placing a 100-point grid over each photo and summing the number of grid points directly overlying each species. Following harvesting, stocking density was reset back to the original treatment level by removing excess biomass. However, the proportional composition of each species in culture was not reset back to equal levels to quantify the on-going change in species competition (dominance) over time. The experiment was re-run for a further two weeks, providing for a total of three harvests.

Total AFDW productivity was calculated for each replicate for each week as described above. To evaluate competition, specific growth rates were calculated for each replicate for *Oedogonium* only, using the formula above where B_f and B_I are the final and initial biomasses of *Oedogonium* within each culture. B_f was calculated by multiplying the total final FW biomass of each replicate by the proportional composition of *Oedogonium* in that replicate. In week 1 B_I was calculated as half or one third of the total initial biomass stocked into bi-cultures and polycultures respectively; in weeks 2 and 3, B_I was calculated by multiplying the total initial FW biomass by the proportional composition of *Oedogonium* in each replicate in the preceding week. Multivariate PERMANOVAs were used to analyse the effect of competition and density on total AFDW productivity, *Oedogonium* specific growth rates and the proportional composition of *Oedogonium* in bi-cultures and poly-cultures over the three week experiment (Table S4).

Results and Discussion

Species Selection Experiment

Productivity, as determined by AFDW, varied significantly between the three species (Fig. 2a). *Oedogonium* was the most productive species across all treatments when grown under high aeration with CO_2 (8.0 g AFDW m^{-2} day^{-1}) and the productivity of *Oedogonium* was at least 20% greater than that of *Cladophora* and *Spirogyra* in all treatment combinations except when grown with low aeration and no CO_2 (Table S3). In contrast to productivity as measured by AFDW, specific growth rate was highest across all treatments for *Cladophora* when grown under low aeration with CO_2 (17.4% day^{-1}). In all treatment combinations, *Cladophora* growth rates were at least 30% higher than *Oedogonium* and, with the exception of the no aeration treatment, *Spirogyra* growth rates were at least 20% higher (Fig. 2b; Table S3). Striking differences in the relative position of the three species in AFDW productivity compared to specific growth rate were driven by differences in their FW:DW ratios and ash contents. FW:DW ratios varied significantly between species (Fig. 2c; Table S3), with the ratio for *Spirogyra* (7.3±0.22 S.E.) being more than double that of *Oedogonium* (3.4±0.04 S.E.). There were also significant differences in ash content between species (Fig. 2d; Table S3). *Oedogonium* ash contents (3–8%) were less than half those of *Cladophora* (11–16%) and *Spirogyra* (12–19%) in every individual treatment combination. Consequently, despite slower growth rates, *Oedogonium* cultures produced larger amounts of dried ash-free biomass - the critical measure for the majority of end-product applications, particularly bioenergy. Rapid growth rates are often used as one of the key desirable characteristics when assessing the suitability of algae for large scale cultivation [33]. However, as has been shown for other macroalgae species [34], our results demonstrate that fast growth rates are not necessarily equivalent to high productivity, providing support to previous assertions that culture productivities should not be extrapolated from growth rates obtained in controlled experiments [17].

The key biological attributes of *Oedogonium* that contributed to its higher AFDW productivity - lower ash content and lower FW:DW - are also important considerations in the evaluation of feedstocks for biomass applications. For example, a higher water content

(high FW:DW values) means higher inputs are required to obtain dried feedstock, which is necessary if the feedstock is to be transported from point of production to a centralised processing location [35]. Similarly, higher ash contents appear to be correlated with high water contents and may negatively influence bioenergy processes such as hydrothermal liquefaction (HTL) and biogas production due to the concentration of mineral salts at higher levels than traditional lignocellulosic feedstocks [3]. Species differences for bioenergy potential were also reflected in the CHONS analysis and higher heating values (Table 1). *Oedogonium* had the highest carbon content (45%) and correspondingly the best higher heating values ($\sim$20 MJ kg^{-1}). These values are comparable to those recorded for terrestrial energy crops of woody plants (16–23 MJ kg^{-1}) [36–38], confirming that *Oedogonium* biomass has high energy potential and would provide a suitable feedstock for bioenergy applications. Furthermore, the consistently high productivity recorded for *Oedogonium* across a range of conditions (e.g. high/low aeration, with/without CO_2) implies that this species can be reliably grown in a variety of cultivation systems, and is also compatible with industrial waste water streams to provide algal bioremediation (e.g. [13,14]).

Cultivation conditions are clearly important for biomass production as all treatments had variable effects on culture productivity, growth rates, FW:DW ratios and ash content over the three experimental weeks (Table S3). In general, cultures without aeration had lower growth rates and AFDW productivity, and higher ash contents relative to treatments with aeration

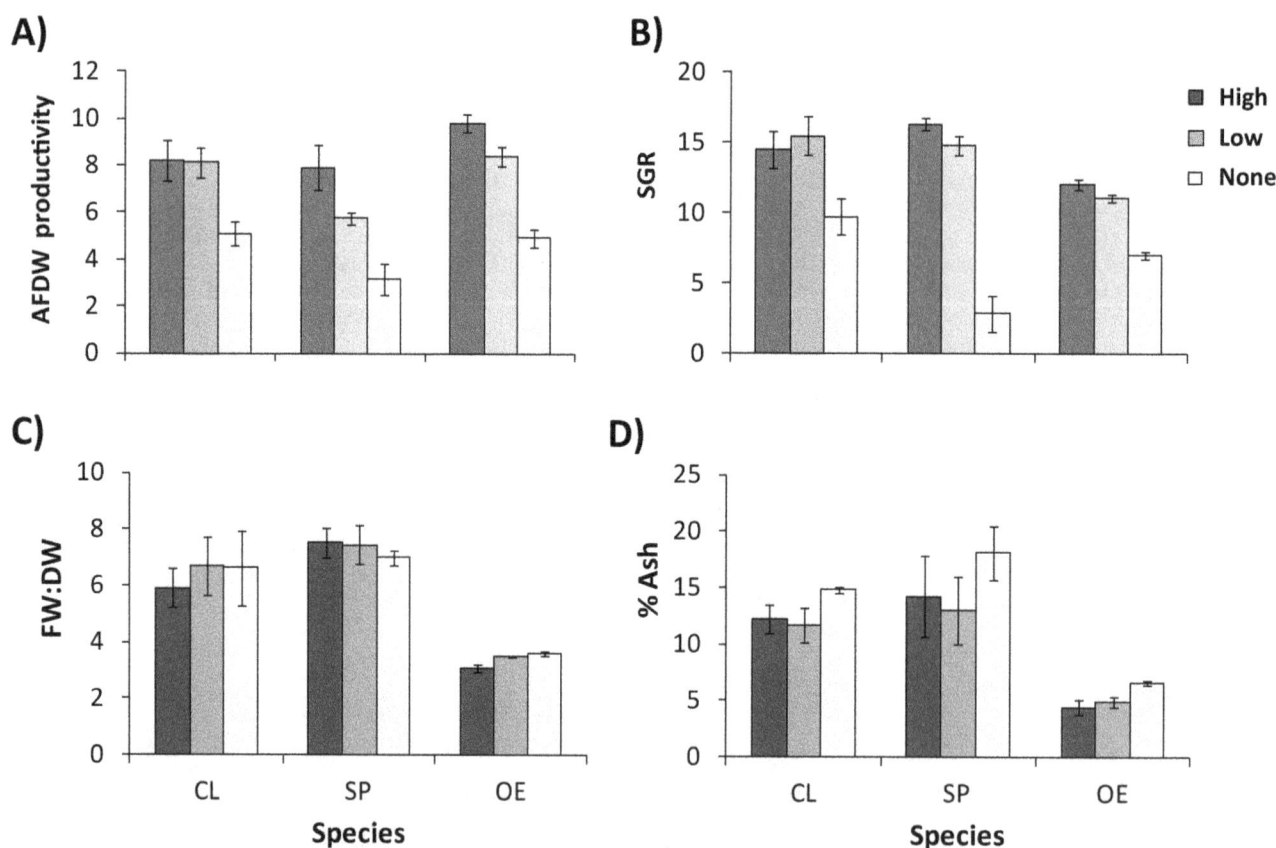

Figure 2. Productivity, growth rates, FW:DW ratios and ash contents of macroalgae cultures. Mean ($\pm$S.E.) ash-free dry weight productivity (g m^{-2} day^{-1}) (A), specific growth rate (SGR, % day) (B), FW:DW ratio (C) and ash content (D) of three macroalgae grown under three aeration levels. CL: *Cladophora*; SP: *Spirogyra*; OE: *Oedogonium*. Data are pooled across CO_2 treatments. Standard errors are calculated as the variation in means between the three weeks of the experiment (n = 3).

Table 1. Ultimate analysis of macroalgae biomass.

Species	CO_2 treatment	Ash	C	H	O	N	S	HHV
Oedogonium	CO_2	2.9 (0.2)	45.3 (0.1)	6.7 (0.1)	38.3 (0.9)	3.5 (0.0)	0.0 (0.0)	19.7 (0.2)
	No CO_2	3.7 (0.5)	45.5 (0.2)	6.9 (0.0)	37.4 (0.6)	3.6 (0.1)	0.1 (0.1)	20.1 (0.1)
Cladophora	CO_2	9.5 (0.7)	43.1 (0.3)	6.2 (0.1)	34.5 (0.9)	4.6 (0.2)	0.3 (0.2)	18.6 (0.2)
	No CO_2	12.1 (2.0)	43.0 (0.5)	6.3 (0.1)	34.3 (1.0)	4.7 (0.1)	0.2 (0.1)	18.6 (0.2)
Spirogyra	CO_2	13.5 (2.1)	42.7 (0.5)	6.3 (0.0)	35.4 (1.2)	4.4 (0.1)	0.0 (0.1)	18.3 (0.4)
	No CO_2	8.7 (0.8)	43.6 (0.1)	6.4 (0.1)	36.8 (0.5)	4.3 (0.1)	0.1 (0.0)	18.7 (0.1)

Ash, ultimate analysis (weight %, on a dry basis) and higher heating value (MJ/kg, on a dry basis) of biomass from three freshwater macroalgae cultured with and without CO_2. Values are means (±S.E.), n = 4, biomass was sampled at the end of the species selection experiment. Note that Cladophora and Spirogyra samples were not pure cultures (see Results and Discussion).

(Figs. 2a,b,d; Table S3). Variation in both FW:DW ratios and ash content was much greater between species than between treatments within each species, and both Cladophora and Spirogyra cultures with high FW:DW ratios consistently had high ash contents (Figs. 2c,d; Table S3). Notably these same species had the highest growth rates and lowest AFDW productivities. In contrast to recent research showing that CO_2 can have pronounced effects on Oedogonium productivity [39], CO_2 had no effect on AFDW productivity or growth rate in the current study (Table S3), suggesting that cultures without additional CO_2 were not limited by the availability of dissolved inorganic carbon. However, as CO_2 was directly bubbled into cultures as a gas and not dissolved in the water, it is also possible that a large proportion of the CO_2 added to cultures was lost to the atmosphere through off gassing [24], resulting in minimal differences in the amount of dissolved inorganic carbon supplied to cultures. Some of the variability in the experiment for Cladophora and Spirogyra was driven by contamination of cultures with other species (predominantly Hydrodictyon species and Stigeoclonium species), resulting from the growth of dormant spores or small contaminant filaments in the biomass when it was first collected. Analysis of the biomass composition at the end of the experiment indicated that contamination was ~80% in Cladophora cultures and ~30% in Spirogyra cultures, inferring that it will be difficult to maintain monocultures of these species over extended periods.

Competition Experiment

In general, the AFDW productivity of mixed species cultures was at least 10% lower than Oedogonium monocultures in the first week of the competition experiment, but there were no differences between cultures in the third week (Fig. 3; Table S4). Changes in culture AFDW productivities between weeks reflect increases in the relative proportions of Oedogonium in bi-cultures and poly-cultures over the course of the three-week experiment (Fig. 4). Although bi-cultures and polycultures were initially stocked with equal proportions of each species, by the end of the third week the proportion of Oedogonium in mixed species cultures was not significantly different (Table S4) and had increased to at least 96% (±7 S.E.) in Oedogonium-Spirogyra bi-cultures, 86% (±16 S.E.) in Oedogonium-Cladophora bi-cultures and 82% (±18 S.E.) in polycultures. These results clearly demonstrate that Oedogonium is competitively dominant and unlikely to become contaminated by other non-target macroalgae species when cultured in "open" systems, providing opportunity for high flow and water exchanges to maximise productivities [30,31].

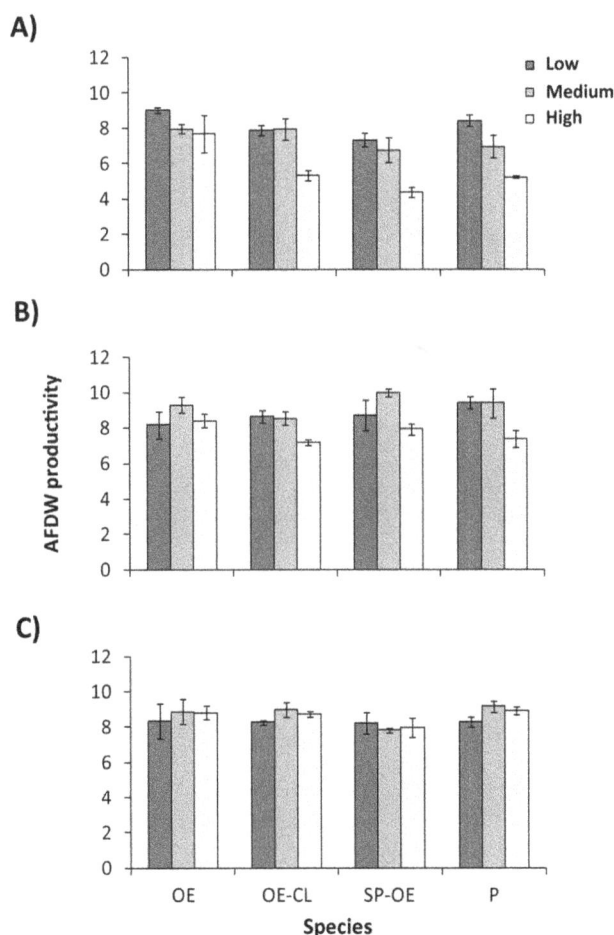

Figure 3. AFDW productivity of mixed species cultures in competition experiment. Mean (±S.E.) total ash free dry weight productivity (g m^{-2} day^{-1}) of monoculture, bi-culture and polyculture combinations of three macroalgae grown under three stocking densities (low, medium, and high) in A) Week 1, B) Week 2 and C) Week 3 of the competition experiment. OE: Oedogonium monoculture (control); CL-OE: Cladophora – Oedogonium bi-culture; SP-OE: Spirogyra – Oedogonium bi-culture; P: Polyculture of all three species.

A)

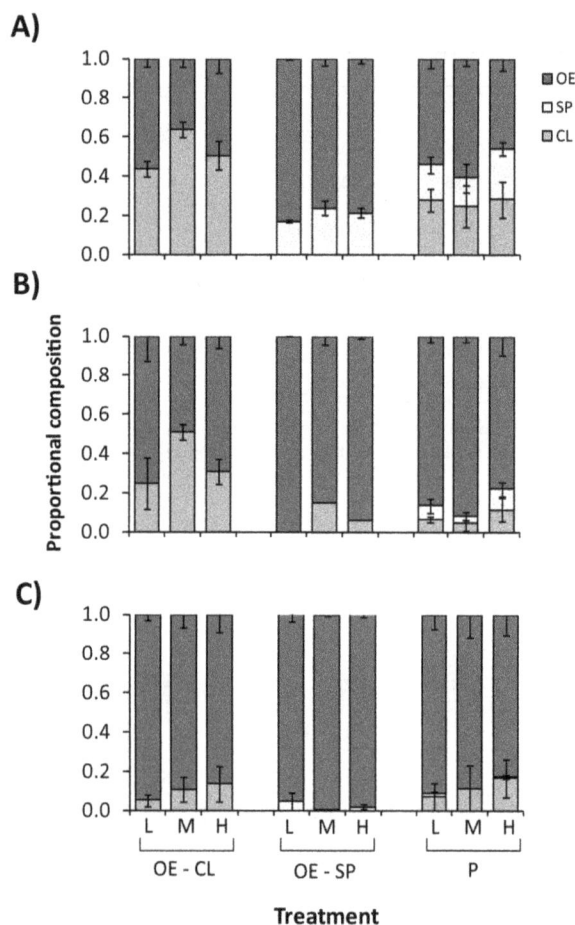

B)

C)

Figure 4. Proportional composition of mixed species cultures. Mean (±S.E.) proportional species composition of bi-culture and polyculture combinations of three macroalgae grown under three stocking densities (low, medium, and high) in A) Week 1, B) Week 2 and C) Week 3 of the competition experiment. Species abbreviations follow Fig. 3.

When selecting algal species for cultivation, fast growth rates are expected to provide a competitive advantage [33]. Yet in contrast to this expectation, the competitively dominant *Oedogonium* had the lowest growth rate of monocultures of all three species in the species selection experiment. However, in the first week of the competition experiment, growth rates of *Oedogonium* were up to 50% higher in mixed species cultures compared to the monoculture. For example, specific growth rates of *Oedogonium* were 12.2% per day (±0.2 S.E.) in the *Spirogyra-Oedogonium* bi-culture under high stocking density, but only 8.1% (±0.8 S.E.) per day in the *Oedogonium* monoculture. These results suggest that *Oedogonium* may increase growth rate as a competitive response to the presence of other species. Regardless, these results demonstrate that inferring competitive abilities based on the growth rates of species in monoculture can be misleading, and likewise inferring bioenergy potential from growth rates could lead to erroneous conclusions about feedstock quality.

The strong competitive response of *Oedogonium* was unaffected by the total stocking density of cultures, with all cultures arriving at greater than 80% *Oedogonium* at the end of the experiment regardless of stocking density treatment (Fig. 3). Similarly, by the third week of the experiment when all mixed species cultures were dominated by *Oedogonium*, stocking density had negligible effects on

AFDW productivity (Fig. 2; Table S4). In contrast, *Oedogonium* growth rates were significantly higher in the low stocking density treatment (23.4% day^{-1}±0.8 S.E.) compared to the medium (16.2% day^{-1}±1.0 S.E.) and high (9.8% day^{-1}±0.8 S.E.) stocking density treatments across all species combinations (Table S4). Macroalgae productivity is generally higher at higher stocking densities [40,41]; although this is not always the case (e.g. [42]) and optimal densities can vary between species [43]. Our results suggest that initially stocking *Oedogonium* cultures at low densities (0.25 g L^{-1}) and harvesting over longer time periods would result in similar productivity to that achieved by stocking cultures at high densities (1 g L^{-1}) and harvesting frequently. This could minimise operational costs associated with harvesting, an important consideration of any aquaculture operation.

Conclusions

For the first time, this study compares the productivity, growth and biochemical composition of freshwater macroalgae in order to identify target species for intensive single species cultivation. *Oedogonium* had the highest AFDW productivity and a consistent biochemical composition, with a high carbon content and bioenergy potential across a range of cultivation conditions. Moreover, *Oedogonium* was competitively dominant in mixed species cultures and quickly overgrew other species within weeks. *Oedogonium* is a cosmopolitan algal genus with a broad geographical distribution. In combination, these factors make *Oedogonium* an ideal freshwater macroalgal target for large-scale production and as a biomass source for bioenergy applications. Our results show that green freshwater macroalgae have much potential for biomass applications but are thus far an under-utilised feedstock. They represent a diverse group of algae for which the greatest opportunity appears to be with small filamentous morphologies, such as *Oedogonium*, that are more cryptic than larger, end succession macroalgae that are apparent in algal blooms (e.g. *Cladophora*, *Spirogyra*).

Supporting Information

Table S1 GenBank accession numbers and results of BLAST searches for *Oedogonium* sequences at four DNA barcode markers.

Table S2 Water nutrient concentrations and productivity of three macroalgae species in nutrient limitation pilot experiments.

Table S3 Results of full factorial permutational analyses of variance (PERMANOVAs) testing the effects of week, species, CO_2 and aeration on productivity as AFDW, specific growth rate, FW:DW ratios and ash content of cultures in the species selection experiment.

Table S4 Results of full factorial multivariate permutational analyses of variance (PERMANOVAs) testing the effects of competition and density on productivity as AFDW, proportional composition of *Oedogonium* and specific growth rate of *Oedogonium* in cultures in the competition experiment.

Text S1 DNA sequencing identification of algae.

Text S2 Pilot experiments to test for nutrient limitation.

Acknowledgments

This research is part of the MBD Energy Research and Development program for Biological Carbon Capture and Storage. We thank Maria Martinez, Amanda Ricketts and Boer Bao for assistance with experiments; Dean Jerry for advice on DNA sequencing; and Good Fortune Bay Fisheries Ltd, Kelso, and the Barramundi Fishing Farm, Townsville, for allowing collection of algae from their ponds.

Author Contributions

Conceived and designed the experiments: RJL RdN NAP. Performed the experiments: RJL. Analyzed the data: RJL RdN NAP. Contributed reagents/materials/analysis tools: RJL RdN NAP. Wrote the paper: RJL RdN NAP. N/A.

References

1. Chopin T, Sawhney M (2009) Seaweeds and their mariculture. The Encyclopedia of Ocean Sciences Elsevier, Oxford: 4477–4487.
2. Bird MI, Wurster CM, De Paula Silva PH, Paul NA, De Nys R (2012) Algal biochar: effects and applications. GCB Bioenergy 4: 61–69.
3. Ross AB, Jones JM, Kubacki ML, Bridgeman T (2008) Classification of macroalgae as fuel and its thermochemical behaviour. Bioresource Technology 99: 6494–6504.
4. Gosch BJ, Magnusson M, Paul NA, Nys R (2012) Total lipid and fatty acid composition of seaweeds for the selection of species for oil-based biofuel and bioproducts. GCB Bioenergy 4: 919–930.
5. Wargacki AJ, Leonard E, Win MN, Regitsky DD, Santos CNS, et al. (2012) An Engineered Microbial Platform for Direct Biofuel Production from Brown Macroalgae. Science 335: 308–313.
6. Rowbotham J, Dyer P, Greenwell H, Theodorou M (2012) Thermochemical processing of macroalgae: a late bloomer in the development of third-generation biofuels? Biofuels 3: 441–461.
7. Nigam PS, Singh A (2011) Production of liquid biofuels from renewable resources. Progress in Energy and Combustion Science 37: 52–68.
8. Pate R, Klise G, Wu B (2011) Resource demand implications for US algae biofuels production scale-up. Applied Energy 88: 3377–3388.
9. Wigmosta MS, Coleman AM, Skaggs RJ, Huesemann MH, Lane LJ (2011) National microalgae biofuel production potential and resource demand. Water Resour Res 47: W00H04.
10. Vorosmarty CJ, McIntyre PB, Gessner MO, Dudgeon D, Prusevich A, et al. (2010) Global threats to human water security and river biodiversity. Nature 467: 555–561.
11. Mehta S, Gaur J (2005) Use of algae for removing heavy metal ions from wastewater: progress and prospects. Critical Reviews in Biotechnology 25: 113–152.
12. Mulbry W, Kondrad S, Pizarro C, Kebede-Westhead E (2008) Treatment of dairy manure effluent using freshwater algae: Algal productivity and recovery of manure nutrients using pilot-scale algal turf scrubbers. Bioresource Technology 99: 8137–8142.
13. Craggs R, Sutherland D, Campbell H (2012) Hectare-scale demonstration of high rate algal ponds for enhanced wastewater treatment and biofuel production. Journal of Applied Phycology 24: 329–337.
14. Saunders RJ, Paul NA, Hu Y, de Nys R (2012) Sustainable sources of biomass for bioremediation of heavy metals in waste water derived from coal-fired power generation. PloS one 7: e36470.
15. Goldman JC, Ryther JH (1975) Mass production of marine algae in outdoor cultures. Nature 254: 594–595.
16. Park J, Craggs R (2011) Algal production in wastewater treatment high rate algal ponds for potential biofuel use. Water Science and Technology 63: 2403–2410.
17. Grobbelaar JU (2010) Microalgal biomass production: challenges and realities. Photosynthesis research 106: 135–144.
18. Francke JA, Den Oude PJ (1983) Growth of *Stigeoclonium* and *Oedogonium* species in artificial ammonium-N and phosphate-P gradients. Aquatic Botany 15: 375–380.
19. Simons J (1994) Field ecology of freshwater macroalgae in pools and ditches, with special attention to eutrophication. Aquatic Ecology 28: 25–33.
20. Cardinale BJ (2011) Biodiversity improves water quality through niche partitioning. Nature 472: 86–89.
21. Entwisle TJ, Skinner S, Lewis SH, Foard HJ (2007) Algae of Australia: Batrachospermales, Thoreales, Oedogoniales and Zygnemaceae. Collingwood, Australia: CSIRO PUBLISHING/Australian Biological Resources Study 200 p.
22. Yee N, Entwisle TJ (2013) ALGKEY website – Interactive Identification of Australian Freshwater Algae. Available: http://203.202.1.217/algkey/. Accessed 2012 October 28.
23. Mei H, Luo W, Liu G, Hu Z (2007) Phylogeny of Oedogoniales (Chlorophyceae, Chlorophyta) inferred from 18 S rDNA sequences with emphasis on the relationships in the genus *Oedogonium* based on ITS-2 sequences. Plant Systematics and Evolution 265: 179–191.
24. Bidwell R, McLachlan J, Lloyd N (1985) Tank Cultivation of Irish Moss, *Chondrus crispus* Stackh. Botanica marina 28: 87–98.
25. Menéndez M, Martínez M, Comín FA (2001) A comparative study of the effect of pH and inorganic carbon resources on the photosynthesis of three floating macroalgae species of a Mediterranean coastal lagoon. Journal of Experimental Marine Biology and Ecology 256: 123–136.
26. Mata L, Silva J, Schuenhoff A, Santos R (2007) Is the tetrasporophyte of *Asparagopsis armata* (Bonnemaisoniales) limited by inorganic carbon in integrated aquaculture? Journal of Phycology 43: 1252–1258.
27. Gonen Y, Kimmel E, Friedlander M (1993) Effect of relative water motion on photosynthetic rate of red alga *Gracilaria conferta*. Hydrobiologia 260: 493–498.
28. Vergara JJ, Sebastian M, Perez-Llorens JL, Hernandez I (1998) Photoacclimation of *Ulva rigida* and *U. rotundata* (Chlorophyta) arranged in canopies. Marine Ecology Progress Series 165: 283–292.
29. Neori A, Chopin T, Troell M, Buschmann AH, Kraemer GP, et al. (2004) Integrated aquaculture: rationale, evolution and state of the art emphasizing seaweed biofiltration in modern mariculture. Aquaculture 231: 361–391.
30. Hurd CL (2000) Water motion, marine macroalgal physiology, and production. Journal of Phycology 36: 453–472.
31. Grobbelaar JU (1994) Turbulence in mass algal cultures and the role of light/dark fluctuations. Journal of Applied Phycology 6: 331–335.
32. Channiwala S, Parikh P (2002) A unified correlation for estimating HHV of solid, liquid and gaseous fuels. Fuel 81: 1051–1063.
33. Borowitzka MA (1992) Algal biotechnology products and processes–matching science and economics. Journal of Applied Phycology 4: 267–279.
34. Lapointe BE, Ryther JH (1978) Some aspects of the growth and yield of *Gracilaria tikvahiae* in culture. Aquaculture 15: 185–193.
35. Richard TL (2010) Challenges in scaling up biofuels infrastructure. Science 329: 793–796.
36. Ebeling J, Jenkins B (1985) Physical and chemical properties of biomass fuels. Transactions of the ASAE (American Society of Agricultural Engineers) 28: 898–902.
37. McKendry P (2002) Energy production from biomass (part 1): overview of biomass. Bioresource Technology 83: 37–46.
38. Cantrell KB, Bauer PJ, Ro KS (2010) Utilization of summer legumes as bioenergy feedstocks. Biomass and Bioenergy 34: 1961–1967.
39. Cole AJ, Mata L, Paul NA, De Nys R (In press) Using CO_2 to enhance carbon capture and biomass applications of freshwater macroalgae. GCB Bioenergy.
40. Nagler PL, Glenn EP, Nelson SG, Napolean S (2003) Effects of fertilization treatment and stocking density on the growth and production of the economic seaweed *Gracilaria parvispora* (Rhodophyta) in cage culture at Molokai, Hawaii. Aquaculture 219: 379–391.
41. Pereira R, Yarish C, Sousa-Pinto I (2006) The influence of stocking density, light and temperature on the growth, production and nutrient removal capacity of *Porphyra dioica* (Bangiales, Rhodophyta). Aquaculture 252: 66–78.
42. Abreu MH, Pereira R, Yarish C, Buschmann AH, Sousa-Pinto I (2011) IMTA with *Gracilaria vermiculophylla*: Productivity and nutrient removal performance of the seaweed in a land-based pilot scale system. Aquaculture 312: 77–87.
43. Mata L, Schuenhoff A, Santos R (2010) A direct comparison of the performance of the seaweed biofilters, *Asparagopsis armata* and *Ulva rigida*. Journal of Applied Phycology 22: 639–644.
44. Saunders GW, Kucera H (2010) An evaluation of rbcL, tufA, UPA, LSU and ITS as DNA barcode markers for the marine green macroalgae. Cryptogamie Algologie 31: 487–528.
45. Hall JD, Fucikova K, Lo C, Lewis LA, Karol KG (2010) An assessment of proposed DNA barcodes in freshwater green algae. Cryptogamie Algologie 31: 529–555.
46. Harrison PJ, Hurd CL (2001) Nutrient physiology of seaweeds: application of concepts to aquaculture. Cahiers de biologie marine 42: 71–82.
47. Anderson MJ, Gorley RN, Clarke KR (2008) PERMANOVA+ for PRIMER: Guide to Software and Statistical Methods. PRIMER-E, Plymouth, UK.

9

Resource Availability and Spatial Heterogeneity Control Bacterial Community Response to Nutrient Enrichment in Lakes

KathiJo Jankowski*, Daniel E. Schindler, M. Claire Horner-Devine

University of Washington, School of Aquatic and Fisheries Sciences, Seattle, Washington, United States of America

Abstract

The diversity and composition of ecological communities often co-vary with ecosystem productivity. However, the relative importance of productivity, or resource abundance, versus the spatial distribution of resources in shaping those ecological patterns is not well understood, particularly for the bacterial communities that underlie most important ecosystem functions. Increasing ecosystem productivity in lakes has been shown to influence the composition and ecology of bacterial communities, but existing work has only evaluated the effect of increasing resource supply and not heterogeneity in how those resources are distributed. We quantified how bacterial communities varied with the trophic status of lakes and whether community responses differed in surface and deep habitats in response to heterogeneity in nutrient resources. Using ARISA fingerprinting, we found that bacterial communities were more abundant, richer, and more distinct among habitats as lake trophic state and vertical heterogeneity in nutrients increased, and that spatial resource variation produced habitat specific responses of bacteria in response to increased productivity. Furthermore, changes in communities in high nutrient lakes were not produced by turnover in community composition but from additional taxa augmenting core bacterial communities found in lower productivity lakes. These data suggests that bacterial community responses to nutrient enrichment in lakes vary spatially and are likely influenced disproportionately by rare taxa.

Editor: Martin Heil, Centro de Investigación y de Estudios Avanzados, Mexico

Funding: University of Washington Keeler Professorship to Daniel Schindler, Seattle ARCS Foundation and EPA STAR Fellowship to KathiJo Jankowski. The funders had no role in study design, data collection and analysis, decision to publish, or preparation of the manuscript.

Competing Interests: The authors have declared that no competing interests exist.

* E-mail: kathijo@u.washington.edu

Introduction

Ecosystem productivity is an important driver of the diversity and composition of ecological communities. Much attention has been given to understanding how communities change with increased productivity, due to the desire to understand how species and their threats are distributed globally [1] and the widespread increase in nutrient enrichment and primary productivity of many ecosystems [2]. Productive ecosystems often support high species richness [3], as evidenced by diversity hotspots in ecosystems such as marine upwelling zones [4] and tend to host distinct communities from low productivity ecosystems. Productivity is thought to promote changes in species richness and composition due to the increased energy available to support the coexistence of multiple species and trophic levels [5,6], as well as by promoting shifts to species that dominate in productive environments. However, productivity is not always a good predictor of species richness [7], and the mechanisms behind observed richness and compositional changes in response to increased ecosystem productivity remain obscure [8].

Spatial or temporal heterogeneity in resource availability can also facilitate the coexistence of species in many environments [9,10], and is commonly used to explain why species richness varies with ecosystem productivity [11–13]. Yet, the relative importance of resource availability and heterogeneity in influencing patterns of species richness and composition in productive

ecosystems remains unclear for many ecological communities [8,14–16], especially for prokaryotes. Bacteria are a fundamental component of food webs and provide the foundation for overall ecosystem functioning, yet we know relatively little about how bacterial communities respond to increases in productivity in most ecosystems [17–19]. In addition, bacteria have unique characteristics, such as metabolic flexibility and dormancy that might make their response to productivity and resource heterogeneity unique. In addition, bacteria can acquire new functional capacities through the exchange of genetic material [20], thus, taxonomic richness may be unresponsive to changes in productivity [21].

Lakes vary widely in productivity and the heterogeneity of resource distribution in response to variation in nutrient loading from human and watershed sources [15,22]. Increased primary production, or trophic status, in lakes is associated with changes in species richness and composition of many ecological communities, including bacteria [23,24]. The richness of macroorganisms often declines at richer trophic state, due to the dominance of phytoplankton that are less palatable or toxic to consumers [25], declines in littoral productivity [26], and changes to the physical and chemical characteristics of the lake environment [22]. Therefore, changes associated with increased lake trophic status often negatively impact diversity of lake communities, change their composition, and lead to the dominance of a few species through homogenization of food resources and reduction in habitat availability [15,27]. Bacterial communities are known to shift in response to increased lake trophic status [28], but the fundamental mechanisms and importance of resource distribution in mediating those changes have not been fully explored.

Nutrient enrichment in lakes tends to magnify the vertical differences in physical and chemical characteristics such as nitrogen (N), phosphorus (P) and dissolved oxygen (DO) among lake strata [22]. However, existing studies of the response of bacterial communities to eutrophication have only evaluated the responses of surface communities or the integrated water column rather than habitat-specific responses [28,29]. In stratified lakes, the surface layer (epilimnion) is typically warm, nutrient-poor, and productive, whereas the deep layer (hypolimnion) is cooler, richer in nutrients, and often low in dissolved oxygen (DO). These differences may be especially important when considering how the response of lake bacteria may differ from eukaryotic communities since vertical differences in physical and chemical conditions are known to structure bacterial communities in stratified lakes [30–32]. For example, low DO in the hypolimnion promotes the use of diverse energy pathways by bacteria such as denitrification and sulfate reduction that are not energetically advantageous in the oxic epilimnion, and therefore, could promote higher diversity of bacterial communities in the entire water column in response to increased trophic status. Therefore, bacterial communities are likely less similar among lake strata in high productivity (eutrophic) than in low productivity (oligotrophic) lakes.

In addition, although several studies have observed changes in bacterial communities with increased lake trophic status [24,29,33], few have identified which type of bacterial taxa are responsible for driving shifts in overall composition [34,35]. For example, while there is increasing evidence that some taxa flourish in high productivity lakes [36], it is unclear whether taxonomic changes result from a complete turnover in the community [37], an increase in the relative abundance of a few key taxa [38], or the increased presence of previously rare or novel taxa that augment a core community of taxa present in low nutrient lakes. For example, a study that evaluated how dominant, common and rare taxa responded to another important disturbance in lakes, lake mixing, found that shifts in the bacterial community were driven by the increased dominance of a few taxa [38].

We evaluated how bacterial abundance, taxonomic richness, and composition changed among and within lakes along a gradient of increasing trophic state. In particular, we quantified the amount of variation in the bacterial response that was explained by trophic state, resource heterogeneity, and their combination. Second, we evaluated whether communities associated with different lake habitats (specifically the epi-, meta- and hypolimnion) responded differently to increased trophic status than communities assessed in the surface layer or integrated at the whole-lake scale. Finally, we evaluated which taxa were responsible for changes in community characteristics; specifically, we asked whether patterns were driven by turnover in the community or by additional taxa augmenting a core community present across all lakes. Thus, in this study we were able to address whether changes in the observed number of taxa and composition of bacterial communities followed the same patterns as eukaryotic communities in response to productivity in lakes and whether the distribution or abundance of resources was more important in shaping those patterns.

Methods

We sampled 21 lakes in the Puget Sound region of western Washington (USA) and southern British Columbia (Canada,

Figure 1) that spanned a large gradient of anthropogenic nutrient loading and productivity [39]. We sampled during the summer-stratified period of July and August 2008. Therefore, our samples reflected the communities that had developed following two to three months of stratified conditions within the water column [31]. As previously described [39], the lakes included in this study were physically similar. Twenty of the 21 lakes were monomictic, and one lake was too shallow to develop thermal stratification. No permissions were required to access 17 of these lakes since they were accessible via a public boat launch. We obtained permission from the University of British Columbia to access the remaining four lakes, which were on the property of their Malcolm Knapp Research Forest. No endangered or protected species were involved in this research.

All bacterial community samples and measurements of lake environmental characteristics were collected over the deepest point in each lake. Water samples for nutrient and chlorophyll a analyses were collected from the epilimnion (surface), metalimnion (thermocline depth), and hypolimnion (within 3 m of the lake bottom) with a van Dorn bottle. Total N (TN) was determined using the perchloric acid digestion method [40] followed by analysis with automated colorimetry on a Lachat autoanalyzer (Lachat Instruments, Loveland, CO, USA). Total P (TP) concentration was determined colorimetrically after persulfate digestion and reaction with molybdate and stannous chloride [40]. Water samples for inorganic N and P determination were pre-filtered through a 0.2 mm Supor filters (Supor-200, Pall Gelman, East Hills, NY) and then analyzed colorimetrically using the same methods as above without a pre-digestion step. Chlorophyll a concentration was determined fluorometrically (Turner Designs, Sunnyvale, California) and used as a surrogate for algal community biomass. Temperature, dissolved oxygen (DO), and pH measurements were taken at 1-m depth intervals with a YSI sonde 6600 (YSI Integrated Systems & Services, Yellow Springs, OH, USA). Other physical lake data such as mean and maximum depth, lake area, and drainage area were obtained from the King County Water and Land Resources Division and the Washington Department of Ecology.

Two water samples for bacterial community analysis were collected from the epilimnion, metalimnion and hypolimnion of each lake with a Van Dorn Bottle. Two 300-mL samples were pooled and bacteria collected on 0.2-μm filters (Supor-200, Pall Gelman, East Hills, NY). Filters were frozen immediately and stored at $-80°C$ until further processing. DNA was extracted from replicate filters using the Qiagen DNEasy Blood and Tissue Mini-kit (Qiagen, Valencia, CA). Samples for bacterial cell enumeration were preserved with 2% formalin, filtered onto a 0.2 μm black polycarbonate filter, stained with 4′, 6-diamidino-2-phenylindole (DAPI), and viewed with a Nikon Eclipse 80i digital microscope at $1000\times$ magnification.

Bacterial community composition and observed richness were assessed using automated ribosomal intergenic spacer analysis [41]. ARISA generates fingerprints of the microbial community based on the length heterogeneity in the intergenic spacer region between the 16S and 23S rRNA genes, which varies among organisms. ARISA has similar limitations as other PCR-based fingerprinting approaches [41] and tends to only survey dominant taxa in a community, thus our assessment of bacterial community composition is really a comparison of the community of dominant taxa among lakes. However, ARISA has been shown to give a robust, high-resolution view of bacterial assemblages in aquatic ecosystems [42,43], to generate results that are consistent with more high resolution techniques [42,44], and can represent species-level taxonomic resolution (98–99% sequence similarity;

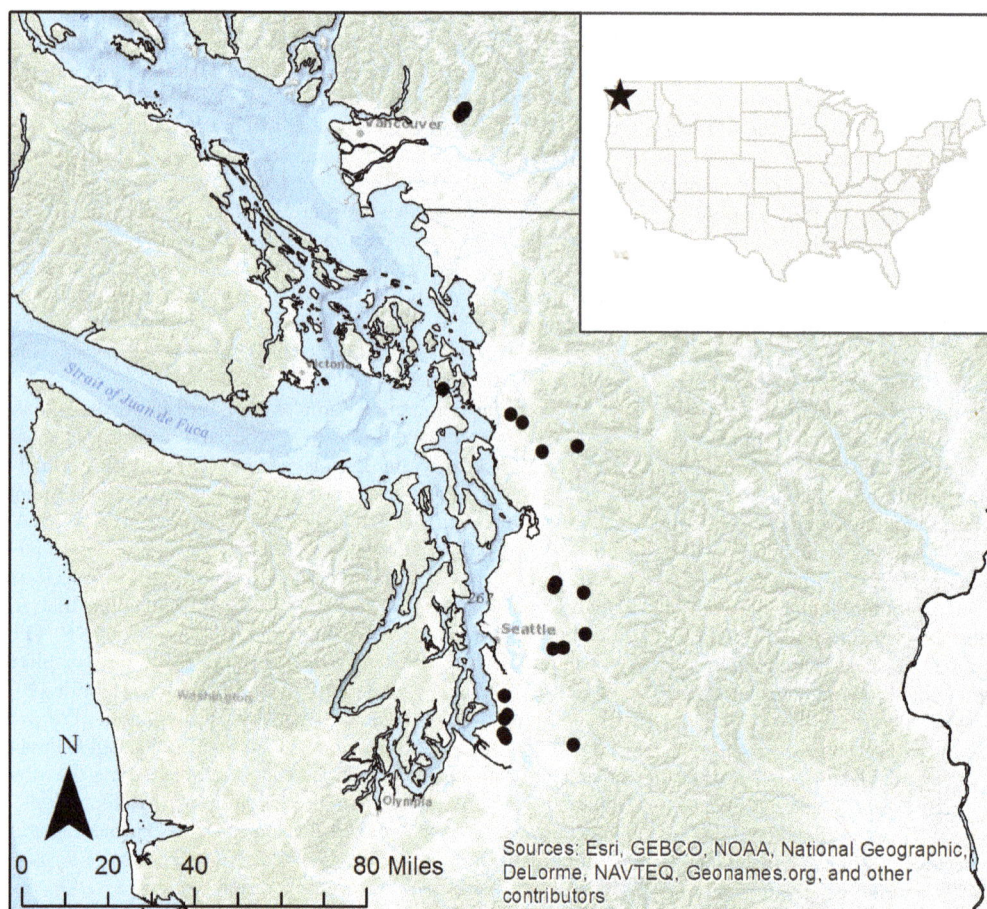

Figure 1. Map of study sites. Lakes included in the study were located in the Puget South Basin in Washington state, USA and British Columbia, Canada. Lakes are indicated by black points.

[42]). The 16S-23S intergenic region was amplified using the polymerase chain reaction (PCR) from the total extracted DNA using 6-FAM-labelled universal 1406-F primer (5′ TGYACA-CACCGCCCGT-3′) and bacterial specific primer 23S-R(5′-GGGTTBCCCCATTCRG-3′) [41,45]. PCRs were conducted on a Mastercycler gradient thermocycler (Eppendorf, New York). PCR products were pooled, quantified, and analyzed on a MegaBACE 1000 automated capillary sequencer (GE Healthcare Corporation, New Jersey). Operational taxonomic units (OTUs) were generated by binning ARISA fragments into successively larger length bins based on their size and eliminating fragments that were <150 and >1300 bp [42]. We used peak area to estimate relative abundance of OTUs in our samples [43], which we considered to be the ratio of the peak area of an OTU in a sample to the total peak area of the sample. We also converted the peak area matrix to presence-absence to assess the composition of bacterial communities in the ARISA profiles by occurrence patterns. We calculated observed richness from ARISA profiles by summing the number of OTUs observed in each sample, hereafter referred to as profile richness. We found no differences in bacterial community patterns using peak height vs. peak area.

Statistical Analyses

We used a principal components analysis (PCA) to summarize physical and chemical variation related to trophic status among lakes. We found that lakes varied little in relevant physical

characteristics (lake area and mean depth), and thus we used the first axis of the resulting PCA as a multivariate proxy for increasing lake trophic state (Figure 2A). Although we did not measure primary productivity directly, other studies have found good agreement between primary productivity and chlorophyll *a* and nutrient concentrations in lakes with similar concentrations as lakes in this study [46].

To quantify vertical heterogeneity in chemical and physical variables within each lake (e.g., TN, TP, temperature, and DO), we used the standard deviation of measurements among lake strata (Table S1). We then performed a PCA that included only the standard deviations of these physical and chemical variables to establish a gradient of resource heterogeneity among lakes.

To compare the influence of increasing trophic state ("trophic status"), depth variation in resource availability ("resource heterogeneity"), and their combination (trophic status and resource heterogeneity) on the bacterial community, we then took the scores from the first principal component (PC 1) of each PCA and regressed them against metrics of bacterial abundance, ARISA profile richness, and an index of community similarity among ARISA profiles (see below for description). The combined trophic status and resource heterogeneity model contained two predictor variables: PC 1 of the trophic status PCA, and PC1 of the resource heterogeneity PCA. All variables were transformed to meet the assumptions of normality prior to the PCA. We evaluated the support for each of the three candidate models describing the

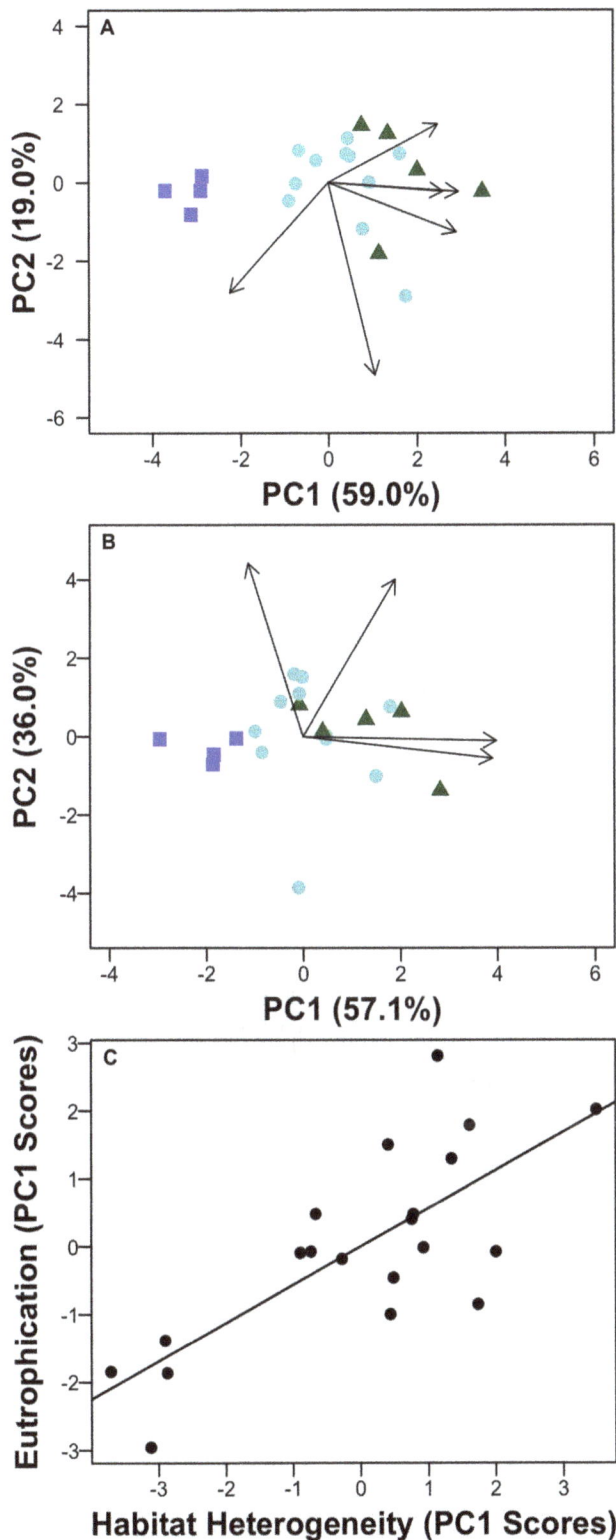

TP, 'HypoTP' = Hypolimnetic TP, 'HypoDO' = hypolimnetic DO, and DO = epilimnetic DO. '.std' indicates that the standard deviation of measurements of the specified variable among layers. Triangles = eutrophic, stars = mesotrophic, and squares = oligotrophic lakes.

relationships between environmental conditions and the bacteria community attributes using Akaike's Information Criteria adjusted for small sample sizes [47]. The model with the lowest AIC_c was considered the best model, and models within 2 AIC_c units of one another were considered to be equally good [47]. In addition, we calculated AIC weights (w_i) for each individual model, which estimates a probability that model i is the best model given the set of models we considered. Finally, to evaluate overall importance of the individual variables, trophic status and resource heterogeneity, we calculated w_i for each term across the three models we compared (Table 1).

Bacterial community similarity among samples was assessed using Sorensen's coefficient for occurrence data [48] and the Chao-Sorensen abundance estimator for relative abundance data [49]. We assessed the overall similarity of communities within a lake by using an average dissimilarity value among ARISA profiles from all two-layer comparisons. We used a constrained analysis of principal coordinates (CAP) to evaluate if changes in community composition were associated with increasing trophic state [50] since it allowed us to use the Chao-Sorensen similarity index.

Finally, we investigated whether changes in the bacterial community with increased lake trophic status were realized by shifts in "widespread" or "narrowly distributed" taxa. We assessed how the relative contribution of widespread OTUs (taxa observed in the majority of lakes) changed with trophic state and habitat heterogeneity. We considered widespread taxa to be those that were observed in 90% of lakes in our study (but see Table S2 for evaluation of different thresholds). We then assessed whether the occurrence and relative abundance of these taxa changed across the lake trophic gradient and with increasing heterogeneity (i.e., PC1 of trophic status and resource heterogeneity PCAs). All analyses were done in R Version 2.14.0 [51] using the vegan package [52].

Results

Lake Characteristics

Productivity-related variables such as TP, TN, and chlorophyll a explained a substantial portion of the environmental variation among lakes in this study (59%; Figure 2, Table S1). Epilimnetic TP concentrations ranged from 4.6 μ g L^{-1} to over 30 μg L^{-1}, and chlorophyll a ranged from 0.23 to 10.2 μg L^{-1}, thus the lakes ranged from oligotrophic to eutrophic [22]. Environmental conditions did not change similarly in each layer with increased trophic state; for example, the epilimnion was less variable among lakes than either the metalimnion or hypolimnion in most environmental characteristics (Table S3). As a result, conditions within the water column were more heterogeneous as trophic state increased ($R^2 = 0.57$, Figure 2C). We observed the most significant differences in TN and TP concentrations among layers as trophic state increased (Figures 2B and C). TN and TP were correlated with the availability of NH$_4$ (r = 0.71) and PO$_4$ (r = 0.86), respectively. Thus, nutrient availability was variable within the water column. Finally, the percent change in nutrient concentrations was greater in the hypolimnion than in either the epi- or metalimnion (Table S3).

Figure 2. Principal component analyses (PCAs) showing environmental variation across lakes in this study. Panel A shows PCA results based on trophic state variables, panel B shows PCA results based on Heterogeneity variables (standard deviation among depths in variables) and panel C shows the correlation between the trophic state PC 1 scores and heterogeneity PC 1 scores ($R^2 = 0.57$). Arrows show significant variables (p<0.05) and values in parentheses show percentages of total environmental variation among lakes explained by each axis. 'EpiTN' = epilimnetic TN, 'EpiTP' = epilimnetic

Table 1. Comparison of models evaluating the effects of trophic state, heterogeneity, and their combination on bacterial communities.

	Model	n	k	R^2	AIC_c	ΔAIC_c	w_i*	w_i of T^+	w_i of H^{++}
ABUNDANCE									
	Trophic State	21	3	0.46	612.2	9.1	0.01	0.34	0.83
	Heterogeneity	21	3	0.66	603.1	0.0	0.50		
	PC1 T+PC1 H	21	4	0.69	603.3	0.0	0.33		
RICHNESS									
	Trophic State	17	3	0.30	129.3	2.9	0.16	0.33	0.84
	Heterogeneity	17	3	0.41	126.4	0.0	0.67		
	PC1 T+PC1 H	17	4	0.40	129.2	2.8	0.17		
DISSIMILARITY									
	Trophic State	17	3	0.12	−23.8	0.9	0.35	0.46	0.65
	Heterogeneity	17	3	0.17	−24.7	0.0	0.54		
	PC1 T+PC1 H	17	4	0.12	−21.5	3.1	0.11		
WIDESPREAD TAXA	Trophic State	18	3	0.29	−36.9	1.5	0.26	0.44	0.74
	Heterogeneity	18	3	0.35	−38.4	0.0	0.56		
	PC1 T+PC1 H	18	4	0.35	−36.1	2.3	0.18		

*AIC_c weight,
+Trophic State,
++Heterogeneity.
Relationship of trophic state, heterogeneity, and their combination with abundance, richness, dissimilarity of bacterial communities among lake habitats ("Dissimilarity"), and the proportional abundance of common taxa. 'n' = sample size and 'k' = number of parameters in each model. 'w_i' refers to the AIC_c weight calculated for each model and the weights for the individual terms 'T' and 'H' across all models.

Did Lake Bacterial Communities Shift in Response to Increasing Trophic State?

Bacterial communities at the whole-lake scale shifted significantly in association with increasing lake trophic state (Figure 3, Figure S1). Average bacterial abundance ($R^2 = 0.46$, Table 1, Figure 2A) and ARISA profile richness increased linearly with our proxy for lake eutrophication ($R^2 = 0.30$, Table 1, Figures 3A & E) and ranged from 66 to 106 OTUs per lake. Our CAP model showed that bacterial community composition shifted with increased lake trophic status and shifts were strongly associated with increasing chlorophyll a ($r = 0.99$) and epilimnetic TN ($r = 0.76$) concentration (Figure S1). The CAP model explained 29% of the total variation in community composition among lakes. The first CAP axis captured the majority of that explained variation (71.4%), indicating that community composition changed in response to increased lake trophic state.

Did the Responses of the Bacterial Community Vary among Habitats in Lakes?

Bacterial communities associated with surface and deep habitats displayed different patterns of abundance, and the richness and composition of ARISA profiles changed significantly as lake trophic state increased (Figure 3, Figure S1). Average bacterial abundance increased with trophic state ($R^2 = 0.46$), was highest in the metalimnion (ANOVA; $F = 5.6$, $p = 0.006$), but increased significantly in all layers across the trophic gradient (Figure 3). The richness of ARISA profiles also varied significantly among layers (ANOVA, $F = 10.6$, $p < 0.001$), but only increased notably in the hypolimnion in response to trophic state ($R^2 = 0.15$, Figure 3H). Profile richness was highest on average in the hypolimnion (55 ± 7 SD), which also had the largest range of observed richness,

ranging from 37 OTUs in Gwendoline Lake, an oligotrophic lake, to 71 OTUs found in more nutrient-rich Geneva Lake. Therefore, increases in the profile richness in the hypolimnion accounted for the increases we observed in overall lake richness ($R^2 = 0.39$).

When all lake communities were considered together, there were significant, but small, compositional differences among epi-, meta- and hypolimnetic communities (ANOSIM, $R = 0.16$, $p = 0.001$), and surface and deep communities shared the fewest taxa (data not shown). Furthermore, surface communities were significantly less variable than deep communities across the trophic gradient (Homogeneity of dispersion, $p < 0.001$), and surface and deep communities within a given lake tended to become less similar to one another as trophic state increased ($R^2 = 0.12$). However, heterogeneity in nutrient concentrations among strata explained slightly more of that variation than trophic state alone ($R^2 = 0.17$, Figure 4, Table 1).

In all cases, the heterogeneity model or the trophic status plus heterogeneity model explained more variation in bacterial communities among lakes than the trophic status model alone (Table 1). We found that greater vertical heterogeneity of nutrient availability (Figure 2) was strongly related to increased abundance, ARISA profile richness, and decreasing similarity of communities among lake strata (Table 1, Figure 4). Total abundance and observed richness were both more strongly related to increases in the heterogeneity of P and N than increases in their concentrations alone or to differences in temperature and DO among strata, which other studies have shown to be associated with heterogeneity in bacterial community composition (Table S1; [31,32]). Therefore, although we saw an increase in observed richness with increased trophic state ($R^2 = 0.30$), observed richness was more closely linked to greater heterogeneity of nutrients within the water column ($R^2 = 0.41$). Overall, the AIC_c shows that the heteroge-

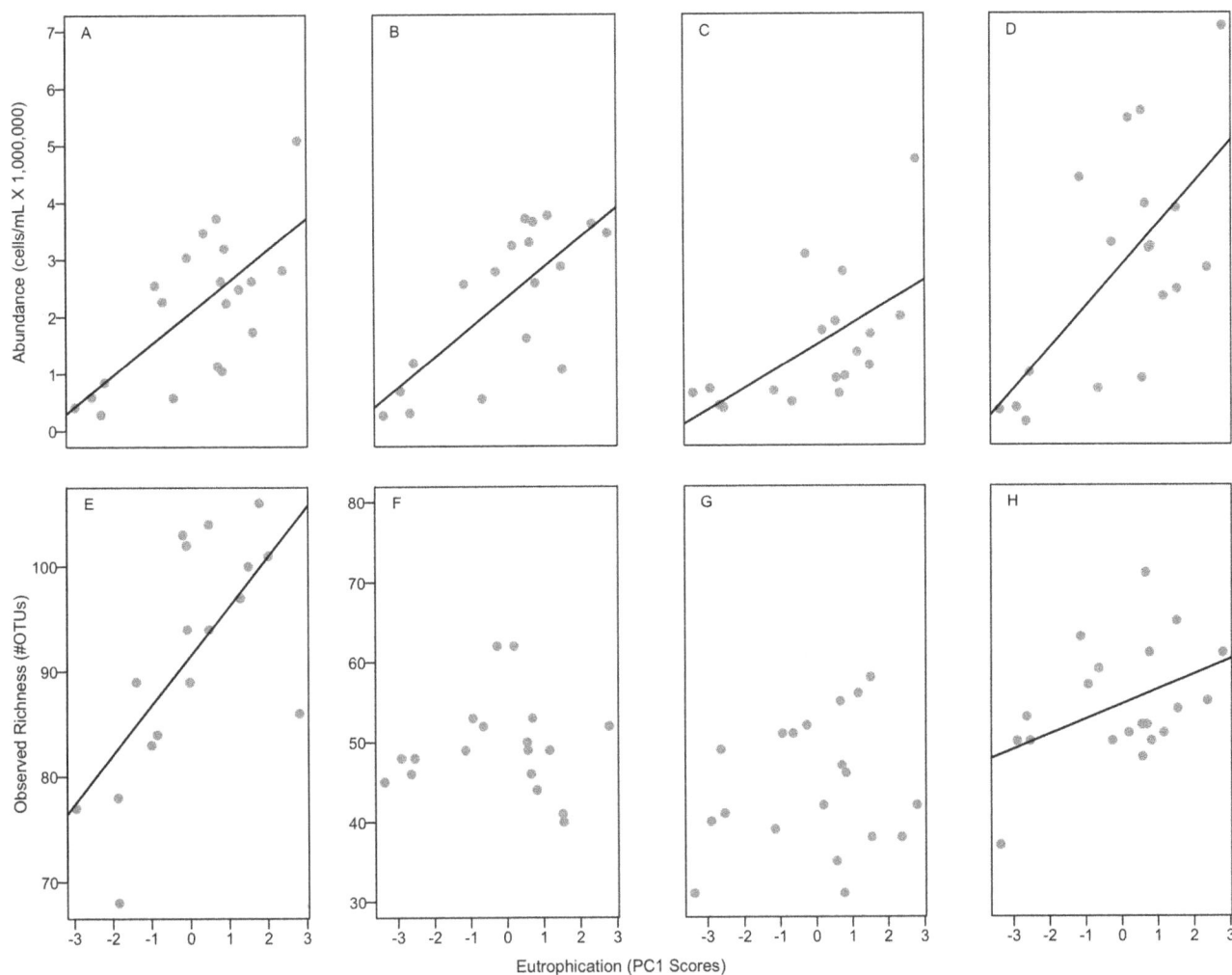

Figure 3. Relationships of increasing trophic state (PC 1 scores) with abundance and richness. A) Whole-lake average abundance ($R^2 = 0.46$, $p<0.001$), B) Epilimnetic abundance ($R^2 = 0.52$, $p<0.001$), C) Metalimnetic abundance ($R^2 = 0.33$, $p = 0.007$), D) Hypolimnetic abundance ($R^2 = 0.40$, $p = 0.003$), E) Whole-lake average richness ($R^2 = 0.30$, $p = 0.003$), F) Epilimnetic richness ($R^2 = -0.06$, $p = 0.98$), G) Metalimnetic richness ($R^2 = -0.03$, $p = 0.49$), and H) Hypolimnetic richness ($R^2 = 0.15$, $p = 0.05$).

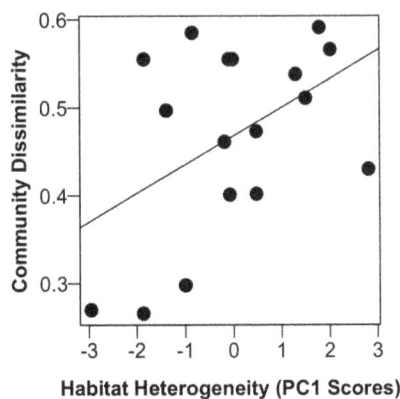

Figure 4. Bacterial communities were less similar among lake strata as chemical heterogeneity increased ($R^2 = 0.17$, $p = 0.05$). Similarity of communities was based on the average Chao Sorensen Dissimilarity from comparisons of relative abundance of taxa among three depth strata.

neity model had the most support, and that, in fact, the heterogeneity term had the most weight across models (Table 1). Thus, as measured here, bacterial communities exhibited habitat-specific responses to lake eutrophication, and spatial variation in resource availability often influenced bacterial community composition more than simple increases in nutrient concentration and productivity (Figures 4 and 5).

Which Taxa Accounted for changes in Community Composition with Eutrophication?

We observed a total of 221 OTUs across all lakes and found that some of those taxa were widespread among lakes. We observed that while a core community of 11 OTUs was present and detected using ARISA in ~90% of lakes in this study ("widespread taxa"), and while still present across lakes, made up a decreasing proportion of both the number ($R^2 = 0.23$) and relative abundance of taxa in lakes as lake trophic state increased ($R^2 = 0.35$, Figure 5). Specifically, while these widespread taxa comprised 35–40% of the relative abundance of the community in the more homogenous oligotrophic lakes, they were less than 15% in the more heterogeneous eutrophic lakes. These trends were

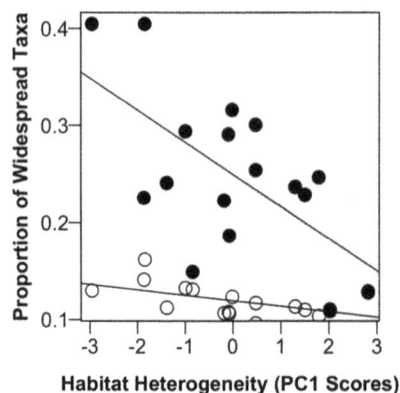

Figure 5. Relationship of habitat heterogeneity with the proportional representation of widespread taxa in each lake. Heterogeneity was measured as the PC 1 scores of 'heterogeneity' model. Filled circles show the total proportion of relative abundance made up by widespread taxa ($R^2 = 0.35$, $p = 0.007$) and open circles show widespread taxa as a proportion of the total number of taxa observed ($R^2 = 0.23$, $p = 0.03$).

robust to using different thresholds to define "widespread" (e.g., present in 70–90% of lakes), and strengthened as we considered thresholds up to 90% (Table S2). We only observed four OTUs that were present in >90% of lakes, which likely were ubiquitous taxa that would be present regardless of lake trophic state [36]. In addition, we found that the increasing representation of previously low abundance or new taxa in response to trophic status was explained most by increasing heterogeneity among habitats (Table 1). Although the AIC_c values for all three models were very close, the AIC value and variable w_i's suggest that heterogeneity explained the declining contribution of the widespread OTUs we observed to the overall lake community (Table 1). This suggests that either novel, previously low abundance, (i.e., below the detection limit of ARISA), or dormant taxa [35] increased in their relative importance as nutrient status of lakes increased, and that chemical heterogeneity in the lake environment most likely facilitated the increased prevalence of these taxa in the lake community.

Discussion

We found that the abundance and richness of bacterial communities increased as lake nutrient status increased, in parallel to what has been reported in other studies [28,33,45]. However, we found even stronger relationships between the overall abundance and number of OTUs observed and the spatial heterogeneity in nutrient conditions among lake layers as trophic status increased. Additionally, because environmental conditions among lake strata diverged and communities associated with these habitats responded differently to lake trophic status (Figure 3), bacterial communities as measured by ARISA fingerprinting were much less similar among lake strata as lake trophic state increased (Figure 4, Table S3). Furthermore, our analyses demonstrated that a core bacterial community of dominant bacteria did not change systematically in high nutrient lakes but rather that increasing habitat heterogeneity, specifically due to large changes in conditions in the deepest layer of lakes, provided additional habitat for previously low abundance or absent taxa that became detectable in more eutrophic conditions. Our study shows that the response of bacterial communities to increased productivity in lakes may differ from that of other lake organisms as a result of

spatial heterogeneity in resources specifically affecting the ecology of bacteria and supports the idea that the recruitment of rare or dormant bacterial taxa in lake communities may facilitate much of this response (e.g., [35]).

Effects of Trophic State on Lake Bacterial Communities

We observed significant increases in bacterial abundance and changes in the richness and composition of ARISA profiles at the whole-lake scale in response to increases in lake trophic status. As expected, abundance was positively related to increases in nutrient availability (TN and TP) and chlorophyll a (Table 1) across the entire range of ecosystem productivity we observed. High abundances of bacteria are often correlated with high rates of bacterial productivity [46], which can be nutrient-limited in oligotrophic lakes. We also observed increased ARISA profile richness as trophic state increased (Figure 3E), suggesting that more taxa coexisted as dominant types (and thus able to be detected by ARISA) as resource availability increased. As has been observed for macroorganisms in lakes [23], other studies have shown lake bacterial richness to exhibit a range of linear and unimodal responses to productivity depending on the spatial and taxonomic resolution of the study [17,18,29]. We likely observed a linear increase in observed richness since our lakes represented a relatively modest gradient in nutrient loading and chlorophyll a compared to the global distribution of lake productivity [22], thus, we may have only captured the initial upward slope of a unimodal response across the global range of lake trophic states. Alternatively, is also possible that we observed a consistently increasing number of taxa rather than a unimodal trend due an effect of increased sampling; in other words, because we took discrete samples from each lake habitat and ran a separate ARISA analysis on each sample rather than on one integrated sample from each lake, we were able to detect more taxa than previous studies [23,29].

Bacterial Community Response to Resource Heterogeneity

Bacterial community shifts were most strongly associated with changes in the heterogeneity of the lake environment in all cases (Table 1). Specifically, we observed more OTUs and decreased similarity among bacterial communities in surface and deep habitats in response to heterogeneity in N and P concentrations (Figures 2B and 3). Environmental conditions within lakes changed to different degrees among the epi-, meta- and hypolimnion (Table S3), which translated to habitat specific responses of bacterial communities in each layer across lakes (Figure 3). For instance, although bacterial abundance increased in all layers, we only observed increases in richness in the hypolimnion with increased trophic status (Figure 3H), which also had the largest range in nutrient concentrations across the trophic gradient (Figure S1). Interestingly, communities in the surface layer varied less across the trophic gradient than communities in the hypolimnion, suggesting that studies that evaluate trends in bacterial communities in response to eutrophication miss an important aspect of the bacterial response.

While our study design does not allow us to definitely conclude whether patterns in bacterial communities were related more to increased trophic state or to habitat heterogeneity because the two were correlated (Figure 2C), we found strong support for our heterogeneity model (Table 1) and patterns suggesting that both were important for bacterial communities. For instance, we observed lower richness (66 OTUs) and higher similarity of communities among lake strata (Sorensen similarity = 0.86) in the only lake in our study that did not stratify into discrete habitats.

Although this lake had similar nutrient concentrations as eutrophic lakes, environmental conditions were more homogeneous within the water column in this lake suggesting that heterogeneity in resource availability influenced bacterial richness more than nutrient concentration alone in all lakes in our study.

While changes in bacterial communities in response to increasing lake trophic status have been widely observed [28,29,45] as have differences in communities among lake strata [30,31,53] previous studies have not linked these observations to evaluate the combined effects of lake resource availability and resource heterogeneity on the richness and composition of lake bacterial communities. Thus, while similar communities may inhabit eutrophic lakes [36], communities become increasingly distinct from one another as surface and deep habitat conditions diverge and biotic interactions change [54]. Furthermore, these findings contrast with studies of other lake organisms, such as zooplankton and fish, which find increased productivity reduces diversity as a result of homogenization of food resources and loss of habitat [15,26,27]. Thus, our study demonstrates that the type of heterogeneity that influences communities varies among macro- and microorganisms and that the response to a large-scale environmental change is expressed differently among habitats within ecosystems.

In addition, our results suggest that changes in bacterial communities with increased trophic status may be more strongly related to vertical differences in nutrient concentrations than dissolved oxygen, temperature, and light, which are typically thought to structure differences among communities in different strata within lakes [24,30,31,53]. We tested this by regressing axis 1 and 2 of our environmental PCA (Figure S1), which represented differences among layers in nutrient concentrations and DO/ temperature, respectively. We found that in all cases axis 2 (variation in DO and temperature) did not have a strong relationship with any bacterial response measure. Few studies have tested for this and those that have, have often focused on a single lake or small set of lakes [30,53]. Thus, these results suggest that increased vertical differences in nutrient concentrations may be more important in structuring the bacterial community than vertical differences in DO and temperature as the trophic state of lakes increases.

Were Widespread or Narrowly Distributed Taxa Responsible for Shifts in Community Composition?

We found that habitat heterogeneity played an important role in shaping the bacterial response to increasing lake trophic status through enabling the increased contribution of new or previously low abundance taxa in lake communities. Furthermore, we found that increased resource heterogeneity simultaneously allowed for the retention of a core group of taxa that were widespread among lakes while also providing new habitat and resources for these previously unobserved or low abundance taxa (Figure 5). Thus, our results support the general notion of the importance of rare taxa in microbial communities [35,55] and suggest that bacteri- ally-relevant habitat heterogeneity may be an important mecha- nism driving the bacterial response to increased lake ecosystem productivity. Additionally, this pattern may be common to many types of planktonic communities [34]. For instance, a study of the response of a lake phytoplankton community to eutrophication and recovery found that there were phytoplankton that were consistently present through time, but that temporally rare taxa were most responsive to changes in lake nutrient status and drove changes in community composition [56].

While the use of ARISA allowed us to screen the bacterial community in a large number of lakes and has been shown to have species-level taxonomic resolution, as with so many methods used to sample bacterial communities, there are limitations associated with using this approach [57]. For example, ARISA underesti- mates the total richness of the bacterial community and has biases such as preferential amplification of abundant organisms [41]. For example, many of the additional taxa that we observed in more eutrophic lakes could have been present at low abundances in oligotrophic lakes, and therefore below the detection limit of ARISA. However, increased detection of these taxa in eutrophic lakes suggests that they are at higher abundances and thus may be more functionally important in those communities. In addition, other studies have shown that ARISA captures similar patterns in diversity among communities as more high-resolution techniques such as clone libraries [58]. Further, although sequencing and clone library techniques would have allowed us to identify specific taxa, the higher costs associated with those techniques would have limited our ability to sample the entire trophic gradient in our study. Finally, our results are comparable to other studies using similar techniques [24,33,35], and recent studies of bacterial responses to nutrient additions that have used high throughput sequencing techniques suggest that using a more thorough sampling approach likely would not likely reveal a different trend in how richness and composition respond to increased ecosystem productivity (e.g., [19,44]). Thus, the fact that we observed such striking trends using this approach suggests that higher resolution sampling would have only strengthened observed patterns.

In summary, we showed that bacterial community composition changed and was richer and more heterogeneous within lakes as trophic status increased. In contrast to trends in macroorganisms whose diversity is often negatively associated with increases in lake productivity [15,23,27], we showed that the high degree of heterogeneity in bacterial resources in eutrophic lakes promoted higher richness as a result of differentiation of bacterial taxa among lake habitats. We found that eutrophication alters the drivers of bacterial community differences within lakes from physical and redox related variables to changes in nutrient availability. Furthermore, our results suggest that rare or dormant taxa may be most responsible for changes in bacterial communities with increased lake trophic state [35,59]. This "seed bank" of taxa has increasingly been recognized to be important in responding to changes in many types of ecosystems [60], and understanding the role of rare and dormant taxa an important frontier for understanding the processes that regulate how microbial commu- nities respond to ecosystem change in general.

Supporting Information

Figure S1　Constrained analysis of principal coordinates (CAP) of bacterial community composition with envi- ronmental variables.

Table S1　Results of eutrophication and heterogeneity Principal Components Analyses (PCAs).

Table S2　Model results from comparing effects of trophic state, heterogeneity, and the combination of the two (T+H) on the relative abundance of widespread taxa in lakes.

Table S3　Means and variation of key environmental variables among and within lakes.

Acknowledgments

The authors would like to acknowledge M. Dyen, H. Bekris, and A. Coogan for extensive help with field data collection and laboratory analyses. We are grateful to R. Lange, J. Griffiths, J. Armstrong and two anonymous reviewers for helpful reviews on the manuscript.

Author Contributions

Conceived and designed the experiments: KJJ DES MCHD. Analyzed the data: KJJ. Contributed reagents/materials/analysis tools: DES MCHD. Wrote the paper: KJJ DES MCHD.

References

1. Reid WV, Mooney HA, Cropper A, Capistrano D, Carpenter SR, et al. (2005) Ecosystems and human well-being: synthesis. Washington, D.C.: Island Press.
2. Carpenter SR, Caraco NF, Correll DL, Howarth RW, Sharpley AN, et al. (1998) Nonpoint pollution of surface waters with phosphorus and nitrogen. Ecological Applications 8: 559–568.
3. Rozenzweig ML (1995) Species diversity in space and time. Cambridge, U.K.: Cambridge University Press.
4. Ribalet F, Marchetti A, Hubbard KA, Brown K, Durkin CA, et al. (2010) Unveiling a phytoplankton hotspot at a narrow boundary between coastal and offshore waters. Proceedings of the National Academy of Sciences of the United States of America 107: 16571–16576.
5. Wright DH (1983) Species-energy theory - An extension of species-area theory. Oikos 41: 496–506.
6. Mittelbach GG, Steiner CF, Scheiner SM, Gross KL, Reynolds HL, et al. (2001) What is the observed relationship between species richness and productivity? Ecology 82: 2381–2396.
7. Adler PB, Seabloom EW, Borer ET, Hillebrand H, Hautier Y, et al. (2011) Productivity Is a Poor Predictor of Plant Species Richness. Science 333: 1750–1753.
8. Cardinale BJ, Hillebrand H, Harpole WS, Gross K, Ptacnik R (2009) Separating the influence of resource 'availability' from resource 'imbalance' on productivity-diversity relationships. Ecology Letters 12: 475–487.
9. Hutchinson GE (1961) The paradox of the plankton. American Naturalist 95: 137–145.
10. Tilman D (1982) Resource Competition and Community Structure. Princeton, New Jersey: Princeton University Press.
11. Chase JM, Leibold MA (2002) Spatial scale dictates the productivity-biodiversity relationship. Nature 416: 427–430.
12. Shurin JB, Winder M, Adrian R, Keller W, Matthews B, et al. (2010) Environmental stability and lake zooplankton diversity - contrasting effects of chemical and thermal variability. Ecology Letters 13: 453–463.
13. Reich PB, Frelich LE, Voldseth RA, Bakken P, Adair EC (2012) Understorey diversity in southern boreal forests is regulated by productivity and its indirect impacts on resource availability and heterogeneity. Journal of Ecology 100: 539–545.
14. Stevens MHH, Carson WP (2002) Resource quantity, not resource heterogeneity, maintains plant diversity. Ecology Letters 5: 420–426.
15. Barnett A, Beisner BE (2007) Zooplankton biodiversity and lake trophic state: Explanations invoking resource abundance and distribution. Ecology 88: 1675–1686.
16. Eilts JA, Mittelbach GG, Reynolds HL, Gross KL (2011) Resource Heterogeneity, Soil Fertility, and Species Diversity: Effects of Clonal Species on Plant Communities. American Naturalist 177: 574–588.
17. Horner-Devine MC, Leibold MA, Smith VH, Bohannan BJM (2003) Bacterial diversity patterns along a gradient of primary productivity. Ecology Letters 6: 613–622.
18. Smith VH (2007) Microbial diversity-productivity relationships in aquatic ecosystems. Fems Microbiology Ecology 62: 181–186.
19. Bowen JL, Ward BB, Morrison HG, Hobbie JE, Valiela I, et al. (2011) Microbial community composition in sediments resists perturbation by nutrient enrichment. Isme Journal 5: 1540–1548.
20. Madigan MT, Martinko JM, Stahl DA, Clark DP (2010) Brock biology of microorganisms. San Francisco: Benjamin Cummings.
21. Green JL, Bohannan BJM, Whitaker RJ (2008) Microbial biogeography: From taxonomy to traits. Science 320: 1039–1043.
22. Wetzel RC (2001) Limnology: lake and river ecosystems. San Diego: Academic Press.
23. Dodson SI, Arnott SE, Cottingham KL (2000) The relationship in lake communities between primary productivity and species richness. Ecology 81: 2662–2679.
24. Longmuir A, Shurin JB, Clasen JL (2007) Independent gradients of producer, consumer, and microbial diversity in Lake Plankton. Ecology 88: 1663–1674.
25. Smith VH (2003) Eutrophication of freshwater and coastal marine ecosystems - A global problem. Environmental Science and Pollution Research 10: 126–139.
26. Vadeboncoeur Y, Jeppesen E, Vander Zanden MJ, Schierup HH, Christoffersen K, et al. (2003) From Greenland to green lakes: Cultural eutrophication and the loss of benthic pathways in lakes. Limnology and Oceanography 48: 1408–1418.
27. Vonlanthen P, Bittner D, Hudson AG, Young KA, Muller R, et al. (2012) Eutrophication causes speciation reversal in whitefish adaptive radiations. Nature 482: 357–U1500.
28. Kolmonen E, Haukka K, Rantala-Ylinen A, Rajaniemi-Wacklin P, Lepisto L, et al. (2011) Bacterioplankton community composition in 67 Finnish lakes differs according to trophic status. Aquatic Microbial Ecology 62: 241–U249.
29. Korhonen JJ, Wang JJ, Soininen J (2011) Productivity-Diversity Relationships in Lake Plankton Communities. Plos One 6.
30. De Wever A, Muylaert K, Van der Gucht K, Pirlot S, Cocquyt C, et al. (2005) Bacterial community composition in Lake Tanganyika: Vertical and horizontal heterogeneity. Applied and Environmental Microbiology 71: 5029–5037.
31. Shade A, Jones SE, McMahon KD (2008) The influence of habitat heterogeneity on freshwater bacterial community composition and dynamics. Environmental Microbiology 10: 1057–1067.
32. Garcia SL, Salka I, Grossart HP, Warnecke F (2013) Depth-discrete profiles of bacterial communities reveal pronounced spatio-temporal dynamics related to lake stratification. Environ Microbiol Rep 5: 549–555.
33. Lindstrom ES (2000) Bacterioplankton community composition in five lakes differing in trophic status and humic content. Microbial Ecology 40: 104–113.
34. Galand PE, Casamayor EO, Kirchman DL, Lovejoy C, Karl DM (2009) Ecology of the rare microbial biosphere of the Arctic Ocean.. Proceedings of the National Academy of Sciences of the United States of America 106: 22427–22432.
35. Jones SE, Lennon JT (2010) Dormancy contributes to the maintenance of microbial diversity. Proceedings of the National Academy of Sciences of the United States of America 107: 5881–5886.
36. Newton RJ, Jones SE, Eiler A, McMahon KD, Bertilsson S (2011) A Guide to the Natural History of Freshwater Lake Bacteria. Microbiology and Molecular Biology Reviews 75: 14–49.
37. Bell T, Bonsall MB, Buckling A, Whiteley AS, Goodall T, et al. (2010) Protists have divergent effects on bacterial diversity along a productivity gradient. Biology Letters 6: 639–642.
38. Shade A, Chiu CY, McMahon KD (2010) Seasonal and Episodic Lake Mixing Stimulate Differential Planktonic Bacterial Dynamics. Microbial Ecology 59: 546–554.
39. Jankowski KJ, Schindler DE, Holtgrieve GW (2012) Assessing non-point source nitrogen loading and nitrogen fixation in lakes using $\delta 15N$ and nutrient stoichiometry. Limnology and Oceanography 57: 671–683.
40. Eaton A, Greenberg AE, Rice EW (2005) Standard Methods for the Examination of Water and Wastewater: American Public Health Association.
41. Fisher MM, Triplett EW (1999) Automated approach for ribosomal intergenic spacer analysis of microbial diversity and its application to freshwater bacterial communities. Applied and Environmental Microbiology 65: 4630–4636.
42. Brown MV, Schwalbach MS, Hewson I, Fuhrman JA (2005) Coupling 16S-ITS rDNA clone libraries and automated ribosomal intergenic spacer analysis to show marine microbial diversity: development and application to a time series. Environmental Microbiology 7: 1466–1479.
43. Yannarell AC, Triplett EW (2005) Geographic and environmental sources of variation in lake bacterial community composition. Applied and Environmental Microbiology 71: 227–239.
44. Logue JB, Langenheder S, Andersson AF, Bertilsson S, Drakare S, et al. (2012) Freshwater bacterioplankton richness in oligotrophic lakes depends on nutrient availability rather than on species-area relationships. ISME J 6: 1127–1136.
45. Yannarell AC, Kent AD, Lauster GH, Kratz TK, Triplett EW (2003) Temporal patterns in bacterial communities in three temperate lakes of different trophic status. Microbial Ecology 46: 391–405.
46. Pace ML, Cole JJ (1994) Comparative and experimental approaches to top-down and bottom-up regulation of bacteria. Microbial Ecology 28: 181–193.
47. Burnham KP, Anderson DR (2002) Model selection and multi-model inference: A practical information-theoretic approach. New York: Springer-Verlag.
48. Legendre P, Legendre L (1998) Numerical Ecology. Amsterdam: Elsevier Science B.V.
49. Chao A, Chazdon RL, Colwell RK, Shen TJ (2005) A new statistical approach for assessing similarity of species composition with incidence and abundance data. Ecology Letters 8: 148–159.
50. Legendre P, Anderson MJ (1999) Distance-based redundancy analysis: Testing multispecies responses in multifactorial ecological experiments. Ecological Monographs 69: 1–24.
51. R Core Team (2012) R: A language and environment for statistical computing. Vienna, Austria: R Foundation for Statistical Computing.
52. Oksanen J, Guillaume Blanchet F, Kindt R, Legendre P, Minchin PR, et al. (2012) vegan: Community Ecology Package. R package version 2.0–5 ed.
53. Dorigo U, Fontvieille D, Humbert JF (2006) Spatial variability in the abundance and composition of the free-living bacterioplankton community in the pelagic zone of Lake Bourget (France). FEMS Microbiol Ecol 58: 109–119.
54. Kent AD, Jones SE, Yannarell AC, Graham JM, Lauster GH, et al. (2004) Annual patterns in bacterioplankton community variability in a humic lake. Microbial Ecology 48: 550–560.
55. Sogin ML, Morrison HG, Huber JA, Mark Welch D, Huse SM, et al. (2006) Microbial diversity in the deep sea and the underexplored "rare biosphere".

Proceedings of the National Academy of Sciences of the United States of America 103: 12115–12120.

56. Schindler DE, Chang GC, Lubetkin S, Abella SEB, Edmonson WT (2003) Rarity and functional importance in a phytoplankton community.. In: Kareiva P, Levin SA, editors. The importance of species: perspectives on expendability and triage. Princeton, New Jersey: Princeton University Press.

57. Bent SJ, Forney LJ (2008) The tragedy of the uncommon: understanding limitations in the analysis of microbial diversity. ISME J 2: 689–695.

58. Kovacs A, Yacoby K, Gophna U (2010) A systematic assessment of automated ribosomal intergenic spacer analysis (ARISA) as a tool for estimating bacterial richness. Res Microbiol 161: 192–197.

59. Lyons KG, Brigham CA, Traut BH, Schwartz MW (2005) Rare species and ecosystem functioning. Conservation Biology 19: 1019–1024.

60. Lennon JT, Jones SE (2011) Microbial seed banks: the ecological and evolutionary implications of dormancy. Nature Reviews Microbiology 9: 119–130.

10

Representation of Ecological Systems within the Protected Areas Network of the Continental United States

Jocelyn L. Aycrigg[1]*, Anne Davidson[1], Leona K. Svancara[2], Kevin J. Gergely[3], Alexa McKerrow[4], J. Michael Scott[5]

1 National Gap Analysis Program, Department of Fish and Wildlife Sciences, University of Idaho, Moscow, Idaho, United States of America, 2 Idaho Department of Fish and Game, Moscow, Idaho, United States of America, 3 United States Geological Survey Gap Analysis Program, Boise, Idaho, United States of America, 4 United States Geological Survey Gap Analysis Program, Raleigh, North Carolina, United States of America, 5 Department of Fish and Wildlife Sciences, University of Idaho, Moscow, Idaho, United States of America

Abstract

If conservation of biodiversity is the goal, then the protected areas network of the continental US may be one of our best conservation tools for safeguarding ecological systems (i.e., vegetation communities). We evaluated representation of ecological systems in the current protected areas network and found insufficient representation at three vegetation community levels within lower elevations and moderate to high productivity soils. We used national-level data for ecological systems and a protected areas database to explore alternative ways we might be able to increase representation of ecological systems within the continental US. By following one or more of these alternatives it may be possible to increase the representation of ecological systems in the protected areas network both quantitatively (from 10% up to 39%) and geographically and come closer to meeting the suggested Convention on Biological Diversity target of 17% for terrestrial areas. We used the Landscape Conservation Cooperative framework for regional analysis and found that increased conservation on some private and public lands may be important to the conservation of ecological systems in Western US, while increased public-private partnerships may be important in the conservation of ecological systems in Eastern US. We have not assessed the pros and cons of following the national or regional alternatives, but rather present them as possibilities that may be considered and evaluated as decisions are made to increase the representation of ecological systems in the protected areas network across their range of ecological, geographical, and geophysical occurrence in the continental US into the future.

Editor: Kimberly Patraw Van Niel, University of Western Australia, Australia

Funding: The National Gap Analysis Program at the University of Idaho is supported by the United States Geological Society Gap Analysis Program under grant #G08A00047. The url: gapanalysis.usgs.gov. The agreement mentioned above supported JA and AD to do the study design, data collection and analysis as well as the decision to publish and preparation of the manuscript. LS was supported by Idaho Department of Fish and Game to help with the study design, data analysis, and preparation of the manuscript. AM and KG were funded by United States Geological Survey GAP to help with study design, data collection and analysis. KG is the program officer for this agreement and he has been involved with the study design, data collection, and data analysis for this manuscript. JMS is retired and he helped with the study design, decision to publish and preparation of the manuscript. The funders had a role in the study design, data collection, and data analysis, but not in the decision to publish or preparation of the manuscript.

Competing Interests: The authors have declared that no competing interests exist.

* E-mail: aycrigg@uidaho.edu

Introduction

Traditionally, a mix of opportunity, available resources, and agency-specific conservation priorities are the foundation upon which networks of protected areas are developed over time [1–4]. This has led to a protected areas network in the continental US cultivated for multiple purposes including protecting biological resources, such as vegetation communities [5–8]. Often, to respond to conservation issues, such as habitat loss, the protected areas network is expanded by establishing new protected areas or enlarging existing ones [9–13]. However, with increasing land-use intensification the opportunities for expanding such networks are dwindling [4,14]. Furthermore, with the imminence of climate change along with increased loss and fragmentation of vegetation communities, the exigency of protecting areas that represent the full suite of vegetation communities and therefore the species found therein, has increased [15–17].

The conservation community has increasingly focused on landscape levels for national decision making, but the lack of relevant and consistent data at a national scale has been an impediment [18–20]. Most public land management agencies, even those with the broadest authorities to protect natural resources have yet to implement ecosystem-scale approaches, perhaps due to lack of relevant data [21,22]. However, the impediment that once prevented a national-scale approach to protected areas management in the continental US has recently been overcome with the availability of national-level data for vegetation communities, classified to ecological systems [23], and a protected areas database for the US [24]. Ecological systems are groups of vegetation communities that occur together within similar physical environments and are influenced by similar ecological processes (e.g., fire or flooding), substrates (e.g., peatlands), and environmental gradients (e.g., montane, alpine or subalpine zones) [23,25]. Ecological systems represent vegetation communities with spatial scales of tens to thousands of hectares and temporal scales of 50–100 years. They represent the habitat upon which vertebrate species rely for survival. The Protected Areas Database of the US (PAD-US) represents public land ownership and conservation lands (e.g., federal and state lands), including privately protected areas that are voluntarily provided (e.g. The Nature Conservancy) [24]. Each land parcel within PAD-US is assigned a protection status that denotes both the intended level of biodiversity protection and indicates other natural, recreational and cultural uses (Table 1) [24]. Together, these databases provide the foundation for assessing the representation of vegetation communities in the continental US within the protected areas network and thereby informing decision making at the national level.

The protected areas network within the continental US is often viewed as one of our best conservation tools for securing vegetation communities and the species they support into the future [26–29]. An inherent assumption behind a network of protected areas is that protection of vegetation communities will also protect the species that rely on them, including invertebrate and vertebrate species, many of which little is known of their life history or habitat requirements [11,30,31]. For our analysis, we narrowly defined a protected area as an area of land having permanent protection from conversion of natural land cover and a mandated management plan in operation to maintain a natural state within which disturbance events may or may not be allowed to proceed without interference and/or be mimicked through management (Table 1) [24]. Furthermore, we defined a protected areas network as a system of protected areas that increase the effectiveness of *in situ* biodiversity conservation [32]. Lastly, we defined biodiversity as a hierarchy from genes to communities encompassing the interdependent structural, functional, and compositional aspects of nature [33].

The questions of how much of a vegetation community to protect and what approach is best for systematically protecting vegetation communities have been discussed at length [34,35]. No single solution or specific amount of area has been established to meet both policy targets and biological conservation needs [35]. Most recently the Convention on Biological Diversity set a target of 17% for terrestrial areas in the Aichi Biodiversity Targets described within the Strategic Plan 2011–2020 [36]. The Aichi Biodiversity Targets also attempt to address biological needs by stating that areas protected should be ecologically representative [36]. Representation of vegetation communities is often put forth as a goal of conservation planning because the aim is to protect something of everything in order to conserve the evolutionary potential of the entire protected areas network [34,37,38]. The US has not explicitly addressed the representation of vegetation communities within the protected areas network; however, Canada has used representation targets to structure their protected areas network [39–41]. Even though climate change will likely alter what is represented within Canada's protected areas network, starting from a representative group of protected vegetation communities provides a foundation for climate change adaptation [40,41].

Numerous assessments of the US protected areas network and its effectiveness at conserving vegetation communities have all concluded the network is falling short [15,20,42–48]. Each assessment used the best data available at the time, but in all cases, extent, resolution, and consistency of the data were limited. Shelford [42] conducted the first assessment of protected areas in the US in 1926. His aim was to study the native biota of North America, which started with inventorying the existing protected areas and how their vegetation communities had been modified from pre-settlement conditions. Later, Scott et al. [15] found that 302 of 499 (~60%) mapped vegetation communities within the US had <10% representation within protected areas. Dietz and Czech [20] found the median percentage of area protected within the continental US was 4% for the ecological analysis units they defined.

We recently have had the opportunity to evaluate the representation (i.e., saving some of everything) and redundancy (i.e., saving more than one of everything) of ecological systems within the existing protected areas network for the continental US. This opportunity was possible because of the availability of a complete ecological systems database for the continental US and a comprehensive database of the current protected areas network. Hence, we can now assess how well the protected areas network

Table 1. Description of protection status categories in the Protected Areas Database for US [24].

Protection status	Description	Example
Lands managed to maintain biodiversity (i.e., protected areas network)	An area of land having permanent protection from conversion of natural land cover and a mandated management plan in operation to maintain a natural state within which disturbance events may or may not be allowed to proceed without interference and/or be mimicked through management.	Yellowstone National Park, Wyoming
Lands managed for multiple-use, including conservation	An area having permanent protection from conversion of natural land cover for the majority of the area, but subject to extractive uses of either a broad low-intensity type (e.g., logging) or localized intense type (e.g., mining). Protection of federally listed endangered and threatened species throughout the area may be conferred.	Kaibab National Forest, Arizona
Lands with no permanent protection from conversion, but may be managed for conservation	An area with no known public or private institution mandates or legally recognized easements or deed restrictions held by the managing entity to prevent conversion of natural habitats to anthropogenic habitat types. Conversion to unnatural land cover throughout is generally allowed and management intent is unknown.	Fort Irwin, California

Protection status denotes the intended level of biodiversity protection and indicates other natural, recreational, and cultural uses. These designations emphasize the managing entity rather than the land owner because the focus is on long-term management intent. Therefore an area gets a designation of permanently protected because that is the long-term management intent.

encompasses the ecological and evolutionary patterns and processes that maintain ecological systems and thereby the species that depend on them [37]. Additionally, based on the Aichi Biodiversity Targets within the Strategic Plan 2011–2020 of the Convention on Biological Diversity, we can evaluate the current protected areas network in the continental US in context of meeting the suggested 17% target for terrestrial areas [36].

If the current protected areas network is falling short of conserving vegetation communities then what potential alternatives might be available to address those shortfalls? One such alternative is to replace protected areas that contribute minimally to conservation of vegetation communities with those with greater conservation value [49]. The goal would be to increase the overall biodiversity protection of the entire protected areas network. This approach proposed by Fuller et al. [49] could be attractive because the sale of protected areas with less conservation value could go towards acquiring new ones. Fuller et al. [49] proposed this approach in Australia where a protected areas network has been systematically designed with broad representation of Australia's vegetation types [49]. The protected areas network in the continental US has not been systematically designed [2,4]. Would this approach be feasible if the criteria for determining the contribution to conservation (i.e., cost-effectiveness analysis) could be agreed upon consistently across the continental US?

Another alternative to address the current protected areas network's shortfall could be to expand the network in area and number of protected areas [9,11,13]. A national assessment would be needed to identify vegetation communities not represented or under-represented within the existing protected areas network and a national conservation plan would be developed to prioritize acquisition of these vegetation communities to increase their representation on protected lands [50,51]. There are approximately 300 million hectares of public and private lands with no permanent protection on which native vegetation communities occur [23,24]. Could the representation of vegetation communities within the protected areas network be increased by prioritizing acquisition within these lands with no permanent protection?

A third alternative for addressing the protected areas network's shortcomings might be to increase the emphasis of maintaining biodiversity on some public and private lands currently managed for multiple-use (Table 1). Swaty et al. [52] found that in addition to the 29% of the continental US land area that has been converted by human use; there were an additional 23% of non-converted lands with altered vegetation structure and composition, which likely are lands managed for multiple-use. The protected areas network is comprised of approximately 50 million hectares in the continental US, while there are about 140 million hectares of public and private lands managed for multiple-use [24]. Vegetation communities that are currently not represented or underrepresented within the current protected areas network may have representation on the approximately 140 million hectares of land managed for multiple-use [20,24]. Could, therefore, an emphasis on maintaining biodiversity on a strategically targeted subset of lands managed for multiple-use be used to effectively expand the representation of vegetation communities within the entire protected areas network?

From a conservation management perspective for the US, the Department of Interior (DOI) has established a framework of Landscape Conservation Cooperatives (LCC) with the mission of landscape-level planning and management [53]. This national framework further supports the need for nationally consistent databases and analyses. We focused our analysis on alternative ways to potentially increase the representation of ecological systems in the protected areas network of the continental US.

Specifically we asked (1) how well are ecological systems represented in the protected areas network relative to their occurrence in the continental US, including with regards to soil productivity and elevation, (2) how alternative approaches may potentially increase the representation of ecological systems in the protected areas network, and (3) how Landscape Conservation Cooperatives (LCC), the new landscape unit for conservation initiatives, can be used to regionally assess conservation status of ecological systems.

Materials and Methods

Data Description

We used the National Gap Analysis Program (GAP) Land Cover [23] and US Geological Survey GAP's (USGS-GAP) Protected Areas Database of the US (PAD-US 1.0) [24] as the national datasets for our analyses. The land cover data contains 3 nested hierarchical levels of vegetation communities. Level I contains 8 groupings, based on generalized vegetative physiognomy (e.g., grassland, shrubland, forest), while Level II has 43 groupings representing general groups of ecological systems based on physiognomy and abiotic factors (e.g., lowland grassland and prairie, alpine sparse and barren). The third hierarchical level contains 551 map classes, including 518 ecological systems. We focused on the non-modified, non-aquatic classes at each level (Level I: 5 classes, Level II: 37 classes, and Level III: 518 ecological systems).

The National GAP Land Cover was compiled from the Southwest, Southeast, Northwest, and California GAP land cover data completed during 2004–2009 [23]. We incorporated data from LANDFIRE (www.landfire.gov) for the Midwest and Northeast. These national land cover data were based on consistent satellite imagery (Landsat Thematic Mapper (TM) and Enhanced Thematic Mapper (ETM)) acquired between 1999 and 2001 in conjunction with digital elevation model (DEM) derived datasets (e.g., elevation, landform) and a common classification system (i.e., ecological systems) to model natural and semi-natural vegetation [54–56]. The resolution is 30-m and typically the minimum mapping unit is 1 ha. Regional accuracy assessments and validations have been conducted and, based on those, in general, forest and some shrub ecological systems typically had higher accuracies than rare and small patch ecological systems, such as wetlands [57,58].

PAD-US (Version 1.0) consists of federal, state, and voluntarily provided privately protected area boundaries and information including ownership, management, and protection status [24]. Protection status is assigned to denote the intended level of biodiversity protection and indicate other natural, recreational, and cultural uses (Table 1) [24]. In assigning protection status, the emphasis is on the managing entity rather than the owner and focuses on long-term management intent instead of short-term processes [11]. The criteria for assigning protection status includes perceived permanence of biodiversity protection, amount of area protected with a 5% allowance of total area for intensive human use, protection of single vs. multiple features, and the type of management and degree to which it is mandated [59]. The protection status ranges from lands managed to maintain biodiversity to lands with little or no biodiversity protection (Table 1). Lands managed for multiple-use, including conservation, are permanently protected, but allow for extractive uses, such as mining and logging. In the continental US, lands with no permanent protection are considered any land parcel not designated either of the other protection status categories. We included only lands permanently protected and managed to maintain biodiversity in our definition of the protected areas network.

We also used elevation data obtained from the National Elevation Dataset (NED) [60] and soil productivity. The National Elevation Dataset, a seamless dataset with a resolution of approximately 30 m, was the best available raster elevation data for the continental US [60]. We divided the National Elevation Dataset into 8 classes ranging from 0 to 4500 meters at 500-meter intervals. Soil productivity classes for the continental US were based on STATSGO data (http://soils.usda.gov/survey/geography/statsgo/). These data were reclassified into 8 soil productivity classes based on land capability classes (http://soils.usda.gov/technical/handbook) and ranged from very high to very low productivity.

To apply our analysis and results to current conservation management in the continental US, we used the LCC framework [53]. LCCs represent large area conservation-science partnerships between DOI and other federal agencies, states, tribes, non-governmental organizations (NGOs), universities, and other public and private stakeholders. Their intent is to inform resource management decisions to address landscape-level stressors, such as land use change, invasive species, and climate change [53].

Data Analysis

The PAD-US 1.0 [24] and LCC data [53] were converted to grids (i.e., 30×30 m cells) and combined with the National GAP Land Cover [23] using ArcGIS 9.3.1 (ESRI, Redlands, CA). To assess the protection of ecological systems relative to their occurrence, we calculated a frequency distribution of protected area sizes within the existing protected areas network. To evaluate how the size range of protected areas would change with the inclusion of land managed for multiple-use, we calculated a frequency distribution of the protected areas network with lands managed for multiple-use added in (Table 1). We also calculated the amount of area of land managed for multiple-use needed to meet the 17% Aichi Biodiversity Target. To assess least protected or most endangered ecosystems, we summarized within each hierarchical level of the National GAP Land Cover (i.e., Levels I, II, and ecological systems) the number, size, protection status, and ownership of land parcels within PAD-US, as well as their distribution among LCCs. At the broadest level (Level I), we calculated percent availability versus percent protected to gain insight into the representation of each system in the protected areas network. We used a comparison index line (i.e., 1:1 line) to indicate the relationship between percent availability and percent protected [61]. Similarly, we calculated the percent area of ecological systems protected (i.e., managed to maintain bio-diversity), managed for multiple-use, and not permanently protected for soil productivity and elevation ranges by combining these data with PAD-US [24] using ERDAS Imagine 9.3 (Table 1).

The diversity of ecological systems across and redundancy within LCCs was calculated by counting the number of ecological systems occurring within each LCC. Diversity was defined as the number of ecological systems within each LCC, while redundancy was defined as the number of LCCs in which a single ecological system occurred [37]. For example, if an ecological system occurred in 2 LCCs, its redundancy value was 2. Unique ecological systems were those that occurred in a single LCC. Furthermore, we calculated the number and percent area protected of ecological systems by each protection status within each LCC. To assess whether lands were being protected at the same rate as those converted to human dominated classes, such as developed areas, cultivated croplands, orchards, vineyards, quarries, mines, gravel pits, oil wells, and pastures, we calculated

the conservation risk index (CRI) for each LCC by dividing percent area converted by percent area managed to maintain biodiversity or percent area managed to maintain biodiversity and for multiple-use [23,62]. Finally, we summarized CRI values by protection status.

Results

The current protected areas network in the continental US covers approximately 10% of the total area in which ecological systems occur. Across about 30,000 protected areas, the mean size of an individual protected area was 1942 ha with a size range of approximately 25–2,500,000 hectares over all protected areas. The analysis of representation of the network shows that the distribution of ecological systems managed to maintain bio-diversity (i.e., the distribution of the protected areas network) is skewed towards high elevation and low productivity soils (Figure 1A). Overall 68% of all 518 ecological systems have <17% of their area protected, which is a target suggested by the Aichi Biodiversity Target of the Convention of Biological Diversity [36] and most of the ecological systems with <17% protected occur at low elevation and in areas with moderate to high productivity soils (Figures 1B and 1C, Table S1).

In examining the percent available versus percent protected for lands managed to maintain biodiversity, only two of the five Level I land cover groups (sparse and barren; riparian and wetland) occurred above the 1:1 line indicating a greater percentage of these groups are protected in relation to their availability (Figure 2). Representation of Level II land cover groups was lowest for lowland grassland and prairie (xeric-mesic), but most groups had <17% protected (Figure 3). Out of 37 Level II groups, 11 fell at or above the 17% Aichi Biodiversity Target [36].

Ecological systems on lands managed for multiple-use and on lands with no permanent protection comprised 29% and 61%, respectively, of the total area of the continental US in which ecological systems occur. When lands managed for multiple-use were included as part of the protected areas network, the overall number of protected areas increased to about 88,000 with a size range of approximately 25–117,757,000 hectares.

When both lands managed to maintain biodiversity and for multiple-use were included all five Level I land cover groups occurred above the 1:1 line and all five occurred at or above the suggested 17% Aichi Biodiversity Target (Figure 2) [36]. The largest increases were within the shrubland, steppe, and savanna group, forest and woodland group, and sparse and barren group. The percent area of Level II land cover groups increased for all 37 groups when lands managed for multiple-use were added to lands managed to maintain biodiversity (Figure 3). The largest increases in percent area occurred within the lowland grassland and prairie (xeric-mesic) and sagebrush dominated shrubland. Out of 37 Level II groups, 33 fell at or above the 17% Aichi Biodiversity Target [36] when both lands managed to maintain biodiversity and multiple-use were included (Figure 3).

To meet the suggested 17% Aichi Biodiversity Target [36], approximately 9 million hectares (6.4%) of the 140 million hectares of public and private lands managed for multiple-use or 34 million hectares (11.3%) of the 300 million hectares of lands with no permanent protection would need to emphasize main-taining biodiversity or be acquired as part of the protected areas network (Table S1). Including lands managed for multiple-use with lands managed to maintain biodiversity, 98% of all ecological systems increased their percent area protected (Table S1). Using the suggested 17% Aichi Biodiversity Target [36], we found 32%

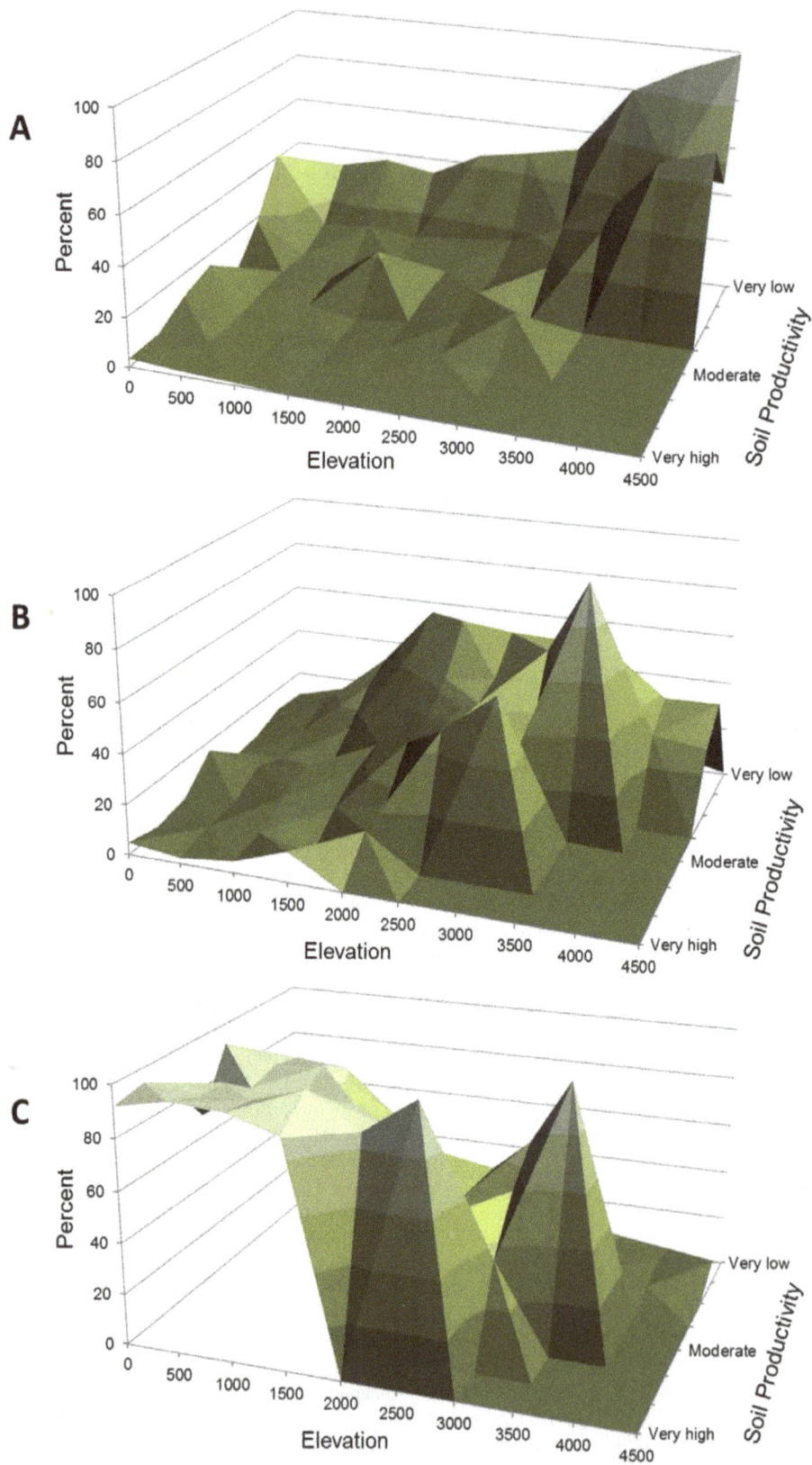

Figure 1. Percent area of ecological systems in relation to elevation, soil productivity, and protection status. Protection status designations include lands managed to maintain biodiversity (A), lands managed for multiple-use (B), and lands that have no permanent protection (C). See Table 1 for protection status descriptions. Percent area of ecological systems determined by combining data for elevation (meters) and soil productivity (http://soils.usda.gov/technical/handbook) with ecological systems grouped by protection status [23,24,60].

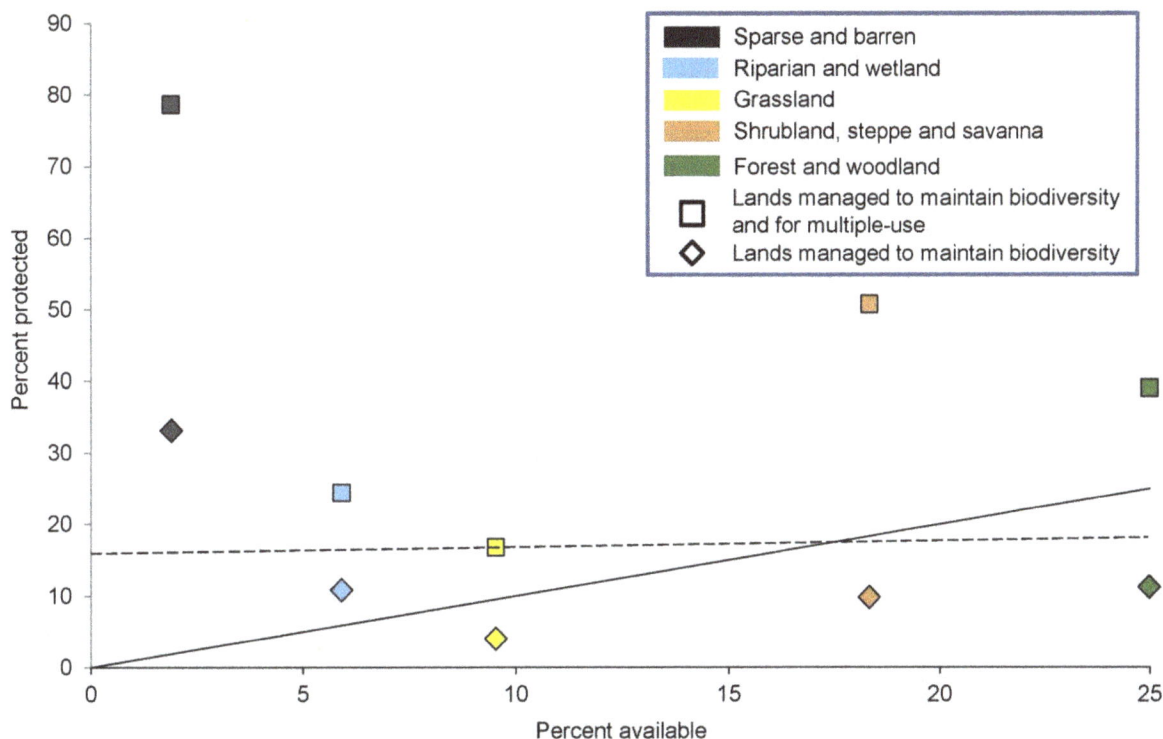

Figure 2. Percent protected and available for each Level I land cover group by protection status. Lands managed to maintain biodiversity (diamonds) are shown relative to lands managed to maintain biodiversity and for multiple-use (squares). See Table 1 for protection status descriptions. A comparison index line is shown, which indicates a 1:1 relation between percent availability and percent protected [61]. A value below the 1:1 line represents a Level I land cover group under-represented in the protected areas network, a value above represents a Level I land cover group well represented in the protected areas network, while a value on the line indicates a Level I land cover group available and protected equally [61]. For example, grassland, a Level I land cover group, has about 4% of its area managed to maintain biodiversity, but that increased to about 17% when lands managed for multiple-use were included [23,24]. A dashed line representing the 17% Aichi Biodiversity Target of the Convention on Biological Diversity is shown [36].

of all ecological systems met that target, but that increased to 68% when lands managed for multiple-use were included (Table S1).

Including lands managed for multiple-use in the protected areas network would result in dramatic geographic changes in the western US, but noticeable changes were also evident in northeastern US, Florida, the Appalachian mountains, and around the Great Lakes (Figure 4). Federal, state, and local governments as well as private entities manage lands to maintain biodiversity and for multiple-use (Figure 5). There are approximately 50 million hectares of lands managed to maintain biodiversity with Bureau of Land Management (BLM) and US Forest Service (USFS) managing about 29 million hectares, which is more than US Fish and Wildlife Service (USFWS), National Park Service (NPS), and all other federal land combined (Figure 5). Approximately 140 million hectares is managed for multiple-use in the continental US with BLM and USFS managing about 100 million hectares (Figure 5, Table S1).

Redundancy values for ecological systems occurring in LCCs ranged from 1–8, with redundancy values higher in LCCs in the west (Figure 6A). Ecological systems were highly diverse in 4 LCCs (Great Northern, Great Basin, Desert, and Gulf Coast Plain and Ozarks); however, only 1 had numerous unique ecological systems (Gulf Coast Plains and Ozarks; Figure 6B and Table 2). When including lands managed for multiple-use in the protected areas network, 7 out of the 16 LCCs in the continental US more than doubled the percent area protected (Table 2). Lands managed to maintain biodiversity represented between 0.6–17.0% of the area

of LCCs, adding lands managed for multiple-use increased that to 1.2–62.9% (Table 2). Eight out of 16 LCCs contained ecological systems that occurred only on lands managed for multiple-use or had no permanent protection (e.g., Great Plains, North Atlantic; Figure 7). The CRI values varied across LCCs with the Eastern Tallgrass Prairie and Big Rivers having the highest value (126.4) because almost 80% of its area was converted to human use (i.e., cultivated cropland) and the Desert and Southern Rockies having the lowest (0.2) because >10% of their area contained lands managed to maintain biodiversity (Figure 8). Including lands managed for multiple-use lowered the CRI for all LCCs and increased the number of LCCs meeting the suggested Aichi Biodiversity Target of 17% target from 1 to 7 (Figure 8) [36].

Discussion

Protection of Ecological Systems Relative to their Occurrence in the Continental US

The existing protected areas network in the continental US would need to capture a more representative complement of ecological systems if the US aims to meet the suggested Aichi Biodiversity Target of 17% for ecologically representative terrestrial areas [36]. The 518 ecological systems mapped in the continental US are disproportionately distributed by number, size, and protection status relative to elevation and soil productivity, which translates to an uneven representation of ecological systems within the protected areas network (Figure 1A) [15,63]. Soils with

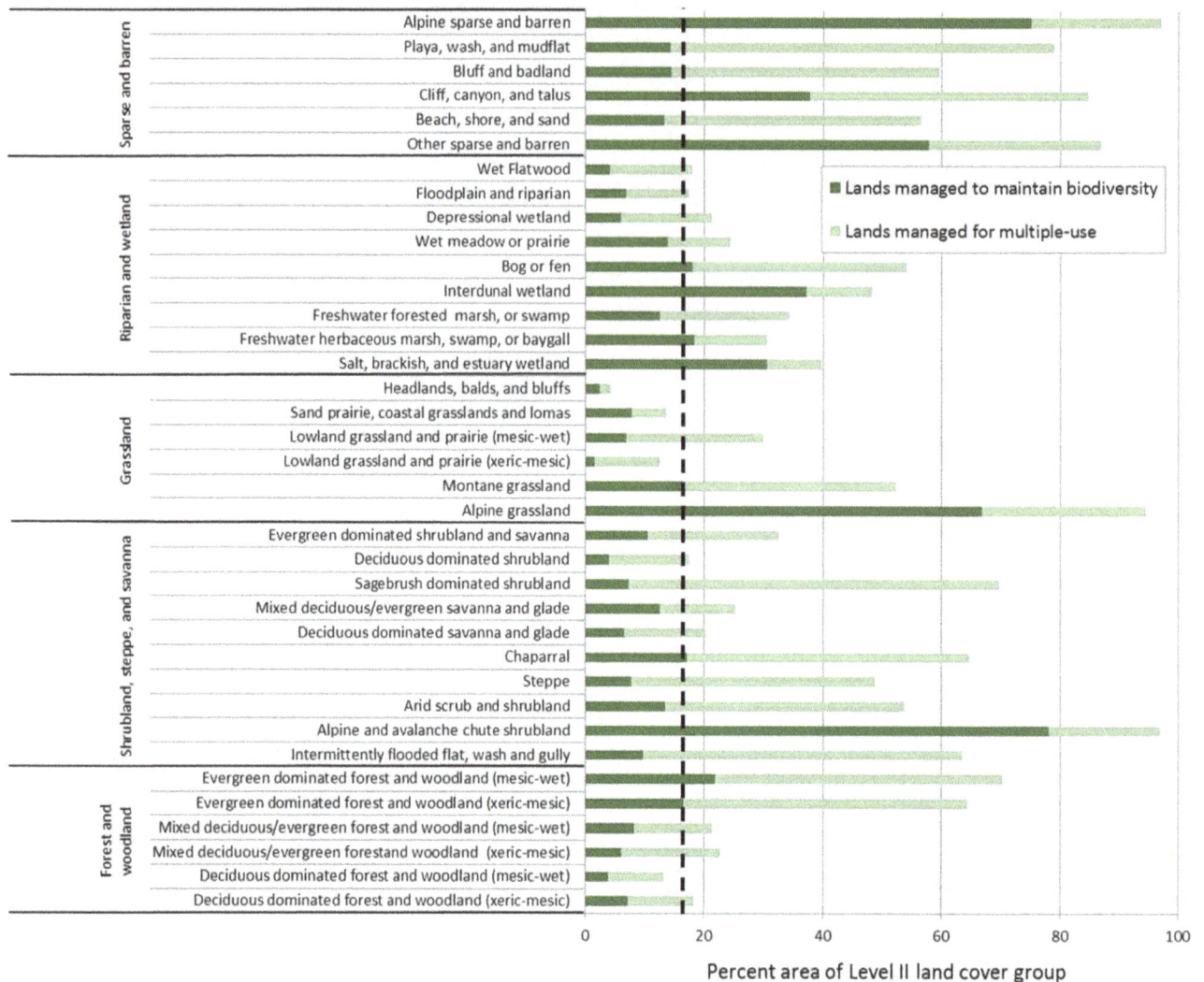

Figure 3. Percent area of Level II land cover groups by protection status. The Level II land cover groups are arranged by Level I land cover groups (see Table S1) [23]. Percent area for both lands managed to maintain biodiversity and lands managed for multiple-use are shown [24]. See Table 1 for protection status descriptions. A dashed line representing the 17% Aichi Biodiversity Target of the Convention on Biological Diversity is shown [36].

low productivity at high elevation are more likely to be found within the protected areas network; therefore ecological systems that occur in those areas are disproportionally represented in the network. Typically, low soil productivity at high elevations occurs in sparse and barren areas and these areas are well represented within the protected areas network (Figure 2) [15]. Capturing a broader range of elevation could be important to spatial patterns of biodiversity because ecological systems might shift with climate change, but the patterns of biodiversity will likely endure with geophysical features, such as elevation range [64]. How can the representation of ecological systems increase within the protected areas network of the continental US?

Alternatives for Increasing Representation and Conservation of Ecological Systems

Many alternatives exist for conserving ecological systems and successful conservation will likely come from employing one or more of them. One approach, presented earlier in the paper, would be to replace protected areas that are minimally contrib-

uting to conservation and have a high cost associated with protecting ecological systems within a specific protected area (i.e., least cost effective) with those having greater conservation value (i.e., more cost effective) to increase the overall biodiversity protection of the entire network [49]. Applying this approach could be challenging because public support for existing protected areas may make it difficult to convince those supporters to relinquish a protected area for the benefit of the entire network [8,65]. This approach, even though controversial because of the concept of giving up protected areas, could play a prominent role in addressing the impacts of climate change because of the potential opportunity to shift the distribution of ecological systems on current protected areas in response to shifts in temperature and precipitation [66,67].

Protected areas have long been downgraded, downsized, delisted, and degazetted and these practices are currently widespread [68,69]. Approximately 60 National Parks have been delisted and downgraded since the establishment of the National Park System in 1916 [68,70,71]. One of the major drivers of protected area degazettement, which is loss of legal protection for

A

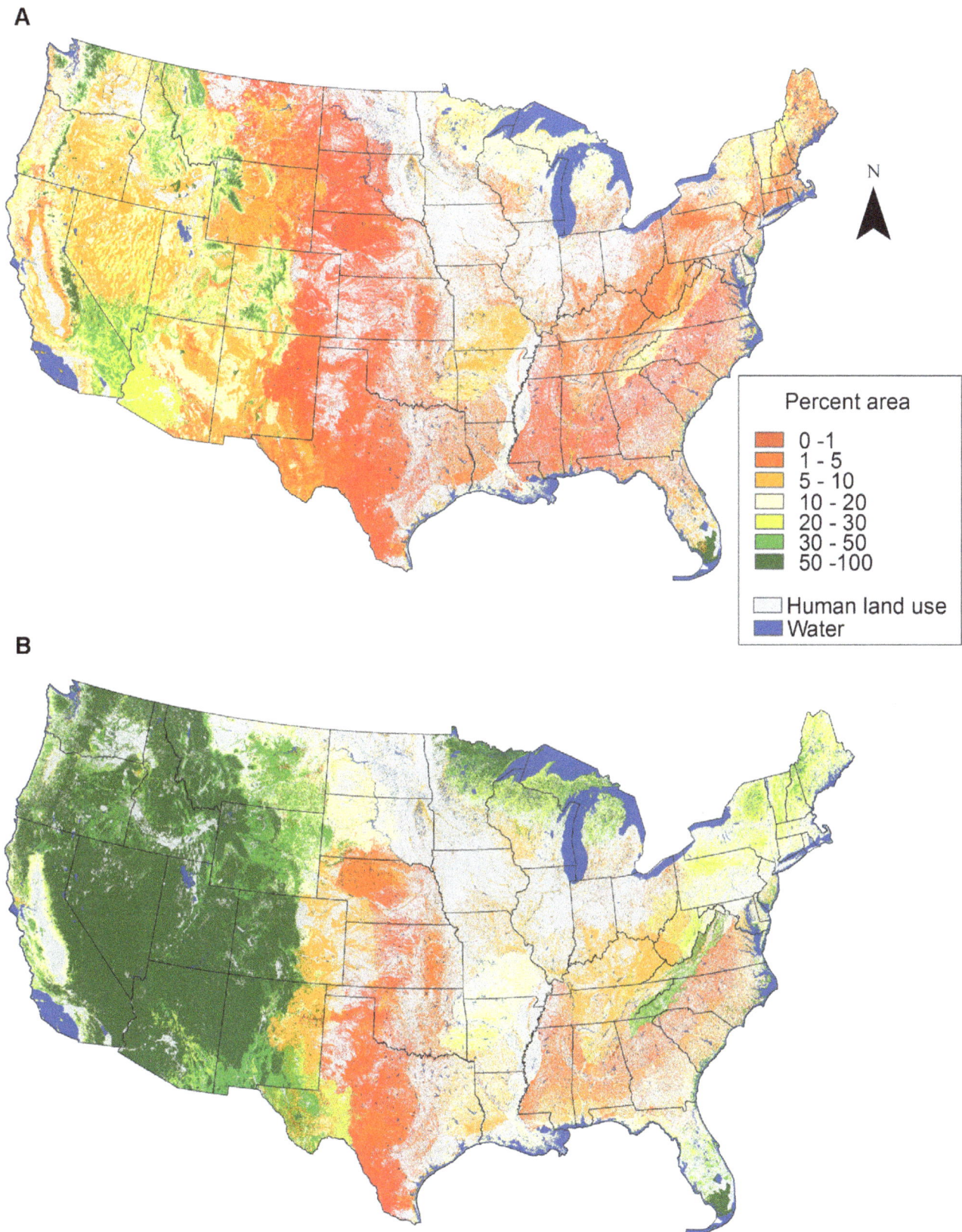

B

Figure 4. Percent area of ecological systems by protection status. Protection status designations are lands managed to maintain biodiversity (A) and lands managed to maintain biodiversity and multiple-use (B) for the continental US. Percent area is based on the area of each ecological system within each protection status divided by the total area of each ecological system [23,24]. See Table 1 for protection status descriptions. Only non-modified, non-aquatic ecological systems were included (n = 518; Table S1).

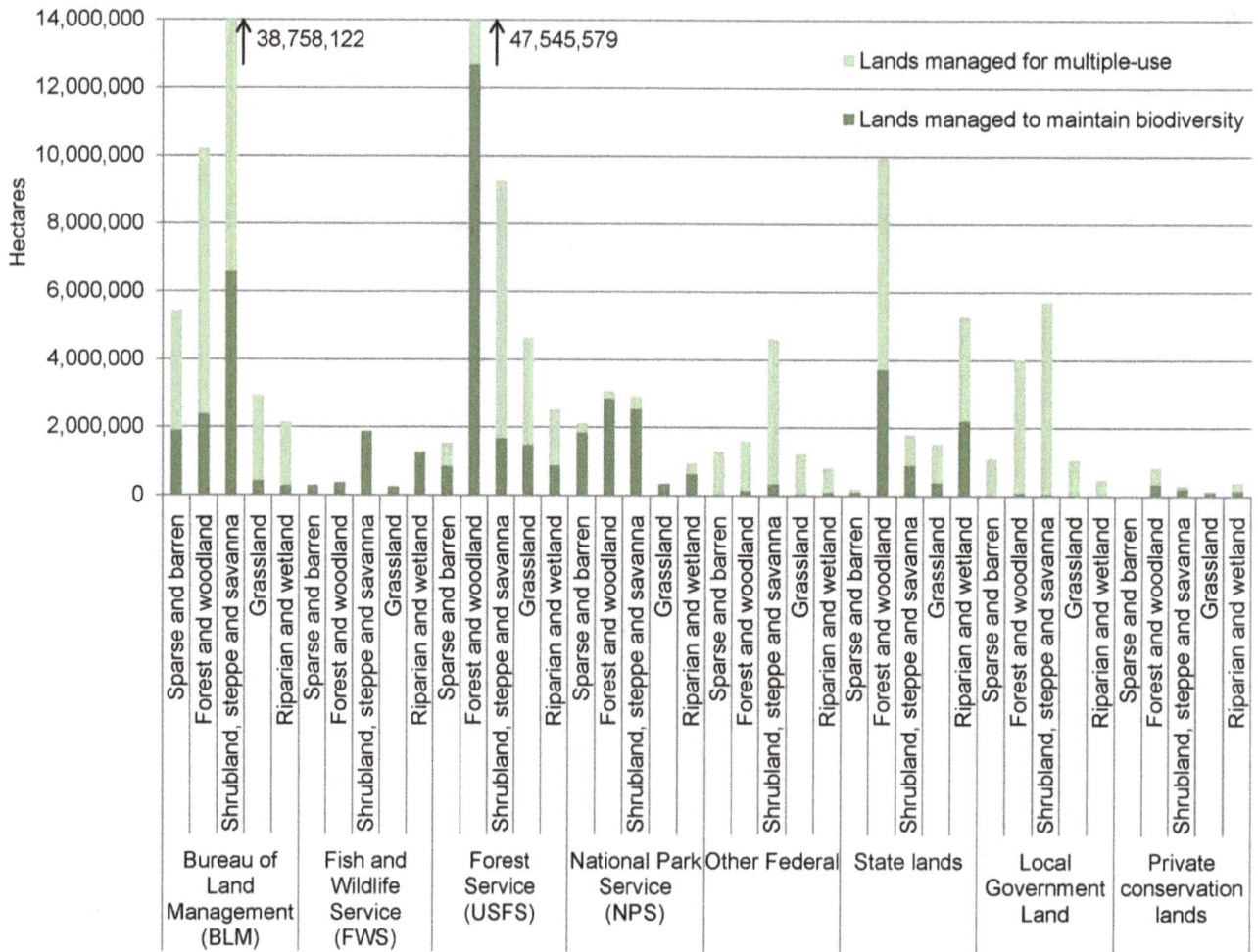

Figure 5. Area (ha) of Level I land cover groups by ownership and protection status. Ownership includes federal, state, and local governments as well as private conservation lands. See Table 1 for protection status descriptions. These values were for the continental US. Both BLM and USFS have areas of Level I land cover groups that fall outside the scale on this graph [23,24]. Values for those Level I land cover groups are shown.

an entire protected area, is access to and use of natural resources (e.g., commodity extraction) [69]. The impact on biodiversity protection because of access and use of natural resources is evident in Midwestern US where a low percent area of land is managed to maintain biodiversity and many areas are mapped as human land use (Figure 4). LCC's in the Midwest (i.e., Plains and Prairie Potholes, Great Plains, and Eastern Tallgrass Prairie and Big Rivers) have low diversity and few unique ecological systems (Figure 6B). A large percent of their area has been converted to human land use, which is reflected in high CRI values (Figure 8). To date, the ecological consequences of degazettement are unclear [69]. Both Fuller et al. [49] and Kareiva [8] believe degazettement would lead to a more dynamic and flexible approach to maintaining the current protected areas network, however it could depend on the level of systematic design used to establish the protected areas network.

Even though we did not specifically assess cost effectiveness of protected areas, our analysis could help inform the approach proposed by Fuller et al. [49]. A cost effectiveness analysis could be based on land ownership, protection status, and percent area converted to human modified systems. For example, the Great Basin LCC has potential for including some of the most cost effective protected areas because it has a low CRI value and

<10% of its area is converted. There is the potential to lower its CRI value and meet the suggested 17% Aichi Biodiversity Target [36] by increasing the percent of area managed to maintain biodiversity by 60% through emphasizing protection of biodiversity (Figure 8). The Great Basin LCC also contains ecological systems that occur only on lands managed for biodiversity (Figure 7) and has a high diversity of ecological systems even though only 1 is unique (Figure 6B). Other factors beyond land ownership, protection status, and percent area converted to human modified systems could be considered in efforts to assess the cost effectiveness of protected areas, such as representation of ecological systems and transaction costs. However, our analysis could help inform a conservation strategy for the continental US if the approach described by Fuller et al. [49] were implemented.

The second alternative for improving the conservation and representation of ecological systems described previously would be to increase the size (i.e., area or number) of our existing protected areas network through acquisition for the least protected, most endangered, or high priority ecological systems [50,51]. If a systematic approach for choosing new protected areas could increase the representation of elevation and soil productivity and thereby ecological systems then the network's ability to respond to varying conditions and future change could be strengthened

A

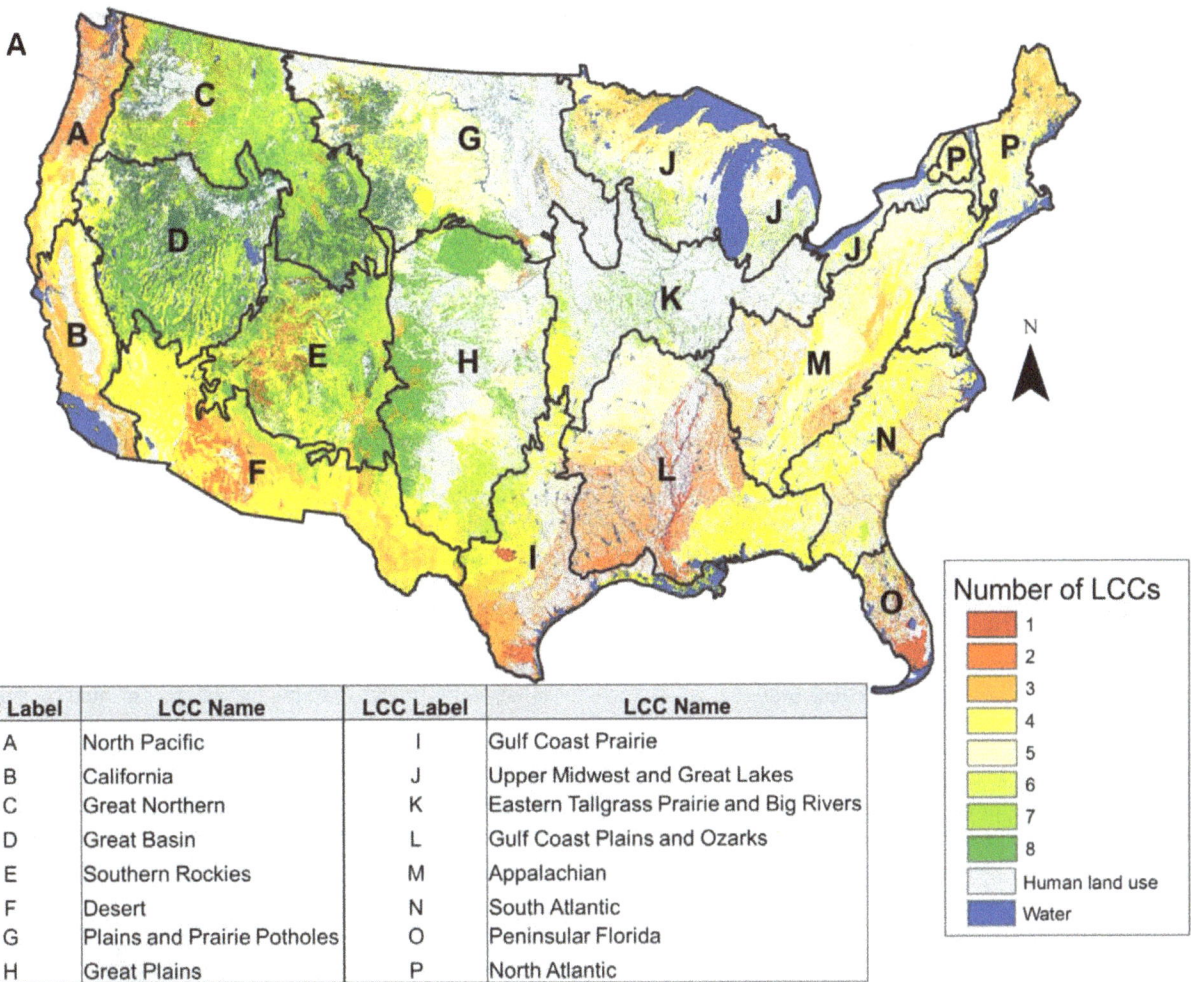

LCC Label	LCC Name	LCC Label	LCC Name
A	North Pacific	I	Gulf Coast Prairie
B	California	J	Upper Midwest and Great Lakes
C	Great Northern	K	Eastern Tallgrass Prairie and Big Rivers
D	Great Basin	L	Gulf Coast Plains and Ozarks
E	Southern Rockies	M	Appalachian
F	Desert	N	South Atlantic
G	Plains and Prairie Potholes	O	Peninsular Florida
H	Great Plains	P	North Atlantic

B

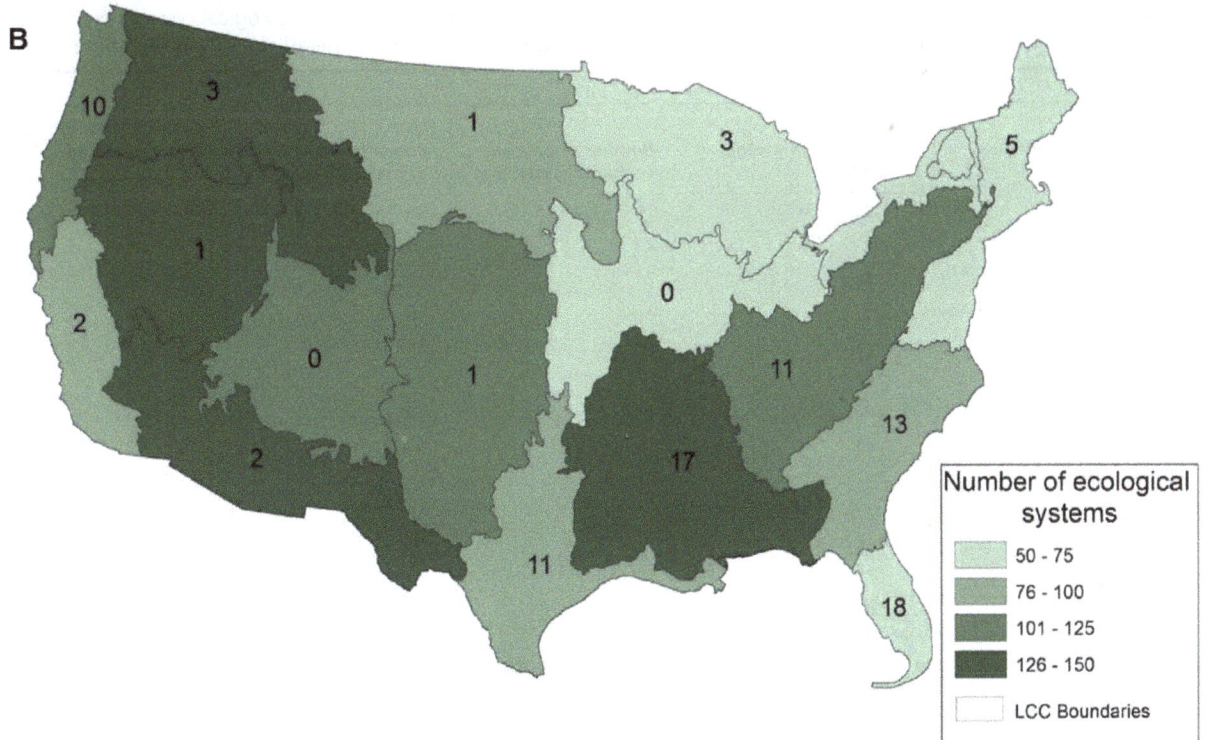

Figure 6. Redundancy, diversity, and uniqueness of ecological systems within Landscape Conservation Cooperatives (LCC). Redundancy measures the number of LCC's in which a single ecological system occurs (A) [23]. The higher the number of LCC's in which an ecological systems occurs the more redundancy displayed by that ecological system. For example, if an ecological system occurs in 2 LCCs, it has a redundancy value of 2. Diversity is the total number of ecological systems occurring with an LCC, which is shown by color shading of LCCs (B). Uniqueness is the number of ecological systems that occur in a single LCC, which is indicated by the number within each LCC (B). For example, the Great Northern LCC encompasses 126–150 ecological systems total, most of these occur in a total of 7 or 8 LCCs, but 3 are unique and only found in this LCC. Only non-modified, non-aquatic ecological systems were included (n = 518; Table S1). Each LCC is assigned a letter, which indicates the name of the LCC.

(Figure 1) [15,63]. Our results were similar to Scott et al. [15] because we found that ecological systems at lower elevations and higher soil productivity were under-represented within the current protected areas network (Figure 1). These areas could be prioritized if acquisition of new protected areas was employed for increasing protection of ecological systems. The least protected ecological systems and potentially most endangered (see Figure 8) are within all the Level I land cover groups except sparse and barren (Figures 2, 3, and 5, Table S1) and are located mostly in the Midwestern US (Figure 4). Prioritizing acquisition of the Level I land cover groups within the Midwestern US would increase the overall representation of ecological systems in the continental US. However, the feasibility of land acquisition for conservation is continually a challenge as resources for obtaining new protected areas are dwindling and competition for undeveloped private land is limiting expansion opportunities [4,14]. Furthermore, the support of policy makers for creating new protected areas could be perceived as ephemeral [72]. The idea of increasing the amount of protected land is attractive in part because of the perceived permanence associated with that protection. In other words, expanding the protected areas network reduces the risk of more land being converted to a state from which it might not recover (i.e., urban development), even though the immediate benefit to conservation is dependent upon management strategies employed.

A third alternative for improving the current protected areas network might be to take stock of our management within the current protected areas network and to evaluate the potential role of lands managed for multiple-use in conserving ecological systems. Our analysis found that increasing the emphasis on maintaining biodiversity on lands currently managed for multiple-use, which are permanently protected, but allow for extractive uses (e.g., mining and logging), offers an alternative for increasing the representation of ecological systems. However, much of the land managed for multiple-use has undergone ecosystem alteration and increased management or restoration may be needed to recover existing ecological systems [52]. If we increased the emphasis on maintaining biodiversity on some public and private lands managed for multiple-use, the total percent area of ecological systems protected could increase up to 39% in the continental US (lands managed to maintain biodiversity: 10%; lands managed for multiple-use: 29%). Geographically, the greatest potential for increased emphasis on maintaining biodiversity on lands managed for multiple-use is in the West, but also in the Northeast, South, and Midwest (Figure 4). To meet the suggested Aichi Biodiversity Target of 17% [36] increased emphasis on maintaining biodiversity would need to occur on 6.4% of the lands managed for multiple-use (Table S1). Even though lands managed for multiple-use occur on both public (i.e., federal, state, and local government) and private (i.e., non-governmental organization) lands, the potential for conservation efforts to increase the protection of

Table 2. Total number and unique number of ecological systems as well as percent area of ecological systems on lands managed to maintain biodiversity and for multiple-use within each Landscape Conservation Cooperative (LCC) in the continental US.

Landscape Conservation Cooperative (LCC)	Number of ecological systems	Number of unique ecological systems	Percent area of lands managed to maintain biodiversity	Percent area of lands managed for multiple-use
Appalachian	103	11	3.5	8.3
California	88	2	10.7	16.3
Desert	133	2	17.0	40.0
Eastern Tallgrass Prairie & Big Rivers	75	0	1.2	1.2
Great Basin	143	1	11.2	62.9
Great Northern	143	3	14.8	39.3
Great Plains	102	1	0.6	2.5
Gulf Coast Plains & Ozarks	148	17	3.5	4.9
Gulf Coast Prairie	95	11	1.3	1.4
North Atlantic	63	5	6.6	8.7
North Pacific	123	10	15.1	25.5
Plains & Prairie Potholes	95	1	2.4	10.6
Peninsular Florida	56	18	8.8	13.1
South Atlantic	97	13	2.8	4.0
Southern Rockies	116	0	14.1	50.6
Upper Midwest & Great Lakes	60	3	5.7	8.3

See Figure for location of LCC. See Table 1 for protection status descriptions. LCCs are listed alphabetically.

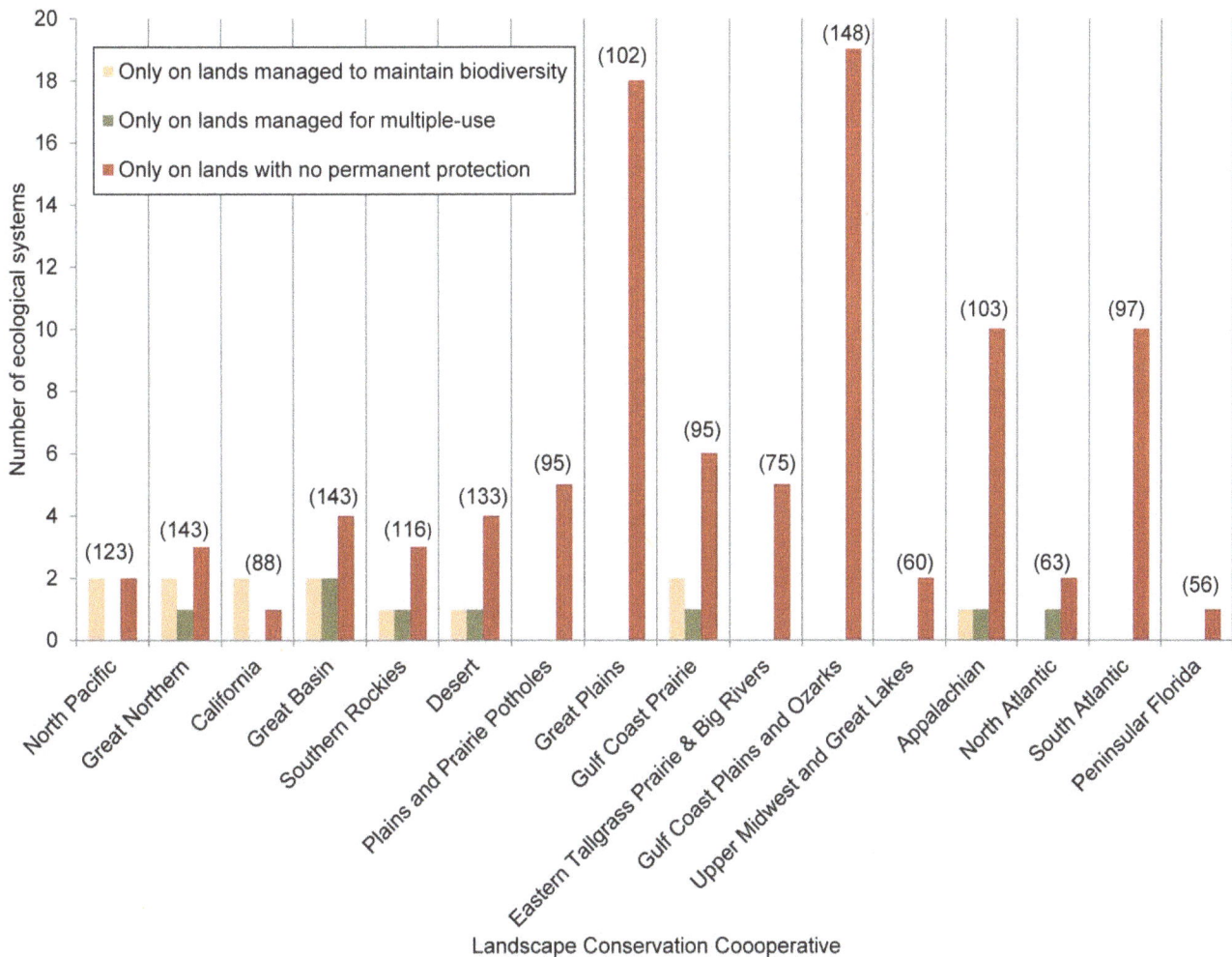

Figure 7. Number of ecological systems occurring only within each protection status by Landscape Conservation Cooperative (LCC). Ecological systems included occur only within the specified protection status [23,24]. The total number of ecological systems within each LCC is shown parenthetically. For example, the Great Plains LCC contains 102 ecological systems with 18 occurring only on lands with no permanent protection and none occurring on lands managed to maintain biodiversity or for multiple-use. See Table 1 for protection status descriptions. Only non-modified, non-aquatic ecological systems are included (n = 518; Table S1).

ecological systems on public lands is greater (i.e., quantitatively and geographically) (Figure 5).

To protect a broad representation of ecological systems within the continental US, opportunities within public land management agencies fall largely on lands managed by BLM and USFS (Figure 5). Both manage lands that maintain biodiversity, but the majority of the lands they manage are for multiple-use (Figure 5). However, if the US is to become less dependent on foreign energy sources and meet its own resource needs within its boundaries, then shifting management focus on even a small portion of lands currently managed for multiple-use could become a public lands dilemma. Lands managed for multiple-use provide multiple public benefits, including domestic energy production. [17,73,74].

In addition to the lands BLM manages for multiple-use, it has also designated 11 million hectares to the National Landscape Conservation System (NLCS), which is a network of conservation areas specifically aimed at conserving biodiversity [75]. The USFS manages over 17 million hectares of land managed to maintain biodiversity, which is more than USFWS, NPS, and other federal land management agencies combined (Figure 5). With BLM and USFS managing millions of hectares of land for maintaining

biodiversity, their role in protecting ecological systems is well established, and there may be potential to expand the protection and representation of ecological systems, for example, through the expansion of NLCS. In the past, administrative jurisdictional land transfers have occurred between land management agencies (e.g., BLM, USFWS, NPS, and USFS) [76–78]. Some of these land transfers have led to more emphasis on maintaining biodiversity.

Landscape Conservation Cooperatives Setting Priorities for Conservation of Ecological Systems

The framework and partnerships of the LCCs informs conservation at the landscape level, which will be needed to implement conservation across jurisdictional boundaries. Our analysis indicates that ecological systems in the East are less redundant and at more risk of conversion than those in the West (Figures 6 and 8). Because of this East-West dichotomy, increased conservation on some public and private lands may be important to the representation of ecological systems in the West, whereas increased public-private partnerships may play an important role in the East to increase the representation of ecological systems (Figures 4, 5, 6, 7, 8).

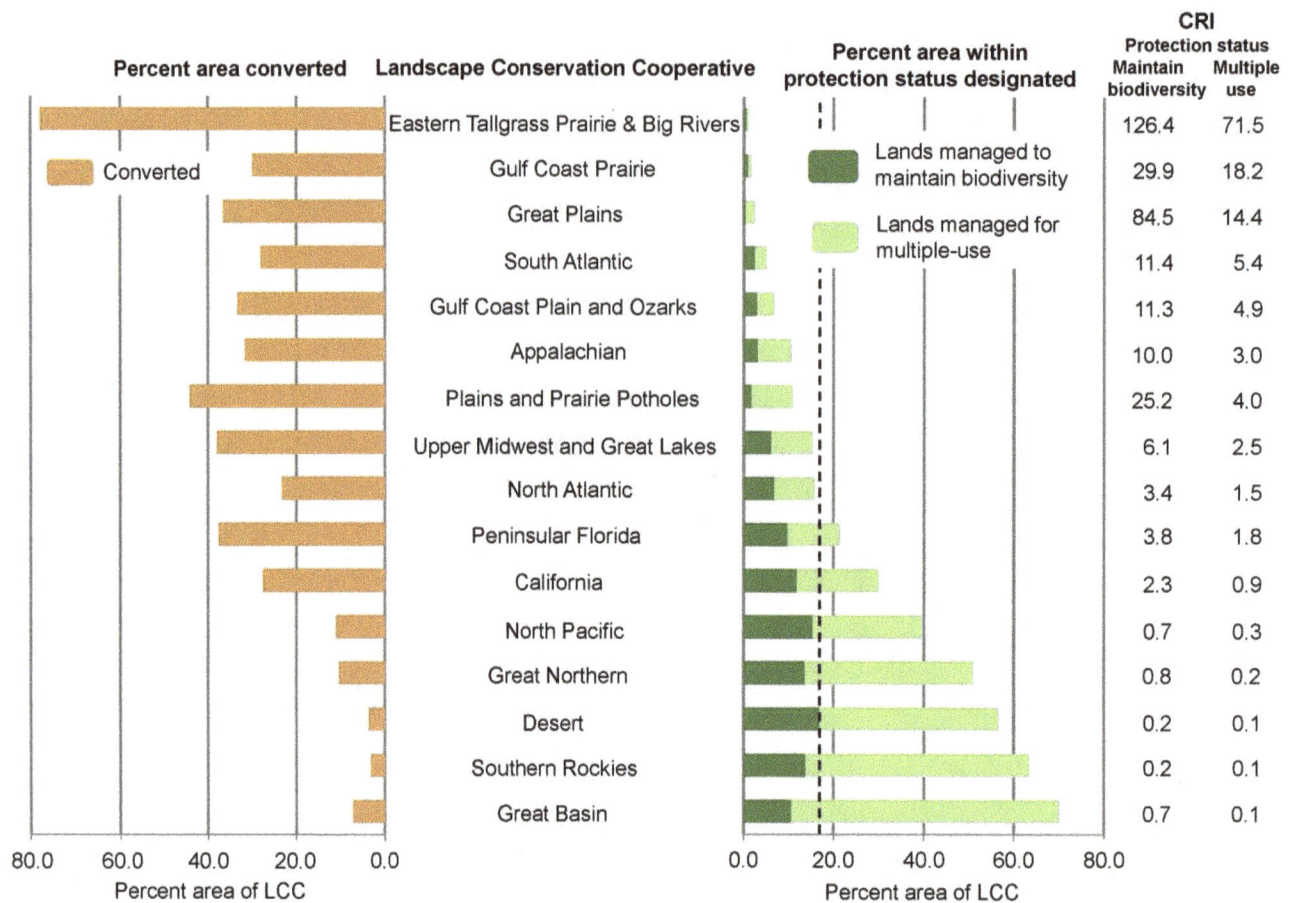

Figure 8. Percent area of Landscape Conservation Cooperative (LCC) protected or converted and its conversion risk index (CRI). CRI for each LCC is calculated by dividing percent area converted by percent area protected [62]. The CRI index is shown for lands managed to maintain biodiversity (i.e., labeled maintain biodiversity) as well as for lands managed to maintain biodiversity and multiple-use (i.e., labeled multiple-use) [23]. The LCCs are ordered by percent area within each protection status. See Table 1 for protection status descriptions. A dashed line representing the 17% Aichi Biodiversity Target of the Convention on Biological Diversity is shown [36].

Our research results highlighting low redundancy and unique ecological systems corroborate results from other studies [13,18]. In particular, the eastern US was identified as an ecoregion with high threats and irreplaceability value with regards to identifying conservation priorities [13,18]. For example, the Gulf Coast Plain and Ozarks LCC in southeastern US has high diversity and uniqueness, but low redundancy and a high conservation risk index (Figures 6 and 8). Within this LCC, there are few opportunities for increasing the representation of ecological systems on lands managed for multiple-use (Table 2, percent protected changes from 3.5% to 4.9%). An initial practical approach for conservation of ecological systems in this LCC, which contains many diverse and unique ecological systems, would be to engage both public and private conservation partners. In this case, our research results could serve as a catalyst for building public and private conservation partnerships. The larger scale perspective of LCCs provides a unique forum that previously did not exist for putting nationwide conservation planning at a scale that allows strategic emphasis on ecological systems that are in most need of added representation and protection.

There are numerous benefits to exploring alternatives for increasing the conservation and representation of ecological systems in the protected areas network. First, we can increase the number and area of ecological systems protected. Ecological

systems represent a range of the habitats upon which many species rely; therefore we are increasing the protection of numerous species, including threatened, endangered, and species of concern. Second, we can increase the adaptability of ecological systems and the protected areas network to climate change impacts [79]. A wider range of environmental variables will enable ecological systems and the vertebrate species that rely on them to have room to shift their ranges in response to changes in climate. Third, we can increase the buffer area for all ecological systems and thereby reduce edge effects and increase the integrity of existing ecological systems. Lastly, we are more likely to capture the ecological processes that drive the pattern of ecological systems that we observe and allow for a more fully functional and robust protected areas network.

The current protected areas network for the continental US does not capture the full range of ecological systems or geophysical features (i.e., elevation and soil productivity). As a consequence, the species that rely on these ecological systems and geophysical features have fewer opportunities to adjust to changing environmental conditions. We have not assessed the pros and cons of using our alternatives for increasing the representation of ecological systems, but rather we have presented them as possibilities that may be considered and evaluated as decisions are made to conserve biodiversity. Each alternative may increase the representation of ecological systems, which can lead to protecting and securing habitat across a broader range of ecological, geographical, and geophysical occurrence of species. And may provide the greatest opportunity for evolutionary processes to persist regardless of imminent changes in the near, intermediate, and long term.

Supporting Information

Table S1 Area (ha) and percent area of ecological systems by protection status nested into Level I and II land cover groups [23,24]. All 5 Level I groups, 37 Level II groups, and 518 ecological systems are listed. See Table 1 for protection status descriptions. Only non-modified, non-aquatic ecological systems are included (n = 518).

References

Acknowledgments

We thank C. Conway, D. Weinstein, R. White, and 2 anonymous reviewers for their comments that improved this manuscript. We also thank M. Croft, L. Duarte, J. Lonneker, K. Mallory, A. Radel, and G. Wilson for their help. Any use of trade, product, or firm names is for descriptive purposes only and does not imply endorsement by the US Government.

Author Contributions

Conceived and designed the experiments: JMS LS JA AD KG. Analyzed the data: AD JA. Wrote the paper: JLA LS AM AD JS KG.

1. Miller KR (1982) Parks and protected areas: considerations for the future. Ambio 11: 315–317.
2. Pressey RL (1994) Ad hoc reservations: forward and backward steps in developing representative reserve systems? Conserv Biol 8: 662–668.
3. Margules CR, Pressey RL (2000) Systematic conservation planning. Nature 405: 243–253.
4. Fairfax SK, Gwin L, King MA, Raymond L, Watt LA (2005) Buying nature: The limits of land acquisition as a conservation strategy, 1780–2004. Cambridge: The MIT Press. 357 p.
5. Pressey RL, Humphries CJ, Margules CR, Vane-Wright RI, Williams PH (1993) Beyond opportunism: key principles for systematic reserve selection. Trends Ecol Evol 8: 124–128.
6. Ando A, Camm J, Polasky S, Solow A (1998) Species distributions, land values, and efficient conservation. Science 279: 2126–2128.
7. van Jaarsveld AS, Freitag S, Chown SL, Muller C, Koch S, et al. (1998) Biodiversity assessment and conservation strategies. Science 279: 2106–2108.
8. Kareiva P (2010) Trade-in to trade-up. Nature 466: 322–323.
9. Harrison J, Miller K, McNeely J (1982) The world coverage of protected areas: development goals and environmental needs. Ambio 11: 238–245.
10. Ehrlich PR, Wilson EO (1991) Biodiversity studies: science and policy. Science 253: 758–762.
11. Scott JM, Davis F, Csuti B, Noss R, Butterfield B, et al. (1993) Gap analysis: a geographic approach to protection of biological diversity. Wildlife Monographs 123: 1–41.
12. Chape S, Blyth S, Fish L, Fox P, Spalding M (2003) 2003 United Nations list of protected areas. Available: http://www.unep.org/pdf/un-list-protected-areas.pdf. Access 29 February 2012.
13. Rodrigues ASL, Akçakaya HR, Andelman SJ, Bakarr MI, Boitani L, et al. (2004) Global gap analysis: priority regions for expanding the global protected-area network. Bioscience 54: 1092–1100.
14. McDonald RI (2009) The promise and pitfalls of systematic conservation planning. Proc Natl Acad Sci U S A 106: 15101–15102.
15. Scott JM, Davis FW, McGhee RG, Wright RG, Groves C, et al. (2001) Nature preserves: do they capture the full range of America's biological diversity? Ecol Appl 11: 999–1007.
16. Baron JS, Griffith B, Joyce LA, Kareiva P, Keller BD, et al. (2008) Preliminary review of adaptation options for climate-sensitive ecosystems and resources. Available: http://library.globalchange.gov/products/assessments/sap-4-4-preliminary-review-of-adaptation-options-for-climate-sensitive-ecosystems-and-resources. Accessed 29 February 2012.
17. Glicksman RL (2008) Ecosystem resilience to disruptions linked to global climate change: An adaptive approach to federal land management. Neb Law Rev 87: 833–892.
18. Brooks TM, Bakarr MI, Boucher T, Da Fonseca GAB, Hilton-Taylor C, et al. (2004) Coverage provided by the global protected-area system: is it enough? Bioscience 54: 1081–1091.
19. Estes JE, Mooneyhan DW (1994) Of maps and myths. Photogramm Eng Remote Sensing 60: 517–524.
20. Dietz RW, Czech B (2005) Conservation deficits for the continental United States: an ecosystem gap analysis. Conserv Biol 19: 1478–1487.
21. Noss RF, Cooperrider AY (1994) Saving Nature's legacy. Washington DC: Island Press. 416 p.
22. The President's Council of Advisors on Science and Technology (PCAST) (2011) Sustaining environmental capital: protecting society and the economy. Available: http://www.whitehouse.gov/administration/eop/ostp/pcast/docsreports. Accessed 2011 July 29.
23. US Geological Survey, Gap Analysis Program (USGS-GAP) (2010) National GAP Land Cover, Version 1. Available: http://gapanalysis.usgs.gov. Accessed 29 July 2011.
24. US Geological Survey, Gap Analysis Program (USGS-GAP) (2010) Protected Areas Database of the United States, Version 1.0. Available: http://gapanalysis.usgs.gov. Accessed 2011 July 29.
25. Comer P, Faber-Langendoen D, Evans R, Gawler S, Josse C, et al. (2003) Ecological systems of the United States: a working classification of US terrestrial systems. Available: http://www.natureserve.org/library/usEcologicalsystems.pdf. Accessed 29 July 2011.
26. Redford KH, Richter BD (1999) Conservation of biodiversity in a world of use. Conserv Biol 13: 1246–1256.
27. Sanderson EW, Jaiteh M, Levy MA, Redford KH, Wannebo AV, et al. (2002) The human footprint and the last of the wild. Bioscience 52: 891–904.
28. Hobbs RJ, Arico S, Aronson J, Baron JS, Bridgewater P, et al. (2006) Novel ecosystems: theoretical and management aspects of the new ecological world order. Global Ecol Biogeogr 15: 1–7.
29. Sodhi NS, Butler R, Laurance WF, Gibson L (2011) Conservation successes at micro-, meso-, and macroscales. Trends Ecol Evol 26: 585–594.
30. Noss RF, LaRoe III ET, Scott JM (1995) Endangered ecosystems of the United States: A preliminary assessment of loss and degradation. Available: http://biology.usgs.gov/pubs/ecosys.htm. Accessed 2010 July 22.
31. Bunce RGH, Bogers MMB, Evans D, Halada L, Jongman RHG, et al. (2012). The significance of habitats as indicators of biodiversity and their links to species. Ecol Indic http://dx.doi.org/10.1016/j.ecolind.2012.07.014. Accessed 2012 August 31.
32. Dudley N (2008) Guidelines for applying protected area management categories. Available: http://data.iucn.org/dbtw-wpd/edocs/paps-016.pdf. Accessed 2012 February 29.
33. Noss RF (1990) Indicators for monitoring biodiversity: a hierarchical approach. Conserv Biol 4: 355–364.
34. Tear TH, Kareiva P, Angermeier PL, Comer P, Czech B, et al. (2005) How much is enough? The recurrent problem of setting measurable objectives in conservation. Bioscience 55: 835–849.
35. Svancara LK, Brannon R, Scott JM, Groves CR, Noss RF, et al. (2005) Policy-driven versus evidence-based conservation: a review of political targets and biological needs. Bioscience 55: 989–995.
36. Convention on Biological Diversity Strategic Plan for Biodiversity 2011–><2020 including Aichi Biodiversity Targets. Available:.Accessed 2012 February 29.
37. Shaffer ML, Stein BA (2000) Safeguarding our precious heritage. In: Stein BA, Kutner LS, Adams JS, editors. Precious heritage: the status of biodiversity in the United States. New York: Oxford University Press. 301–321.
38. Groves CR (2003) Drafting a conservation blueprint: a practioner's guide to planning for biodiversity. Washington DC: Island Press. 458 p.
39. Scott D, Malcolm JR, Lemieux C (2002) Climate change and modelled biome representation in Canada's national park system: implications for system planning and park mandates. Global Ecol Biogeogr 11: 475–484.
40. Lemieux CJ, Scott DJ (2005) Climate change, biodiversity conservation and protected area planning in Canada. Can Geogr 49: 384–399.
41. Lemieux CJ, Beechey TJ, Gray PA (2011) Prospects for Canada's protected areas in an era of rapid climate change. Land use policy Available: doi:10.1016/j.landusepol.2011.03.008. Accessed 2011 April 29.
42. Shelford VE (1926) Naturalist's guide to the Americas. Baltimore: The Williams and Wilkins Company. 761 p.
43. Crumpacker DW, Hodge SW, Friedley D, Gregg WP (1988) A preliminary assessment of the status of major terrestrial and wetland ecosystems on Federal and Indian lands in the United States. Conserv Biol 2: 103–115.
44. Caicco SL, Scott JM, Butterfield B, Csuti B (1995) A gap analysis of the management status of the vegetation of Idaho (USA). Conserv Biol 9: 498–511.
45. Davis FW, Stine PA, Stoms DM, Borchert MI, Hollander AD (1995) Gap analysis of the actual vegetation of California 1. The Southwestern Region. Madroño 42: 40–78.

46. Stoms DM, Davis FW, Driese KL, Cassidy KM, Murray MP (1998) Gap analysis of the vegetation of the intermountain semi-desert ecoregion. Great Basin Nat 58:199–216.

47. Scott JM, Murray M, Wright RG, Csuti B, Morgan P, et al. (2001) Representation of natural vegetation in protected areas: capturing the geographic range. Biodivers Conserv 10: 1297–1301.

48. Wright RG, Scott JM, Mann S, Murray M (2001) Identifying unprotected and potentially at risk plant communities in the western USA. Biol Conserv 98: 97–106.

49. Fuller RA, McDonald-Madden E, Wilson KA, Carwardine J, Grantham HS, et al. (2010) Replacing underperforming protected areas achieves better conservation outcomes. Nature 466: 365–367.

50. Langhammer PF, Bakarr MI, Bennun LA, Brooks TM, Clay RP, et al. (2007) Identification and gap analysis of key biodiversity areas: targets for comprehensive protected area systems. Available: data.iucn.org/dbtw-wpd/edocs/pag-015.pdf. Accessed 2012 February 29.

51. Kark S, Levin N, Grantham HS, Possingham HP (2009) Between-country collaboration and consideration of costs increase conservation planning efficiency in the Mediterranean Basin. Proc Natl Acad Sci U S A 106: 15368–15373.

52. Swaty R, Blankenship K, Hagen S, Fargione J, Smith J, et al. (2011) Accounting for ecosystem alteration doubles estimates of conservation risk in the conterminous United States. PLoS ONE 6: 1–10. DOI: 10.1371/journal.pone.0023002.

53. Millard MJ, Czarnecki CA, Morton JM, Brandt LA, Shipley FS, et al. (2012) A national geographic framework for guiding conservation on a landscape scale. Journal of Fish and Wildlife Management 3: 175–183.

54. Lowry Jr JH, Ramsey RD, Boykin K, Bradford D, Comer P, et al. (2007) Land cover classification and mapping. In: Prior-Magee JS, et al., editors. Southwest Regional Gap Analysis Final Report. Available: http://fws-nmcfwru.nmsu.edu/swregap/report/swregap%20final%20report.pdf. Accessed 2011 November 16.

55. Zhu Z, Vogelmann J, Ohlen D, Kost J, Chen X, et al. (2006) Mapping existing vegetation composition and structure for the LANDFIRE Prototype Project. In: Rollins MG, Frame CK, editors. The LANDFIRE prototype project: nationally consistent and locally relevant geospatial data for wildland fire management Available: http://www.treesearch.fs.fed.us/pubs/24700. Accessed 2011 January 18.

56. Rollins MG (2009) LANDFIRE: a nationally consistent vegetation, wildland fire, and fuel assessment. International Journal of Wildland Fire 18: 235–249.

57. Sanborn (2006) GAP zone 1 vegetation mapping final report. Available: http://gap.uidaho.edu. Accessed 2011 July 29.

58. Lowry J, Ramsey RD, Thomas K, Schrupp D, Sajwaj T, et al. (2007) Mapping moderate-scale land-cover over very large geographic areas within a collaborative framework: a case study of the Southwest Regional Gap Analysis Project (SWReGAP). Remote Sens Environ 108: 59–73.

59. Crist PJ, Prior-Magee JS, Thompson BC (1996) Land management status categorization in gap analysis: a potential enhancement. In Brackney ES, Jennings MD, editors. Gap Analysis Bulletin 5. Available: http://www.gap.uidaho.edu/bulletins/5/LMSCiGA.html. Accessed 2010 March 29.

60. U. S. Geological Survey (2006) National elevation dataset. Available: http://ned.usgs.gov/. Accessed 2011 July 29.

61. Hazen HD, Anthamatten PJ (2004) Representation of ecological regions by protected areas at the global scale. Physical Geography 25: 499–512.

62. Hoekstra JM, Boucher TM, Ricketts TH, Roberts C (2005) Confronting a biome crisis: global disparities of habitat loss and protection. Ecol Lett 8: 23–29.

63. Groves CR, Kutner LS, Stoms DM, Murray MP, Scott JM, et al. (2000) Owning up to our responsibilities: who owns land important for biodiversity? In: Stein BA, Kutner LS, Adams JS, editors. Precious heritage: The status of biodiversity in the United States. New York: Oxford University Press. 399 p.

64. Andersen MG, Ferree CE (2010) Conserving the stage: climate change and the geophysical underpinnings of species diversity. PLoS ONE 5: 1–10. DOI:10.737/journal.pone.0011554.

65. Tversky A, Kahneman D (1974) Judgment under uncertainty: heuristics and biases. Science 185: 1124–1131.

66. Parmesan C (2006) Ecological and evolutionary responses to recent climate change. Annu Rev Ecol Evol Syst 37: 637–669.

67. Parmesan C, Yohe G (2003) A globally coherent fingerprint of climate change impacts across natural systems. Nature 421: 37–42.

68. Retti DR (1995) Our National Park System: caring for America's greatest natural and historic treasures. Urbana: University of Illinois Press. 293 p.

69. Mascia MB, Pailler S (2011) Protected area downgrading, downsizing, and degazettement (PADDD) and its conservation implications. Conservation Letters 4: 9–20.

70. Hogenauer AK (1991) Gone, but not forgotten: The delisted units of the US National Park System. The George Wright Forum 7: 2–19.

71. Hogenauer AK (1991) An update to "Gone, but not forgotten: The delisted units of the US National Park System. The George Wright Forum 8: 26–28.

72. US Department of Interior and US Department of Agriculture (2005) National Land Acquisition Plan. Available: http://www.fs.fed.us/land/staff/LWCF/Final%20DOI-USDA%20Land%20Acquisition%20Report%20to%20Congress.pdf. Accessed 2012 September 5.

73. Loomis JB (1993) Integrated public lands management. New York: Columbia University Press 474 p.

74. Thomas JW, Sienkiewicz A (2005) The relationship between science and democracy: public land policies, regulation, and management. Public Land and Resources Law Review 26: 39–69.

75. Darst CR, Huffman KA, Jarvis J (2009) Conservation significance of America's newest system of protected areas: National Landscape Conservation System. Natural Areas Journal 29: 224–254.

76. Towns E, Cook JE (1998) USDA, Forest Service, USDI, National Park Service: Notice of Transfer of Administrative Jurisdiction, Coconino National Forest and Walnut Canyon National Monument. Available: http://www.gpo.gov/fdsys/pkg/FR-1998-08-25/pdf/98-22723.pdf. Accessed 2012 August 31.

77. Stobaugh J (2003) Notice of Proposed Withdrawal and Opportunity for Public Meeting: Nevada. Available: http://www.gpo.gov/fdsys/pkg/FR-2003-07-09/pdf/03-17392.pdf. Accessed 2012 August 31.

78. Allred CS (2007) Public Land Order No. 7675: Transfer of Administrative Jurisdiction, Petrified Forest National Park Expansion, Arizona. Available: http://www.gpo.gov/fdsys/pkg/FR-2007-05-18/pdf/E7-9586.pdf. Accessed 2012 August 31.

79. Kujala H, Araújo MB, Thuiller W, Cabeza M (2011) Misleading results from conventional gap analysis – messages from the warming north. Biol Conserv 144: 2450–2458.

11

Legacy Effects of Canopy Disturbance on Ecosystem Functioning in Macroalgal Assemblages

Leigh W. Tait[¤]*, David R. Schiel

Marine Ecology Research Group, School of Biological Sciences, University of Canterbury, Christchurch, New Zealand

Abstract

Macroalgal assemblages are some of the most productive systems on earth and they contribute significantly to nearshore ecosystems. Globally, macroalgal assemblages are increasingly threatened by anthropogenic activities such as sedimentation, eutrophication and climate change. Despite this, very little research has considered the potential effects of canopy loss on primary productivity, although the literature is rich with evidence showing the ecological effects of canopy disturbance. In this study we used experimental removal plots of habitat-dominating algae (Order Fucales) that had been initiated several years previously to construct a chronosequence of disturbed macroalgal communities and to test if there were legacy effects of canopy loss on primary productivity. We used *in situ* photo-respirometry to test the primary productivity of algal assemblages in control and removal plots at two intertidal elevations. In the mid tidal zone assemblage, the removal plots at two sites had average primary productivity values of only 40% and 60% that of control areas after 90 months. Differences in productivity were associated with lower biomass and density of the fucoid algal canopy and lower taxa richness in the removal plots after 90 months. Low-shore plots, established three years earlier, showed that the loss of the large, dominant fucoid resulted in at least 50% less primary productivity of the algal assemblage than controls, which lasted for 90 months; other smaller fucoid species had recruited but they were far less productive. The long term reduction in primary productivity following a single episode of canopy loss of a dominant species in two tidal zones suggests that these assemblages are not very resilient to large perturbations. Decreased production output may have severe and long-lasting consequences on the surrounding communities and has the potential to alter nutrient cycling in the wider nearshore environment.

Editor: Simon Thrush, National Institute of Water & Atmospheric Research, New Zealand

Funding: This work was funded by a University of Canterbury scholarship, the Andrew W Mellon Foundation of New York, and the New Zealand Foundation for Research, Science and Technology (Coasts and Oceans OBI, grants C01X0307 and C01X0501). The funders had no role in study design, data collection and analysis, decision to publish, or preparation of the manuscript.

Competing Interests: The authors have declared that no competing interests exist.

* E-mail: taitl@science.oregonstate.edu

¤ Current address: Department of Zoology, Oregon State University, Corvallis, Oregon, United States of America

Introduction

Macroalgae play a critical role in primary production [1] and habitat provision [2], and are commonly recognized as key species in structuring the biodiversity within communities in temperate marine systems [3,4]. It has been widely documented that there is a current global decline in rocky shore habitat-forming macroalgal species from a wide range of stressors [5–7]. These species are often in high abundance and biomass, but susceptible to loss through both acute and diffuse stressors placed on coastal ecosystems such as increased coastal run-off and sedimentation [8–10], coastal development [11], impaired water quality [10], increased temperatures [12–14] and changes to the wave climate [15]. The loss of these habitat-forming species typically shifts systems that are structured in multiple dimensions (canopy, subcanopy and basal layers) to one dominated by low-lying, turf forming, filamentous or ephemeral algae [7,16,17]. The susceptibility of macroalgae to disturbance and their critical role in structuring communities makes them a prime candidate to test the long-term consequences of species loss on ecosystem function.

Although the long-term consequences of losing habitat-forming species, particularly fucoid algae, on community structure and composition are increasingly understood [18–22], there is relatively little information on the potential consequences to a critical ecosystem function, that of primary productivity. Macroalgal diversity has been shown to enhance primary productivity [23] and nitrogen uptake [24], at least in some circumstances, and the loss of these species could have significant impacts on the stability of nearshore ecosystems. Furthermore, there may be very little functional replacement of canopy forming macroalgae within intertidal assemblages [4], potentially making these assemblages vulnerable to prolonged shifts in community composition.

Chronosequences have been extensively used by ecologists to examine successional patterns where the long life span of species precludes time series observations of the entire successional sequence [25]. A chronosequence can be referred to as a mosaic of patches that have been developing for various lengths of time following a known disturbance. Such observations have been useful in formulating models of succession in terrestrial plant communities [25] and seagrass beds [26]. After a major disturbance, in certain cases, the community may not reach what is considered its 'climax state,' even after an extended period of time [16,18]. Long-term shifts in community composition have the potential to drastically alter the functioning of macroalgal assemblages. While chronosequences have been used to elucidate patterns of succession, they may also be useful in examining changes in primary productivity over time.

Here we use a series of patches, disturbed experimentally at known times, and field-based photo-respirometry to test the legacy effects of the loss of canopy-forming algae on primary productivity. *In situ* photo-respirometry can measure primary productivity without altering target assemblages [27], thereby allowing them to recover along their natural successional trajectory. Understanding the direct and indirect effects of canopy loss on primary productivity over ecologically significant time-scales will allow insight into the role of macroalgal succession in the variability of primary productivity. The aim of our work, therefore, was to test the ability of macroalgal assemblages to recover to control (or pre-disturbed) levels of primary productivity and to determine the association of primary productivity with community composition. We test the null hypothesis that primary productivity of mid and low shore assemblages is unaffected by canopy removal and time since removal.

Methods

Chronosequence experiment, standardization and species cover

To test the long-term consequences of disturbance to dominant canopies on primary productivity, a combination of old and new removal plots were used. Canopy-forming algal species of the Order Fucales were removed from areas at two tidal heights (low and mid shore). In mid-shore algal assemblages (tidal elevation of 0.8–0.9 m), the dominant fucoid *Hormosira banksii* was removed from 3 m ×3 m experimental plots in June 2002 at two sites (North Reef Moeraki 45°11'S, 170°98'E and Wairepo Reef Kaikoura, 42°25'S, 173°42'E) to test its role in ameliorating stress to the mid-shore assemblages of subcanopy algae and invertebrates [16,21]. *Hormosira* reaches lengths of around 40 cm, with densities of hundreds of plants per m^2 and a biomass of around 7 kg per m^2 [4]. In the low shore zone (tidal elevation of 0–0.25 m), the large fucoid *Durvillaea antarctica* was removed from 2 m ×2 m plots (n = 4) at North Reef, Moeraki and Oaro Reef, Kaikoura (42°30'S, 173°30'E) in March 2006 to examine the role of *D. antarctica* in driving community composition. *D. antarctica* adults can reach 10 m in length, densities of several plants per m^2, and a biomass of 70 kg per m^2. Because the very uneven limestone and conglomerate reef at Oaro Reef rendered it impossible to attach the incubation chambers effectively (see below), this site could not be used for photo-respirometry. In both mid and low intertidal assemblages, therefore, we used these plots to gauge the long-term consequences of canopy loss to primary productivity.

To gauge shorter term effects, we then added new canopy removal treatments in the mid-shore during March 2008 (Moeraki) and January 2010 (Kaikoura) and in the low-shore (Moeraki) in September 2009. New controls with *H. banksii* (mid-shore) and *D. antarctica* (low shore) canopies intact (n = 3), were also established, to produce a balanced design along with the old controls, which remained with *H. banksii* (mid-shore) and *D. antarctica* (low shore) canopies intact [16]. This chronosequence of canopy disturbance enabled us to test the trajectory of recovery of primary productivity following single episodes of major disturbance.

Incubations measuring primary productivity were done using custom-designed photo-respirometry chambers [27]. These were sealed around target assemblages immediately prior to incubations and removed after incubations to limit any long-term disturbance to the target assemblage. This allowed assemblages to be exposed to natural conditions and sampled over time with minimal impact [27]. Chambers were 25 cm in diameter, had a volume of 14.7 L, and covered 491 cm^2 of substratum. Before attachment of chambers, a two-compound epoxy resin was used to fill in deeper cracks within the substratum so that chambers could be sealed, but care was taken not to change the reef composition such that pooling of water occurred. Before incubations were done, all visible invertebrates were removed from the assemblages to limit the effects of heterotrophic respiration. On the flat reef surfaces, the only manipulation of the reef was the drilling of holes for rawl plugs for attachment of chambers. Oxygen concentrations were sampled using a Hach LDO meter at intervals no greater than 20 minutes to avoid super-saturation of oxygen [27]. The change in dissolved oxygen over time was then converted to changes in carbon uptake using a P:Q (photosynthetic quotient) ratio of 1:1 and scaled up to carbon uptake per m^2 of reef surface (g C m^{-2} h^{-1}). During incubations, irradiance was measured using HOBO irradiance and temperature loggers (cross-calibrated with Li-Cor LI 192 quantum sensor) and averaged for the period of the incubation. Plots were sampled across a range of natural irradiance intensities (from 0–2100 µmol m^{-2} s^{-1}), with single plots often taking multiple days to sample. Primary productivity for each treatment was then taken as the maximum productivity (P_{max}) per sample area and adjusted to g C m^2 hr^{-1}; it was also standardized as productivity per dry biomass of algal material mg C gDW^{-1} h^{-1} to take account of the changing biomass per area in the experimental plots. Because treatment and control plots were re-visited throughout the years of the experiment, harvesting plots for biomass was not plausible, but dry biomass was estimated using length/mass relationships of *D. antarctica* and percent cover/mass relationships for *H. banksii*.

In addition to assessing primary productivity, the cover and diversity of species were measured at each sampling time. This was done so that primary productivity could be related to recovery and succession through time as the species composition changed. During primary productivity sampling, all macroalgal species within the areas covered by the chambers were recorded and their percentage cover estimated using a gridded quadrat. Multi-dimensional scaling plots (MDS) using PRIMER were used to visualize the variation and recovery trajectory in community structure between treatments through time. Differences between treatment plots through time were analyzed using PERMANOVA (using 9999 simulation permutations; [28]) analysis. For analysis, all replicate plots (n = 3) were plotted independently, but for graphical presentation replicate plots were averaged for each time period and only the centroid of each treatment at each sampling time was plotted. The data for percent cover of *H. banksii* and *D. antarctica* were removed from MDS analysis and PERMANOVA analysis, as is standard procedure in removal-type experiments because these distort community analyses between treatments.

Sequence of primary productivity

One incubation chamber was set up within each of the old experimental treatments on the mid-shore. When we began the primary productivity studies, these plots were 78 (Moeraki) and 90 (Kaikoura) months old (6.5 and 7.5 years respectively). In the new removal treatments, the *H. banksii* canopy was removed from three 0.5 ×0.5 m areas, within which the incubation chambers were situated. Moeraki plots were sampled in spring 2008, autumn 2009 and spring 2009 giving the corresponding chronosequence of 0, 6, 12, 78, 84 and 90 months after canopy disturbance. Kaikoura plots were sampled in austral summer 2010 and winter 2010, giving the corresponding chronosequence of 0, 6, 90 and 96 months after canopy disturbance.

Similarly on the low shore, one incubation chamber was set up within each of the old experimental removal and control plots. These plots were 30 months old when we began the primary productivity studies. In the new removal plots, *D. antarctica* adults were removed from three 0.5×0.5 m areas, within which the incubation chambers were situated. Incubations testing primary productivity in all treatments were done during spring 2009 and autumn 2010. This gave a chronosequence of 0, 6, 30 and 36 months after canopy disturbance. Because of the large sizes that *D. antarctica* plants can reach, controls were set up around moderate sized plants (no taller than 50–60 cm) because larger ones would not fit inside chambers.

Results

Sequence of primary productivity in mid-shore communities

The loss of the *H. banksii* canopy had a major impact on the productivity of the mid- shore assemblages of both sites (Fig. 1). The immediate effect was a reduction of primary productivity from 2.1 to 0.6 g C m^{-2} h^{-1} at Moeraki (Fig. 1A) and 2.9 to 0.4 g C m^{-2} h^{-1} at Kaikoura (Fig. 1C). Primary productivity results were similar when standardized to biomass, with a c. 50% reduction in removal treatments at both sites (Figs. 1B and D). In all cases, the lower values in the *Hormosira* removal plots represented the productivity of the remaining understory algal assemblage and indicate that both on a per-area and per-biomass

basis, the loss of the canopy fucoid resulted in a significant loss of primary productivity. The differences between treatments and sites persisted throughout the first year at Moeraki and at least for 6 months at Kaikoura (per-area Treatment×Site interaction at 6 months: $F_{8,11} = 10.22$, p = 0.013). The long-term plots, however, showed increasing differences between the two sites. After 90 months, primary productivity in removal plots at Moeraki did not recover beyond initial levels, but those at Kaikoura had recovered much of their primary productivity (per-area Treatment×Site interaction: $F_{8,11} = 5.57$, p = 0.046). We were able to do an additional set of incubations at Kaikoura. This showed that after 8 years (96 months) there was finally a convergence of treatment and control plots in primary productivity (per-area: $F_{1,4} = 1.96$, p = 0.23). Multi-dimensional scaling plots illustrated the large variation in assemblage structure in the removal treatments over time at both sites (Fig. 2). Treatments diverged greatly after six months, as the remaining algal assemblages in removal plots responded to the loss of the canopy and increasing stress in the mid tidal zone. PERMANOVA analysis of community composition at Moeraki showed a significant difference between treatments ($F_{1,24} = 16.8$, p<0.01), times ($F_{5,24} = 7.6$, p<0.01) and a significant interaction (treatment×time, $F_{5,24} = 2.1$, p<0.05). The poor recovery at this site was indicated by the continued divergence of the community, even after 90 months (Fig. 2A). Furthermore, although *H. banksii* cover had recovered somewhat, it was still lower than controls (95% cover in controls, 78% cover in removals). The control plots had more fucoid algae, especially

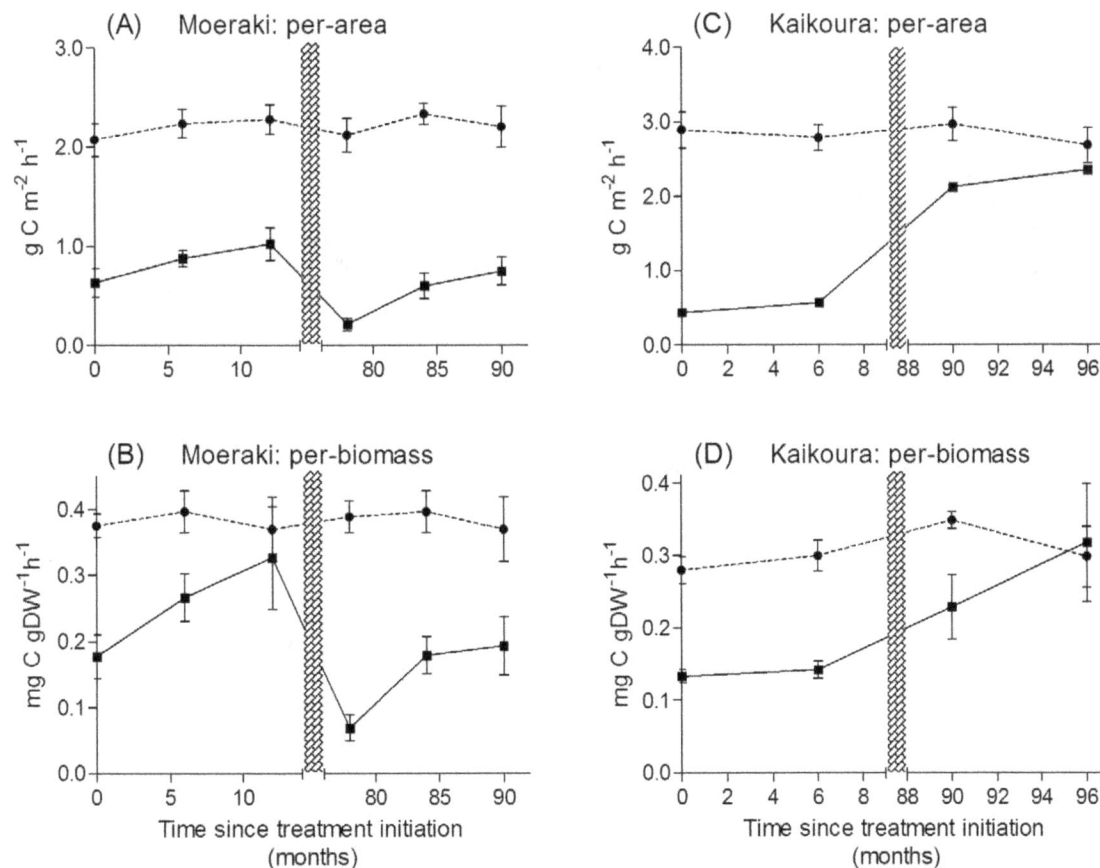

Figure 1. Primary productivity (±SE) in mid intertidal zone control (dashed lines) and canopy (*Hormosira banksii*) removal (solid lines) plots in two sites through time, standardized on a per-area basis (Moeraki (A) and Kaikoura (C)) and per-biomass basis (Moeraki (B) and Kaikoura (D)). Shaded bar indicates the change from new to old treatments (see text).

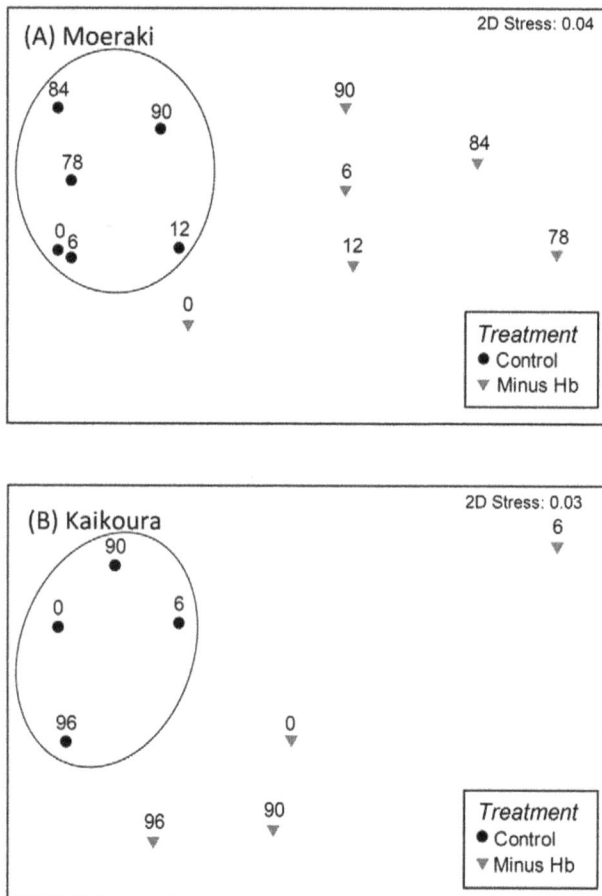

Figure 2. MDS plots of changing community composition in mid intertidal zone control (intact assemblages), and canopy (*Hormosira banksii*) removal plots through time at Moeraki (A) and Kaikoura (B). Each point represents the centroid of three replicate plots. Numbers above symbols indicate the number of months since the initiation of treatments.

Cystophora spp, and branching red algae (*Lophothamnion hirtum*). Removal plots had more turfing coralline algae, especially a greater cover of *C. officinalis* and *Jania micrathrodia*, and more ephemeral species. The cover of corallines varied considerably over time due to burn-off during the summer months. After 6 months, *H. banksii* had started recruiting into removal plots but their cover and biomass were far less than in controls. Ninety months after initial removal, there was a similar species composition in control and removal plots, with very similar cover of *Corallina officinalis* and encrusting corallines. However, *H. banksii* density and biomass remained less than in controls [29], which translated into the large differences seen in primary productivity.

At Kaikoura, the algal community composition of removal plots was very different to controls as early as 6 months after canopy removal (Fig. 2B). This was caused by the die-back of corallines and the abundance of ephemeral, disturbance-oriented species such as *Ulva*, *Colpomenia* and *Adenocystis*. PERMANOVA analysis showed a significant effect of treatment ($F_{1,16} = 6.5$, p<0.05) and time ($F_{3,16} = 2.5$, p<0.01), but no interaction. Throughout the time series, there were generally fewer fucoids, more corallines and more ephemerals in removal plots relative to controls. After 96 months, removal treatments were similar to controls (Fig. 2 B). The *H. banksii* canopy recovery was much quicker at Kaikoura

than at Moeraki, and there was around a 90% canopy cover of recruits after approximately 2 years, which grew into full cover of large plants. At both sites, however, species richness and the abundance of mid-canopy fucoids were still lower in removal plots than controls after 90 months at both Kaikoura and Moeraki [29].

Sequence of primary productivity in low shore communities

Removal of the *Durvillaea antarctica* canopy had a large initial impact on per-area productivity ($F_{1,4} = 17.2$, p = 0.014), which persisted at the same level for at least 6 months ($F_{1,4} = 37.0$, p = 0.004; Fig. 3A). The controls had an average of around 10 g C m^{-2} h^{-1} compared to the removals at about 1.3 g C m^{-2} h^{-1}. The older removal treatments showed that recovery of plots had begun by 30 months after canopy removal but significant differences remained after 36 months ($F_{1,4} = 9.99$, p = 0.034), with controls at an average of 10.4 g C m^{-2} h^{-1} and removal plots at 5.4 g C m^{-2} h^{-1}. These levels of primary productivity in *Durvillaea* rival that of other large macroalgae such as *Macrocystis pyrifera* [30]. Although less pronounced, there was also a significant effect of *D. antarctica* removal on a per dry biomass basis ($F_{1,4} = 8.24$, p = 0.041, Fig. 3B). Between 0–6 months after canopy removal, per-biomass productivity in removal plots was approximately 50% of the control treatment (compared to c. 15% when standardized by reef area), which indicates the large per-capita contribution that *Durvillaea* makes to productivity. On both a per-area and per-biomass basis, therefore, its loss greatly affected productivity.

MDS plots of community composition show an obvious separation between treatments in multi-dimensional space, which increased through 6 and 30 months and began converging by 36 months (Fig. 4). PERMANOVA analysis showed there were significant differences between treatments ($F_{1,16} = 6.0$, p<0.05) and time ($F_{3,16} = 1.8$, p<0.05). These differences involved a) the shifting proportions of assemblages between *D. antarctica* and *Corallina officinalis* in control plots, and b) expanded cover of corallines, incursions of the invasive kelp *Undaria pinnatifida*, and recruitment of the low-shore fucoid *Cystophora torulosa* in removal plots. Removal plots had a 30% increase in the cover of coralline turf (*Haliptilon roseum* and *C. officinalis*) in the first 6 months, but an overall decline in algal diversity. After 30 months, the removal plots had a large variation in species diversity and cover, with plots dominated by *C. torulosa*, *U. pinnatifida*, *C. officinialis* and the subtidal fucoid *Xiphophora gladiata*, but with increasing cover of juvenile *D. antarctica* by 36 months. In control plots, the canopy cover of *D. antarctica* was almost 100% at all sampling intervals, although there was some natural variation in the subcanopy composition.

Discussion

The loss or severe reduction of canopy-forming algae in the nearshore zone has produced changes in communities worldwide, usually as a result of multiple stressors [31]. Where fucoid algae are involved, the trajectory of recovery is often slow, sometimes over decades [18–22]. The consequences of these changes on one of the most important ecosystem functions, that of primary productivity, are generally not known. Here, however, we show that reduced productivity occurs for several years in both the mid and low intertidal assemblages, primarily because of the slow pace of recovery of canopy species [29] and the advent of lower-lying turfs and fleshy algae.

Community changes following fucoid canopy impacts are complex and varied among different regions. On British shores, for example, loss of a dominant may allow other fucoids to establish, involve grazer dynamics and eventual slow re-establish-

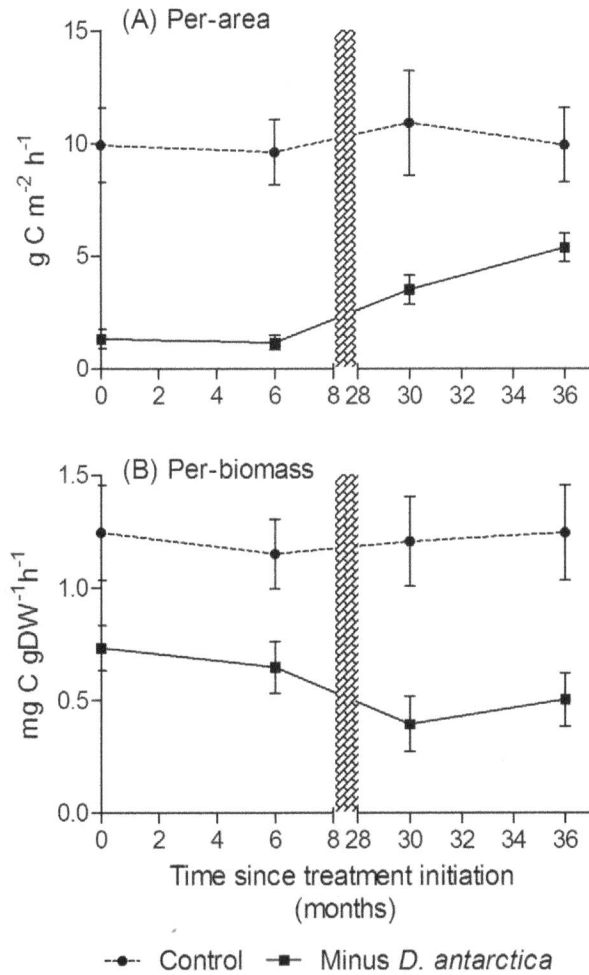

Figure 3. Primary productivity (±SE) in low intertidal zone control (dashed lines) and canopy (*Durvillaea antarctica*) removal (solid lines) plots at Moeraki, standardized on a per-area basis (A) and per-biomass basis (B). Shaded bar indicates the change from new to old treatments (see text).

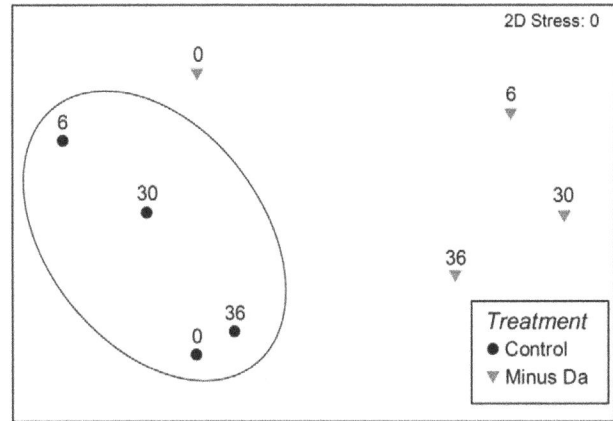

Figure 4. MDS plots of changing community composition in low intertidal zone control (intact assemblages), and canopy (*Durvillaea antarctica*) removal in plots through time at Moeraki. Each point represents the centroid of three replicate plots. Numbers above symbols indicate the number of months since the initiation of treatments.

ment of the original dominant species [32]. In many areas where stressors such as coastal sedimentation persist, there is an increased abundance of low-lying turf species [5–7]. Although it was not tested in this study, higher amounts of sedimentation, associated with greater coralline cover [29], and an increase in the wave climate [15] at the southern site of our study may have slowed recruitment of *Hormosira banksii* compared to the northern site of Kaikoura. Along the shores of southern New Zealand and southeastern Australia, the loss or reduction of the mid intertidal dominant, *Hormosira banksii*, produces variable responses. For example, Underwood [18,33] showed that a severe storm removed large patches of this species and that its recovery over several years was affected by both direct and indirect effects of grazers. Bellgrove et al [34] speculated for shores further south in Australia that patches of *Hormosira* and coralline turfs represented alternate states of intertidal communities. Schiel [15], however, showed that over 17 years, *Hormosira* canopies came and went in response to storms and wave damage, and that canopy losses in the mid intertidal zone triggered a series of cascading losses of assemblage structure, as newly exposed understory species became desiccated and died off. These were often replaced for several years by coralline turfs and seasonally blooming ephemeral species before

eventual recovery of the *Hormosira* canopy. It was experimentally shown, however, that even after 8 years, there were still differences in biomass and diversity between canopy-impacted areas and controls [29]. Our study shows the resulting compromised primary productivity over several years.

In both the mid and low intertidal zones, the removal of the dominant fucoid canopy immediately resulted in a large fall in primary productivity. This is hardly surprising as both the dominant biomass and much of the areal cover of algae were lost. Biomass has often been used as a surrogate for rates of primary productivity in macroalgae [30]. Our study showed that biomass was a key determinant of overall per-area primary productivity, but when productivity was expressed per-biomass there was still significantly lower productivity in removal treatments, indicating that biomass alone does not entirely explain the loss of productivity and that certain taxa are disproportionately important. In particular, the dynamics of light use by the species that dominated canopy removal areas play a key role in this response. For example, on both a per-area and per-biomass basis, corallines are some of the least productive algae. Their calcium carbonate structure makes them heavy relative to their photosynthetic capability [35]. In contrast to corallines, the ephemeral species that bloomed in removal areas had high productivity on a per-biomass basis, but were patchy and did not have great productivity on a per-area basis. Species such as sea lettuce (*Ulva* spp), for example, in a full light environment can have productivity as high as 10 mg C gDW^{-1} h^{-1} [35], compared to corallines at 0.3 mg C gDW^{-1} h^{-1} [27]. Similar high levels of per-biomass productivity occur for other ephemeral algae [35]. In intact assemblages with a full fucoid canopy, these species remain in the understory, along with several species of low-shore fucoids, and contribute to the overall assemblage primary productivity at high light intensities [36]. However, when fully exposed to a high light environment following canopy removal, they can undergo photo-inhibition [37], and often desiccate and bleach. In contrast to the loss of smaller fucoids higher up on the shore, the loss of the much larger "bull kelp" *Durvillaea antarctica* on the low shore removes the abrasion of the substratum by fronds [38], and results in elongation of corallines and the recruitment of lower biomass species with far less areal productivity. Even an incursion by the invasive annual kelp *Undaria* did not compensate in primary productivity for the loss of *Durvillaea*.

If the results seen in our study apply to assemblages elsewhere, where the loss of dominant algal canopies has led to long-term changes in communities, then there are potentially large consequences to benthic primary productivity for algal communities worldwide. Given the susceptibility of large, canopy-forming macroalgae to disturbance [8–14], understanding how their loss affects important ecosystem functions such as primary productivity is essential. Major alterations to carbon fixation through primary productivity have the potential to alter the functioning of ecosystems and the functioning of global biogeochemical cycles [39]. Decreased carbon output could have far-reaching consequences, with stable isotope signatures indicating the importance of macroalgae well beyond the nearshore environment [40,41]. The decline of dominant canopy-forming macroalgae worldwide

[5] has the potential, therefore, to alter carbon fixation of nearshore ecosystems, with consequences reaching far up the food chain.

Acknowledgments

We thank the Marine Ecology Research Group for help in the field, particularly Stacie Lilley and Paul South, Dr Malcolm Forster for continued mentoring, Dr Ian Hawes for useful discussion, and two anonymous reviewers for helpful and constructive comments.

Author Contributions

Conceived and designed the experiments: LWT DRS. Performed the experiments: LWT DRS. Analyzed the data: LWT DRS. Contributed reagents/materials/analysis tools: LWT DRS. Wrote the paper: LWT DRS.

References

1. Mann KH (2000) Ecology of coastal waters. Blackwell, Malden, Masaschusetts, USA.
2. Bertness MD, Leonard GH, Levine JM, Schmidt PR, Ingraham AO (1999) Testing the relative contribution of positive and negative interactions in rocky intertidal communities. Ecology 80: 2711–2726.
3. Bruno JF, Stachowicz JJ, Bertness MD (2003) Inclusion of facilitation into ecological theory. Trends in Ecology and Evolution 18: 119–125.
4. Schiel DR (2006) Rivets or bolts? When single species count in the function of temperate rocky reef communities. J Exp Mar Biol Ecol 338: 233–252.
5. Airoldi L, Beck MW (2007) Loss, status and trends for coastal marine habitats of Europe. Oceanography and Marine Biology: An Annual Review 45: 345–405.
6. Connell SD, Russell BD, Turner DJ, Shepherd SA, Kildea T, et al. (2008) Recovering a lost baseline: missing kelp forests from a metropolitan coast. Mar Ecol Prog Ser 360: 63–72.
7. Perkol-Finkel S, Airoldi L (2010) Loss and recovery potential of marine habitats: an experimental study of factors maintaining resilience in subtidal algal forests at the Adriatic Sea. PLoS one 5: e10791.
8. Irving AD, Balata D, Colosio F, Ferrando GA, Airoldi L (2009) Light, sediment, temperature, and the early life-history of the habitat alga *Cystoseira barbata*. Marine Biology 156: 1223–1231.
9. Shepherd SA, Watson JE, Womersley HBS, Carey JM (2009) Long-term changes in macroalgal assemblages after increased sedimentation and turbidity in Western Port, Victoria, Australia. Botanica Marina 52: 195–206.
10. Foster MS, Schiel DR (2010) Loss of predators and the collapse of southern California kelp forests (?): alternatives, explanations and generalizations. J Exp Mar Biol Ecol 393: 59–70.
11. Mangialajo L, Chiantore M, Cattaneo-Vietti R (2008) Loss of fucoid algae along a gradient of urbanisation, and structure of benthic assemblages. Mar Ecol Prog Ser 358: 63–74.
12. Schiel DR, Steinbeck JR, Foster MS (2004) Ten years of induced ocean warming causes comprehensive changes in marine benthic communities. Ecology 85: 1833–1839.
13. Connell SD, Russell BD (2010) The direct effects of increasing CO2 and temperature on non-calcifying organisms: increasing the potential for phase shifts in kelp forests. Proceedings of the Royal Society B 277: 1409–1415.
14. Wernberg T, Thomsen MS, Tuya F, Kendrick GA, Staehr PA, et al. (2010) Decreasing resilience of kelp beds along a latitudinal temperature gradient: potential implications for a warmer future. Ecology Letters 13: 685–694.
15. Schiel DR (2011) Biogeographic patterns and long-term changes on New Zealand coastal reefs: Non-trophic cascades from diffuse and local impacts. J Exp Mar Biol Ecol 400: 33–51.
16. Lilley SA, Schiel DR (2006) Community affects following the deletion of habitat-forming alga from rocky marine shores. Oecologia 148: 672–681.
17. Airoldi L, Balata D, Beck MW (2008) The grey zone: relationships between habitat loss and marine diversity and their applications in conservation. J Exp Mar Biol Ecol 366: 8–15.
18. Underwood AJ (1998) Grazing and disturbance: an experimental analysis of patchiness in recovery from a severe storm by the intertidal alga *Hormosira banksii* on rocky shores in New South Wales. J Exp Mar Biol Ecol 231: 291–306.
19. Jenkins SR, Norton TA, Hawkins SJ (2004) Long term effects of *Ascophyllum nodosum* canopy removal on mid shore community structure. Journal of the Marine Biological Association of the United Kingdom 84: 327–329.
20. Cervin G, Aberg P, Jenkins SR (2005) Small-scale disturbance in a stable canopy dominated community: implications for macroalgal recruitment and growth. Mar Ecol Prog Ser 305: 31–40.
21. Schiel DR, Lilley SA (2007) Gradients of disturbance to an algal canopy and the modification of an intertidal community. Mar Ecol Prog Ser 339: 1–11.

22. Ingolfsson A, Hawkins SJ (2008) Slow recovery from disturbance: a 20 year study of Ascophyllum canopy clearances. Journal of the Marine Biological Association of the United Kingdom 88(4): 689–691.
23. Bruno JF, Boyer KE, Duffy JE, Lee SC, Kertesz JS (2005) Effects of macroalgal species identity and diversity on primary productivity in benthic marine communities. Ecology Letters 8: 1165–1174.
24. Bracken MES, Friberg SE, Gonzalez-Dorantes CA, Williams SL (2008) Functional consequences of realistic biodiversity changes in a marine ecosystem. Proc Natl Acad Sci USA 105: 924–928.
25. Wardle DA, Walker LR, Bardgett RD (2004) Ecosystem properties and forest decline in a contrasting long-term chronosequence. Nature 305: 509–513.
26. Peterson BJ, Rose CD, Rutten LM, Fourqurean JW (2002) Disturbance and recovery following catastrophic grazing: studies of a successional chronosequence in a seagrass bed. Oikos 97: 361–370.
27. Tait LW, Schiel DR (2010) Primary productivity of intertidal macroalgal assemblages: comparison of laboratory and in situ photorespirometry. Mar Ecol Prog Ser 416: 115–125.
28. Anderson MJ (2001) A new method for non-parametric multivariate analysis of variance. Austral Ecology 26: 32–46.
29. Schiel DR, Lilley SA (2011) Impacts and negative feedbacks in community recovery over eight years following removal of habitat-forming macroalgae. J Exp Mar Biol Ecol 407: 108–115.
30. Reed DC, Rassweiler A, Arkema KK (2008) Biomass rather than growth rate determines variation in net primary productivity by giant kelp. Ecology 89: 2493–2505.
31. Schiel DR (2009) Multiple stressors and disturbances: When change is not in the nature of things. In: Wahl M, ed. Marine Hard Bottom Communities, Ecological Studies, 2009, Volume 206, Part 4. pp 281–294.
32. Jenkins SR, Coleman RA, Della Santina P, Hawkins SJ, Burrows MT, et al. (2005) Regional scale differences in the determinism of grazing effects in the rocky intertidal. Mar Ecol Prog Ser 287: 77–86.
33. Underwood AJ (1999) Physical disturbances and their direct effect on an indirect effect: responses of an intertidal assemblage to a severe storm. J Exp Mar Biol Ecol 232: 125–140.
34. Bellgrove A, McKenzie PF, McKenzie JL, Sfiligoj BJ (2010) Restoration of the habitat-forming fucoid alga Hormosira banksii at effluent-affected sites: competitive exclusion by coralline turf. Mar Ecol Prog Ser 419: 47–56.
35. Littler MM, Arnold KE (1982) Primary productivity of marine algal functional-form groups from south-western North America. Journal of Phycology 18: 307–311.
36. Tait LW (2010) Primary production of intertidal marine macroalgae: factors influencing primary production over wide spatial and temporal scales. PhD Thesis. University of Canterbury.
37. Tait LW, Schiel DR (2011) Dynamics of productivity in naturally structured macroalgal assemblages: importance of canopy structure on light-use efficiency. Mar Ecol Prog Ser 421: 97–107.
38. Santelices B, Ojeda FP (1984) Recruitment, growth and survival of *Lessonia nigrescens* (Phaeophyta) at various tidal levels in exposed habitats of central Chile. Mar Ecol Prog Ser 19: 73–82.
39. Falkowski PG, Barber RT, Smetacek V (1998) Biogeochemical controls and feedbacks on ocean primary production. Science 281: 200–206.
40. Fischer G, Wiencke C (1992) Stable carbon isotope composition, depth distribution and fate of macroalgae from the Antarctic Peninsula region. Polar Biology 12: 341–348.
41. Hill JM, McQuaid CD, Kaehler S (2006) Biogeographic and nearshore-offshore trends in isotope ratios of intertidal mussels and their food sources around the coast of southern Africa. Mar Ecol Prog Ser 318: 63–73.

12

Isolation and Identification of *Oedogonium* Species and Strains for Biomass Applications

Rebecca J. Lawton[1]*, Rocky de Nys[1], Stephen Skinner[2], Nicholas A. Paul[1]

1 School of Marine and Tropical Biology, James Cook University, Townsville, Queensland, Australia, **2** Molonglo Catchment Group, Fyshwick, Australia

Abstract

Freshwater macroalgae from the genus *Oedogonium* have recently been targeted for biomass applications; however, strains of *Oedogonium* for domestication have not yet been identified. Therefore, the objective of this study was to compare the performance of isolates of *Oedogonium* collected from multiple geographic locations under varying environmental conditions. We collected and identified wild-type isolates of *Oedogonium* from three geographic locations in Eastern Australia, then measured the growth of these isolates under a range of temperature treatments corresponding to ambient conditions in each geographic location. Our sampling identified 11 isolates of *Oedogonium* that could be successfully maintained under culture conditions. It was not possible to identify most isolates to species level using DNA barcoding techniques or taxonomic keys. However, there were considerable genetic and morphological differences between isolates, strongly supporting each being an identifiable species. Specific growth rates of species were high ($>26\%$ day^{-1}) under 7 of the 9 temperature treatments (average tested temperature range: 20.9–27.7°C). However, the variable growth rates of species under lower temperature treatments demonstrated that some were better able to tolerate lower temperatures. There was evidence for local adaptation under lower temperature treatments (winter conditions), but not under higher temperature treatments (summer conditions). The high growth rates we recorded across multiple temperature treatments for the majority of species confirm the suitability of this diverse genus for biomass applications and the domestication of *Oedogonium*.

Editor: Jonathan H. Badger, J. Craig Venter Institute, United States of America

Funding: This research is supported by the Australian Government through the Australian Renewable Energy Agency and the Advanced Manufacturing Cooperative Research Centre (AMCRC), funded through the Australian Government's Cooperative Research Centre Scheme. The funders had no role in study design, data collection and analysis, decision to publish, or preparation of the manuscript.

Competing Interests: Author Stephen Skinner as he is employed by the commercial company Molonglo Catchment Group.

* E-mail: rebecca.lawton@jcu.edu.au

Introduction

Freshwater macroalgae have diverse applications as targets for biofuels [1], the bioremediation of waste waters [2–4], fertiliser and soil conditioners [5]and as a tool for carbon sequestration [1]. But despite their potential for biomass applications, freshwater macroalgae have thus far been under-utilised as a feedstock. To date, only a single study has compared the performance of freshwater macroalgae in order to identify target species for biomass applications. Lawton et al. [6] identified the cosmopolitan genus *Oedogonium* as a target for biomass applications due to its high productivity, favourable biochemical composition, cosmopolitan distribution and competitive dominance over other algal species. However, these findings were based on the performance of a single wild-type strain of *Oedogonium*. In order to realise the full potential of this alga for biomass applications, high productivity strains of *Oedogonium* need to be identified and domesticated.

Domestication and selective breeding of agricultural food crops has resulted in large gains in productivity compared to wild-type varieties and generated the high yielding strains which are cultivated today [7].For example, worldwide production of cereals more than doubled between 1960 and 2000 as a result of selective breeding and optimisation of cultivation techniques [8].Applying the same processes and principles used in past agricultural

domestications to new target species, biofuel crops for example, will enable efficient identification of strains for domestication [9]. As a first step towards domestication, the natural variability of wild-type strains for production related traits needs to be determined and the degree to which different genotypes perform best under different environments assessed [9]. Performance tests of different provenances and strains over a range of conditions are necessary in order to reveal possible genotype by environment (G X E) interactions and determine appropriate strains to domesticate [10]. If there are strong genotype by environment interactions, some strains may be adapted to local conditions at their collection locations and may not perform well under other conditions [11]. Therefore, the best candidates for domestication of freshwater macroalgae should be strains with naturally high productivities under a range of conditions, particularly if this productivity has a genetic rather than environmental basis.

A wide range of environmental conditions can also influence the growth and productivity of macroalgae [12]. Conditions such as nutrient availability and water flow rate can be easily manipulated in intensive large scale cultivation systems; however, it is often difficult and costly to manipulate temperature. Therefore, the effect of temperature on productivity should be a key consideration when selecting strains for domestication. In general, photosynthesis and growth rates of macroalgae, and therefore productivity, increase

Table 1. Sample information.

Isolate	Date	Location	Accession number[2]	Provisional species identification	Rationale
Riv1	13/11/2012	Yass, NSW ,34°30'08"S, 146°13'56"E	KF606971	*Oedognoium* sp., belonging to capillare or crassum group.	Not possible to assign a species name based on morphological characteristics. ITS sequence does not form a clade with any other *Oedogonium* species.
Riv2	13/11/2012	Yenda, NSW,34°15'05"S, 146°12'54"E	KF606972	*Oedogonium* sp.aff.*pringsheimii Cramer ex.Hisn*	Identification based on morphological characteristics. Doe0 snot show the distinctly narrower males of *O. pringsheimii*. ITS sequence does not form a clade with any other *Oedogonium* species.
Riv3	14/11/2012	Jerilderie, NSW,35°21'16"S, 145°43'29"E	U/R	*Oedogonium* sp.	Not possible to assign a species name based on morphological characteristics. Could not obtain readable ITS sequence.
Riv4	14/11/2012	Rutherglen, VIC, 36°02'15"S, 146°23'41"E	KF606973	*Oedogonium implexum*	Identification based on morphological characteristics. ITS sequence does not form a clade with any other *Oedogonium* species.
Riv5	14/11/2012	Chiltern, VIC, 36°09'28"S, 146°34'36"E	U/R	*Oedogonium undulatum var Wissmanii*	Identification based on morphological characteristics. Could not obtain readable ITS sequence.
Tar1	31/05/2012	Tarong, QLD, 26°46'02"S, 151°55'26"E	KF606974	*Oedogonium* sp.	Not possible to assign a species name based on morphological characteristics. ITS sequence does not form a clade with any other *Oedogonium* species.
Tar2	11/10/2012	Tarong, QLD, 26°46'01"S, 151°54'56"E	KF606975	*Oedogonium* sp.	Not possible to assign a species name based on morphological characteristics. ITS sequence does not form a clade with any other *Oedogonium* species.
Tar3	11/10/2012	Tarong, QLD, 26°46'01"S, 151°54'56"E	U/R	*Oedogonium* sp.	Not possible to assign a species name based on morphological characteristics. Could not obtain readable ITS sequence.
Tar4	11/10/2012	Tarong, QLD, 26°46'01"S, 151°55'12"E	KF606976	*Oedogonium* sp.	Not possible to assign a species name based on morphological characteristics. ITS sequence does not form a clade with any other *Oedogonium* species.
Tsv1	-[1]	Townsville, QLD, 19°19'45"S, 146°45'41"E	KC701473	*Oedogonium* sp.	Not possible to assign a species name based on morphological characteristics. ITS sequence does not form a clade with any other *Oedogonium* species.
Tsv2	15/07/2012	Townsville, QLD, 19°19'57"S, 146°45'33"E	KF606977	*Oedogonium cf intermedium*	Most species defining morphological characteristics were not present. ITS sequence does not form a clade with any other *Oedogonium* species.

[1]This strain has been maintained in continuous culture at James Cook University for >2 years.
[2]U/R - unreadable sequence.
List of isolates used in this study, collection date and location, GenBank accession number for ITS sequences and provisional species identification based on morphological characteristics and phylogenetic trees constructed using ITS sequence data.

with increasing temperature up to an optimum point and then rapidly decline near an upper critical temperature [13,14]. The exact parameters of these curves and temperature optima vary both between and within species; however, maximal growth rates and tolerance to temperature variation are typically correlated with the temperature regime in the local habitat of an algal species [12,13]. Consequently, strains of the same algal species inhabiting different environments may have very different optimum temperatures for growth and tolerances to temperature variation.

The objective of the current study was to compare the performance of isolates of *Oedogonium* collected from multiple geographic locations under varying environmental conditions. (We use the term isolates here to refer to individual samples or variants within a species, including strains, ecotypes or genotypes).Our specific aims were to 1) collect and identify wild-type isolates of *Oedogonium* from three geographic locations; and 2) measure the

growth of these isolates under a range of temperature treatments corresponding to ambient conditions in each geographic location. Ideally, we wanted to identify isolates that exhibit naturally high growth rates under a range of temperature conditions. Assessing the performance of a variety of *Oedogonium* isolates collected from a range of environments will also provide insights as to whether the high growth rates of *Oedogonium* recorded by Lawton et al. [6] are unique to the single wild-type isolate tested in their study or are characteristic of the genus.

Methods

Sample collection and isolation

Samples of freshwater macroalgae were collected from naturally occurring water bodies, irrigation channels and wetland areas in three distinct geographic regions of Australia– Riverina (35°S,

145°E), Tarong (26°S, 151°E) and Townsville (19°S, 146°E).Permission was obtained from owners and local authorities where appropriate to collect samples. These locations were chosen as they encompass a broad range of environmental conditions (Supporting information, Fig. S1) and they are focal areas for a range of industries with potential biomass applications for freshwater algae. The Riverina region has a cool temperate climate and is a centre for agricultural production; Tarong has a warm temperate climate and is the location of a coal fired power station which produces large amounts of complex industrial waste effluents; and Townsville has a tropical climate and is a centre for biofuels research. Twenty samples were collected from the Riverina, 5 samples from Tarong and 3 samples from Townsville (including Tsv1, the original isolate tested by Lawton et al. [6]). Samples were transported in water taken at the collection site back to James Cook University, Townsville, where they were identified to genus using a compound microscope. Any samples containing *Oedogonium* were maintained in nutrient enriched autoclaved freshwater in a temperature and light controlled laboratory (12:12 light:dark cycle, 50 μmol photons $m^{-2} s^{-1}$, 23°C), all other samples were discarded. Stock cultures of *Oedogonium* isolates were created by isolating individual filaments of *Oedogonium* from each sample and maintaining these in sterile petri dishes with nutrient enriched (Guillards F/2 medium; 12.3 mg L^{-1} nitrogen, 1.12 mg L^{-1}phosphorus) autoclaved freshwater in culture cabinets at 24.5°C with 12 hour light: 12 hour dark cycles and a light level of 50 μmol $m^{-2} s^{-1}$. These conditions correspond to the middle temperature treatment (Tarong) of the constant temperature experiment (see below). In some cases multiple isolates of *Oedogonium* were isolated from the same sample. In total, 26 isolates of *Oedogonium* were isolated, of which 11 were successfully scaled up into stock cultures – 5 isolates from Riverina, 4 isolates from Tarong and 2 isolates from Townsville (Table 1). Stock cultures of each isolate were maintained for at least 2 months prior to the start of each experiment to allow acclimation to the culture system and ensure that all algae were pre-exposed to identical conditions.

Isolation and identification

Isolates of *Oedogonium* were identified using DNA barcoding and taxonomic keys. Samples of each isolate were examined under dissecting and compound light microscopes and their morphological characteristics were recorded. Where possible, each sample was identified to species using taxonomic keys [15]. For the DNA barcoding, DNA sequences from the internal transcribed spacer (ITS) region of the ribosomal cistron were used to assign species names to isolates of *Oedogonium*. This marker has been widely used in species level phylogenetic studies of green algae and was the marker of choice in a recent phylogenetic study of *Oedogonium* [16].Genomic DNA was isolated from fresh tissue samples of each isolate using a Qiagen DNEasy Plant Mini Kit following the manufacturer's instructions. The ITS region was amplified using the primers ITS1 [17] and G4 [18]. Polymerase chain reaction (PCR) amplifications were performed in a 25 μL reaction mixture containing 1.5 U of MyTaq HS DNA polymerase (Bioline), 5× MyTaq reaction buffer, 0.4 μM each primer, and 1 μL of genomic DNA (25–30 ng). Amplifications were performed on a BioRad C1000 Thermal Cycler with a touchdown PCR cycling profile (cycling parameters: 5 min at 94°C, 30 cycles of 30 s denaturing at 95°C, 45 s annealing at 56°C with the annealing temperature decreasing by 0.5°C each cycle, 60 s extension at 72°C, and a final extension at 72°C for 5 min). PCR products were column purified using Sephadex G25 resin and sequenced in both directions by the Australian Genome Research Facility (Brisbane, Australia). If sequences were unreadable or contaminated a second PCR

attempt was made and sequenced. If these sequences were also unreadable or contaminated, then DNA was re-extracted from a fresh sample and further PCRs and sequencing attempts were made. However, despite these steps we were not able to obtain readable sequences for three isolates – Riv3, Riv5 and Tar3. Sequences were edited using Bioedit [19] and submitted to GenBank under the accession numbers given in Table 1.

Isolates were identified based on their DNA sequences by constructing phylogenetic trees using sequences downloaded from GenBank. All publically available *Oedogonium* ITS sequences were downloaded. Duplicate sequences were removed from each dataset and then all remaining sequences were aligned with ours and trimmed to a standard length in MEGA 5.0 [20]. The ITS dataset included 32 sequences, 24 of which were retrieved from GenBank and the alignment consisted of 766 positions. Maximum likelihood (ML) phylogenetic trees were constructed in MEGA using a *Bulbochaete rectangularis* sequence (AY962677) as an outgroup. jModelTest 2.1 [21,22]showed that the SYM+I+G model of molecular evolution best fitted the data. However, as this model was not available in MEGA we used the simple Kimura two-parameter model to estimate genetic distance [23] as this is the standard model of molecular evolution used in barcoding studies [24]. The reliability of tree topologies was estimated using bootstrapping (1,000 replicates). Pairwise differences between all sequences used in the analysis were generated in MEGA using a maximum composite likelihood model.

Constant temperature experiments

To determine which isolates of *Oedogonium* would be suitable for targets for biomass applications, growth trials were conducted on all isolates under three temperature treatments that were kept constant at all times. Nine filaments of each isolate were cut to a standardised length of 6 mm. Three filaments from each isolate were then grown under each of three constant temperature treatments (21.3°C, 24.5°C, and 27.7°C) in culture cabinets (Sanyo MLR-351) with 12 hour light: 12 hour dark cycles and a light level of 50 μmol $m^{-2} s^{-1}$ (Philips TLD 36W/850 Daylight) for 7 days. These temperatures correspond to the mean annual 3pm air temperature in Wagga Wagga – the central point of our Riverina sampling locations (21.3°C), Kingaroy – the closest weather station to Tarong (24.5°C) and Townsville (27.7°C) (Table 2).This range of temperatures also represents the lower range of Townsville (21.3°C) and the upper range of Wagga Wagga and Kingaroy (27.7°C). Each individual filament was maintained in a sterile 60 mm petri dish with nutrient enriched autoclaved freshwater and photographed under a dissecting microscope (Olympus model SZ61) at the start and end of the 7 day period (Supporting information, Fig. S2). The 2-dimensional surface area of filaments was determined using ImageJ [25]. Specific growth rates were calculated for each individual replicate of each isolate using the equation SGR (% day^{-1}) = $Ln(B_f/B_i)/T*100$, where B_f and B_i are the final and initial surface areas (mm^2) and T is the number of days in culture. The experiment was repeated a further two times using new filaments from stock cultures at the start of each 7 day period to give a total of 3 replicate weeks of data. Permutational analysis of variance (PERMANOVA) was used to analyse the effects of isolate, temperature (both fixed effects) and week (random effect) on the specific growth rate of isolates. Analyses were conducted in Primer v6 (Primer-E Ltd, UK) using Bray-Curtis dissimilarities on fourth root transformed data and 9999 unrestricted permutations of raw data [26].

Table 2. Temperature treatments for growth experiments.

Region	Weather station	Constant	Summer variable			Winter variable		
			Min	Average	Max	Min	Average	Max
Riverina	Wagga Wagga AMO	21.3	16.3	20.9	32.2	2.9	5.8	12.9
Tarong	Kingaroy airport	24.5	18.0	21.8	30.9	3.4	8.1	19.4
Townsville	Townsville aero	27.7	24.7	26.7	31.7	14.0	17.3	25.3

Temperature treatments (°C) used in growth experiments. Temperatures for each experiment were based on those recorded by the Australian Bureau of Meteorology between 1981 and 2010 at the following weather stations: Wagga Wagga AMO for the Riverina region (35.16°S, 147.46°E), Kingaroy airport for the Tarong region (26.57°S, 151.84°E) and Townsville aero for the Townsville region (19.25°S, 146.77°E). Treatments for the constant temperature experiment correspond to the mean annual 3pm temperature in each location. Treatments for the summer variable temperature experiment correspond to the minimum and maximum mean temperatures recorded in January for each location. Treatments for the winter variable temperature experiment correspond to the minimum and maximum mean temperatures recorded in July for each location. Average daily values for each temperature profile in the variable temperature experiments are given. See methods and Figures 1B and 2B for more detail on the variable temperature treatments.

Variable temperature experiments

To provide further insights into the performance of isolates under different temperature conditions, growth trials were also conducted on isolates using a summer variable temperature treatment and a winter variable temperature treatment. In these variable temperature experiments, temperature treatments increased and decreased during the day to represent natural variations recorded in summer (January) and winter (July) at each location (Table 2; Figs. 1B and 2B). Temperatures were maintained at the minimum summer or winter temperature from 6 pm until 6 am, then increased in equal increments every 30 minutes to reach the maximum summer or winter temperature at 12 pm. Temperatures were maintained at the maximum between 12 pm and 2 pm, then decreased in equal increments every 30 minutes from 2 pm to reach the minimum summer or winter temperature at 6pm. These profiles resulted in average daily temperatures of 20.9°C for the Riverina treatment, 21.8°C for the Tarong treatment and 26.7°C for the Townsville treatment in the summer variable temperature experiment; and 5.8°C for the Riverina treatment, 8.1°C for the Tarong treatment and 17.3°C for the Townsville treatment in the winter variable temperature experiment (Table 2). The methods used were identical to the constant temperature experiment except that each variable temperature experiment was run for a single period of 7 days. Permutational analysis of variance (PERMANOVA) was used to analyse the effects of isolate and temperature (both fixed effects) on the specific growth rate of isolates. Data for the summer variable temperature experiment and winter variable temperature experiment were analysed separately. Analyses were conducted in Primer v6 (Primer-E Ltd, UK) using Bray-Curtis dissimilarities on fourth root transformed data and 9999 unrestricted permutations of raw data [26].

Results

Strain identification

Specific morphological characteristics of species were not visible in most isolates (Supporting information, Text S1). Consequently, we were only able to assign species names to 2 of our isolates based on morphological characteristics – isolate Riv4 was identified as *Oedogonium implexum* and isolate Riv5 as *Oedogonium undulatum* var *Wissmanii* (Table 1). The DNA sequence analysis was similarly inconclusive in terms of matching isolates with extant species of *Oedogonium*. None of our isolates had identical ITS sequences to any of the *Oedogonium* sequences downloaded from GenBank. Furthermore, none of the isolates formed tight clades in the

phylogenetic tree with any GenBank sequences (Fig. 3). Therefore it was not possible to assign species names to any isolates based on the ITS phylogenetic tree. Pairwise differences between GenBank samples included in the analysis ranged from 0.002 to 0.181. As the smallest pairwise distance between a sequence from our isolates and any other sequence was 0.005 (between isolates Tar4 and Tar1), it was also not possible to use pairwise distances to infer that any of our isolates were the same species as other isolates or GenBank samples included in the analysis. However, the higher pairwise differences recorded for our isolates (all >0.005), and the fact that all isolates formed distinct clades on the phylogenetic tree and did not group with any other sequences, demonstrate that isolates of *Oedogonium* are genetically distinct and therefore provides strong support that each isolate is a genetically differentiated species.

Constant temperature experiments

Average specific growth rates were high across all isolates ranging from 38.5% day^{-1} (±2.9 S.E.) at 21.3°C (Riverina treatment) to 45.3% day^{-1} (±2.3 S.E.) at 27.7°C (Townsville treatment) in the constant temperature experiment (Fig. 4). Specific growth rates varied significantly between isolates of *Oedogonium*, however this effect was not consistent between the three replicate weeks of the experiment (Table 3). There were significant differences in growth rates between weeks for all isolates except Riv5 and Tar4 ($P<0.05$) (Supporting information, Table S1). Most isolates had lower specific growth rates in week 2 of the experiment compared to week 1 and 3. There was no obvious effect of collection location on specific growth rate (Fig. 4), with isolates collected from the same location attaining their highest growth rates under different temperature treatments. For example, growth of Riverina isolate Riv2 was highest at 21.3°C (Riverina treatment), while that of Riverina isolate Riv3 was highest at 27.7°C (Townsville treatment).

Variable temperature experiments

Growth rates in the summer variable temperature experiment were similar to those seen in the constant temperature experiment. Average specific growth rates varied across all isolates, ranging from 31.9% day^{-1} (±5.4 S.E.) under the Riverina temperature treatment, to 42.3% day^{-1} (±6.8 S.E.) under the Tarong temperature treatment (Fig. 1A). However, neither isolate nor temperature had any significant effect on growth in this experiment (Table 3). In the winter variable temperature experiment, growth rates were much lower overall compared to the summer variable temperature experiment and the constant

A) Growth experiment results

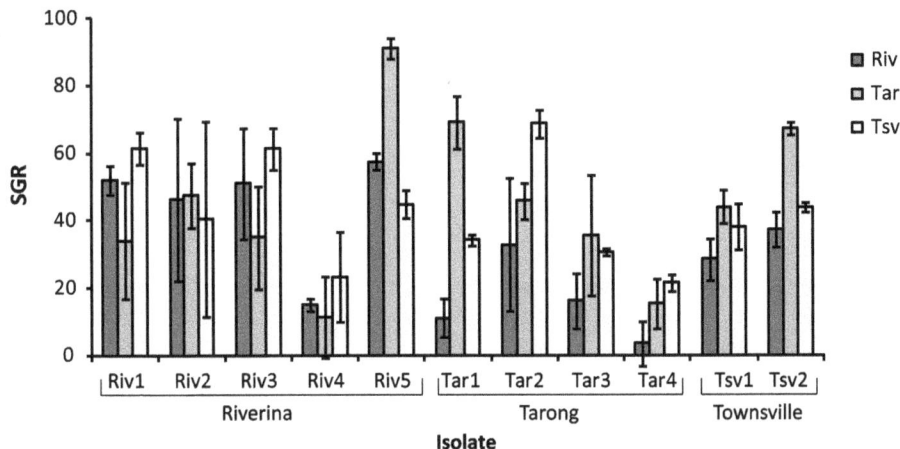

B) Temperature treatment profile

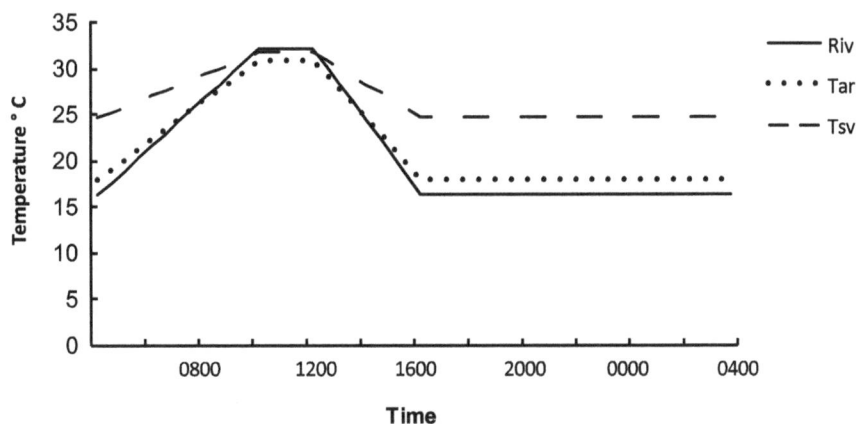

Figure 1. Growth rates of *Oedogonium* isolates and temperature profiles for the summer variable temperature experiment. A) Mean (±S.E.) specific growth rates (% day^{-1}) of isolates of *Oedogonium* collected from the Riverina (Riv), Tarong (Tar) and Townsville (Tsv) and grown under three variable temperature treatments and B)temperature profiles for Riverina, Tarong and Townsville treatments in the summer variable temperature experiment. These profiles result in average daily temperatures of 20.9°C for the Riverina treatment, 21.8°C for the Tarong treatment and 26.7°C for the Townsville treatment.

temperature experiment. Across all isolates, average specific growth rates ranged from 2.6% day^{-1} (±2.1 S.E.) under the Riverina temperature treatment to 51.7% day^{-1} (±5.3 S.E.) under the Townsville temperature treatment (Fig. 2A). Temperature had a significant effect on the growth rate of isolates (Table 3) and growth was significantly higher under the Townsville temperature treatments compared to the Tarong and Riverina treatment ($P<0.001$, PERMANVOA post hoc tests for main effect of temperature). Specific growth rates were <10%day^{-1} for all isolates except Riv2 under the Riverina temperature treatment and <10% day^{-1} for all isolates except Riv2, Riv3 and Riv5 under the Tarong temperature treatment. In contrast, growth rates ranged from 26–85% day^{-1} for all isolates under the Townsville temperature treatment.

Discussion

Determining the natural variability of wild-type isolates for production related traits and the degree to which different isolates perform best under different environments is an essential starting point for the domestication of new target species [9]. Our survey of natural populations in three geographic regions of Australia identified 11isolates of *Oedogonium* that were successfully maintained under culture conditions. Despite being collected from a range of environments and comprising multiple species of *Oedogonium*, supported by molecular data, the majority of these isolates had high specific growth rates across a range of temperature treatments (>30% day^{-1} in the constant temperature experiment and summer variable temperature experiment). These specific growth rates are considerably higher than those recorded for marine macroalgae in similar studies [27,28]. The high performance of the majority of isolates (species) across the range of temperature treatments tested here further confirms the findings of Lawton et al. [6] that species of *Oedogonium* are ideal targets for biomass applications and should be a focus for domestication efforts.

Temperature is a major factor controlling the growth of macroalgae [12,13,29]. However, temperature treatment only had a strong effect on *Oedogonium* growth rates in the winter variable temperature experiment. Growth rates across all treat-

A) Growth experiment results

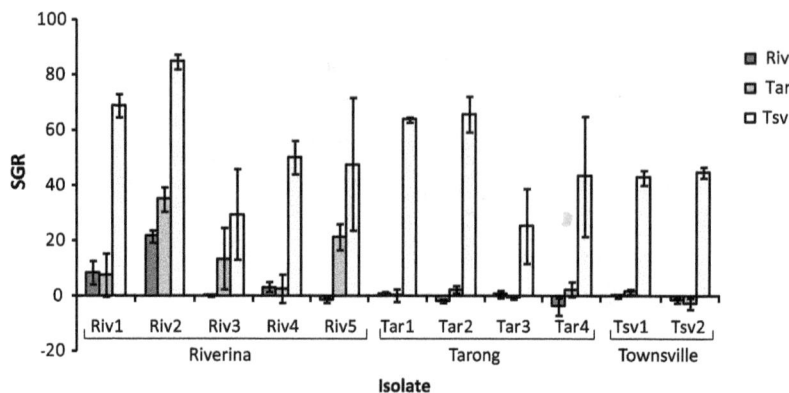

B) Temperature treatment profile

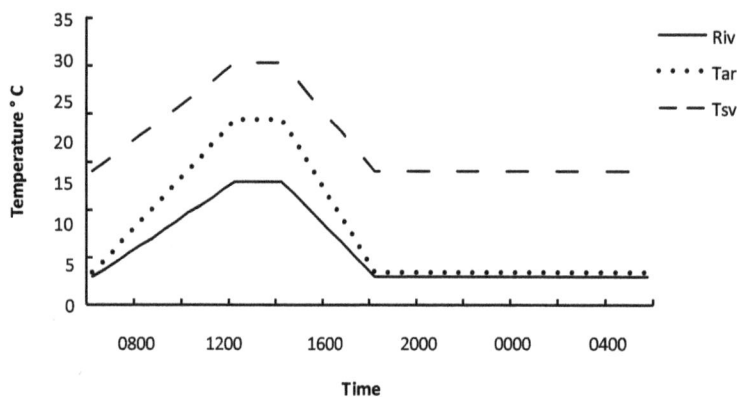

Figure 2. Growth rates of *Oedogonium* isolates and temperature profiles for the winter variable temperature experiment. A) Mean (±S.E.) specific growth rates (% day^{-1}) of isolates of *Oedogonium* collected from the Riverina (Riv), Tarong (Tar) and Townsville (Tsv) and grown under three variable temperature treatments and B)temperature profiles for Riverina, Tarong and Townsville treatments in the winter variable temperature experiment. These profiles result in average daily temperatures of 5.8°C for the Riverina treatment, 8.1°C for the Tarong treatment and 17.3°C for the Townsville treatment.

ments were lower in this experiment compared to the constant temperature experiment and the summer variable temperature experiment, and most isolates had very low growth rates (<10% day^{-1}) under the treatments with lowest temperatures (i.e. under Riverina and Tarong conditions). While the cause of these low growth rates is unknown, sub-lethal effects may have accumulated over the 7-day experimental period as a result of the low temperatures, limiting growth in some isolates. Low temperatures limit metabolic and photosynthetic rates in macroalgae as some of the enzymes involved in these processes are temperature dependent [12,14]. Low temperatures also impair the synthesis and functioning of photosynthetic pigment proteins [13,14] and damage cells in macroalgae by limiting electron transport and photon capture [12]. While it was not a goal of this study to quantify temperature thresholds for growth, it is possible that the growth rates recorded under the lowest temperature treatments represent the response of isolates to a shock treatment or acute stress rather than to low temperatures *per se*. In the winter variable temperature experiment filaments were transferred directly from stock cultures maintained at 24.5°C to each treatment. For the Tarong and Riverina treatments, this represented an immediate temperature change of at least 5 to 11°C; however, it is notable that the daily temperature swings for both treatments are greater

than this at 15–20°C. The rate of cooling is an important factor affecting plant responses to low temperatures [30] and could potentially be used as a strategy to ease domesticated cultures into winter conditions to enhance annual production yields.

Irrespective of the cause of low growth rates in the winter variable temperature experiment, our results demonstrate that some isolates are better able to tolerate lower temperatures than others. For example, under all three winter temperature treatments isolate Riv2 maintained specific growth rates >20% day^{-1}. Similarly, some isolates were more susceptible to low temperatures than others. For instance, isolate Tsv2 had negative growth rates under the Tarong and Riverina winter temperature treatments. Most macroalgae are capable of acclimating to variable temperatures to some degree; however, this ability is expected to be higher in algae native to habitats with large annual variations in temperature compared to those from habitats with more stable climates [12,13]. In agreement with these predictions, the best performing isolates (Riv2, Riv3 and Riv5) under the lower winter temperature treatments were all collected from the Riverina – the sampling location with the greatest annual variations in temperature. These results suggest that domestication efforts should further target isolates from regions with large annual variations in temperature as these isolates are likely to have broad

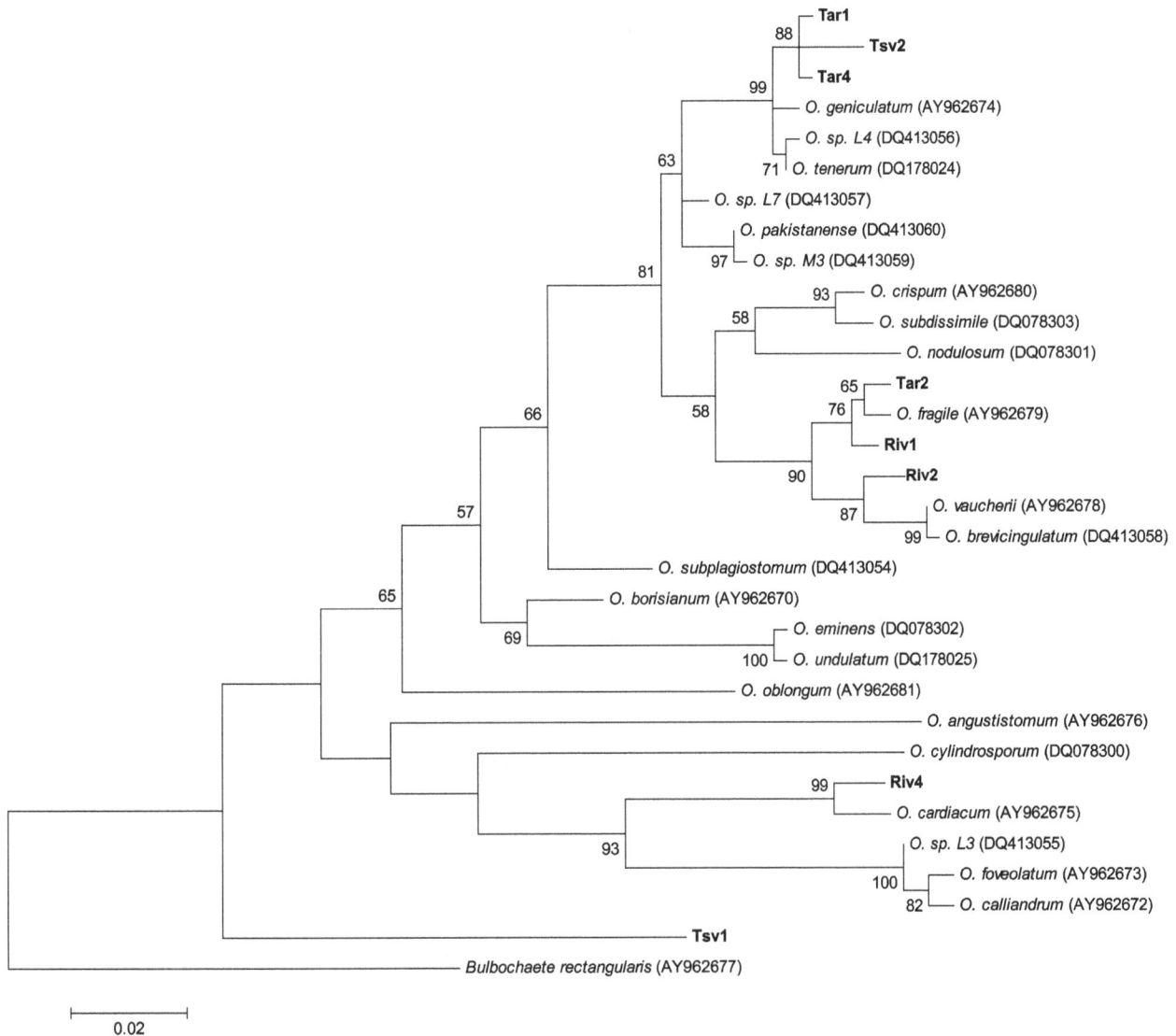

Figure 3. Maximum likelihood tree of *Oedogonium* internal transcribed spacer (ITS) sequence data. Maximum likelihood tree of *Oedogonium* internal transcribed spacer (ITS) sequence data (scale at bottom). Numbers near each node refer to bootstrap support values, nodes with <50% bootstrap support are not labelled. Isolates collected in this study shown in bold. Numbers accompanying the species names are GenBank accession numbers for the sequences used in the analysis.

temperature tolerances and therefore perform well across a wide range of conditions. The next steps in assessing the suitability of isolates for domestication and biomass applications should be the measurement of areal productivity in larger outdoor cultures [1,6]. Subsequent tests of other traits of interest between isolates, such as biochemical profiles for bioenergy potential [31], bioremediation ability [4] and carbon sequestration [1], can then be evaluated in the scaled biomass.

The lower growth rates recorded for most *Oedogonium* isolates under the Riverina and Tarong winter temperature treatments indicate that biomass production of *Oedogonium* is likely to be low over the winter period in these regions and other locations that experience similar temperatures. Estimating growth and expected production across multiple seasons is important when considering potential biomass applications for *Oedogonium*, particularly for locations where seasonal temperatures vary significantly, as use of data from a single season may result in erroneous estimates of annual productivity [12]. In contrast to growth rates recorded

under the Riverina and Tarong treatments, *Oedogonium* growth rates were high (over 45% day^{-1}) under the Townsville treatment in the winter variable temperature experiment. These high growth rates indicate that Townsville, and other locations with similar temperature profiles, are target locations for *Oedogonium* cultivation as it will be possible to maintain high biomass productivity year round under ambient conditions in these locations using any of the domesticated strains.

Conclusion

This study compares, for the first time, the growth rates of multiple isolates of *Oedogonium* under a range of temperature regimes in order to identify isolates that exhibit naturally high growth rates under varying environmental conditions. The high growth rates recorded across multiple temperature treatments for the majority of isolates tested here confirm the suitability of this genus for biomass applications and demonstrate that most

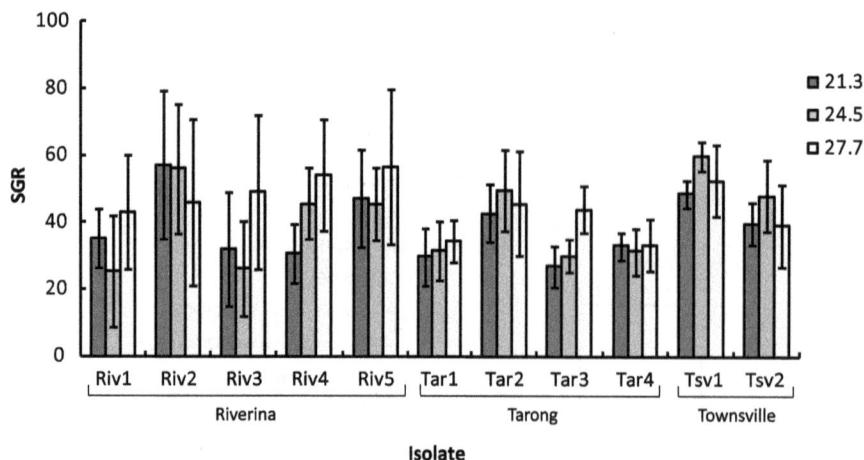

Figure 4. Growth rates of *Oedogonium* isolates for the constant temperature experiment. Mean ($\pm$S.E.) specific growth rates (% day^{-1}) of isolates of *Oedogonium* collected from three locations (Riverina, Tarong, Townsville) and grown under three constant temperature treatments. Temperatures are average daily temperatures for each profile (°C).

naturally occurring isolates of *Oedogonium* can be domesticated. All isolates tested here had similar growth rates to isolate Tsv1, the original isolate tested by Lawton et al. [6] which outperformed other genera of freshwater macroalgae across a range of metrics. Our results confirm that the genus *Oedogonium* contains many species that have a superior performance and provide a vital first step to determining the suitability of *Oedogonium* isolates for biomass applications prior to measuring areal productivity and assessing biochemical composition in large outdoor cultures. Worldwide production of food crops is based around a small number of target species. Similarly, production of marine macroalgae is dominated by only six genera [32]. Future production of freshwater macroalgae is also likely to follow these trends and rely on a select group of target genera. Our findings provide strong rationale for the inclusion of *Oedogonium* as a key target genus for the

development of this industry worldwide. Our approach of collecting isolates of *Oedogonium* from multiple geographic locations and comparing their performance under a range of conditions provides a template that can be applied to other novel target species as a first step towards domestication.

Supporting Information

Figure S1 Annual temperature profiles for sampling locations. Average 9am monthly temperatures recorded in Wagga Wagga (Riverina region), Kingaroy (Tarong region) and Townsville. Error bars show mean maximum and mean minimum monthly temperatures. Data from the Australian Government Bureau of Meteorology.

Figure S2 Before/after photographs of *Oedogonium* growth. Microscope photos of single *Oedogonium* filament at the start and end of 7 day growth period in the constant temperature experiment.

Table S1 PERMANOVA post hoc tests for constant temperature experiment. Results of PERMANOVA post hoc tests on main effect of Week x Isolate (We x Is) in the constant temperature experiment. *P* values for each test are presented, significant terms shown in bold.

Text S1 Morphological characteristics of *Oedogonium* isolates.

Acknowledgments

This research is part of the MBD Energy Research and Development program for Biological Carbon Capture and Storage. We thank Maria Martinez, Amanda Ricketts and Boer Bao for assistance with experiments and Dean Jerry for advice on molecular analysis.

Author Contributions

Conceived and designed the experiments: RL NP RdN. Performed the experiments: RL SS. Analyzed the data: RL SS. Contributed reagents/materials/analysis tools: RL SS RnD NP. Wrote the paper: RL RnD NP.

Table 3. Results of permutational analyses of variance (PERMANOVAs) testing the effects of temperature (Te), isolate(Is) and week (We) on specific growth rates of *Oedogonium* in the constant temperature experiment; and the effects of temperature (Te) and isolate (Is) on specific growth rates of *Oedogonium* in the summer and winter variable temperature experiments.

Source	df	Constant		Summer variable		Winter variable	
		F	P	F	P	F	P
Te	2	0.71	0.575	1.89	0.159	**26.09**	**<0.001**
Is	10	1.09	0.374	1.55	0.124	1.83	0.069
We[1]	2	**24.23**	**<0.001**				
Te x Is	20	0.94	0.539	1.02	0.449	1.29	0.215
Te x We[1]	4	1.46	0.204				
Is x We[1]	20	**3.47**	**<0.001**				
Te x Is x We[1]	40	1.11	0.297				

[1]Week was not a factor in the summer and winter variable temperature experiments.
Pseudo F (F) and P values are presented, significant terms shown in bold.

References

1. Cole AJ, Mata L, Paul NA, De Nys R (In press) Using CO_2 to enhance carbon capture and biomass applications of freshwater macroalgae. GCB Bioenergy. DOI: 10.1111/gcbb.12097

2. Kebede-Westhead E, Pizarro C, Mulbry W (2006) Treatment of swine manure effluent using freshwater algae: Production, nutrient recovery, and elemental composition of algal biomass at four effluent loading rates. J Appl Phycol 18: 41–46.

3. Mulbry W, Kondrad S, Pizarro C, Kebede-Westhead E (2008) Treatment of dairy manure effluent using freshwater algae: Algal productivity and recovery of manure nutrients using pilot-scale algal turf scrubbers. Bioresour Technol 99: 8137–8142.

4. Saunders RJ, Paul NA, Hu Y, de Nys R (2012) Sustainable sources of biomass for bioremediation of heavy metals in waste water derived from coal-fired power generation. PloS one 7: e36470.

5. Bird MI, Wurster CM, de Paula Silva PH, Paul NA, de Nys R (2012) Algal biochar: effects and applications. GCB Bioenergy 4: 61–69.

6. Lawton RJ, de Nys R, Paul NA (2013) Selecting Reliable and Robust Freshwater Macroalgae for Biomass Applications. PloS one 8: e64168.

7. Evenson RE, Gollin D (2003) Assessing the Impact of the Green Revolution, 1960 to 2000. Science 300: 758–762.

8. Tilman D, Cassman KG, Matson PA, Naylor R, Polasky S (2002) Agricultural sustainability and intensive production practices. Nature 418: 671–677.

9. Yan J, Chen W, Luo F, Ma H, Meng A, et al. (2012) Variability and adaptability of Miscanthus species evaluated for energy crop domestication. GCB Bioenergy 4: 49–60.

10. Achten WMJ, Nielsen LR, Aerts R, Lengkeek AG, Kjær ED, et al. (2009) Towards domestication of Jatropha curcas. Biofuels 1: 91–107.

11. Kawecki TJ, Ebert D (2004) Conceptual issues in local adaptation. Ecol Lett 7: 1225–1241.

12. Lobban CS, Harrison P (2000) Seaweed ecology and physiology. Cambridge University Press, Cambridge. pp. 366.

13. Eggert A (2012) Seaweed Responses to Temperature. In: Wiencke C, Bischof K, editors. Seaweed Biology: Springer Berlin Heidelberg. pp. 47–66.

14. Davison IR (1991) Environmental effects on algal photosynthesis: temperature. J Phycol 27: 2–8.

15. Entwisle TJ, Skinner S, Lewis SH, Foard HJ (2007) Algae of Australia: Batrachospermales, Thoreales, Oedogoniales and Zygnemaceae. Collingwood, Australia: CSIRO PUBLISHING/Australian Biological Resources Study 200 p.

16. Mei H, Luo W, Liu G, Hu Z (2007) Phylogeny of Oedogoniales (Chlorophyceae, Chlorophyta) inferred from 18S rDNA sequences with emphasis on the relationships in the genus Oedogonium based on ITS-2 sequences. Plant Syst Evol 265: 179–191.

17. Bakker FT, Olsen JL, Stam WT (1995) Evolution of nuclear rDNA its sequences in the Cladophora albida/sericea clade (Chlorophyta) J Mol Evol 40: 640–651.

18. Harper JT, Saunders GW (2001) The application of sequences of the ribosomal cistron to the systematics and classification of the florideophyte red algae (Florideophyceae, Rhodophyta). Cah Biol Mar 42: 25–38.

19. Hall TA. (1999) BioEdit: a user-friendly biological sequence alignment editor and analysis program for Windows 95/98/NT; 1999. pp. 95–98.

20. Tamura K, Peterson D, Peterson N, Stecher G, Nei M, et al. (2011) MEGA5: molecular evolutionary genetics analysis using maximum likelihood, evolutionary distance, and maximum parsimony methods. Mol Biol Evol 28: 2731–2739.

21. Guindon S, Gascuel O (2003) A simple, fast, and accurate algorithm to estimate large phylogenies by maximum likelihood. Syst Biol 52: 696–704.

22. Darriba D, Taboada GL, Doallo R, Posada D (2012) jModelTest 2: more models, new heuristics and parallel computing. Nat Meth 9: 772–772.

23. Kimura M (1980) A simple method for estimating evolutionary rates of base substitutions through comparative studies of nucleotide sequences. J Mol Evol 16: 111–120.

24. Hebert PDN, Ratnasingham S, de Waard JR (2003) Barcoding animal life: cytochrome c oxidase subunit 1 divergences among closely related species. Proceedings of the Royal Society of London Series B: Biological Sciences 270: S96–S99.

25. Schneider CA, Rasband WS, Eliceiri KW (2012) NIH Image to ImageJ: 25 years of image analysis. Nat Meth 9: 671–675.

26. Anderson MJ, Gorley RN, Clarke KR (2008) PERMANOVA+ for PRIMER: Guide to Software and Statistical Methods. PRIMER-E, Plymouth, UK.

27. Lawton RJ, Mata L, De Nys R, Paul NA (2013) Algal bioremediation of waste waters from land-based aquaculture using Ulva: selecting target species and strains. PLoS ONE 8(10): e77344. doi:10.1371/journal.pone.0077344

28. de Paula Silva PH, McBride S, de Nys R, Paul NA (2008) Integrating filamentous 'green tide' algae into tropical pond-based aquaculture. Aquaculture 284: 74–80.

29. Raven JA, Geider RJ (1988) Temperature and algal growth. New Phytol 110: 441–461.

30. Minorsky PV (1989) Temperature sensing by plants: a review and hypothesis. Plant, Cell & Environment 12: 119–135.

31. Ross AB, Jones JM, Kubacki ML, Bridgeman T (2008) Classification of macroalgae as fuel and its thermochemical behaviour. Bioresour Technol 99: 6494–6504.

32. Paul NA, Tseng CK, Borowizka M (2012) Seaweed and microalgae. In: Lucas JS, Southgate PC, editors. Aquaculture: Farming aquatic animals and plants. Sussex, U.K.: Blackwell Publishing Ltd.

13

Effects of Increased Nitrogen Deposition and Rotation Length on Long-Term Productivity of *Cunninghamia lanceolata* Plantation in Southern China

Meifang Zhao[1], Wenhua Xiang[1]*, Dalun Tian[1], Xiangwen Deng[1], Zhihong Huang[1], Xiaolu Zhou[2], Changhui Peng[2,1]

1 Faculty of Life Science and Technology, Central South University of Forestry and Technology, Changsha, Hunan, People's Republic of China, 2 Institute of Environment Sciences, Department of Biological Sciences, University of Quebec at Montreal, Montreal, Quebec, Canada

Abstract

Cunninghamia lanceolata (Lamb.) Hook. has been widely planted in subtropical China to meet increasing timber demands, leading to short-rotation practices that deplete soil nutrients. However, increased nitrogen (N) deposition offsets soil N depletion. While long-term experimental data investigating the coupled effects related to short rotation practices and increasing N deposition are scarce, applying model simulations may yield insights. In this study, the CenW3.1 model was validated and parameterized using data from pure *C. lanceolata* plantations. The model was then used to simulate various changes in long-term productivity. Results indicated that responses of productivity of *C. lanceolata* plantation to increased N deposition were more related to stand age than N addition, depending on the proportion and age of growing forests. Our results have also shown a rapid peak in growth and N dynamics. The peak is reached sooner and is higher under higher level of N deposition. Short rotation lengths had a greater effect on productivity and N dynamics than high N deposition levels. Productivity and N dynamics decreased as the rotation length decreased. Total productivity levels suggest that a 30-year rotation length maximizes productivity at the 4.9 kg N ha^{-1} year^{-1} deposition level. For a specific rotation length, higher N deposition levels resulted in greater overall ecosystem C and N storage, but this positive correlation tendency gradually slowed down with increasing N deposition levels. More pronounced differences in N deposition levels occurred as rotation length decreased. To sustain *C. lanceolata* plantation productivity without offsite detrimental N effects, the appropriate rotation length is about 20–30 years for N deposition levels below 50 kg N ha^{-1} year^{-1} and about 15–20 years for N deposition levels above 50 kg N ha^{-1} year^{-1}. These results highlight the importance of assessing N effects on carbon management and the long-term productivity of forest ecosystems.

Editor: Bruno Hérault, Cirad, France

Funding: This study was financially supported by the Key Program of State Forestry Special Fund for Public Welfare Industry of China (number 2011432009), the Program of Introducing Advanced Technology (948 program) from the China State Forestry Administration (number 2010-4-03), and the Furong Scholar Program. The funders had no role in study design, data collection and analysis, decision to publish, or preparation of the manuscript.

Competing Interests: The authors have declared that no competing interests exist.

* E-mail: xiangwh2005@163.com

Introduction

Nitrogen (N) is the element that has the greatest limiting effect on plantation productivity [1]. However, in many areas, forest ecosystems have experienced increased atmospheric N deposition in recent years [2–5]. For example, in most forests in subtropical China, the current rate of N deposition ranges from 18 to 73 kg N ha^{-1} year^{-1}, and this is expected to increase in coming decades [6–9]. The increase in N that is deposited in forests will increase soil N content and, consequently, stimulate increased productivity over the short term [10]. Although N deposition could meet the N requirements of these forests, excess N will result in a nutrimental burden and offsite environmental effects [5,11,12]. Balancing the N supply and the demand for tree growth in these forests is critical for optimizing productivity, sustainability, and environmental protection [13,14].

N cycling in forest ecosystems is a complex process because of the numerous soil–plant interactions. Nutrient pools and fluxes, as well as the nutrient requirements of the forest trees, vary as the stands develop [15]. To determine the effects of management practices on nutrient cycling and the responses in forest productivity that occur following changes in N input, long-term observational data sets are required. Previous studies have manipulated N levels to examine the short-term responses in tree growth and assess N requirements. However, these studies did not consider variations in the long term productivity and the N requirements of forest stands as they develop. Mechanistic models can accurately predict long-term changes in productivity and are widely used in N fertilizer applications, resulting in increased N-use efficiency and reduced pollution [15,16]. Some mechanistic models have been used to analyze the effects of rotation length on tree and soil carbon (C) stocks and the quality of wood in different European forests [17], the effects of harvesting intensity and rotation length on long-term soil N and C dynamics in boreal forests in central Canada [18], the effects of N deposition, climate change, and rotation length on forest carbon sequestration and the harvesting of Scots pine forests in southern Finland [19], and to compare the ecological impact of natural disturbances and harvesting [20].

Cunninghamia lanceolata (Lamb.) Hook. is the third most commonly planted tree species in plantations in the world [21], and its total plantation area is now 9.21 million hectares in China [22]. Due to its high value in terms of timber quality and versatility, *C. lanceolata* has mainly been planted for timber production, but its bark and roots are usually harvested for local construction purposes and fuel [23,24]. Currently, most *C. lanceolata* plantations are monocultures and successive rotation planting is commonly practiced [24,25]. Increased timber demand and improvements in processing techniques have resulted in shorter harvesting cycles of *C. lanceolata* plantations (originally 30 years to now ≤20 years). Soil N depletion due to timber removal has been recognized as the major factor responsible for the decline in yields of *C. lanceolata* plantations [26]. Therefore, sustaining the productivity of *C. lanceolata* plantations over successive rotations is a concern of numerous researchers [27–29] and has been identified as a priority for plantation management in China [4]. Regarding *C. lanceolata* plantations, various models have been used to simulate age-related C dynamics [30,31], yield decline due to management activities and degradation of soil fertility [28], and C sequestration at N deposition levels <1–50 kg N ha^{-1} year^{-1} at various rotation cycles [32]. However, the simultaneous effects of increased N deposition and rotation lengths with successive planting on long-term production and management in *C. lanceolata* plantations have not been studied so far.

This study uses the CenW3.1 model [33,34] to simulate long-term production and N requirement dynamics in response to N deposition and different rotation lengths and establish sound management practices for the sustainable productivity of *C. lanceolata* plantations. The major parameters of production and N dynamics include stem wood biomass, net primary production (NPP), annual N requirement (ANR) for tree growth, soil available N for growth, and soil organic C (SOC). Specifically, the purposes of this study are to (1) determine how increased N deposition affects forest productivity as stand development; (2) examine the interactive effects of N deposition and rotation length on long-term forest productivity; (3) investigate whether practicing increased atmosphere N with shorter rotation that can avert N saturation damage in the study region.

Materials and Methods

Site Description

The *C. lanceolata* plantations that were evaluated in this study are located at the experimental area of the Huitong National Forest Ecosystem Research Station (NFERS) (lat 26°50′N, long 109°45′E), Hunan Province, southern China [31]. The region is characterized by a humid mid-subtropical monsoon climate. The annual mean temperature is 16.5°C, which ranges from an average of 4.3°C during the coolest month (January) to 29.4°C during the warmest month (July). The annual rainfall is approximately 1270–1650 mm, mainly occurring between April and August. The soil is a subgroup of clay loam red soil that is formed from shale and slate parent materials (Table 1).

Eight *C. lanceolata* plantation watersheds were established within the research station in 1984. Automatic weather stations were installed within the plantations to collect climatic data. The first-rotation *C. lanceolata* plantations were established on a clear-cut of natural forest in 1966. After the slashed understory plants and harvesting residues had been burned, the soil was prepared by digging holes and 1-year-old *C. lanceolata* seedlings were planted. No weed control, fertilization or other treatment was employed.

The establishment, survival, and growth of the plantation have been described previously [31]. This study was carried out at two watersheds (II and III), situated within NFERS. The second successive rotation plantation was established in watershed III in February 1987 and in watershed II in March 1995 with an initial planting density of 3318 trees ha^{-1} using identical planting methods as used in the first rotation. The ecological factors and characteristics of the *C. lanceolata* plantations related to this study were previously described [31]. The biomass and stand characters, including height (H, m), diameter at breast height (DBH, cm), and stand density (trees ha^{-1}), were measured at different growing-stages beginning at age 7. Total volume (under bark) and dry matter biomass of the wood, bark, branches, and needles were calculated. Biomass, litter, forest floor, and understory samples were collected at an interval of five years for nutrient concentration and growth analysis. Linear regression was used to test the relationships between soil bulk density and soil C content.

CenW Model

The CenW3.1 model was developed as a comprehensive monoculture forest model that runs in daily time-step with the C gain that is calculated from light absorption and can be modified by taking into account temperature, soil water status, and foliage N concentration [34]. CenW allows the explicit modeling of both C and N pools as well as their fluxes, thereby integrating tree growth parameters, environmental factors, and management treatments. The projection of stand growth and N dynamics are based on the rates of key ecological processes regulating the availability of and competition for light and nutrient resources. Net primary productivity (NPP, t ha^{-1} year^{-1}, represents the remains of stand-level photosynthesis after subtracting losses of stand-level respiration and litterfall) can be reduced through a number of processes, such as increasing respiration, increasing senescence and mortality losses, immobilisation of nutrients and unfavourable shifts in biomass allocation. Fixed photosynthate is used for plant growth, with allocation to different plant organs determined by plant nutrient status, tree height and species-specific allocation factors. Stem wood biomass (t DM ha^{-1}) includes heartwood and sapwood. Tree death is estimated as a simple daily fractional mortality rate. It is assumed that the ratio of above- to below-ground allocation increases with foliar N concentration. Foliar N concentrations are essentially determined through the relative rates of C and N uptake. It is also assumed that 25% of foliar N is relocated to other plant part before litter fall. Water use is calculated using the Penman-Monteith equation, with canopy resistance given by the inverse of stomatal conductance, which, in turn, is linked to calculated photosynthetic carbon gain. Water is lost through transpiration and soil evaporation, and gained by rainfall or irrigation which together determines soil water status for the following day. Stand density was fully independently modeled as a result of self-thinning. The nutrient cycle is considered through litter production by the shedding of plant parts, such as roots, bark, branches and foliage. Litter is assumed to be produced as a constant fraction of live biomass pools. In addition, foliage is shed during drought or when canopies become too dense. Litter is then added to the organic matter pools from where C is eventually lost and N becomes available again as inorganic mineral N. N can come from external addition (atmospheric deposition and fertilizer addition) or mineralization during the decomposition of soil organic matter. Plant available N depends on is current soil available N status plus additions from atmospheric deposition. Soil decomposition rate is determined by temperature, soil water status and soil organic matter quality. Input litter tends to have a wide C:N ratio so that some C needs to be lost through decomposition

Table 1. Soil characteristics of the *C. lanceolata* plantations at the Huitong National Forest Ecosystem Research Station, Hunan Province, China.

Characteristic	Value	Source
Soil pH	4.32–4.86	[60]
Bulk density (g cm^{-3})	1.09–1.7	[61]
Porosity (% of the total volume)	52.1–62.5	[62]
Permeable velocity (cc min^{-1})	2.83–10.1	[62]
Maximum moisture holding capacity (% of the total volume)	36.1–40.4	[62]
Moisture content (%) for most of the year	>30	[63–64]
Total N (g kg^{-1})	1.876–2.478	[65]
NH$_4^+$-N (mg kg^{-1})	6.67–9.1	[66]
NO$_3$-N (mg kg^{-1})	11.2–24.3	[67]
Total P (g kg^{-1})	0.92–1.42	[68]
Total K (g kg^{-1})	4.27–5.13	[69]
TOC (g kg^{-1})	21.1–33.4	[70]
Soil C:N ratio	13–16	[64]

before critical C:N ratio is reached in organic matter and excess N can be mineralized. The model also offers the potential to manipulate forests in a variety of experimental ways and predict outputs. Detailed descriptions of model features, structures, mathematical representations, sensitivity analyses, and building strategies have been described [33,34].

Model Calibration and Validation

The calibration and validation of CenW3.1 in *C. lanceolata* plantations in this study followed two steps. The first step for model calibration was to select the data from watershed II to calibrate CenW3.1 model. The purpose of this step was to determine the model parameters to ensure simulation accuracy as much as possible. The second step for model validation was to evaluate the model against the independent watershed III to test the model accuracy. The data used in the model simulation included environmental factors, tree growth parameters, tree-N response data, forest floor decomposition rates, and N inputs. The primary parameters and values for productivity, above-ground biomass, decomposition, and soil C were previously described in detail [31] (see Appendix S1).

For this study, the CenW3.1 simulations were run for a period of 120 years using climate records that were obtained during the period from 1987 to 2010. The historic records of daily maximum and minimum values of air temperature, daily sum of radiation (MJ m^{-2} day^{-1}), and daily precipitation (mm) over the period 1987–2010 at the study site were used for climate inputs. For future periods without observed climate data, we randomly assign each year's climate condition with detrended climate data [35].

Tree growth parameters, tree-N response data, forest floor decomposition rates, and N inputs were obtained from the literature or field measurements (see Appendix S1). The rates of these processes were calculated from a combination of historical bioassay data (biomass accumulation in component pools, stand density) and measures decomposition rates and photosynthetic saturation curves by relating 'biologically active' biomass components (foliage and small roots) to calculate nutrient uptake, the capture of light energy, and net primary production. Then, the model generated a suite of growth properties for each tree components to be represented based on the historical data (i.e.

growth and yield tables or long-term permanent plots), which was used to calibrate tree growth. These growth properties were subsequently used to simulate plant growth as a function of resource availability and management practices. Calibration data were assembled that describe the accumulation of tree biomass (above and below-ground components) as stands developed. Aging was taking into account by using calibration data pairs for different stand ages (i.e. age vs. tree size, age vs. stem density, age vs. biomass, etc.). Tree biomass and stand self-thinning rate data were estimated from the height, DBH and stand density as output of traditional empirical growth and yield models in conjunction with species-specific component biomass allometric equations. To calibrate the nutritional aspects of the model, data describing the concentration of nutrients in the various biomass components were required. The data required in the model on the degree of shading produced by different quantities of foliage and the response of foliage to different light levels were derived from literature values, field measurements, or simulation models. Lastly, decomposition rates of various litter types and soil organic matter were required for the model to simulate nutrient cycling. Fertilizer applications, irrigation management techniques, weather conditions (such as sunny, cloudy, and etc.), incidences of insect problems or disease, plowing, and fire were excluded in the simulations.

Moreover, the annual N requirement (ANR) for tree growth was defined as the year-to-year change in amounts of N required for biomass accumulated during the year (expressed in kg N ha^{-1} year^{-1}). This was estimated using the following equation:

$$ANR_t = \sum_{i=1}^{n} (B_{i(t)} \times C_{i(t)} - B_{i(t-1)} \times C_{i(t-1)}) \qquad (1)$$

where ANR_t is the annual N requirement at stand age of year t; $B_{i(t)}$ and $B_{i(t-1)}$ represent the accumulated biomass of tree compartment i at stand age of year t and $t-1$, respectively; $C_{i(t)}$ and $C_{i(t-1)}$ represent the N concentration in tree compartment i at stand age of year t and $t-1$; and n is the number of compartments (including sapwood, heartwood, bark, branches, leaves, fine root, coarse root, flower, and fruits) within the stand. B_i was quantified as the amount of biomass in each compartment i, plus the biomass returned to the soil as litter, predicted by CenW3.1 for each stand during each rotation cycle.

Linear regressions of the observed data obtained from the study site were compared with the simulated average stand height (H), average diameter at breast height (DBH), stand density (stems per hectare), and stem woody dry matter biomass that were calculated using CenW3.1. These comparisons were used to validate the CenW3.1 model for long-term predictions (Fig. 1). For a better quantification of the error in fitting low flows, the efficiency proposed by Nash and Sutcliffe (NSE) [36], defined as one minus the sum of the absolute squared differences between the predicted and observed values normalized by the variance of the observed values during the period under investigation, was used to tested modeling performances.

The H value (Fig. 1A) that was calculated by the model fitted well with the observations across different climatic conditions ($R^2 = 0.77$, NSE = 0.72). The predicted and observed data were closely matched for DBH (Fig. 1B; $R^2 = 0.96$, NSE = 0.96), stand density (Fig. 1C; $R^2 = 0.99$, NSE = 0.98), and stem woody dry matter mass (Fig. 1D; $R^2 = 0.99$, NSE = 0.98) in *C. lanceolata* plantations during the period 1990–2010. The validation results, therefore, confirm the reliability of the applied model for further investigations into the long-term effects of N deposition and

rotation lengths on the productivity and N dynamics of *C. lanceolata* plantations.

Model Simulation Runs

Five dynamic variables including stem wood biomass, NPP, ANR, soil available N for growth (kg N ha^{-1}), and soil organic C (SOC, t C ha^{-1}), were consider in this study. However, these variables are sometimes not sufficient to account for the full magnitude of changes. This study combines that response with simple assumptions about external N depositions and current rotation length practices. The atmospheric N depositions (kg N ha^{-1} year^{-1}) refers to external input rates as wet and dry atmospheric deposition. For effects of increased N deposition on long-term forest productivity, five N deposition levels e.g. 4.9, 18, 30, 50, 70 and 90 kg N ha^{-1} year^{-1} were applied during the simulations. Based on the reported N depositions in China, the initial N deposition in 1987 is approximately estimated to be 4.9 kg N ha^{-1} year^{-1} (N4.9) at the study site [37], which represents the average deposition rate in areas with human population without industrial development. A level of 18 kg N ha^{-1} year^{-1} (N18) is the current deposition rate in southern China [8]. Whereas in the polluted areas in Hunan Province, levels around 30 kg N ha^{-1} year^{-1} (N30) is the most common, but

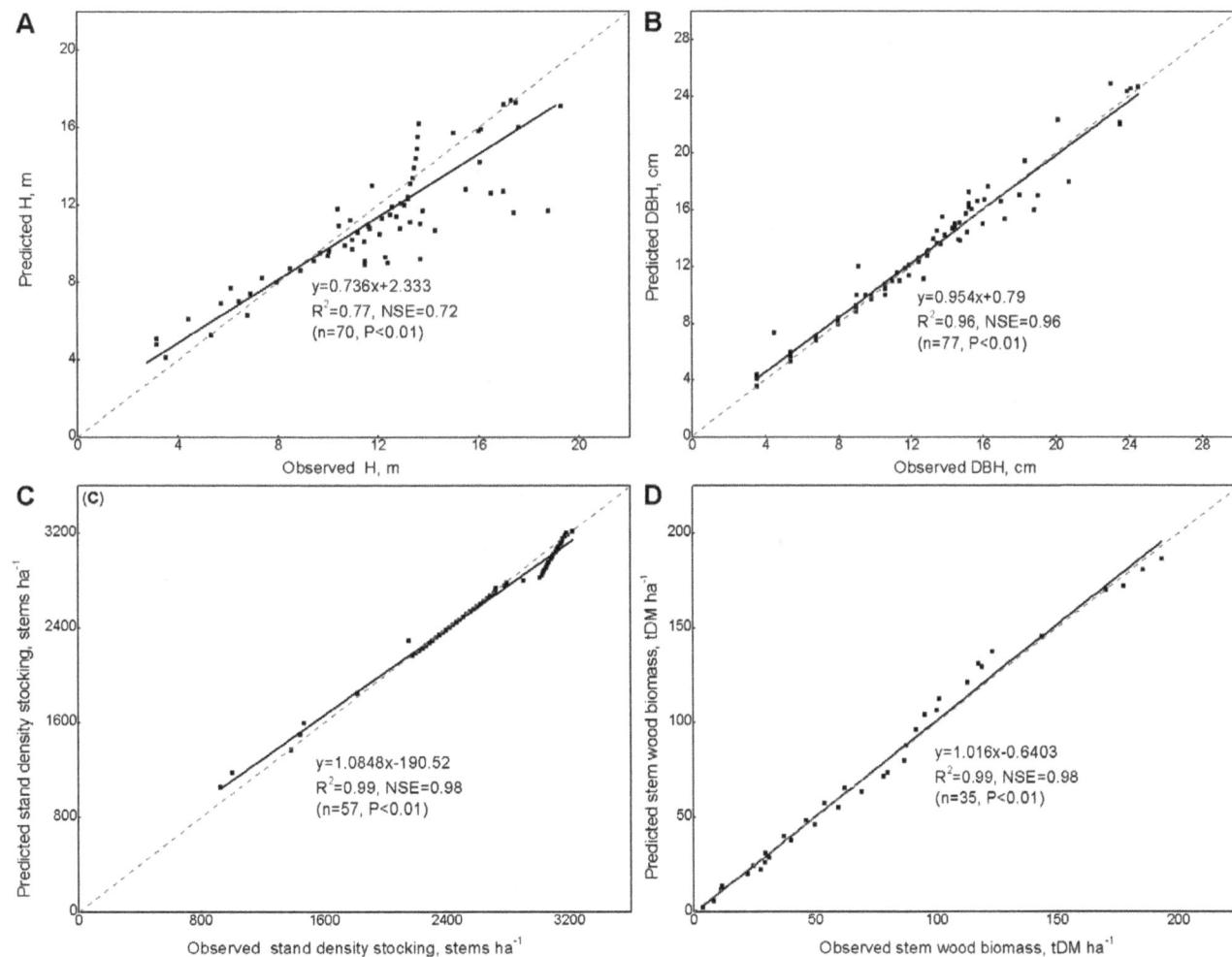

Figure 1. The model was validated by fitting the predicted and observed values. (A) Stand height (H, m). (B) Diameter at breast height (DBH, cm). (C) Stand stocking (stems per ha). (D) Stem woody biomass in tone dry matter per hector (t DM ha^{-1}). Gray dashed lines represent a 1:1 relationship; solid lines are linear regressions. NSE for "Nash-Sutcliffe efficiency".

reaches up to 50 kg N ha^{-1} year^{-1} (N50) [8]. Higher levels over 70 kg N ha^{-1} year^{-1} (N70) in the large industrialized urban areas have been reported. It could further increases up to 90 kg N ha^{-1} year^{-1} (N90) for near future [4]. In the model, it was taken to be a constant rate throughout the life of the stand evenly distributed into every day. Assuming no major climatic changes, the atmospheric N depositions alone is relatively constant for a given site over time. We will also assume that it is constant within a given rotation and over successive rotations. Harvesting and thinning affect C and N dynamics in different ways, depending on the harvesting intensity and the rotation length. In this study, C and N dynamics were simulated and compared using rotation lengths of 120 (RL120), 40 (RL40), 30 (RL30), 20 (RL20) and 15 (RL15) years over a period of 120 years. All rotations started with the planting of 3318 seedlings per hectare. At the end of each rotation period, the stands were clear-felled, and bole only harvesting was adopted, the stem wood at harvest was considered as removed from the site. When a bole harvesting only event occurs, 94% the wood and bark N and volume were removed from the site. The remaining 6% is transferred to the forest floor. The rate of removal of foliage and branch N and volume from the site remains at zero, since the entire amount is transferred to the forest floor as slash. The long-term growth was seldom affected by residue treatment and field observations generally confirm this [38].

The model was used to simulate growth (physiological) constrains on long-term productivity in the study site over a period of 120 years without the effects of rotation (RL120) under initial level of N deposition (N4.9). The growth responses to increased N deposition were re-expressed as growth-response (ecophysiological) constraints on long-term productivity over a period of 120 years without rotation (RL120) under each increased deposition level (N18, N30, N50, N70 and N90). The model was then used to investigate the effects of rotation lengths (RL40, RL30, RL20, and RL15) on long-term forest productivity for initial (N4.9) or increased (N18, N30, N50, N70 and N90) N deposition. We examine the changes in long-term forest in response to different rotation length and increased N deposition, and figure out an appropriate rotation length for maximizing total stem biomass and average NPP. This approach did not address the issues of biological and genetic diversity conservation, or wider socioeconomic aspects of sustainability, such as energy costs associated with fertilizer inputs, harvest operations, transport and processing of wood products.

Results

Effects of Increased N Deposition

The simulated productivity and N dynamics (Panel A, B, C in Fig. 2) of a *C. lanceolata* plantation, in terms of stand growth under the baseline (N4.9 scenario; Fig. 2a0, b0, c0) and increased N deposition impacts (N18, N30, N50, N70 and N90 scenarios; Fig. 2a1–5, b1–5, c1–5) for a period of 120 years, are presented in Fig. 2.

The baseline simulation (Fig. 2a0, b0, c0) generated a realistic sigmoid function of age in the general pattern of stand growth in which the total dry matter of stem wood biomass continuously increased with stand development, peaked at 255.84 t DM ha^{-1} at a stand age of 53 years, followed by a gentle decline as the stands grew, and yielded at 226.35 t DM ha^{-1} at the end of 120-year period (Fig. 2a0). The increase in wood production with stand age was corresponding to the "growth expansion first and decline later" pattern of NPP (Fig. 2a0). The decline trends as stand grows after 53 years can be explained from ANR (<0 after 54 years) (Fig. 2b0). NPP initially exhibited a steep increase during the first

10 years, approximated a maximum rate (here 22.08 t DM ha^{-1} year^{-1}) at 13 years, and then gradually declined with growth. Thereafter, NPP decreased to nearly 0 (here 0.3 t DM ha^{-1} year^{-1}) at 60 years. Decreased nutrient availability and enhanced stomatal limitation were the major causes for NPP decline with stand age. Possible shifts in allocation to belowground components also may contribute to the apparent decline in productivity and biomass accumulation rate of stem wood. The average NPP over the 120-year period was 5.78 t ha^{-1} year^{-1} (Appendix S2).

Annual N requirement (ANR) demonstrated a drastically different pattern compared with production as the stand developed, formed an age-related "concave up beginning convex later" trends (Fig. 2b0). In terms of increases in foliage and fine root pools, ANR increased sharply from 9 kg N ha^{-1} year^{-1} at 2 years to a maximum value of 109.6 kg N ha^{-1} year^{-1} at 16 years. Then caused by self-thinning, ANR generally decreased between 16 and 23 years of age, corresponded to NPP fluctuates at this period. During the period from 23 to 29 years, ANR increased from 70.7 kg N ha^{-1} year^{-1} to a maximum of 227.9 kg N ha^{-1} year^{-1}. After 29 years, ANR decreased and fell to <0 after 54 years, which indicates any increase in stand level respiration, hydraulic resistance, and perhaps plant tissue maturation as age grew. The average ANR over the 120-year period was 32.4 kg N ha^{-1} year^{-1} (Appendix S2).

Soil available N (Fig. 2c0) decreased from 55.9 kg N ha^{-1} at the time when the seedlings were planted to 21.5 kg N ha^{-1} at a stand age of 5 years (Fig. 2c0). As the stand developed, soil available N increased to 32.6 kg N ha^{-1} during the period from 5 to 15 years. Thereafter, soil available N decreased to 1.5 kg N ha^{-1} at 120 years of age. Similar to soil available N, SOC (Fig. 2c0) varied over the first few years after harvest and then declined by about 10% for the next 5 years (Fig. 2c0). SOC dropped from 48.3 t C ha^{-1} at the time when the seedlings were planted to 43.6 t C ha^{-1} at a stand age of 5 years, and then returned to the baseline level by a stand age of 15 years. Then, SOC increased reaching a peak value of 57.4 t C ha^{-1} at an age of 35 years. Thereafter, SOC decreased to 37.8 t C ha^{-1} at 120 years.

Fig. 2 a1–5, b1–5, c1–5 show the variation of stem wood biomass (t DM ha^{-1}), NPP (t DM ha^{-1} year^{-1}), ANR (kg N ha^{-1} year^{-1}), soil available N (kg N ha^{-1}) and SOC (t C ha^{-1}) with stand ages, predicted by the model under increased N deposition rate (without the influence of harvesting regimes) over the 120 year study period. The results showed that productivity of *C. lanceolata* plantation in response to increased N deposition was more related to age than N addition depending on the proportion and age of stand development. Increased N deposition has shown a rapid peak in growth and N dynamics. Increasing N depositions increased average NPP to 5.95, 6.032, 6.05, 6.06, and 6.06 t DM ha^{-1} year^{-1} under the N deposition rates of 18, 30, 50, 70 to 90 kg N ha^{-1} year^{-1}, respectively (Appendix S2). Increased available N, decreased stomatal limitation and early arrival of a large leaf area were the major causes for the growth peaks decline with stand age.

Effects of Rotation Lengths Under the N4.9 Scenario

Under the null level of N deposition (N4.9; gray solid lines in Fig. 3, solid squares in Fig. 4, and pink solid line with hollow squares in Fig. 5), the effects of rotation lengths on the long-term productivity of *C. lanceolata* plantation are shown in Figs. 3,4,5 and Appendix S2.

Over 120 year period, rotation length resulted in the changes in the productivity and N dynamic of *C. lanceolata* plantation with successive rotations (Fig. 3A, B, C, D, E). Soil available N decreased with the increase in successive rotations (Fig. 3D). First

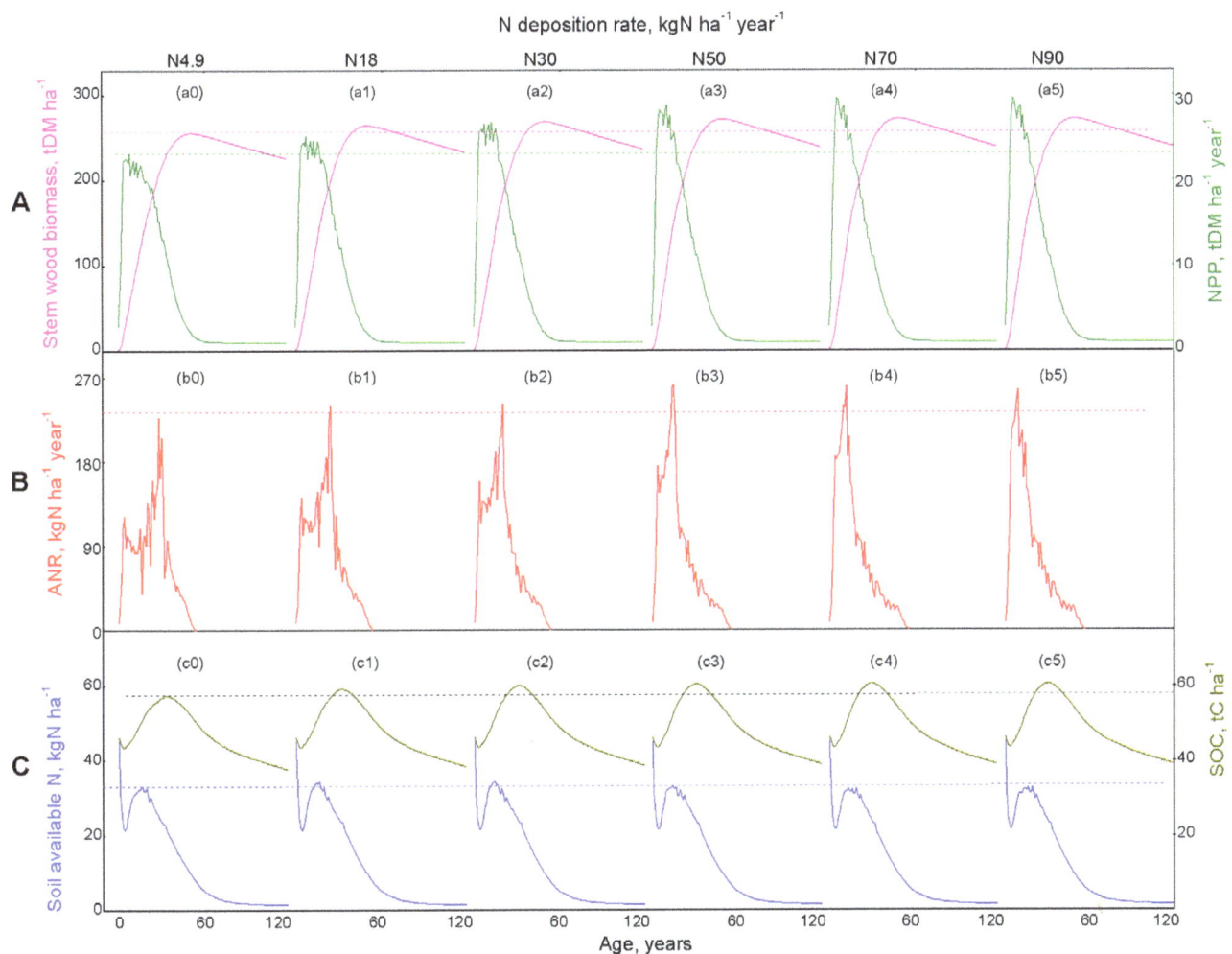

Figure 2. Simulated increased N deposition impacts on forest productivity as stand development. (Panel A) The magenta solid line stands for the stem wood biomass production in tone dry matter per hector (t DM ha^{-1}), and green solid line refers to the net primary production in tone dry matter per hector (NPP, t DM ha^{-1} year^{-1}). (Panel B) Annual increment of ANR (red solid line, ANR, kg N ha^{-1} year^{-1}). (Panel C) The blue solid line represents current status of soil available N for growth in kg N per hector (kg N ha^{-1}), olive solid line refers to soil organic C (SOC, t C ha^{-1}) of *C. lanceolata* plantation in southern China from seedling to 120 years of age. In panels A, B, and C, a0, b0 and c0 represent growth trends from the basic, under initial level of N deposition (N4.9) and without rotation practice (RL120), simulation that made up the baseline data set; and a1–5, b1–5, as well as c1–5 represent predicted growth response trends generated under increased N deposition. The dashed lines represent growth constraint equilibrium state.

rotation plantation exhibited conservative N-cycling properties. As successive rotation practiced, N-cycling properties recovered with increasing availability of soil nitrate relative to ammonium. The dominance of a conservative N cycle for typical mature lowland tropical forests re-emerged [39]. All rotation lengths led to a yield decline in stem wood biomass accumulation (Fig. 3A), annual dry matter biomass increment (NPP, Fig. 3B), and annual increment of ANR (ANR, Fig. 3C) at harvest time over successive rotations. The shorter the rotation length, the greater the decrease in yields at harvest time (Figs. 3). For a certain rotation length (Fig. 4), the successive decrease was greatest for the first 1–3 rotations, usually by about 5–25%, and tended to be modest thereafter (Fig. 4). However, stem wood biomass demonstrated an increase during rotations 5–6 of the RL20 and 6–8 of the RL15 treatments, respectively (Fig. 4A). On the long-term (Fig. 5), in general, over the 120-year period, total productivity (total stem biomass and average NPP) demonstrated "increase first decrease later" patterns as rotation length shortened (Fig. 5A, B), while the trends of

average ANR peaked at RL40 (Fig. 5C), and total stem wood biomass and average NPP peaked at RL30 (Fig. 5A, B). Differences in total production between RL30, RL20, and RL15 gradually increased as the rotation length shortened, and this result was particularly evident in terms of the timber harvest. A rotation length of 30 years was optimal for sustainable management at the N4.9 deposition level.

Interactive Effects of Rotation Length and Increased N Deposition

Rotation length clearly affected the productivity and N dynamics of the simulated *C. lanceolata* plantation over the 120-year period from perspective of growth response constraints (Figs. 3), successive effects (Fig. 4, Appendix S2), and long-term outcomes (Fig. 5, Appendix S2).

For different rotation lengths (RL40, RL30, RL20, and RL15) and the simulated productivity and N dynamic values of the *C. lanceolata* plantation over the 120-year cultivation period are

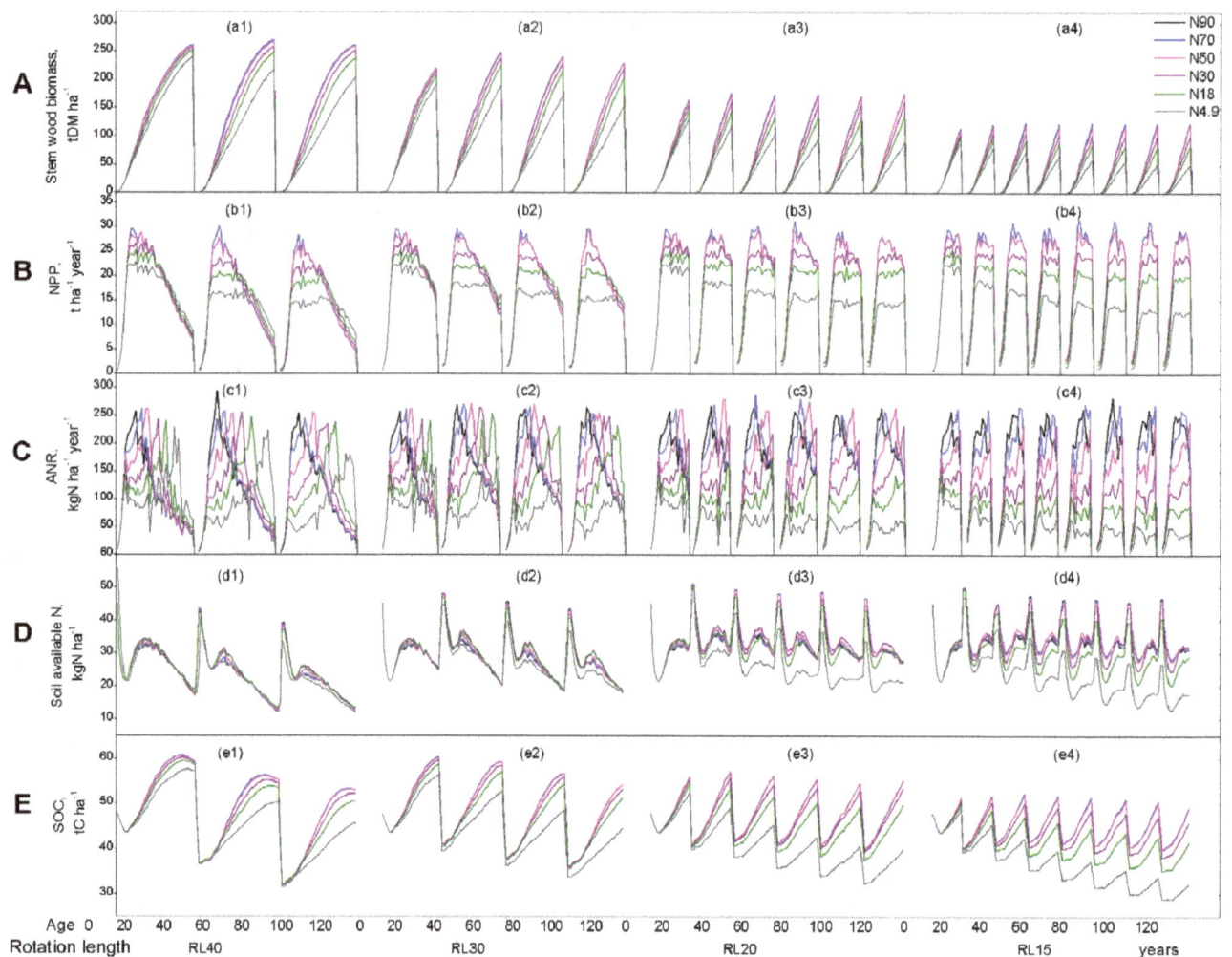

Figure 3. Simulated interactive effects of increased N deposition and short-rotation management as stand development. (A) The stem wood biomass production in tone dry matter per hector (t DM ha^{-1}). (B) The net primary production in tone dry matter per hector (NPP, t ha^{-1} year^{-1}). (C) Annual increment of ANR (kg N ha^{-1} year^{-1}). (D) The current status of soil available N for growth in kg N per hector (soil available N, kg N ha^{-1}). (E) Soil organic C (SOC, t C ha^{-1}). Over the 120-year period coupling chronic levels of atmosphere N depositions of N4.9, N18, N30, N50, N70, and N90 with different rotation intervals (RL40, RL30, RL20, and RL15). RL40, RL30, RL20, and RL15 represent rotation cycles with intervals of 40, 30, 20, and15 years, respectively. N4.9 (black solid line) = 4.9 kg N ha^{-1} year^{-1}, N18 (gray solid line) = 18 kg N ha^{-1} year^{-1}, N30 (light gray solid line) = 30 kg N ha^{-1} year^{-1}, N50 (black dash) = 50 kg N ha^{-1} year^{-1}, N70 (gray dash) = 70 kg N ha^{-1} year^{-1}, and N90 (light gray dash) = 90 kg N ha^{-1} year^{-1}.

presented in Fig. 3, 4 and 5. In general, on the long-term (Fig. 5), short rotation length had a more significant effect on productivity and N dynamics than high N deposition. Under all rotation length cycles, increasing N deposition generally led to increases in both productivity and N dynamics. However, the sensitivity of N deposition for a given rotation length decreased with increasing amounts of N deposition. Over the 120-year period, the total stem biomass, average NPP, and average ANR did not sensitively response to increase in N deposition level higher than N50 (50 kg N ha^{-1} year^{-1}). This result indicated that a N deposition level of 50 kg N ha^{-1} year^{-1} was the upper limit level of thresholds for N deposition impacts for *C. lanceolata* plantations with rotation length of 30 and 40 years (Fig. 5). Similarly, N70 (70 kg N ha^{-1} year^{-1}) was the upper limit level of thresholds for N deposition impacts under a rotation length of 15 and 20 years (Fig. 5).

For all of the increased N deposition scenarios larger than 50 kg N ha^{-1} year^{-1}, all indicators increased as the rotation length was shortened from RL40 to RL20 over the entire 120-year simulated period and exhibited moderate variations in average ANR when

the rotation length was changed from RL20 to RL15 (Fig. 5). These results indicate that the largest increase in total stem wood biomass and average NPP occurred when using the RL20 cycle and smaller differences in total stem wood biomass and average NPP occurred between RL15 and RL30 (Fig. 5A, B). At a specific N deposition level, over the 120-year period, a rotation length of 20–30 years would be the most sustainable practice in terms of total stem wood biomass, average NPP and average ANR under the N18 scenario (Fig. 5A, B, C), while under the N30 scenario, a rotation length of 20 years would be most sustainable in terms of total stem wood biomass and average NPP (Fig. 5A, B), and for N50, N70 and N90 scenarios, the sustainable rotation length would be 15–20 years in terms of total stem wood biomass and average NPP (Fig. 5A, B).

At the same time, a shorter rotation length resulted in a major decline in forest floor biomass. A change in the rotation length from RL40 to RL30 increased SOC by 0.15% under N18, 0.18% under N30, and 0.015% under N50, but decreased SOC by 15% under N70 and N90. Therefore, for N deposition levels higher

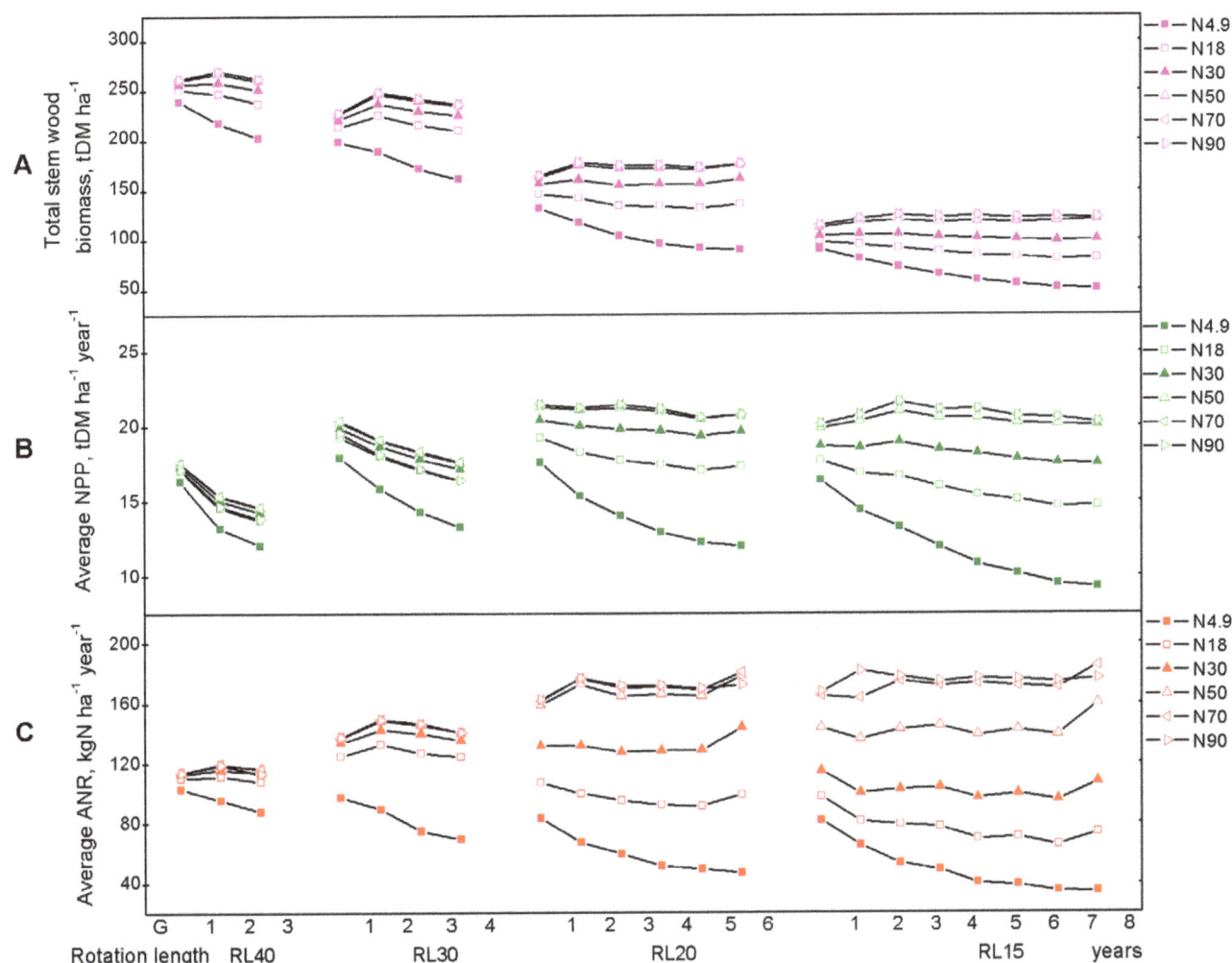

Figure 4. The summary of simulated interactive effects of increased N deposition and short-rotation management over successive rotations. It displays the accumulative statues, (A) Stem wood biomass at the end-of-rotation, and rotation-average values for flux or rates, (B) NPP and (C) ANR. RL40, RL30, RL20, and RL15 represent rotation cycles with intervals of 40, 30, 20, and 15 years, respectively. The effects of atmosphere N deposition levels: N4.9 (solid squares) = 4.9, N18 (hollow squares) = 18, N30 (solid upward triangles) = 30, N50 (hollow upward triangles) = 50, N70 (hollow leftward triangles) = 70, and N90 (hollow rightward triangles) = 90 kg N ha^{-1} year^{-1}.

than N18, rotation lengths <30 years lead to 2.5–8.2% losses in SOC (Fig. 3E).

Discussion

Uncertainties of Model Performance

The results of this study may be limited by four aspects of uncertainties: (1) resulted from complex interactions among the different components of the forest, varied in different forest regions, elevations, among different temporal and spatial scales by comprising multi-level of eco-biological complexity, the response of a forest ecosystem to climate change can have very different magnitudes and even different directions [40–42]. Available studies suggest that forest responses to climate change factors (such as increasing temperature and CO_2) will be limited by competition, disturbance, and nutrient limitations [43–44]. (2) CenW model was unable to acquire sufficient field measurements of reference parameters from literature, pertaining to model parameterization as it relates to *C. lanceolata* stands, fixed parameters of default values or average values from literature or field measurements were used in this study (see Appendix S1). (3)

Our study concentrated on how N deposition and climate would affect the growth for *C. lanceolata* plantation. As mentioned before, we didn't consider land use change resulted from human activities, disturbances such as fires and environmental pollution as aerosols, atmospheric CO_2 and O_3, which are important factors affecting forest productivity and N dynamics [45–46]. (4) Although our results agree with many other studies at realistic temporal scales, but long-term productivity modeling still has limited knowledge of the forest responses to increased N deposition in the field on the long-term observation.

Model Simulation of Effects of Increased N Deposition

The simulated NPP of the *C. lanceolata* plantation demonstrated a steep increase during the first 10 years, reaching its maximum growth rate at the age of 13 years, which then declined thereafter. This growth pattern is in good agreement with the measured peak NPP values of evergreen coniferous forests in tropical and subtropical zones [47]. The simulated average NPP values during the first 60 years under the N4.9 scenario (4.9 kg N ha^{-1} year^{-1}) and RL120 length (11.34 t DM ha^{-1} year^{-1}) in this study are

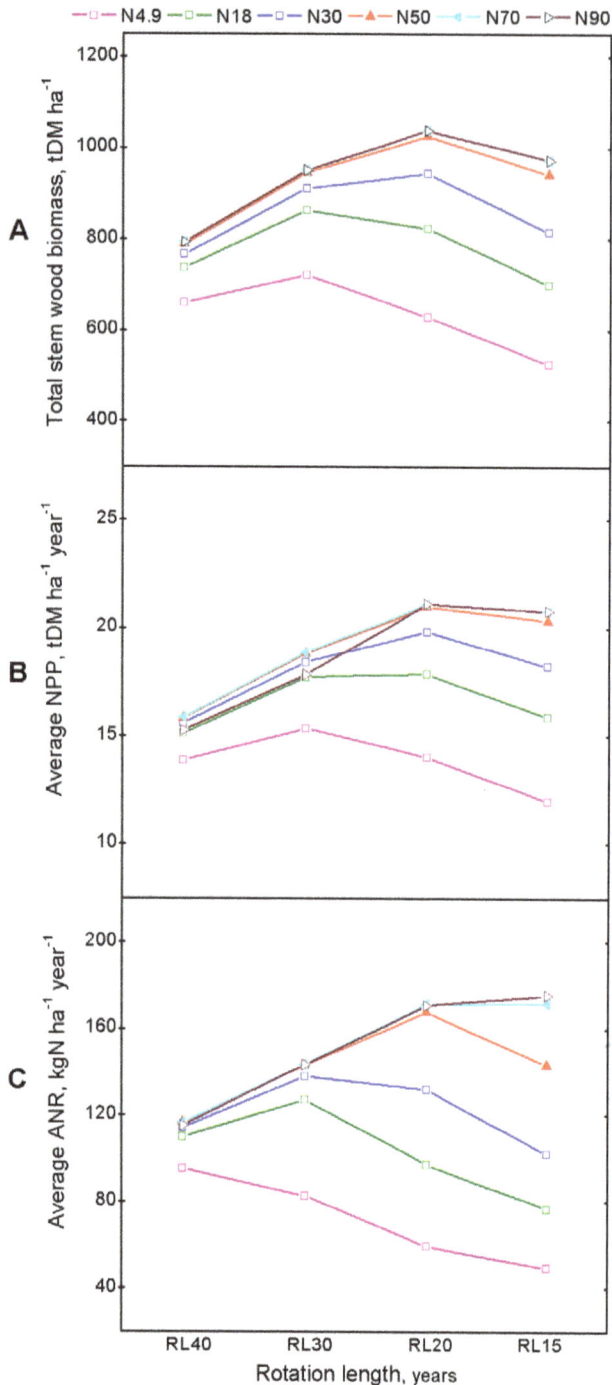

Figure 5. The long-term (total 120-year period) summary of interactive effects of increased N deposition and short-rotation management over successive rotations. It shows the sum of accumulative statues at the end-of-rotations, (A) stem wood biomass, and 120-year-average values for flux or rates, (B) NPP and (C) ANR.

C. lanceolata is a fast growing tree species with a high production value. Despite some uncertainties existed in this study, the simulated results for initial N deposition level (N4.9) could be explained by the five factors reviewed by Ryan et al. [52], including changes in photosynthesis rate as stand development, nutrient supply, respiration, C allocation, and hydrological function. Our study also implied that a decline in stand leaf area usually accompanies a decline in aboveground wood growth. The decline in NPP with stand age can be explained as a combined effect of declining photosynthesis efficiency and declining N-availability for tree growth [53].

The results showed that of the productivity of *C. lanceolata* plantation in response to increased N deposition were more related to stand age than N addition depending on the proportion and age of growing forests. Increased N deposition has shown to raise the equilibrium state and constraint curve. Increased rates of N deposition can affect site nutrient stores and dynamics, microclimate, and tree growth. Recent evidence suggests that sustained growth response is unlikely to be realized as a cumulative effect of N fertilization, but such a possibility might exist with other forms of fertilization, if they become widespread [54]. Based on the assumption that duration of these effects is short, cumulative nutrient mineralization and/or leaching should be unlikely to occur as cumulative effects over time on single site. However, duplication of applications across landscape areas might hypothetically lead to cumulative increases in "baseline" nutrient mobilization and/or leaching for the area as a whole.

Interactive Effects of Increased N Deposition and Short-rotation Length

Maintaining long-term productivity in forests is a fundamental goal of ecologically sound forest management [55]. Factors that may be detrimental to the sustainability of plantations include the removal of a large proportion of material from the site, short-rotation length, and increased site disturbances [18]. Of particular concern is the sustainability of the physical properties of the soil, soil horizons, and complex biogeochemical cycles that are constantly at work within the forest system [28]. A number of studies have investigated the effects of harvesting regimes on the rotation cycle, growth rate, and C storage [18,28,32,56]. Conventional stem-only harvesting is accepted as a sustainable forestry practice [57].

In this study, rotation length was an important factor that affects long-term productivity and N dynamics of *C. lanceolata* plantations if the N deposition level is low. As expected, total productivity over the 120-year simulation period demonstrated a nonlinear response pattern: specifically, an initial increase in productivity, which declined in accordance with a shorter rotation length. The highest value (15.25 t DM ha^{-1} year^{-1}) was achieved using a 30-year rotation cycle (RL30) (Fig. 5B). Therefore, under current climatic conditions and an N deposition level of 4.9 kg N ha^{-1} year^{-1} (N4.9), a reasonable rotation cycle was determined to be 30 years in terms of sustainable productivity over the 120-year simulated period (Fig. 5B).

The rotation length also strongly affects the productivity and N dynamics over successive rotations at *C. lanceolata* plantations, which mostly monocultures that are currently managed using rotation lengths <20 years [28]. Short rotation cycles and intensive forest practices in the past have been linked to yield declines, soil depletion, and reductions in productivity over successive rotations in *C. lanceolata* plantations [4,28,30]. Significant declines in the yield (10–40%) have been reported [25,27]. In this study, the simulated soil available N was highest in the second rotation for rotation lengths <40 years (RL40) (Fig. 3D). The

within the range of the average NPP (9.02–15.47 t DM ha^{-1} year^{-1}) measured in *C. lanceolata* plantations at the same site [48]. These results are close to those values measured in subtropical coniferous forests in China [49,50]. Over the simulated 120-year period, the simulated average NPP (5.78 t DM ha^{-1} year^{-1}) is within the range of values measured in the forests of China (approximately 4.8–6.22 t DM ha^{-1} year^{-1}) [51], indicating that simulated stem wood biomass, NPP, ANR, and SOC values were highest during the first rotation regardless of the rotation cycle, then, decreased as the number of rotation generations increased (Fig. 3A, B, C, E). The decreasing rate of SOC varied over successive rotations, demonstrating a sharp decrease by about 5–25% over the first 1–3 generations, but only a modest successive decrease was demonstrated thereafter. However stem wood biomass increased after the sixth RL20 rotation and the eighth RL15 rotation (Fig. 4A). The decline in yield over successive rotations, as observed in the model simulation, is consistent with the results of Zhang et al. [30], who reported that increments in the stand biomass were reduced by an average of 24% from the first to the second rotation and by a further 40% from the second to the third rotation. SOC was reduced by 10% between the first and the second rotations and by 15% between the second and the third rotations.

Changes in the composition and biomass of forest floors, associated with decomposition, nutrient mobilization, and re-distribution from vegetation and forest floors to mineral soil horizons should occur commonly as a cumulative effect in managed forests, as these trends are influenced in similar or complementary ways by several forestry practices, including harvest, site preparation, fertilization and herbicide application. The long-term significance of these trends is unknown. Short-rotation management aimed to increase wood production may affect forest sustainability depending, among other factors, on initial site conditions and treatment regimes. Excessive removal of C and nutrients and soil disturbance are two of the major factors that may cause site degradation and reduction in forest pro-ductivity [58]. Short-rotation practices can increase productivity above base level and reduce or eliminate negative impacts. Currently, there is a little direct evidence to support that harvest removals in themselves lead to soil depletion over several succeeding rotations. Short-term productivity declines should be due to severe associated effects such as compaction, erosion or loss of organic layers.

The simulation results of this study indicate that the rotation length and N deposition level interactively affect the productivity and N dynamics of C. lanceolata plantations. Even though rotation length is an important factor that affects the productivity and N dynamics of C. lanceolata plantations, the offsite effects of the shorter rotation length decreased as the amount of N deposition increased. A high level of N deposition led to increased production and N accumulation in the plantation ecosystem, but the increasing effects of production due to high N deposition increased as the rotation length was shortened. These phenomena could be interpreted by the fact that high N deposition surpassed the loss of N in the soil due to the removal of harvesting materials and leaching due to runoff.

Continuously increasing atmospheric N deposition will result in soil acidification, N leaching, nutrient imbalances, and NPP depression in forests [4,59]. The results of this study imply that the offsite detrimental effects of chronic atmospheric N deposition could be ameliorated through appropriate forest management initiatives such as rotation length selection. For example, for a given level of N deposition, the responses of NPP to a high level of N deposition were strongly correlated with stand age (Fig. 2). Over the 120-year period, the total stem biomass, average NPP, and average ANR did not sensitively response to increase in N deposition level higher than N50 (50 kg N ha^{-1} year^{-1}). This result indicated that N deposition level of 50 kg N ha^{-1} year^{-1} was the upper limit level of thresholds for nitrogen deposition impacts for C. lanceolata plantations with as rotation length of 30 and 40 years (Fig. 5). Similarly, N70 (70 kg N ha^{-1} year^{-1}) was

the upper limit level of thresholds for nitrogen deposition impacts under a rotation length of 15 and 20 years (Fig. 5). The NPP sensitive to the N deposition level of 18–50 kg N ha^{-1} year^{-1} (N18, N30 and N50) occurred when the stand age was 6–20 years, while the sensitive of NPP to high N deposition, such as N70 (70 kg N ha^{-1} year^{-1}) or N90 (90 kg N ha^{-1} year^{-1}), occurred when the stand age was 6–19 years. Therefore, the sensitive stand age of N increasing was around 20 years for atmospheric N deposition >18 kg N ha^{-1} year^{-1} input to the C. lanceolata plantation. After the stand age of N sensitive, N deposition would exceed N uptake by the trees and the retention capacity, consequently resulting in negative environmental impacts. The sensitive stand age of N increasing for a specific N deposition level may be the appropriate rotation length to sustain productivity without offsite environmental effects in the C. lanceolata plantation.

The results of this study also indicate that increasing atmospheric N deposition might promote improvements in tree growth and site quality at C. lanceolata plantations. While very short rotation lengths reduced the capacity of the plantation ecosystem to accumulate C and N, particularly in sites with initially poor levels of nutrients, these reductions could have been averted by higher atmospheric N deposition. The obvious positive responses in terms of productivity and reduced N-use efficiency in C. lanceolata plantations due to increasing N deposition are consistent with another study on C. lanceolata plantations in southeastern China, which simulated results using the FORECAST model [32]. For example, increasing N deposition from 4.9 kg N kg N ha^{-1} year^{-1} (N4.9) to 70 kg N ha^{-1} year^{-1} (N70) resulted in higher productivity and timber biomass, increased forest N retention, and increased N requirements for tree growth, but lower SOC accumulation and decreased soil available N in C. lanceolata plantations. Further increasing N deposition to N90 (90 kg N ha^{-1} year^{-1}) or higher did not increase productivity, N retention, or affect SOC or soil available N. Therefore, N deposition>N70 (70 kg N ha^{-1} year^{-1}) leads to the loss of N through leaching and offsite effects.

Conclusions

Case studies that use modeling are needed to improve our understanding of the effects of the rotation cycle and N deposition on the long-term productivity of plantations. In terms of how the model simulations performed when they incorporated variables from a C. lanceolata plantation in subtropical China, our results show that CenW3.1 demonstrated considerable potential for this particular application. Not only can the model be used to integrate data from various forest studies in a dynamic manner, thus enabling a more thorough examination of study data, it can also be applied to predict the potential long-term effects of management practices and N deposition on the functions (such as C and N processes) of a forest ecosystem. Over the long term, a combination of indicators, including NPP, stem wood biomass and ANR indicate that 20–30 years is a sustainable rotation cycle for subtropical C. lanceolata plantations assuming an N deposition level of 4.9–50 kg N ha^{-1} year^{-1}; shorter rotation lengths (15–20 years) may be sustainable for N deposition rates between 50 and 70 kg N ha^{-1} year^{-1}, to avoid leaching and offsite effects.

Supporting Information

Appendix S1 Definitions, symbols, values, units, and sources of parameters used in CenW3.1 modeling of a C. lanceolata plantation. For the "Value sources" column, [8,23,31,37,65–72] are refer-ences citations, D = default, F = fitted, O = observed, and A = as-sumed.

Appendix S2 The summarized values for each rotation and over 120-year period (All) in *C. lanceolata* plantation under different N deposition levels (N4.9, N18, N30, N50, N70 and N90 for 4.9, 18, 30, 50, 70, and 90 kg N ha^{-1} year^{-1}) and rotational lengths (RL120, RL40, RL30, RL20 and RL15 for no harvesting, 40, 30, 20, and 15-year rotation).

Acknowledgments

Special thanks to the CenW model group (Dr Miko U.F. Kirschbaum) for free access (http://www.kirschbaum.id.au/) to the CenW 3.1 modeling software.

Author Contributions

Conceived and designed the experiments: WHX MFZ CHP. Performed the experiments: MFZ DLT XWD ZHH CHP. Analyzed the data: MFZ WHX XLZ. Contributed reagents/materials/analysis tools: CHP. Wrote the paper: MFZ WHX CHP.

References

1. Heilman P, Norby R (1998) Nutrient cycling and fertility management in temperate short rotation forest systems. Biomass and Bioenergy 14(4): 361–370.
2. Moffat AS (1998) Global nitrogen overload problem grows critical. Science 279, 988–989.
3. Chen XY, Mulder J, Wang YH, Zhao DW, Xiang RJ (2004) Atmospheric deposition, mineralization and leaching of nitrogen in subtropical forested catchments, south China. Environ. Geochem. Health 26: 179–186.
4. Ma XQ, Heal KV, Liu AQ (2007) Nutrient cycling and distribution in different-aged plantations of Chinese fir in southern China. For. Ecol. Manag. 243(1): 61–75.
5. Lu XK, Mo JM, Gundersern P, Zhu WX, Zhou GY, et al. (2009) Effect of simulated N deposition on soil exchangeable cations in three forest types of subtropical China. Pedosphere 19(2): 189–198.
6. Ma XH (1989) Effects of rainfall on the nutrient cycling in man-made forests of *Cunninghamia lanceolata* and *Pinus massoniana*. Acta Ecologica Sinica 9: 15–20. (in Chinese).
7. Zhou GY, Yan JH (2001) The influence of region atmospheric precipitation characteristics and its element inputs on the existence and development of Dinghushan forest ecosystems, Acta Ecologica Sinica 21: 2002–2012. (in Chinese).
8. Chen XY, Mulder J (2007) Atmospheric deposition of nitrogen at five subtropical forested sites in South China. Sci. Total Environ. 378: 317–330.
9. Hu ZY, Xu CK, Zhou LN, Sun BH, He YQ, et al. (2007) Contribution of atmospheric nitrogen compounds to N deposition in a broadleaf forest of southern China. Pedosphere 17(3): 360–365.
10. Fang YT, Zhu WX, Mo JM, Zhou GY, Gundersen P (2006) Dynamics of soil inorganic nitrogen and their responses to nitrogen additions in three subtropical forests, South China. J. Environ. Sci. 18: 756–763.
11. Maskell LC, Smart SM, Bullock JM, Thompson K, Stevens CJ (2010) Nitrogen deposition causes widespread loss of species richness in British habitats. Glob. Chang. Biol. 16: 671–679.
12. Zaehle S, Friedlingstein P, Friend AD (2010) Terrestrial nitrogen feedbacks may accelerate future climate change. Geophys. Res. Lett. 37: L01401.
13. Clein JS, McGuire AD, Zhang X, Kicklighter DW, Melillo JM, et al. (2002) Historical and projected carbon balance of mature black spruce ecosystems across North America: The role of carbon-nitrogen interactions. Plant Soil 242(1): 15–32.
14. Galloway JN, Aber JD, Erisman JW, Seitzinger SP, Howarth RW, et al. (2003) The nitrogen cascade. Bioscience 53: 341–356.
15. Schoenholtz SH, Van Miegroet H, Burger JA (2000) A review of chemical and physical as indicators of forest soil quality: challenges and opportunities. For. Ecol. Manag. 138: 335–356.
16. Hynynen J, Ahtikoski A, Siitonen J, Sievänen R, Liski J (2005) Applying the MOTTI simulator to analyse the effects of alternative management schedules on timber and non-timber production. For. Ecol. Manag. 207: 5–18.
17. Kaipainen T, Liski J, Pussinen A, Karjalainen T (2004) Managing carbon sinks by changing rotation length in European forests. Environ. Sci. Policy 7:205–219.
18. Peng CH, Jiang H, Apps MJ, Zhang YL (2002) Effects of harvesting regimes on carbon and nitrogen dynamics of boreal forest in central Canada: a process model simulation. Ecol. Model. 155: 177–189.
19. Pussinen A, Karjalainen T, Mäkipää R, Valsta L, Kellomäki S (2002) Forest carbon sequestration and harvests in Scots pine stand under different climate and nitrogen deposition scenarios. For. Ecol. Manag. 158: 103–115.
20. Wei X, Kimmins JP, Zhou G (2003) Disturbances and the sustainability of long-term site productivity in lodgepole pine forests in the central interior of British Columbia: an ecosystem modeling approach. Ecol. Model. 164: 239–256.
21. FAO (2006) Global planted forest thematic study: results and analysis. In: Lugo AD, Ball J, Carle J, editors. Planted Forests and Trees Working Paper. Rome. FP38E.
22. Lei JF (2005) Forest Resources in China. Beijing: China Forestry Publish House. 172 p. (in Chinese).
23. Wu ZL (1984) Chinese Fir. Beijing: China Forestry Publish House. (in Chinese).
24. Guo JF, Yang YS, Liu LZ, Zhao YC, Chen Z, et al. (2009) Effect of temperature on soil respiration in a Chinese fir forest. J. For. Res. 20: 49–53.
25. Tian DL, Xiang WH, Chen XY, Yan WD, Fang X, et al. (2011) A long-term evaluation of biomass production in first and second rotations of Chinese fir plantations at the same site. Forestry 84(4): 411–418.
26. Sheng WT, Fan SH (2005) Long Term Productivity of Chinese Fir Plantations. Beijing: Science Press. (in Chinese).
27. Ding YX, Chen JL (1995) Effect of continuous plantation of Chinese fir on soil fertility. Pedosphere 5: 57–66.
28. Bi J, Blanco JA, Seely B, Kimmins JP, Ding Y, et al. (2007) Yield decline in Chinese-fir plantations: a simulation investigation with implications for model complexity. Can. J. For. Res. 37(9): 1615–1630.
29. Zhang J, Wang SL, Feng ZW, Wang QK (2009) Stability of soil organic carbon changes in successive rotations of Chinese fir (*Cunninghamia lanceolata* (Lamb.) Hook) plantations. J. Environ. Sci. 21: 352–359.
30. Zhang XQ, Kirschbaum MUF, Hou ZH, Guo ZH (2004) Carbon stock changes in successive rotations of Chinese fir (*Cunninghamia lanceolata* (Lamb) hook) plantations. For. Ecol. Manag. 202: 131–147.
31. Zhao MF, Xiang WH, Peng CH, Tian DL (2009) Simulating age-related changes in carbon storage and allocation in a Chinese fir plantation growing in southern China using the 3-PG model. For. Ecol. Manag. 257: 1520–1531.
32. Wei XH, Blanco JA, Jiang H, Hamish Kimmins JP (2012) Effects of nitrogen deposition on carbon sequestration in Chinese fir forest ecosystems. Sci. Total Environ. 416: 351–361.
33. Kirschbaum MUF (1999) Modelling forest growth and carbon storage with increasing CO_2 and temperature. Tellus B 51: 871–888.
34. Kirschbaum MUF, Paul KI (2002) Modelling C and N dynamics in forest soils with a modified version of the CENTURY model. Soil Biol. Biochem. 34: 341–354.
35. Running SW, Ramakrishna RN, Hungerford RD (1987) Extrapolation of synoptic meteorological data in mountainous terrain and its use for simulating forest evapotranspiration and photosynthesis. Can. J. For. Res. 17: 472–483.
36. Nash JE, Sutcliffe JV (1970) River flow forecasting through conceptual models. J. Hydrol. 10: 282–290.
37. Chen XY (1989) Dynamic characteristics of nitrogen in a Chinese fir plantation ecosystem. Journal of Central South Forestry College 9 (Sup.): 66–75. (in Chinese).
38. Yang YS, Guo JF, Chen GS, He ZM, Xie JS (2003) Effects of slash burning on nutrient removal and soil fertility in Chinese fir and evergreen broadleaved forests of mid-subtropical China. Pedosphere 13 (1), 87–96.
39. Davidson EA, Reis de Carvalhohttp://www.nature.com/nature/journal/v447/n7147/full/nature05900.html - a2 CJ, Figueirahttp://www.nature.com/nature/journal/v447/n7147/full/nature05900.html - a3 FY, Ometto JPHB, et al. (2007) Recuperation of nitrogen cycling in Amazonian forests following agricultural abandonment. Nature 447: 995–998.
40. Pastor J, Post WM (1988) Response of northern forests to CO2-induced climate change. Nature 334:55–58.
41. Overpeck JT, Rind D, Goldberg R (1990) Climate-induced changes in forest disturbance and vegetation. Nature 343:51–53.
42. Aber JD, McDowell W, Nadelhoffer K, Magill A, Berntson G, et al. (1998) Nitrogen saturation in northern forest ecosystems, hypotheses revisited. Bioscience 48: 921–934.
43. Galloway JN, Cowling EB (2002) Reactive nitrogen and the world: 200 years of change. Ambio 31: 64–71.
44. Davidson EA (2009) The contribution of manure and fertilizer nitrogen to atmospheric nitrous oxide since 1860. Nat. Geosci. 2: 659–662.
45. Ren W, Tian H, Tao B, Chappelka A, Sun G, et al. (2011) Impacts of tropospheric ozone and climate change on net primary productivity and net carbon exchange of China's forest ecosystems. Glob. Ecol. Biogeogr 20:391–406.
46. Sitch S, Cox PM, Collins WJ, Huntingford C (2007) Indirect radiative forcing of climate change through ozone effects on the land-carbon sink. Nature 448:791–794.
47. Wang SQ, Zhou L, Chen JM, Ju WM, Feng XF, et al. (2011) Relationships between net primary productivity and stand age for several forest types and their influence on China's carbon balance. J. Environ. Manag. 92(6): 1651–1662.
48. Tian DL, Pan WC, Zhang CJ, Lei ZX (1993) The Characteristics of Biomass in Chinese Fir Plantation Ecosystem. In: Liu XZ, editor. Long-term Located Research on Forest Ecosystem. Beijing: China Forestry Publish House. 221–227 p. (in Chinese).

49. Wen DZ, Wei P, Kong GH (1999) The productivity and turnover of fine roots of southern subtropical forest in Dinghu Mountains. Acta Botany Sinica 23: 361–369. (in Chinese).

50. Yang YS, Chen GS, He ZM, Chen YX, Guo JF (2002) Production, distribution and nutrient return of fine roots in a mixed and a pure forest in subtropical China. Chinese Journal of Applied Environment and Biology 8: 2230–233. (in Chinese).

51. Fang JY, Liu GH, Xu SL (1996) Biomass and net production of forest vegetation in China. Acta Ecologica Sinica 16 (5): 497–508. (in Chinese).

52. Ryan M G (1991) The effects of climate change on plant respiration. Ecol. Appl. 1: 157–167.

53. Murty D, McMurtrie RE (2000) The decline of forest productivity as stands age: a model-based method for analyzing causes for the decline. Ecol. Model. 134(2/3): 185–205.

54. Morford SL, Houlton BZ, Dahlgren RA (2011) Increased forest ecosystem carbon and nitrogen storage from nitrogen rich bedrock. Nature 477: 78–81.

55. Helms JA (1998) Dictionary of forestry, Society of American Foresters and CABI Publishing. 210 p.

56. Xin ZH, Jiang H, Jie CY, Wei XH, Blanco J, et al. (2011) Simulated nitrogen dynamics for a *Cunninghamia lanceolata* plantation with selected rotation ages. Journal of Zhejiang Agriculture and Forestry University 28(6): 855–862. (in Chinese).

57. Hakkila P (2002) Operations with reduced environmental impact. In: Richardson J, Bjorheden R, Hakkila P, Lowe AT, Smith CT, editors. Bioenergy from sustainable forestry: Guiding principles and practice. Dordrecht: Klewer Academic Publishers. 244–261 p.

58. Miller RE, Colbert SR, Morris LA (2004) Effects of heavy equipment on physical properties of soils and on long-term productivity. A review of literature and current research. National Council for Air and Stream Improvement Tech. Bull. No. 887. Research Triangle Park, NC. 76 p.

59. Weetman GF, McWiliams ERG, Thompson WA (1992) Nutrient management of Douglas-fir and western hemlock stands: The issues. In: Chappell HN, Weetman GF, Miller RE, editors. Forest Fertilization: Sustaining and improving nutrition and growth of western forests. Institute of Forest Resources, College of Forest Resources, University of Washington, Seattle, WA, Contribution No. 73. 17–27 p.

60. Fang X, Tian DL, Cai BY, Gao YM (2002) Absorption, accumulation and transportation of N, P, K elements in the second-generation Chinese fir plantation. Journal of Central South Forestry University 22(2): 1–6. (in Chinese).

61. Tian DL, Kang WX (1993) Study on the Chinese Fir Plantation Adjustment Function to Water Cycle. In: Liu XZ, editor. Long-term Located Research on Forest Ecosystem. Beijing: China Forestry Publish House. 203–208 p. (in Chinese).

62. Wen SZ, He BF (1993) Studies on Runoff Law in Different interference of Chinese Fir Plantation Ecosystem. In: Liu XZ, editor. Long-term Located Research on Forest Ecosystem. Beijing: China Forestry Publish House. 221–227 p. (in Chinese).

63. Tian DL, Xiang WH (1993) Study on Dynamics Regularity of Soil Moisture in Chinese Fir Plantation. In: Liu XZ, editor. Long-term Located Research on Forest Ecosystem. Beijing: China Forestry Publish House. 209–215 p. (in Chinese).

64. Chen R, Xiang WH, Xu X, Tian DL, Liu J, et al. (2010) Speed of soil nitrogen mineralization and effect of fertilizer combined with warming on it in different ages Chinese fir plantations. Hunan Agricultural Sciences 5: 125–129. (in Chinese).

65. Tian DL, Kang WX, Chen XY, Wen SZ, Wei XM (1989) Study on Microclimatic Characteristics of Chinese Fir Plantation Ecosystem in Small Watersheds. Journal of Central South Forestry College 9(Sup.): 29–37. (in Chinese).

66. Xue L, He YJ, Qu M, Xu Y (2005) Water holding characteristics of litter in plantations in south China. Chinese Journal of Plant Ecology 29(3): 415–421. (in Chinese).

67. Kang WX, Zhao ZH, Deng XW (2007) Study of the Dynamic Effects and the Law of Kinetic Energy Transmission in the Canopy of Chinese fir Plantation Ecosystems. Journal of Central South University of Forestry & Technology 27(2): 1–6. (in Chinese).

68. Yin H (2009) Improvement of Crown Interception Model Based on Vegetation Structural Parameters. Beijing: Beijing Forestry University.

69. Pan WC, Tian DL, Chen XY, Wen SZ (1989) Hydrological process and nutrient dynamics of a subtropical Chinese fir plantation ecosystem. Journal of Central South Forestry College 9 (Sup.): 1–10. (in Chinese).

70. Gu FX, Yu GR, Wen XF, Tao B, Li KR, et al. (2008). Drought effects on carbon exchange in a subtropical coniferous plantation in China. Chinese Journal of Plant Ecology 32(5), 1041–1051. (in Chinese).

71. Qian N (1989) The relationship of leaf and foliage biomass to sapwood area in Chinese Fir. Journal of NanJing Forestry University 13(4): 75–80. (in Chinese).

72. Liu XZ, Kang WX, Wen SZ (1993) Studies on energy balance on canopy in a Chinese fir plantation. In: Liu XZ, editor. Long-term Located Research on Forest Ecosystem. Beijing: China Forestry Publish House. 221–227. (in Chinese).

14

Nutrient Limitation on Ecosystem Productivity and Processes of Mature and Old-Growth Subtropical Forests in China

Enqing Hou[1,2,4], Chengrong Chen[2]*, Megan E. McGroddy[3], Dazhi Wen[1,4]*

1 Key Laboratory of Vegetation Restoration and Management of Degraded Ecosystems, South China Botanical Garden, Chinese Academy of Sciences, Guangzhou, China, **2** Environmental Futures Centre, Griffith School of Environment, Griffith University, Nathan, Queensland, Australia, **3** Department of Environmental Sciences, NASA/University of Virginia, Charlottesville, Virginia, United States of America, **4** University of Chinese Academy of Sciences, Beijing, China

Abstract

Nitrogen (N) is considered the dominant limiting nutrient in temperate regions, while phosphorus (P) limitation frequently occurs in tropical regions, but in subtropical regions nutrient limitation is poorly understood. In this study, we investigated N and P contents and N:P ratios of foliage, forest floors, fine roots and mineral soils, and their relationships with community biomass, litterfall C, N and P productions, forest floor turnover rate, and microbial processes in eight mature and old-growth subtropical forests (stand age >80 yr) at Dinghushan Biosphere Reserve, China. Average N:P ratios (mass based) in foliage, litter (L) layer and mixture of fermentation and humus (F/H) layer, and fine roots were 28.3, 42.3, 32.0 and 32.7, respectively. These values are higher than the critical N:P ratios for P limitation proposed (16–20 for foliage, ca. 25 for forest floors). The markedly high N:P ratios were mainly attributed to the high N concentrations of these plant materials. Community biomass, litterfall C, N and P productions, forest floor turnover rate and microbial properties were more strongly related to measures of P than N and frequently negatively related to the N:P ratios, suggesting a significant role of P availability in determining ecosystem production and productivity and nutrient cycling at all the study sites except for one prescribed disturbed site where N availability may also be important. We propose that N enrichment is probably a significant driver of the potential P limitation in the study area. Low P parent material may also contribute to the potential P limitation. In general, our results provided strong evidence supporting a significant role for P availability, rather than N availability, in determining ecosystem primary productivity and ecosystem processes in subtropical forests of China.

Editor: Sandra Maria Feliciano de Oliveira Azevedo, Federal University of Rio de Janeiro, Brazil

Funding: This study was supported by National Natural Science Foundation of China (No. 31070409), Strategic Priority Research Program - Climate Change: Carbon Budget and Relevant Issues of the Chinese Academy of Sciences (No. XDA05050205 and the Australian Research Council (FT0990547). The support from China Scholarship Council through an overseas joint doctoral fellowship to Enqing Hou is also kindly acknowledged. The funders had no role in study design, data collection and analysis, decision to publish, or preparation of the manuscript.

Competing Interests: The authors have declared that no competing interests exist.

* E-mail: c.chen@griffith.edu.au (CC); dzwen@scbg.ac.cn (DW)

Introduction

Nitrogen (N) and phosphorus (P) have both been shown to control the rates of ecosystem processes and primary productivity in both aquatic and terrestrial ecosystems [1–3]. Global pattern analysis of carbon (C):N:P stoichiometry in foliage and litter supports the hypothesis that N is the major limiting nutrient in temperate regions, while P tends to limit ecosystem productivity and processes in the tropical regions [4–7]. These analyses are generally consistent with the nutrient addition experiments or C:N:P stoichiometry studies at a local or regional scales [8–12], and well explained by variation in climate conditions (e.g. temperature) and soil types [7,12–14]. According to this global pattern, subtropical forests are likely to be co-limited by N and P. However, this supposition has rarely been tested.

Since the beginning of the industrial revolution, human activities (N fertilizer application and burning of fossil fuels) have doubled the N input into the terrestrial ecosystems [15,16]. Although anthropogenic P inputs (mainly as fertilizers) to the biosphere also increased fourfold in the period from 1950s to

1980s and remained more or less constant since 1989, the primary P inputs are mostly confined in agricultural soils and tend to remain and accumulate in crop soils [17]. The greater mobility and biological availability of N in the atmosphere are causing the imbalance supply between N and other mineral nutrients (especially P) in natural ecosystems [17,18], which is likely to transform N-limited ecosystems to P-limited ecosystems [16,19]. In a comprehensive study of nutrients on phytoplankton nutrient limitation in high- and low-N deposition lakes in Norway, Sweden, and Colorado, United States, Elser et al. (2009) found that continued anthropogenic N input increased the stoichiometric ratio of N and P in these lakes, resulting in a shift from N-limitation to P-limitation in high-N deposition lakes [16]. The imbalance of nutrient supply is likely to affect ecosystem productivity and processes and the carbon sequestration potential of terrestrial ecosystems [17,19,20].

China has 0.97 million km^2 of subtropical and tropical forests, which represent 62% of the country's total forested area, and play an important role in maintaining biodiversity and ecological equilibrium, sequestering atmospheric C, and providing important ecological services for social development [21,22]. However, these tropical and subtropical forests in the southern part of China, are generally close to or surrounded by large industrial and/or economic zones. Annual N deposition rate ranging from 18 to 53 kg N ha^{-1} yr^{-1} were reported at several long-term monitoring stations in tropical and subtropical forests [22], comparable to the highest levels of N deposition occurring in Europe [23,24]. Recent studies found that the understory plants generally showed no or even negative responses to experimental N additions (50, 100 and 150 kg N ha^{-1} yr^{-1}) in three mature and old-growth forests at the Dinghushan Biosphere Reserve, south China [25,26]. Nutrients other than N, were proposed as the primary constraint on plant growth at these forest sites with P being the mostly likely candidate [26]. However, direct evidence is still lacking, though one recent study reported a significant increase in litterfall production after experimental P addition (150 kg P ha^{-1} yr^{-1}) at these three forests [27].

While fertilization studies are the gold standard for determining the nature of nutrient limitation [28,29], they are difficult to do well in forest ecosystems [2,30,31] and, in many cases, after several years of study the results are still unclear [31–33]. It may be a question of how much fertilizer to add, as Chapin (1986) suggested [30], or if the nutrient limitation is ultimate, it may take decades or more for species replacement to happen in forest ecosystems and thus delaying measurable results [18,31]. The critical N:P ratio for biomass was shown to work well indicating the limiting nutrient in European wetland ecosystems [34], but is poorly supported in some other terrestrial ecosystems [35,36].

In this study, we investigated the N and P status of foliage, forest floors, fine roots and mineral soil, as well as microbial properties of the forest floors of eight forests in subtropical China. We used regression analysis, to study the relationships between rates of ecosystem productivity and nutrient cycling and N and P availability in these forests. We hypothesized that these selected parameters were more strongly related to P availability than N availability, due to the historically high rates of atmospheric N deposition [22].

Methods

Site Description

The research was conducted in the Dinghushan Biosphere Reserve, located in the middle of Guangdong province in southern China (112°31' E to 112°34' E, 23°09' N to 23°12' N; Figure 1). The Reserve covers an area of 1155 ha, and has a typical subtropical monsoon climate. The entire Reserve has 1843 plant species identified and documented [37]. Mean annual temperature at the site is 21°C, and mean annual precipitation is 1900 mm [38]. Nearly 80% of the precipitation falls in the wet season (from April to September) and 20% in the dry season (from October to March) [38]. Elevation ranges from 10 to 1000 m above sea level. The forest soil has developed from Devonian sandstone and shale during the Holocene (<15 kyr) [39]. Soils are Ferralsols according to the FAO classification, with a pH value ranging from 3.8 to 4.9 [40,41]. The annual rate of atmospheric N deposition was approximately 46 kg N ha^{-1} yr^{-1} between 1989 and 2007 [42].

The basic site information for the eight selected forest communities is summarized in Table S1. These eight communities cover all major forest types in DHSBR, representing five typical forest types in subtropical China [43]. They differ in tree species composition, stand age and topography (Table S1). Four of them (PF: pine forest; PBM1, 2 and 3, pine and broadleaved mixed forest 1, 2 and 3) are mature forests (about 80 years old), while the other four communities (REB1 and 2, ravine evergreen broad-

leaved forest 1 and 2; MEB, monsoon evergreen broadleaved forest; and MTEB, mountainous evergreen broadleaved forest) are old-growth forests (>100 year old) and were regional or topographical climax forests (Table S1) [44]. Seven of the forests have been protected from human disturbance since their establishment. The PF is the exception, it had been disturbed mainly by the harvest of understory vegetation and litter for fuel by local residents between 1950s and 1990s [45], whilst the community has remained and dominated by *Pinus massoniana*, of which biomass was about 90% of the total community biomass [46].

Sampling and Sample Preparation

During 15th – 18th October, 2010, we sampled foliage, forest floors, fine roots and soil at each of the eight study sites. At each site, mature and healthy foliage was sampled from three major tree species which were listed in Table S1. For each species, we collected foliar samples from four individuals.

At each location, four subplots (20×20 m^2) were randomly set up with a distance of at least 10 m between them. In each subplot, 3 small sampling areas (20×20 cm^2) were randomly located with the constraints that they were 1–2 m away from the nearest tree (diameter at breast height ≥5 cm) and at least 5 m from its nearest neighbor. All fine forest floor materials within the sampling area were collected, including leaf litter, and senesced branches, bark, flowers and fruits with diameters ≤1 cm. Forest floor materials were carefully separated into two layers (L layer, litter layer; F/H layer, mixture of fermentation layer and humus layer) in the field. After forest floor materials were sampled, a soil profile was excavated at the same area. Mineral soil from the 0–15 cm depth was sampled by 3 successive cutting rings (Height 5 cm, Volume 100 cm^3) from top to bottom (each 5 cm depth by one cutting ring). By using a cutting ring to sample soil, we also measured the bulk density at the same time of soil sampling. The three forest floor samples from the same layer in each subplot were bulked together as one composite sample (one composite L layer sample and one composite F/H layer sampler per subplot). Nine soils (3 cutting rings per area×3 areas) of each subplot were bulked together as one mineral soil sample. Both forest floor and soil samples were stored at 4°C in the refrigerator within 4 h after sampling.

Leaves were directly oven-dried at 65°C for 72 h prior to grinding for determination of total N and P concentrations. For forest floor materials, the fresh weight (w1, unit: g) was recorded and then the sample was mixed well. A subsample was oven-dried at 65°C for 72 h for the determination of dry weight transfer coefficient (t, proportion of dry weight over the fresh weight). The forest floor biomass was calculated by the equation followed:

L layer (or F/H layer) forest floor biomass (g/cm^2) = (w1×t)/(400 cm^2×3).

The unit was converted later to Mg/ha and shown in the results. After t was determined, the oven-dried sample was ground for the determination of total N and P concentrations. Another subsample was taken and cut into 2–4 mm pieces and stored at 4°C prior to the determination of microbial biomass C concentration, respiration and β-glucosidase activity.

For soil samples, fresh weight (W1, unit: g) was recorded and the sample was mixed well. Stones with diameter >4 mm were picked out during the sieving (4 mm mesh) and weighed (W2, unit: g). A subsample of the sieved soil was air-dried for 2 weeks prior to grinding and determination of soil nutrient concentration. A subsample was used for the soil dry weight transfer coefficient (T; proportion of dry weight after over-dried in fresh weight)

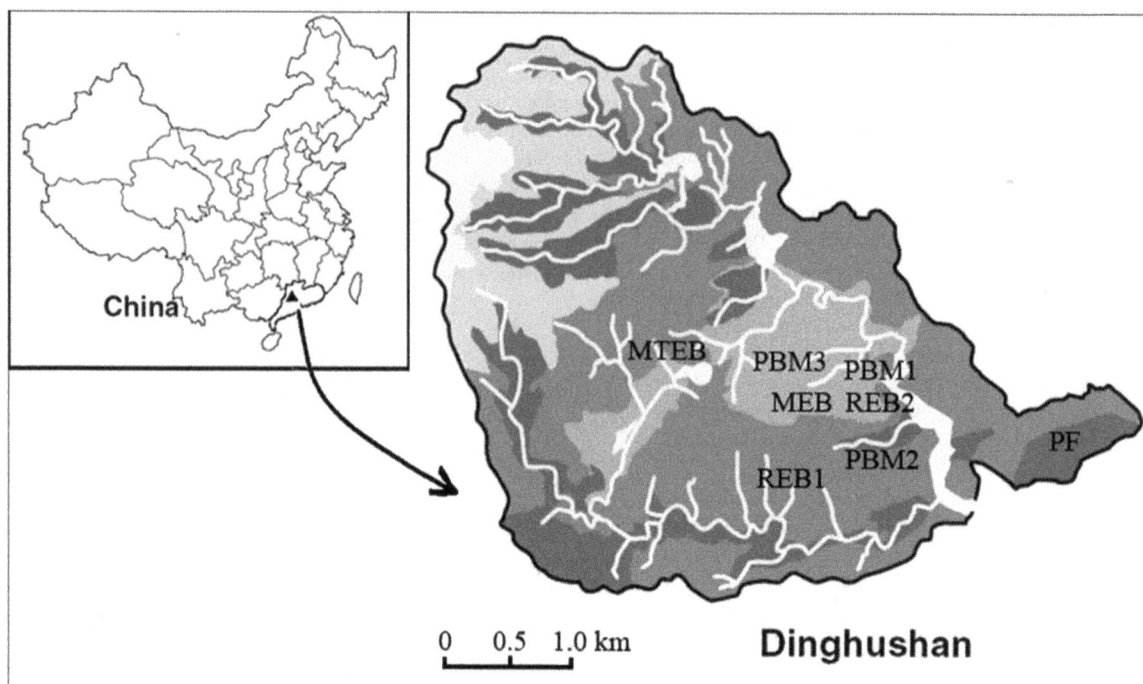

Figure 1. Location of eight study forest sites.

determination by oven-dried at 105°C for 72h. The bulk density was calculated by equations followed:

Bulk density (g/cm^3) = W2+ (W1–W2)×T/(100 cm^3×9).

The remaining soil was weighed and stored at 4°C for the determination of microbial properties (data not shown here) and fine roots (diameter ≤2 mm) collection. Fine roots retained on a 0.6 mm screen were collected and dried at 65°C for 72 h and then weighed for the fine root biomass calculation (data not shown here). Fine roots were finely ground for measurements of total N and P concentrations.

Analytical Methods

Microbial biomass C concentration in the L and F/H layers was determined by a fumigation-extraction (1:25) method using an E_C factor of 2.64 [47]. Microbial respiration was measured by using the incubation method. In brief, 2 g of fresh forest floor materials with moisture adjusted to 60% of the field capacity was incubated aerobically in a 1–L sealed plastic jar at room temperature (ranging from approximate 15°C at night time to 25°C at day time). All CO$_2$ evolved was trapped in 0.1 M NaOH and measured by acid titration (0.1 M HCl) after 1, 3, 7, 14, 21 and 28 days. The activity of β-glucosidase (EC 3.2.1.21) was analyzed following the procedure of Alef and Nannipieri (1995) [48], except that a fresh weight of 0.5 g was used for our forest floor samples.

Total N concentrations of foliage, L and F/H layers, fine roots and soil were all determined using an Isoprime isotope ratio mass spectrometer with a Eurovector elemental analyzer (Isoprime-Euro EA 3000). Total P concentrations were all measured using a nitric acid/perchloric acid digestion, followed by the molybdate blue method [49] using a UV–Vis spectrometer (UV1800, Shimadzu, Japan). Soil extractable N concentration was measured as the hot water extractable total N concentration according to method described by Sparling (1998) [50], which was found to be a simple and useful predictor of mineralizable N and plant available N [51,52]. In brief, 4.0 g air-dried soil was incubated

with 20 ml water in a capped test tube at 70°C for 18 h. The test tube was then shaken on an end-to-end shaker for 5 min, and filtered through Whatman 42 filter paper. Total N concentration of the extract was analyzed using a SHIMADZU TOC-$_{VCPH/CPN}$ analyser (Kyoto, Japan). Soil extractable P concentration was calculated as the sum of inorganic P concentration sequentially extracted by 1.0 M NH$_4$Cl, 0.1 M NH$_4$F and 0.1 M NaOH following the P fractionation scheme of McDowell and Condron (2000) [53]. Concentrations of inorganic P in the extracts of 0.1 M NH$_4$F and 0.1 M NaOH were determined by the molybdate blue method [49] using the same UV–Vis spectrometer mentioned above; while concentrations of inorganic P in the extracts of 1.0 M NH$_4$Cl were too low for the molybdate blue method and thus determined by the malachite green method [54], which works well for the determination of low concentrations of P in soil extracts [54], using the same UV–Vis spectrometer, too.

Community Biomass and Litterfall C, N and P Productions

Community biomass reported by Liu et al. (2007) [55] and litterfall production reported by Zhou et al. (2007) [44] and Yan et al. (2009) [56] from these forest communities were used to investigate the relationships of community biomass and litterfall C, N and P productions with soil nutrient pools in this study. Litterfall C, N and P productions were calculated by multiplying litterfall production by the C, N and P concentrations in the L layer. Community biomass includes dry weight of whole plant of all trees and shrubs with diameter at breast height ≥1 cm (Table S1). Litterfall production was available for seven of the eight study forests, while not for the PBM3 (Table S1). Methods of the data collection or calculation are described in details in Table S1.

Forest Floor Turnover Rate

Jenny et al. (1949) [57] and subsequently Olson (1963) [58] proposed that the rate of change in the forest floor biomass (or biomass C) could be used to determine nutrient transfers from the forest floor to the mineral soil in (near-) equilibrium forests. Forests selected in this study are all in or near an equilibrium status except for the PF site, as suggested by the study of long-term change of litterfall production [44]. Here, we calculated the forest floor turnover rate according to Olson (1963) [58]. The calculation formula is:

Forest floor turnover rate (yr^{-1}) = litterfall C production/forest floor biomass C.

Forest floor biomass C was the sum of L layer and F/H layer biomass C.

Statistical Analyses

Since the data are mostly (near-) normal distributed, Pearson correlation analysis and Pearson linear regression technique were used throughout the manuscript. All N:P ratios shown in this study were calculated on a mass basis. All analyses were performed using SPSS version 16.0. Pearson correlation was used to investigate the correlations between N and P concentration and N:P ratio for all plant and soil samples, and was also used to investigate the correlations between nutrient measures of the plant samples and those of the soil samples. Pearson linear regression technique was used to examine the relationships between community biomass, litterfall C, N and P productions, forest floor turnover rate and forest floor microbial properties with nutrient measures of the soil or the forest floors.

According to the foliar N concentration, foliage was divided into two groups (see Figure S1). One was the high N group with foliar N concentration higher than 25 mg/g (sample number = 20), including species of *Gironniera subaequalis* (from the REB2), *Ormosia fordiana* (REB1), *Caryota ochlandra* (REB2), *Euodia lepta* (PF) and *Sterculia lanceolata* (REB2); the other was low N group with foliar N concentration lower than 25 mg/g (sample number = 76), including the other eight species (Table S4 and Figure S1). For plants of the same species, only four species with a sample number ≥8 were selected for the correlation analysis. To be consistent with respect to units, when community biomass and litterfall C, N and P productions were regressed against the soil nutrients, pools were used; while for all other correlation and regression analysis, concentrations were used.

Results

Community Biomass and Litterfall Productions in Relation to the Nutrient Measures

Community biomass was positively related to both soil total P $(R^2 = 0.39, P<0.001;$ Figure 2B) and extractable P pools $(R^2 = 0.59, P<0.001;$ Figure 2E). The relationships with soil total N $(R^2 = 0.18, P<0.05;$ Figure 2A) and extractable N $(R^2 = 0.20, P<0.05;$ Figure 2D) pools were also significant but mainly caused by the inclusion of the prescribed PF site. When the PF site was excluded, community biomass was not related to either the soil total N $(R^2 = 0.04, P>0.05;$ Figure 2A) or the extractable N pool $(R^2 = 0.02, P>0.05;$ Figure 2D). Community biomass tended to increase with increasing soil total N:P ratio when community biomass was low (<300 Mg/ha), while increasing with decreasing soil total N:P ratio when community biomass was high (≥300 Mg/ha; Figure 2C). Overall, community biomass was not related to the soil total N:P ratio $(R^2<0.01, P>0.05;$ Figure 2C). Community biomass was negatively related to soil extractable N:P ratio when the PF site was excluded from the analysis $(R^2 = 0.54, P<0.001;$ Figure 2F).

Litterfall C, N and P productions were all positively related to the soil extractable P pool $(R^2 = 0.18, 0.57$ and 0.76, respectively, $P<0.05;$ Figures 3B, D and F). Similar to the community biomass,

the relationships between litterfall C, N and P productions and the soil extractable N pool were significant for all sites $(R^2$ was 0.21, 0.33 and 0.23, respectively, $P<0.05;$ Figures 3A, C and E), but none was significant if the PF site was excluded $(R^2 = 0.03, 0.02$ and 0.05, respectively, $P>0.05;$ Figures 3A, C and E). The patterns for litterfall C, N and P productions with soil total fractions and soil extractable fractions were similar (data not shown). Moreover, litterfall C, N and P productions were not related to the soil total N:P ratio (data not shown).

Forest Floor Turnover Rates in Relation to the Nutrient Measures

Forest floor turnover rate (yr^{-1}) ranged from 0.10 to 1.75 among the seven study communities for which it was calculated (no data for PBM3), with an average value of 0.86 (Table S2). As summarized in the Table 1, forest floor turnover rate was positively related to the P concentration in both L and F/H layers (L layer: $R^2 = 0.35, P<0.001;$ F/H layer: $R^2 = 0.40, P<0.001),$ while only weakly related to the N concentration in the L layer $(R^2 = 0.18, P<0.05).$ It was also negatively related to the N:P ratio in both L and F/H layers (L layer: $R^2 = 0.19, P<0.05;$ F/H layer: $R^2 = 0.40, P<0.001).$ When one outlier was excluded, the relationship of forest floor turnover rate with the F/H layer N:P ratio was even stronger $(R^2 = 0.53, P<0.001;$ Figure 4).

Microbial biomass C concentration was not related to any nutrient measures in the L layer, but related to the N and P concentrations in the F/H layer, with a stronger relationship with the P concentration $(R^2 = 0.34, P<0.001;$ Table 1) than with the N concentration $(R^2 = 0.16, P<0.05;$ Table 1). Microbial respiration was positively related to the P concentration $(R^2 = 0.48, P<0.001;$ Table 1) and also negatively related to the N:P ratio in the L layer $(R^2 = 0.37, P<0.001;$ Table 1), while not related to any nutrient measures in the F/H layer (Table 1). In the L layer, β-glucosidase activity was more strongly related to the P concentration $(R^2 = 0.44, P<0.001;$ Table 1) than the N concentration $(R^2 = 0.14, P<0.05;$ Table 1), and negatively related to the N:P ratio $(R^2 = 0.23, P<0.01;$ Table 1). In the F/H layer, β-glucosidase activity was poorly related to both N and P concentration $(R^2 = 0.17$ and 0.16, respectively, $P<0.05;$ Table 1) while not related to the N:P ratio.

Correlations between Nutrient Measures

For both low N and high N groups of foliage, N and P concentrations were poorly correlated with each other (Table 2 and Figure S1A), and the P concentration was more strongly correlated with N:P ratio than the N concentration (Table 2 and Figures S1B and C). Within a species, N:P ratio was negatively correlated with the P concentration in all species $(P≤0.033;$ Table 2), but was not correlated with the N concentration in any species $(P≥0.072;$ Table 2). The relationship between foliar N and P concentration of plants of the same species was only significant for one species (*Castanea henryi*; Table 2). For the forest floors, fine roots and soil extractable fraction, N and P concentrations were all correlated with each other $(r = 0.385–0.782, P≤0.030;$ Table 2), and N:P ratios were all negatively correlated with the P concentrations $(r = -0.610– -0.798, P<0.001),$ while only poorly correlated with the N concentration in the F/H layer $(r = 0.380, P = 0.032;$ Table 2). Soil total N:P ratios were positively correlated with soil total N concentration $(r = 0.450, P = 0.010;$ Table 2) but not correlated with soil total P concentration $(r = -0.201, P = 0.270;$ Table 2). However, the plot of soil total N:P ratio against soil total P concentration showed that the points representing the PF site were distinct from other sites (Figure

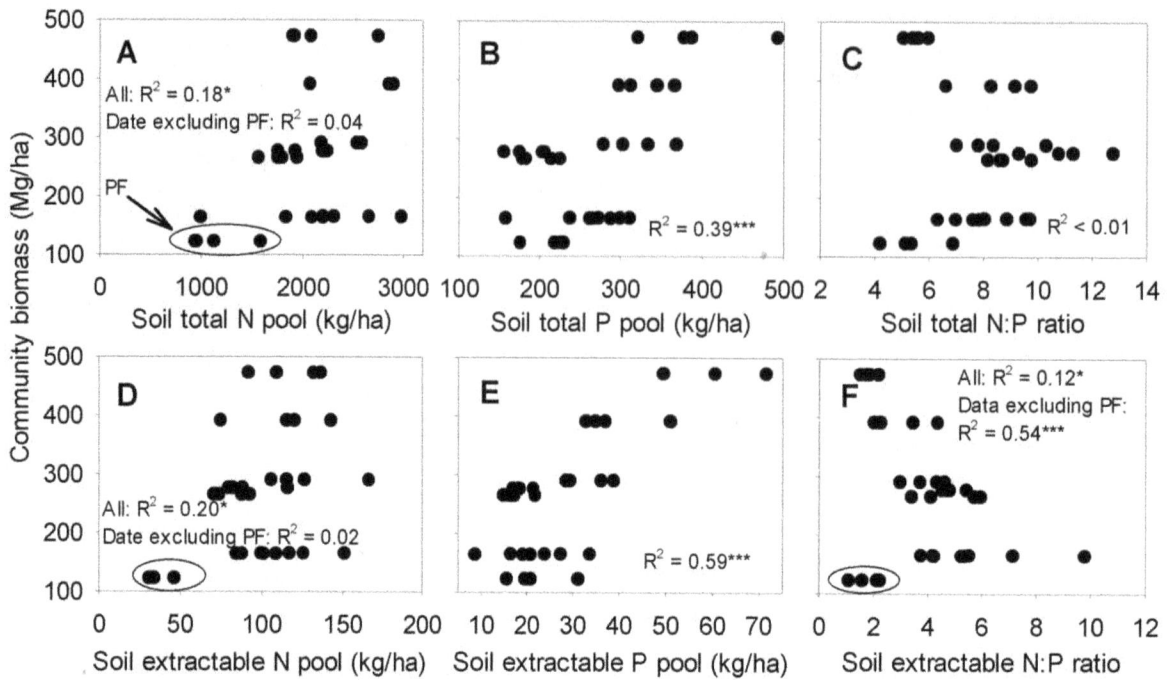

Figure 2. Relationships between community biomass and soil nutrient measures. PF indicates the pine forest. * $P<0.05$; *** $P<0.001$.

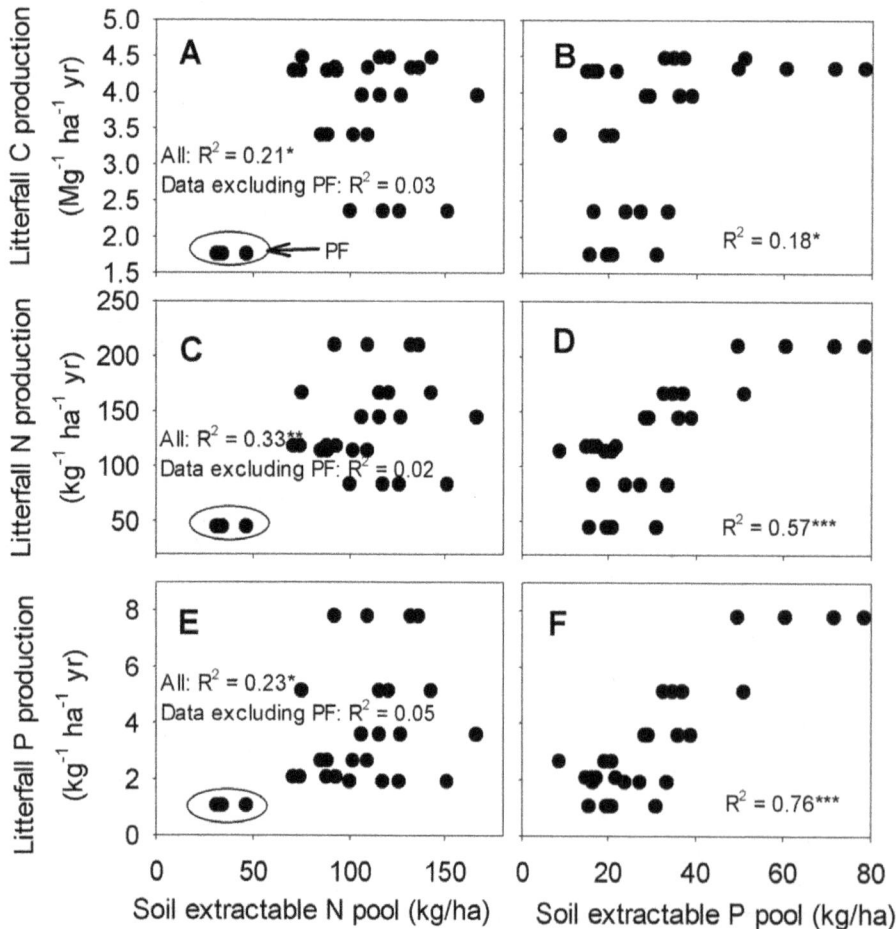

Figure 3. Relationships between litterfall C, N and P productions and soil nutrient measures. PF indicates the pine forest. ** $P<0.01$; *** $P<0.001$.

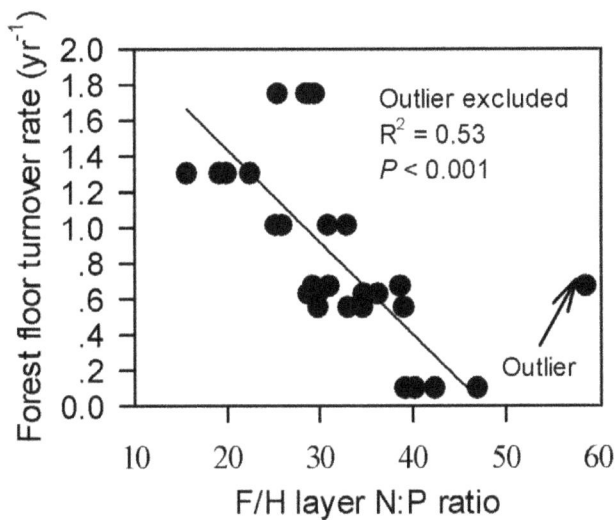

Figure 4. Relationships between forest floor turnover rate with F/H layer N:P ratio.

S2B). Excluding the PF site from analysis, the correlations between the soil total fractions were similar to those of forest floors, fine roots and soil extractable fraction (Table 3 and Figure S2).

As summarized in the Table 3, nutrient measures of all plant materials were generally more strongly correlated with the soil total P and extractable P concentrations than with soil total N and extractable N concentrations. Nitrogen: P ratios of forest floors and fine roots were all negatively correlated with soil total P and extractable P concentrations, while either not or negatively correlated with soil total N or extractable N concentration.

Nitrogen and P contents and N:P ratio

Soil total N:P ratios and extractable N:P ratios varied across the eight study sites, both were lower at the PF site (5.4 and 1.8, respectively) and REB1 site (5.5 and 1.8, respectively) than at other sites (7.2 to 11.0 and 3.0 to 6.1, respectively; Table 4). The PF site, with a long history of human disturbance, was distinct from other sites, with higher bulk density and lower organic C, total N and extractable N pools in the 0–15 cm mineral soil (Table 4). Soil total P and extractable P pools at the PF site were also lower than at the old-growth forest sites (REB1 and 2, MEB and MTEB), but

Table 1. Summary of regressions of forest floor turnover rates against nutrient measures of forest floors.

Parameter	Nutrient measure	Regression	R^2	Significance
Forest floor turnover rate (yr^{-1})				
L layer	N conc. (mg/g)	Rate = 0.067(N conc.) −0.226	0.18	<0.05
	P conc. (mg/g)	Rate = 1.729(P conc.) +0.012	0.35	<0.001
	N:P ratio	Rate = −0.019(N:P ratio) +1.638	0.19	<0.05
F/H layer	N conc. (mg/g)			NS
	P conc. (mg/g)	Rate = 2.70(P conc.) −0.36	0.4	<0.001
	N:P ratio	Rate = −0.036(N:P ratio) +2.025	0.4	<0.001
Microbial biomass C conc. (mg/g)				
L layer	N conc. (mg/g)			NS
	P conc. (mg/g)			NS
	N:P ratio			NS
F/H layer	N conc. (mg/g)	Microbial C = 0.94(N conc.) − 1.46	0.16	<0.05
	P conc. (mg/g)	Microbial C = 30.7(P conc.) − 2.25	0.34	<0.001
	N:P ratio			NS
Microbial respiration (µg CO$_2$-C g^{-1} h^{-1})				
L layer	N conc. (mg/g)			NS
	P conc. (mg/g)	Respiration = 24.6(P conc.) +21.1	0.48	<0.001
	N:P ratio	Respiration = −0.32(N:P ratio) +45.0	0.37	<0.001
F/H layer	N conc. (mg/g)			NS
	P conc. (mg/g)			NS
	N:P ratio			NS
β-glucosidase activity (µg p-nitrophenol g^{-1} h^{-1})				
L layer	N conc. (mg/g)	β-glucosidase = 19.3(N conc.) − 78.9	0.14	<0.05
	P conc. (mg/g)	β-glucosidase = 626.5(P conc.) − 29.5	0.44	<0.001
	N:P ratio	β-glucosidase = −6.8(N:P ratio) − 518.4	0.23	<0.01
F/H layer	N conc. (mg/g)	β-glucosidase = 8.1(N conc.) +45.0	0.17	<0.05
	P conc. (mg/g)	β-glucosidase = 176.5(P conc.) +75.8	0.16	<0.05
	N:P ratio			NS

n = 32. NS indicates statistically not significant at the level of $P<0.05$.

Table 2. Correlations between nutrient measures in the plant and soil samples.

Sample	n	N conc. vs. P conc.		N:P ratio vs. N conc.		N:P ratio vs. P conc.	
		r	P	r	P	r	P
Foliage							
Low N group (<25 mg/g)	76	0.336	0.003	0.435	<0.001	−0.682	<0.001
High N group (≥25 mg/g)	20	−0.449	0.047	0.709	<0.001	−0.929	<0.001
C. henryi	16	0.709	0.002	0.085	0.753	−0.634	0.008
C. concinna	8	0.250	0.550	0.455	0.257	−0.747	0.033
P. massoniana	16	0.458	0.075	0.397	0.128	−0.627	0.009
S. superba	20	0.302	0.196	0.411	0.072	−0.740	<0.001
Forest floor							
L layer	32	0.630	<0.001	−0.111	0.545	−0.798	<0.001
F/H layer	32	0.385	0.030	0.380	0.032	−0.661	<0.001
Fine roots	32	0.782	<0.001	−0.094	0.610	−0.660	<0.001
0–15 cm mineral soil							
Total fraction	32	0.747	<0.001	0.450	0.010	−0.201	0.270
Total fraction excluding the PF site	28	0.617	<0.001	0.156	0.428	−0.647	<0.001
Extractable fraction	32	0.530	0.002	0.194	0.288	−0.610	<0.001

For soil total fraction, inclusion and exclusion of the pine forest (PF) site showed distinct results and thus, correlations for all data and data excluding PF were both shown.

comparable to those at other mature forest sites (PBM1, 2 and 3; Table 4).

Nitrogen and P concentrations varied widely in all plant materials, with average N concentrations of 21.0 mg/g, 16.1 mg/g, 13.2 mg/g and 13.8 mg/g, respectively in the foliage, L and F/H layers and fine roots, while corresponding values for P in these materials were 0.77 mg/g, 0.41 mg/g, 0.43 mg/g and 0.44 mg/g, respectively (Table 5). Site averages of foliar N concentration were all above 15 mg/g (Table S3), and species averages were all higher than 15 mg/g as well, except for the *Pinus massoniana* (14.6 mg/g; from PF site) and *Rhododendron henryi* (11.2 mg/g; from MTEB site; Table S4). Although N:P ratio also varied widely in all plant

Table 3. Correlations of nutrient measures between the plant and soil samples.

Plant sample	0–15 cm mineral soil					
	Total N conc.	Total P conc.	Total N:P ratio	Extractable N conc.	Extractable P conc.	Extractable N:P ratio
Foliage						
N conc.	0.235	0.419	−0.313	0.126	0.548	−0.669
P conc.	0.277	0.357	−0.172	0.137	0.442	−0.538
N:P ratio	−0.253	−0.024	−0.399	−0.159	−0.105	−0.073
L layer forest floor						
N conc.	0.502**	0.668***	−0.074	0.607***	0.708***	−0.147
P conc.	0.218	0.661***	−0.426*	0.377*	0.833***	−0.562**
N:P ratio	0.040	−0.387*	0.482**	−0.076	−0.523**	0.578**
F/H layer forest floor						
N conc.	0.101	0.256	−0.377*	−0.018	0.042	−0.127
P conc.	0.349	0.782***	−0.468**	0.387*	0.787***	−0.604***
N:P ratio	−0.303	−0.550**	0.118	−0.381*	−0.646***	0.477**
Fine roots						
N conc.	0.632***	−0.614***	0.070	0.635***	0.618***	−0.286
P conc.	0.392*	0.697***	−0.326	0.484**	0.812***	−0.558**
N:P ratio	0.050	−0.442*	0.655***	−0.027	−0.489**	0.547**

Data are correlation coefficients. n = 8 for foliage; n = 32 for L and F/H layers and fine roots. * $P<0.05$; ** $P<0.01$; *** $P<0.001$.

Table 4. Selected characteristics of the 0–15 cm mineral soil.

Site	Bulk density (g/cm³)	Soil organic C pool (Mg/ha)	Soil total N pool (kg/ha)	Soil total P pool (kg/ha)	Soil total N:P ratio	Soil extractable N pool (kg/ha)	Soil extractable P pool (kg/ha)	Soil extractable N:P ratio
PF	1.35(0.02)	17(2)	1148(150)	213(13)	5.4(0.6)	36(3)	22(3)	1.8(0.3)
PBM1	0.92(0.03)	35(1)	1759(77)	200(11)	8.8(0.3)	81(5)	17(2)	4.8(0.6)
PBM2	1.02(0.07)	31(5)	1774(271)	243(29)	7.2(0.3)	96(6)	17(3)	6.1(1.2)
PBM3	0.98(0.04)	33(2)	2023(115)	185(12)	11.0(0.7)	91(8)	19(1)	4.9(0.2)
REB1	1.09(0.04)	28(2)	2156(201)	394(36)	5.5(0.2)	117(10)	65(6)	1.8(0.1)
REB2	0.89(0.04)	43(4)	2789(266)	330(15)	8.4(0.7)	113(14)	39(4)	3.0(0.5)
MEB	0.84(0.04)	40(4)	2682(268)	321(19)	8.4(0.7)	128(13)	33(3)	3.9(0.4)
MTEB	0.88(0.03)	44(3)	2533(178)	280(16)	9.0(0.4)	123(11)	25(4)	5.1(0.8)

All data are means (±1 SE), n = 4. The corresponding full names of eight study sites are listed in Table S1.

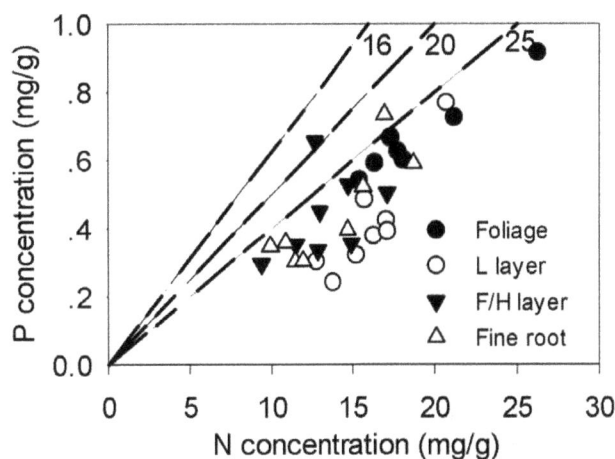

Figure 5. Mean N and P concentrations of plant materials of eight study forests. Dashed lines depict N:P ratios of 16, 20 and 25 on a mass basis. Ratios of 16 and 20 are P limitation thresholds of plant growth proposed by Koerselman and Meuleman (1996) [34] and Güsewell (2004) [31], respectively; ratio of 25 is the critical N:P ratio that indicates P limitation on litter decomposition proposed by Güsewell and Verhoeven (2006) [61].

materials, the values (15.6–72.0) were generally high (Table 5), with 30 of the 32 site averages higher than 25 and other two values around 16 and 20 (Figure 5 and Table S3).

Discussion

Overall Relationship Patterns

Our results revealed consistent relationship patterns of community biomass, litterfall C, N and P productions, forest floor turnover rate and microbial properties with measures of N and P as well as N:P ratios (Figures 2, 3 and 4 and Table 1) that generally suggested that P availability played a more significant role than N availability in determining ecosystem primary productivity and nutrient cycling at the study sites. Considering that ecosystem productivity increases with increasing supply of a limiting nutrient, while showing no or even a negative response to the increased supply of a non-limiting nutrient [30,59,60], our results suggest that P availability was one of the limiting factors of plant growth and nutrient cycling at the study sites.

The idea is supported by the high N:P ratios of foliage, forest floors and fine roots at these sites. Site averages of the N:P ratios of these plant materials (30 of the 32 values higher than 25; Figure 5 and Table S3) were mostly higher than the proposed breakpoints of P limitation on plant growth by both Koerselman and Meuleman (1996) [34] (N:P ratio = 16) and Güsewell (2004) [31] (N:P ratio = 20), and also higher than the critical ratio that indicate P limitation on the decomposition of graminoid leaf litter (N:P ratio = 25) [61]. Moreover, average N:P ratios of foliage (28.3) and forest floors (L layer 42.3; F/H layer 32.0) in this study were much higher than the global averages of both temperature (broadleaf foliage 12.8, coniferous foliage 9.8; broadleaf litter 13.2, coniferous litter 11.8) and tropical regions (foliage 19.6, litter 28.4) [6]. These results generally suggest strong P limitation on primary productivity and litter decomposition at the study sites.

Our results are consistent with previous studies in which experimental N additions (50, 100 and 150 kg N ha^{-1} yr^{-1}) generally did not increase (and in some cases actually decreased) the understory vegetation biomass, litter decomposition rate and

Table 5. Statistical characteristics of nutrient measures of selected plant materials.

| Material | N concentration (mg/g) | | P concentration (mg/g) | | N:P ratio | |
	Mean	Range	Mean	Range	Mean	Range
Foliage	21.0(0.8)	10.7–43.9	0.77(0.04)	0.37–1.97	28.3(0.6)	15.8–41.4
L layer	16.1(0.5)	11.8–25.0	0.41(0.03)	0.21–0.89	42.3(2.0)	23.2–72.0
F/H layer	13.2(0.5)	8.2–19.5	0.43(0.02)	0.27–0.70	32.0(1.5)	15.6–58.7
Fine roots	13.8(0.7)	8.9–21.8	0.44(0.03)	0.24–0.88	32.7(1.2)	21.5–46.1

Data in the brackets are SE; n = 96 for foliage, = 32 for L and F/H layers and fine roots.

soil respiration rate at three (PF, PBM2 and MEB) of the eight study sites [25,26,62,63]. Experimental P addition (150 kg P ha^{-1} yr^{-1}) significantly increased soil microbial biomass C concentration, soil respiration and litterfall production at the old-growth MEB site, and also significantly increased the litterfall production at the other two sites (PF, PBM2) [27].

Different Relationship Patterns for Different Study Sites and Ecosystem Compartments

Although the relationship patterns suggested significant P limitation at the study sites in general, ecosystem productivity or processes may differ in the extent of P limitation or may be co-limited by N at different sites. Fertilization studies showed that at the old-growth MEB site, understory vegetation biomass, litter decomposition rate and soil respiration all responded negatively to experimental N additions [62,26], and soil microbial biomass and respiration and litterfall production showed positive responses to experimental P addition [27]. In contrast, at the PF site, understory vegetation biomass, litter decomposition and soil respiration at the PF site did not respond to the N additions [25,62], soil microbial biomass and respiration also did not respond to the P addition [27]. The responses of the PBM2 site to fertilization additions generally fell between the MEB site and PBM2 site [25–27,62].

The different nutrient limitation patterns of different sites were also reflected by our plots in this study. For example, plot of community biomass against soil extractable N:P ratio showed a distinct pattern at the PF site as compared to the other sites (Figure 2F). For sites with similar community biomass, soil extractable N:P ratio of the PF site was much lower than that of other sites. The distinct pattern was largely attributed to the markedly lower soil extractable N pool of the PF site than that of other sites (Table 4 and Figure 2D). Similarly, the plot of total N:P ratio against total P concentration in the soil also showed a distinct pattern of the PF site from other sites (Figure S2). For soils with similar total P concentration, total N:P ratio of the PF site was much lower than that of other sites due to its lower soil total N concentration compared with other sites. These distinct patterns of the PF site were due to its low soil N availability that probably because of the continuous removal of a large amount of nutrients, particular of N, by the prescribed over 40 years (1950s–1990s) of understory and litter harvest at this site [45].

Several recent studies have found that different ecosystem compartments or processes may differ in response to the addition of the same nutrient [27,64,65]. Nutrient limitation of one ecosystem compartment or process cannot be simply predicted from nutrient limitation of another ecosystem compartment or process [60,64,65]. In this study, community biomass was more strongly related to soil extractable P pool than litterfall C

production (Figures 2E and 3B), suggesting a greater constraint of P availability on community biomass than on litterfall C production. This was probably because of the indirect impact of P availability on litterfall C production via its impact on community biomass, as supported by significant relationship of litterfall C production with community biomass (R^2 = 0.88, P = 0.014, n = 7; data not shown).

Similarly, the relationships of litterfall C, N and P productions with soil P pools were also different. Stronger relationships of litterfall P and N productions with soil extractable P pool (R^2 = 0.76 and 0.47, respectively; Figures 3D and F) than litterfall C production (R^2 = 0.18; Figure 3B) suggest a stronger impact of soil P availability on litter chemistry than on litter quantity. Significant relationships between N and P concentrations and N:P ratio of forest floors and soil P concentrations (Table 2) also suggested the significant impacts of P availability on litter chemistry. These results are consistent with several other studies which revealed that forests on low-fertility soils tended to produce similar quantities of litters as forests on high-fertility soils nearby, but of lower quality [4,5,66–68].

Climate conditions (e.g. evapotranspiration) are likely to be the major factor affecting litter decomposition at a global scale, while at a local scale litter chemistry is always the major factor affecting litter decomposition [14,69,70]. In this study, forest floor turnover rate, microbial biomass C concentration, microbial respiration and β-glucosidase activity were all significantly related to the nutrient measures of the forest floors, with stronger relationships with the P concentration than with the N concentration and negative relationships with the N:P ratio in either L layer or F/H layer or both layers. As for the previous patterns these results suggest P availability rather than N availability plays an important role in controlling litter decomposition rates at all levels from ecosystem, to microbes and enzymes.

Constraint of Ecosystem Development by Physical Environment

The species composition of natural plant communities develops over a long period of succession. During succession, the community interacts with the physical environment (e.g. light, temperature and soil fertility) and finally reaches a relative steady status [71–73]. Physical environments may determine how far a community goes [71,73,74]. At the global or climate scale, primary productivity of forest ecosystems is largely determined by the climate conditions (e.g. precipitation) [75,76]. While at a local scale, if light, temperature and rainfall are relatively consistent across different terrestrial communities, soil fertility may be the major physical condition determining the mature stage of succession [1,2,14]. In this study, the close relationships of community biomass with soil nutrient pools (both total and extractable) did support this hypothesis. Soil total and extractable P pool explained 39% and 59% of the community biomass variation, respectively (Figures 2B and E).

Despite strong theoretical predictions suggesting the existence of relationship between community biomass and soil fertility [14,71,74], the relationships of community biomass and productivity with soil nutrients have been frequently poor in natural ecosystems [77,78], particular in forest ecosystems with high plant diversity as our study forests [14,79]. We proposed that the strong relationship of community biomass and litterfall N and P productions with soil P pools were probably because of markedly P limitation and N enrichment at the study sites compared with many other areas.

In addition to the high foliar and litter N:P ratios, mean N concentrations of the foliage (21.0 mg/g) and L layer (16.1 mg/g) were both ca. 2 times higher than the global averages for evergreen trees and shrubs (foliage 13.7 mg/g; litter 7.3 mg/g) [80], while mean P concentrations (foliage 0.77 mg/g; L layer 0.41 mg/g) were lower than reported global averages (foliage 1.02 mg/g and litter 0.50 mg/g) [80]. Moreover, foliar N concentrations of 11 of the 13 species in this study (range 17.1–42.1 mg/g, the other two values were 11.2 mg/g and 14.6 mg/g; Table S4) were in the upper range of species-specific foliar N concentrations of the tropical rain forests (mostly in the range of 10–30 mg/g) [35], and L layer N concentrations at six of the eight study sites (range of 15.2–20.7 mg/g, the other two values were 12.7 mg/g and 13.7 mg/g; Table S3) were in the upper ranges of site averages of litter N concentration of the tropical rain forests (mostly in the range of 10–19 mg/g) [67]. These results all suggest relatively high N availability and low P availability at our study sites compared with many other areas, which may underlie the strong relationships found between community biomass and litterfall N and P productions and soil P pools in this study.

How do these Consistent Patterns Occur?

As proposed above, strong limitation of P and enrichment of N might be the main causes of the consistent patterns observed in this study. The causes of P limitation in terrestrial ecosystems can be complicated. Here we only address three of six pathways proposed by Vitousek et al. (2010). First is the pathway of depletion that is likely to occur during millions of years of soil development [18,81,82]. Soils at the study sites developed during the Holocene (<15 ky) [39]. The soil ages are comparable to the Laupahoehoe site (20 kyr) in Hawaii that was found to be co-limited by N and P in fertilization studies [82]. The Laupahoehoe site is at comparable latitude (20°N), with similar mean annual temperature (16?), precipitation (2500 mm) and elevation (1170 m) [83] as our study sites (latitude 23°N; mean annual temperature 21?; precipitation 1900 mm; and elevation 50–600 m). Therefore, soil age is not likely to be the major cause of P limitation at the study sites. The second proposed pathway is low-P parent material that can cause P limitation developed quickly and persist over a long timescale [18]. Phosphorus concentrations in the C layer of the study area are approximately 0.40 mg/g [84], which are lower than the average in continental crust (0.70 mg/g, range of 0.04–3.00 mg/g) [18]. Therefore, low P concentration in the parent rock appears to contribute to the P limitation at the study sites but is not likely to be a major cause either.

The third possible pathway is anthropogenic P limitation by enhanced supply of other resources (especially N) [16–18]. This pathway is probably the most important cause of P limitation at the study sites, since N availability at the study sites was high compared with many other areas in the world as discussed above. The high N availability may be mainly attributed to two pathways:

accumulation of N through biological fixation during long-term (>80 years old) forest development [28,72], and high N deposition at the study sites [42,85]. Given an annual atmospheric N input of 40 kg N ha^{-1} yr^{-1} [42,85], atmospheric N input have accounted for 16% to 47% (mean 27%) of the annual total N inputs to the soils (atmospheric N input plus litterfall N production) at the study sites during the last two decades. A recent study reported the gradual increase of vegetation N:P ratios from 21 to 28 at the MEB site during the last three decades and a significant increase of foliar N:P ratio after 100 and 150 kg ha^{-1} yr^{-1} N added [42]. Together these lines of evidence suggested the possibility that the sites were driven to P limitation by excessive N inputs at the study sites.

Conclusions

Our study has revealed consistently stronger relationships between measures of P availability and community biomass, litterfall C, N and P productions, forest floor turnover rates and litter chemistry than we found for measures of N availability, indicating a significant role of P in determining ecosystem primary productivity and processes at the study sites. Vegetation N:P ratios indicated strong P limitation at all sites in this study. The results also showed that different ecosystem compartments or processes may differ in the extent of the nutrient limitation at different sites that depend on the soil nutrient status and land use history of these sites. In general, these results suggested constraint of ecosystem development by soil P availability at the study sites. We proposed that markedly N enrichment was probably a significant driver of the strong P limitation at these study sites. Low P parent material may also partly contribute to the P limitation. Further study is warranted on the mitigation of P limitation of ecosystem productivity and processes in the tropical and subtropical China.

Supporting Information

Figure S1 Correlations between N and P concentrations and N:P ratio of the foliage samples.

Figure S2 Correlations between total N and total P concentrations and total N:P ratio of the 0–15 cm mineral soil.

Table S1 Characteristics of eight study forest sites at Dinghushan Biosphere Reserve, China.

Table S2 Litterfall C production, forest floors biomass C, and forest floor turnover rate of eight study forests.

Table S3 Site averages of N and P concentrations and N:P ratio of foliage, L and F/H layers and fine roots of eight study forests.

Table S4 Species averages of foliar N and P concentrations and mass-based N:P ratio of 13 tree species selected from eight study forests.

Acknowledgments

The authors thank Marijke Heenan, Xuejin Wang, and Yujin Zhang for assistance in field samples collection and some laboratory analyses.

Author Contributions

Conceived and designed the experiments: EH CC DW MEM. Performed the experiments: EH CC DW. Analyzed the data: EH CC DW MEM. Contributed reagents/materials/analysis tools: DW CC. Wrote the paper: EE CC DW MEM.

References

1. Chapin FS (1980) The mineral nutrition of wild plants. Annu Rev Ecol Syst 11: 233–260.
2. Aerts R, Chapin FS (2000) The mineral nutrition of wild plants revisited: A re-evaluation of processes and patterns. In: Fitter AH, Raffaelli DG, editors. Advances in Ecological Research 30. 1–67.
3. Elser JJ, Bracken MES, Cleland EE, Gruner DS, Harpole WS, et al. (2007) Global analysis of nitrogen and phosphorus limitation of primary producers in freshwater, marine and terrestrial ecosystems. Ecol Lett 10: 1135–1142.
4. Vitousek PM (1984) Litterfall, nutrient cycling, and nutrient limitation in tropical Forests. Ecology 65: 285–298.
5. Vitousek PM, Sanford RL (1986) Nutrient cycling in moist tropical forest. Annu Rev Ecol Syst 17: 137–167.
6. McGroddy ME, Daufresne T, Hedin LO (2004) Scaling of C:N:P stoichiometry in forests worldwide: Implications of terrestrial redfield-type ratios. Ecology 85: 2390–2401.
7. Reich PB, Oleksyn J (2004) Global patterns of plant leaf N and P in relation to temperature and latitude. Proc Natl Acad Sci 101: 11001–11006.
8. Kenk G, Fischer H (1988) Evidence from nitrogen fertilisation in the forests of Germany. Environ Pollut 54: 199–218.
9. Herbert DA, Fownes JH (1995) Phosphorus limitation of forest leaf area and net primary production on a highly weathered soil. Biogeochemistry 29: 223–235.
10. Tanner EVJ, Vitousek PM, Cuevas E (1998) Experimental investigation of nutrient limitation of forest growth on wet tropical mountains. Ecology 79: 10–22.
11. Cleveland CC, Townsend AR, Schmidt SK (2002) Phosphorus limitation of microbial processes in moist tropical forests: evidence from short-term laboratory incubations and field studies. Ecosystems 5: 680–691.
12. Tian HQ, Chen GS, Zhang C, Melillo JM, Hall CAS (2010) Pattern and variation of C:N:P ratios in China's soils: a synthesis of observational data. Biogeochemistry 98: 139–151.
13. Aerts R (1989) The effect of increased nutrient availability on leaf turnover and aboveground productivity of two evergreen ericaceous shrubs. Oecologia 78: 115–120.
14. Cleveland CC, Townsend AR, Taylor P, Alvarez-Clare S, Bustamante MMC, et al. (2011) Relationships among net primary productivity, nutrients and climate in tropical rain forest: a pan-tropical analysis. Ecol Lett 14: 939–947.
15. Vitousek PM, Mooney HA, Lubchenco J, Melillo JM (1997) Human domination of earth's ecosystems. Science 277: 494–499.
16. Elser JJ, Andersen T, Baron JS, Bergström AK, Jansson M, et al. 2009. Shifts in lake N:P stoichiometry and nutrient limitation driven by atmospheric nitrogen deposition. Science 326: 835–837.
17. Peñuelas J, Sardans J, Rivas A, Janssens IA (2012) The human induced imbalance between C, N and P in earth's life-system. Glob Chang Biol 18: 3–6.
18. Vitousek PM, Porder S, Houlton BZ, Chadwick OA (2010) Terrestrial phosphorus limitation: mechanisms, implications, and nitrogen-phosphorus interactions. Ecol Appl 20: 5–15.
19. Vitousek PM, Aber JD, Howarth RW, Likens GE, Matson PA, et al. (1997) Human alteration of the global nitrogen cycle: Sources and consequences. Ecol Appl 7: 737–750.
20. Gruber N, Galloway JN (2008) An earth-system perspective of the global nitrogen cycle. Nature 451: 293–296.
21. Zhao TQ, Ouyang ZY, Zheng H, Wang XK, Miao H (2004) Forest ecosystem services and their valuation in China. J Nat Resour 19: 480–491.
22. Liu XJ, Duan L, Mo JM, Du EZ, Shen JL, et al. (2011) Nitrogen deposition and its ecological impact in China: An overview. Environ Pollut 159: 2251–2264.
23. MacDonald JA, Dise NB, Matzner E, Armbruster M, Gundersen P, et al. (2002) Nitrogen input together with ecosystem nitrogen enrichment predict nitrate leaching from European forests. Glob Chang Biol 8: 1028–1033.
24. Aber JD, Goodale CL, Ollinger SV, Smith ML, Magill AH, et al. (2003) Is nitrogen deposition altering the nitrogen status of northeastern forests? BioScience 53: 375–389.
25. Lu XK, Mo JM, Gilliam FS, Yu GR, Zhang W, et al. 2011. Effects of experimental nitrogen additions on plant diversity in tropical forests of contrasting disturbance regimes in southern China. Environ Pollut 159: 2228–2235.
26. Lu XK, Mo JM, Gilliam FS, Zhou GY, Fang YT (2010) Effects of experimental nitrogen additions on plant diversity in an old-growth tropical forest. Glob Chang Biol 16: 2688–2700.
27. Liu L, Gundersen P, Zhang T, Mo JM (2012) Effects of phosphorus addition on soil microbial biomass and community composition in three forest types in tropical China. Soil Biol Biochem 44: 31–38.
28. Vitousek PM, Howarth RW (1991) Nitrogen limitation on land and in the sea - how can it occur. Biogeochemistry 13: 87–115.
29. Van Duren IC, Pegtel DM (2000) Nutrient limitations in wet, drained and rewetted fen meadows: evaluation of methods and results. Plant Soil 220: 35–47.
30. Chapin FS, Vitousek PM, Vancleve K (1986) The nature of nutrient limitation in plant-communities. Am Nat 127: 48–58.
31. Güsewell S (2004) N : P ratios in terrestrial plants: variation and functional significance. New Phytol 164: 243–266.
32. Lipson DA, Bowman WD, Monson RK (1996) Luxury uptake and storage of nitrogen in the rhizomatous alpine herb, Bistorta bistortoides. Ecology 77: 1277–1285.
33. Campo J, Dirzo R (2003) Leaf quality and herbivory responses to soil nutrient addition in secondary tropical dry forests of Yucatán, Mexico. J Trop Ecol 19: 525–530.
34. Koerselman W, Meuleman AFM (1996) The vegetation N:P Ratio: a new tool to detect the nature of nutrient limitation. J Appl Ecol 33: 1441–1450.
35. Townsend AR, Cleveland CC, Asner GP, Bustamante MMC (2007) Controls over foliar N : P ratios in tropical rain forests. Ecology 88: 107–118.
36. von Oheimb G, Härdtle W, Friedrich U, Power SA, Boschatzke N, et al. (2010) N:P Ratio and the nature of nutrient limitation in Calluna -Dominated Heathlands. Ecosystems 13: 317–327.
37. Peng SL, Zhang ZP (1995) Biomass, productivity and energy use efficiency of climax vegetation on Dinghu Mountains, Guangdong, China. Sci China B 38: 67–73.
38. Liu JC, Zhou GY, Zhang DQ (2007) Simulated effects of acidic solutions on element dynamics in monsoon evergreen broad-leaved forest at Dinghushan, China. Part 1: Dynamics of K, Na, Ca, Mg and P. Environ Sci Pollut Res 14: 123–129.
39. Shen CD, Yi WX, Sun YM, Xing CP, Yang Y, et al. (2001) Distribution of C-14 and C-13 in forest soils of the Dinghushan Biosphere Reserve. Radiocarbon 43: 671–678.
40. Lu XK, Mo JM, Gundersern P, Zhu WX, Zhou GY, et al. (2009) Effect of simulated N deposition on soil exchangeable cations in three forest types of subtropical China. Pedosphere 19: 189–198.
41. Zhou GY, Wei XH, Wu YP, Liu SG, Huang YH, et al. (2011) Quantifying the hydrological responses to climate change in an intact forested small watershed in Southern China. Glob Chang Biol 17: 3736–3746.
42. Huang WJ, Zhou GY, Liu JX (2011) Nitrogen and phosphorus status and their influence on aboveground production under increasing nitrogen deposition in three successional forests. Acta Oecol (in press).
43. Wang ZH, He DQ, Song SD, Chen SP, Chen DR, et al. (1982) The vegetation of Ding hu shan Biosphere Reserve. In: Ding Hu Shan Ecosystem Stationary Academia Sincia, editor. Tropical and subtropical forest ecosystem. Beijing: Science Press. 77–141.
44. Zhou GY, Guan LL, Wei XH, Zhang DQ, Zhang QM, et al. (2007) Litterfall production along successional and altitudinal gradients of subtropical monsoon evergreen broadleaved forests in Guangdong, China. Plant Ecol 188: 77–89.
45. Mo JM, Brown S, Lenart M, Kong GH (1995) Nutrient dynamics of a human-impacted pine forest in a MAB reserve of subtropical China. Biotropica 27: 290–304.
46. Fang YT, Mo JM (2002) Study on carbon distribution and storage of a pine forest ecosystem in Dinghushan Biosphere Reserve. Guihaia 22: 305–310.
47. Vance ED, Brookes PC, Jenkinson DS (1987) An extraction method for measuring soil microbial biomass-C. Soil Biol Biochem 19: 703–707.
48. Alef K, Nannipieri P (1995) β-Glucosidase activity. In: Alef K, Nannipieri P, editors. Methods in applied soil microbiology and biochemistry. London: Academic Press. 350–351.
49. Murphy J, Riley JP (1962) A modified single solution method for the determination of phosphate in natural waters. Anal Chim Acta 27: 31–36.
50. Sparling G, Vojvodic-Vukovic M, Schipper LA (1998) Hot-water-soluble C as a simple measure of labile soil organic matter: The relationship with microbial biomass C. Soil Biol Biochem 30: 1469–1472.
51. Chen CR, Xu ZH, Keay P, Zhang SL (2005) Total soluble nitrogen in forest soils as determined by persulfate oxidation and by high temperature catalytic oxidation. Aust J Soil Res 43: 515–523.
52. Curtin D, Wright CE, Beare MH, McCallum FM (2006) Hot water-extractable nitrogen as an indicator of soil nitrogen availability. Soil Sci Soc Am J 70: 1512–1521.
53. McDowell RW, Condron LM (2000) Chemical nature and potential mobility of phosphorus in fertilized grassland soils. Nutr Cycl Agroecosys 57: 225–233.
54. Ohno T, Zibilske LM (1991) Determination of low concentrations of phosphorus in soil extracts using malachite green. Soil Sci Soc Am J 55: 892–895.
55. Liu S, Luo Y, Huang YH, Zhou GY (2007) Studies on the community biomass and its allocations of five forest types in Dinghushan Nature Reserve. Ecol Sci 26: 387–393.
56. Yan JH, Zhang DQ, Zhou GY, Liu JX (2009) Soil respiration associated with forest succession in subtropical forests in Dinghushan Biosphere Reserve. Soil Biol Biochem 41: 991–999.
57. Jenny H, Gessel SP, Bingham FT (1949) Comparative study of decomposition rates of organic matter in temperate and tropical regions. Soil Sci 68: 419–432.

58. Olson JS (1963) Energy storage and the balance of producers and decomposers in ecological systems. Ecology 44: 322–331.

59. Güsewell S, Koerselman M (2002) Variation in nitrogen and phosphorus concentrations of wetland plants. Perspect Plant Ecol Evol Syst 5: 37–61.

60. Wright SJ, Yavitt JB, Wurzburger N, Turner BL, Tanner EVJ, et al. (2011) Potassium, phosphorus, or nitrogen limit root allocation, tree growth, or litter production in a lowland tropical forest. Ecology 92: 1616–1625.

61. Güsewell S, Verhoeven JTA (2006) Litter N:P ratios indicate whether N or P limits the decomposability of graminoid leaf litter. Plant Soil 287: 131–143.

62. Mo JM, Brown S, Xue JH, Fang YT, Li ZA (2006) Response of litter decomposition to simulated N deposition in disturbed, rehabilitated and mature forests in subtropical China. Plant Soil 282: 135–151.

63. Mo JM, Zhang W, Zhu WX, Gundersen P, Fang YT, et al. (2008) Nitrogen addition reduces soil respiration in a mature tropical forest in southern China. Glob Chang Biol 14: 403–412.

64. Hobbie SE, Vitousek PM (2000) Nutrient limitation of decomposition in Hawaiian forests. Ecology 81: 1867–1877.

65. Sundareshwar PV, Morris JT, Koepfler EK, Fornwalt B (2003) Phosphorus limitation of coastal ecosystem processes. Science 299: 563–565.

66. Edwards PJ (1982) Studies of mineral cycling in a montane rain forest in New Guinea: V. rates of cycling in throughfall and litter fall. J Ecol 70: 807–827.

67. Wood TE, Lawrence D, Clark DA (2006) Determinants of leaf litter nutrient cycling in a tropical rain forest: soil fertility versus topography. Ecosystems 9: 700–710.

68. Wood TE, Lawrence D, Clark DA, Chazdon RL (2009) Rain forest nutrient cycling and productivity in response to large-scale litter manipulation. Ecology 90: 109–121.

69. Aerts R (1997) Climate, leaf litter chemistry and leaf litter decomposition in terrestrial ecosystems: A triangular relationship. Oikos 79: 439–449.

70. Wieder WR, Cleveland CC, Townsend AR (2009) Controls over leaf litter decomposition in wet tropical forests. Ecology 90: 3333–3341.

71. Odum EP (1969) The strategy of ecosystem development. Science 164: 262–270.

72. Vitousek PM, Reiners WA (1975) Ecological succession and nutrient budgets - a hypothesis. BioScience 25: 376–381.

73. Wang SQ, Zhou L, Chen JM, Ju WM, Feng XF, et al. (2011) Relationships between net primary productivity and stand age for several forest types and their influence on China's carbon balance. J Environ Manage 92: 1651–1662.

74. Kerkhoff AJ, Enquist BJ, Elser JJ, Fagan WF (2005) Plant allometry, stoichiometry and the temperature-dependence of primary productivity. Glob Ecol Biogeogr 14: 585–598.

75. Schuur EA, Matson PA (2001) Net primary productivity and nutrient cycling across a mesic to wet precipitation gradient in Hawaiian montane forest. Oecologia 128: 431–442.

76. Schuur EA (2003) Productivity and global climate revisited: The sensitivity of tropical forest growth to precipitation. Ecology 84: 1165–1170.

77. DeLaune RD, Buresh RJ, Patrick WH (1979) Relationship of soil properties to standing crop biomass of Spartina alterniflora in a Louisiana marsh. Estuar Coast Mar Sci 8: 477–487.

78. Vermeer JG, Berendse F (1983) The relationship between nutrient availability, shoot biomass and species richness in grassland and wetland communities. Plant Ecol 53: 121–126.

79. Laurance WF, Fearnside PM, Laurance SG, Delamonica P, Lovejoy TE, et al. (1999) Relationship between soils and Amazon forest biomass: a landscape-scale study. Forest Ecol Manage 118: 127–138.

80. Aerts R (1996) Nutrient resorption from senescing leaves of perennials: are there general patterns? J Ecol 84: 597–608.

81. Walker TW, Syers JK (1976) The fate of phosphorus during pedogenesis. Geoderma 15: 1–19.

82. Vitousek PM, Farrington H (1997) Nutrient limitation and soil development: Experimental test of a biogeochemical theory. Biogeochemistry 37: 63–75.

83. Crews TE, Kitayama K, Fownes JH, Riley RH, Herbert DA, et al. (1995) Changes in soil phosphorus fractions and ecosystem dynamics across a long chronosequence in Hawaii. Ecology 76: 1407–1424.

84. Liu JX, Chu GW, Yu QF, Zhang DQ, Zhou GY (2002) The responses of soil chemical properties in different forest types to altitude at Dinghushan. In: Ding Hu Shan Forest Ecosystem Stationary Academia Sinica, editor. Tropical and subtropical forest ecosystem. Beijing: Science Press. 125–131.

85. Fang YT, Gundersen P, Mo JM, Zhu WX (2008) Input and output of dissolved organic and inorganic nitrogen in subtropical forests of South China under high air pollution. Biogeosciences 5: 339–352.

15

Changes in the Abundance of Grassland Species in Monocultures versus Mixtures and Their Relation to Biodiversity Effects

Elisabeth Marquard[1]*, Bernhard Schmid[2], Christiane Roscher[3], Enrica De Luca[2], Karin Nadrowski[4], Wolfgang W. Weisser[5], Alexandra Weigelt[4,6]

1 UFZ – Helmholtz Centre for Environmental Research, Department of Conservation Biology, Leipzig, Germany, 2 Institute of Evolutionary Biology and Environmental Studies and Zurich-Basel Plant Science Centre, University of Zurich, Zurich, Switzerland, 3 UFZ – Helmholtz Centre for Environmental Research, Department of Community Ecology, Halle, Germany, 4 Institute of Biology, University of Leipzig, Leipzig, Germany, 5 Terrestrial Ecology/Department of Ecology and Ecosystem Management, Technische Universität München, Freising-Weihenstephan, Germany, 6 German Centre for Integrative Biodiversity Research (iDiv) Halle-Jena-Leipzig, Leipzig, Germany

Abstract

Numerous studies have reported positive effects of species richness on plant community productivity. Such biodiversity effects are usually quantified by comparing the performance of plant mixtures with reference monocultures. However, several mechanisms, such as the lack of resource complementarity and facilitation or the accumulation of detrimental agents, suggest that monocultures are more likely than mixtures to deteriorate over time. Increasing biodiversity effects over time could therefore result from declining monocultures instead of reflecting increases in the functioning of mixtures. Commonly, the latter is assumed when positive trends in biodiversity effects occur. Here, we analysed the performance of 60 grassland species growing in monocultures and mixtures over 9 years in a biodiversity experiment to clarify whether their temporal biomass dynamics differed and whether a potential decline of monocultures contributed significantly to the positive net biodiversity effect observed. Surprisingly, individual species' populations produced, on average, significantly more biomass per unit area when growing in monoculture than when growing in mixture. Over time, productivity of species decreased at a rate that was, on average, slightly more negative in monocultures than in mixtures. The mean net biodiversity effect across all mixtures was continuously positive and ranged between 64–217 g per m^2. Short-term increases in the mean net biodiversity effect were only partly due to deteriorating monocultures and were strongly affected by particular species gaining dominance in mixtures in the respective years. We conclude that our species performed, on average, comparably in monocultures and mixtures; monoculture populations being slightly more productive than mixture populations but this trend decreased over time. This suggested that negative feedbacks had not yet affected monocultures strongly but could potentially become more evident in the future. Positive biodiversity effects on aboveground productivity were heavily driven by a small, but changing, set of species that behaved differently from the average species.

Editor: Jon Moen, Umea University, Sweden

Funding: The Jena Experiment is funded by the German Research Foundation (FOR 456, FOR 1451) and supported by the Friedrich Schiller University of Jena and the Max Planck Institute for Biogeochemistry, Jena. Additional support was provided by the Swiss National Science Foundation (grant no. 31–65224–01 to B.S.). The funders had no role in study design, data collection and analysis, decision to publish, or preparation of the manuscript.

Competing Interests: The authors have declared that no competing interests exist.

* E-mail: lisa.marquard@ufz.de

Introduction

Numerous biodiversity experiments suggest that, all else being equal, plant communities are more productive when they contain higher numbers of species [1]. Commonly, these studies analysed differences in the performance between high- and low-diversity plant communities and in some experiments these differences were observed over several years [2,3,4,5,6,7]. Monocultures usually provide the baseline for such studies but they have not yet been the focus of interest. However, being the reference for comparisons and analytical tools in the context of biodiversity–productivity relationships, understanding the performance of monocultures over time is of critical importance for interpreting biodiversity effects in plant communities. Long-standing agricultural knowledge suggests that a single plant species is likely to decline in its yield if grown in monoculture at the same site for multiple years

[8,9]. Obviously, plants have the potential to influence the biotic or abiotic conditions they experience. For example, plants may change the soil in which they grow for the worse by an imbalanced depletion of resources [10], the release of toxic compounds [11,12] or the accumulation of soil-borne pathogens over time [13,14,15,16,17,18]. Such interactions are also known as negative plant–soil feedbacks [12,13,19,20,21,22,23]. Similarly, host-specific foliar pathogens may accumulate in monocultures if they respond positively to host density [24]. Therefore, it is conceivable that positive plant species richness–productivity relationships are largely due to negative feedbacks in monocultures and low-diversity mixtures rather than to complementary resource-use among species in high-diversity mixtures [13,16,18,23,25,26,27]. On the other hand, some plant species have the potential to improve the conditions of their environment, possibly by accumulating beneficial soil biota [28,29]. Such positive plant–soil

feedbacks have mainly been studied in the context of plant invasions (see e.g. [30,31,32]).

Interestingly, monoculture performance over time has rarely been studied in the context of biodiversity–ecosystem functioning relationships. Some previous studies conducted in biodiversity experiments compared the performance of particular species, but this was mostly done only at a single point in time (e.g. in monoculture: [33,34], across a diversity gradient: [33,35,36,37,38,39,40,41,42]). Where biodiversity experiments were used to compare the performance of species over multiple years the focus was usually on the biodiversity–stability relationship [6,43,44,45]. To our knowledge, no study has so far explicitly addressed the question how much a potential decline of plant monocultures over time contributes to the common phenomenon of overyielding mixtures in biodiversity experiments.

Here, we present a detailed analysis of the temporal dynamics in aboveground biomass production ("productivity") that occurred over a period of 9 years in monocultures of 60 different grassland species belonging to a large scale biodiversity experiment (Jena Experiment). Given the continuous nutrient export caused by regular mowing, a general decline of plant biomass was expected in the Jena Experiment. Therefore, we could not assess the performance of our monocultures in absolute terms. Instead, we compared the performance of our plant species in monoculture to the performance of populations of the same species within plant mixtures of the same experiment.

Specifically, we tested the following hypotheses:

1) Monocultures produce on average less aboveground community biomass than plant mixtures, due to imbalanced resource depletion and/or the accumulation of detrimental agents such as pathogens or toxins (all these mechanisms are referred to as "negative feedbacks" hereafter) and the lack of mechanisms such as complementary resource-use, facilitation or sampling in monocultures. This results in a positive net biodiversity effect.

2) On average, individual plant species' populations produce less aboveground biomass when growing in monocultures than when growing in mixtures.

3) Over time, the productivity of individual species' populations decreases comparatively more in monoculture than in mixture as all plots suffer from nutrient export but monocultures suffer additionally from negative feedbacks. At the species level, relative biomass change rates are therefore less positive or more negative in monocultures than in mixtures.

4) The net biodiversity effect measuring the difference between monocultures and mixtures increases over time due to a gradual augmentation of positive multi-species interactions such as complementary resource-use or facilitation in mixtures as well as negative feedbacks in monocultures that lead to their deterioration. Over time, the deterioration of monocultures becomes increasingly important for explaining positive changes in the net biodiversity effect.

Methods

Ethics Statement

All samples were taken on the field site of the Jena Experiment. This field site is a former arable land which the research group (represented by the University of Jena) rented from the land owner for the duration of the research grant. The land owner gave the permission to conduct this study on this site. No specific permissions were required for the field work and the data

collection that the current manuscript is based on. The field studies did not involve endangered or protected species.

Field Site and Biomass Sampling

The Jena Experiment is a grassland biodiversity experiment located in the floodplain of the river Saale near Jena, Germany (50°55′ N, 11°35′ E, 130 m above see level). The experiment was established on a former arable field with loamy soil (Eutric Fluvisol) that had received high fertilizer inputs for about four decades. In May 2002, 198 experimental plant communities containing 1, 2, 4, 8, or 16 species were sown (1000 viable seeds/ m^2). The field site had been kept fallow in the year before sowing, harrowed bimonthly, and treated with glyphosate (Roundup, Monsanto, St. Louis, Missouri, USA) in July 2001. Plot size was either 3.5×3.5 m (120 small monocultures during 2003–2007, reduced to 69 plots in 2008; i.e. two replicates per species during 2003–2007 and one replicate per species during 2008–2011; for nine of our species that were also part of another experiment not reported here, we kept the two replicates until 2011) or 20×20 m (16 large monocultures and 62 mixtures with 2–16 species). The small plots were downsized from 3.5×3.5 m to 1×1 m in 2009 and the large plots were downsized from 20×20 m to 6×6 m in 2010. The species pool composed for this experiment contained 60 common Central European grassland species typical for the regional alluvial plains. Based on a cluster analysis of ecological and morphological traits, these species had been assigned to four functional groups prior to the set-up of the experiment. According to this clustering, the species pool was composed of 16 grasses, 12 small herbs, 20 tall herbs, and 12 legumes. Species composition of each large plot was determined by a constrained random draw from the species pool. Constrains were imposed to combine the gradient in plant species richness with a gradient in the number of functional groups as orthogonally as possible. In mixtures, all species were sown with equal proportions. Plots were not fertilized but mown and weeded twice a year. The field site was divided into four blocks, to account for gradually changing characteristics of the floodplain soil. Each block contained four large plots of the species richness levels 1, 2, 4 and 8, three or four 16-species mixtures and 30 monocultures of small plot size. As such, all 60 species were present in small monoculture plots (twice until 2008) and a random subsample of 16 species (four per functional group) was present in large monoculture plots. For more details about the design, establishment and maintenance of the Jena Experiment, see Roscher et al. [46] and Fig. S1.

From 2003–2011, aboveground plant biomass was harvested on all experimental plots twice per year (in late May and in late August). For all harvests, the vegetation was clipped at 3 cm above ground in four (2003–2004, August 2005, 2006–2007), three (May 2005, 2008–2009) or two (2010–2011) randomly placed sampling frames of 0.2×0.5 m per large plot and in two (2003–2009) or one (2010–1011) randomly placed sampling frame(s) of 0.2×0.5 m per small monoculture. The harvested biomass was sorted into species and dried at 70°C for at least 48 h. Part of these data (2003–2008) and more detailed information about data collection have been published by Weigelt et al. [47].

Data Analysis

As a measure of aboveground productivity of our experimental communities and their component species we used the peak standing biomass harvested in May (averaged across all sampling frames per plot). We assessed whether, on average, positive biodiversity effects occurred and whether they increased over time by plotting the median community biomass as well as the median

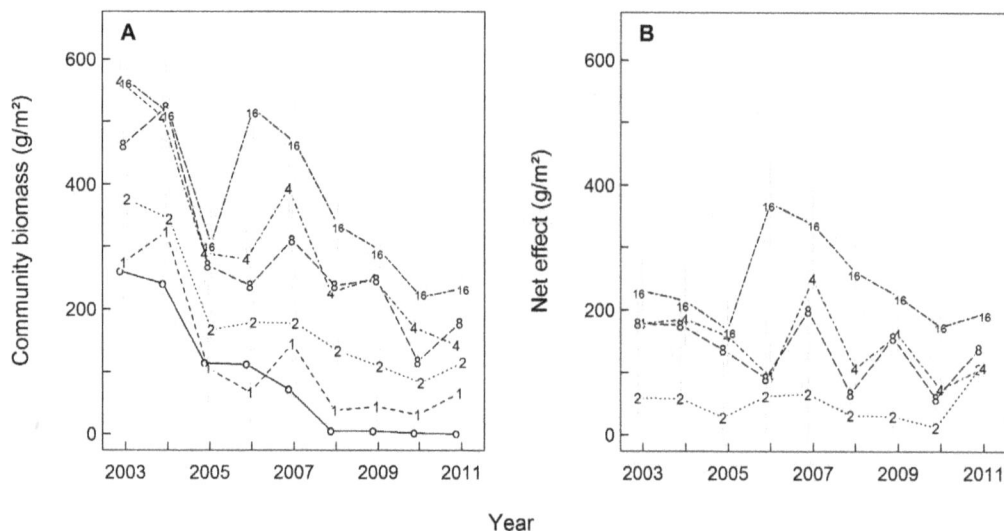

Figure 1. Aboveground community biomass (A) and net biodiversity effect (B) during 2003–2011. Symbols indicate medians per species richness level ±1 standard error (in A, "o" indicates monocultures of small plot size). Symbols were slightly jittered to improve visualization.

net biodiversity effect (see below) per species richness level over time (Fig. 1).

To analyse differences in aboveground productivity and in temporal biomass dynamics between monocultures and mixtures at the species level we plotted the biomasses of the individual species' populations against time (Fig. 2). For this direct comparison of yields of the individual species' populations in monocultures and mixtures, we multiplied the species specific biomasses by the number of species that was sown into the plot on which the respective biomass was measured to correct for differences in the amounts of seeds originally used per species at the different species richness levels. A constant (0.005) was added to the resulting biomass data (corrected for sown diversity) which was then log-transformed (base 10) to improve the residual distribution.

Furthermore, we calculated relative biomass change rates (RBR) for each species in each experimental community as follows:

$$RBR_{ij} \text{ for year } x \text{ to year } (x+1) = \log_{10}(Y_{ij} \text{ in year } (x+1))$$
$$- \log_{10}(Y_{ij} \text{ in year } x),$$

where Y_{ij} = the biomass of species i (in g/m^2) in community j.

This metric had the advantage of being independent of sown diversity. We tested for changes in the species' specific productivity and in RBR over time as well as for differences between monocultures and mixtures by fitting linear mixed-effects models. For each response variable (productivity and RBR) we first determined an appropriate random effects structure by comparing models with the same fixed effects, namely the monoculture-mixture contrast (MMC), the time passed since plants were sown (time), and the interaction of these two terms (MMC×time), using restricted maximum likelihood estimation. The selected random effects structure included main effects of species and harvest event (time as factor), and variable slopes for species over time. For productivity, it also included the main effect of plot as well as the interaction between MMC and time. In a second step we compared models with different fixed effects structures (built from MMC, time, and their interaction) using maximum likelihood

estimation. The resulting models were compared on the basis of the Akaike information criterion, AIC, and significance was determined using likelihood ratio tests of nested models. A summary of the results of these analyses for productivity and RBR at the species level is given in Table 1. The more detailed outputs obtained by fitting the selected models to the data (including the variance components for the random terms and the estimates for the fixed terms) are given in Table S1 for productivity and in Table S2 for RBR. For productivity, the data and the model output are visualized in Fig. S2.

For every year, we calculated the net biodiversity effect for every experimental community as the difference between its observed yield (i.e., the biomass measured at the community level) and its expected yield (i.e., the average of the reference monoculture yields of the composing species, see also [48]), using the following formula:

$$NE_j \text{ in year } x = \sum Ymix_{ij} \text{ in year } x$$
$$- \left(\sum Ymono_i \text{ in year } x\right) * SR_j^{-1},$$

where NE_j = net biodiversity effect of a particular multi-species community j; $Ymix_{ij}$ = the biomass of species i (in g/m^2) in community j; $Ymono_i$ = the biomass of species i (in g/m^2) in its reference monoculture(s) (when two small plots were maintained for a particular species, their yields were averaged to obtain a single reference per species per year); and SR_j = the number of species within j. We excluded three outliers from the data set (two net biodiversity effect values from 2005 and one from 2006, as described by Marquard et al. [49]).

To evaluate the temporal trends in the mean net biodiversity effect and whether the assumed positive changes were increasingly driven by species with deteriorating monocultures but productive populations in plant mixtures we calculated mean annual differences in the net biodiversity effect (averages across all mixtures, dNE) and analyzed how they were impacted by mean annual changes in the monoculture yield of a species (i.e., more precisely, in the expected yield of the species, $\Delta \hat{Y}E_i$) and mean annual changes in the mixture yield of the same species (i.e., in the terminology used here, in the observed yield of the species, $\Delta \hat{Y}O_i$;

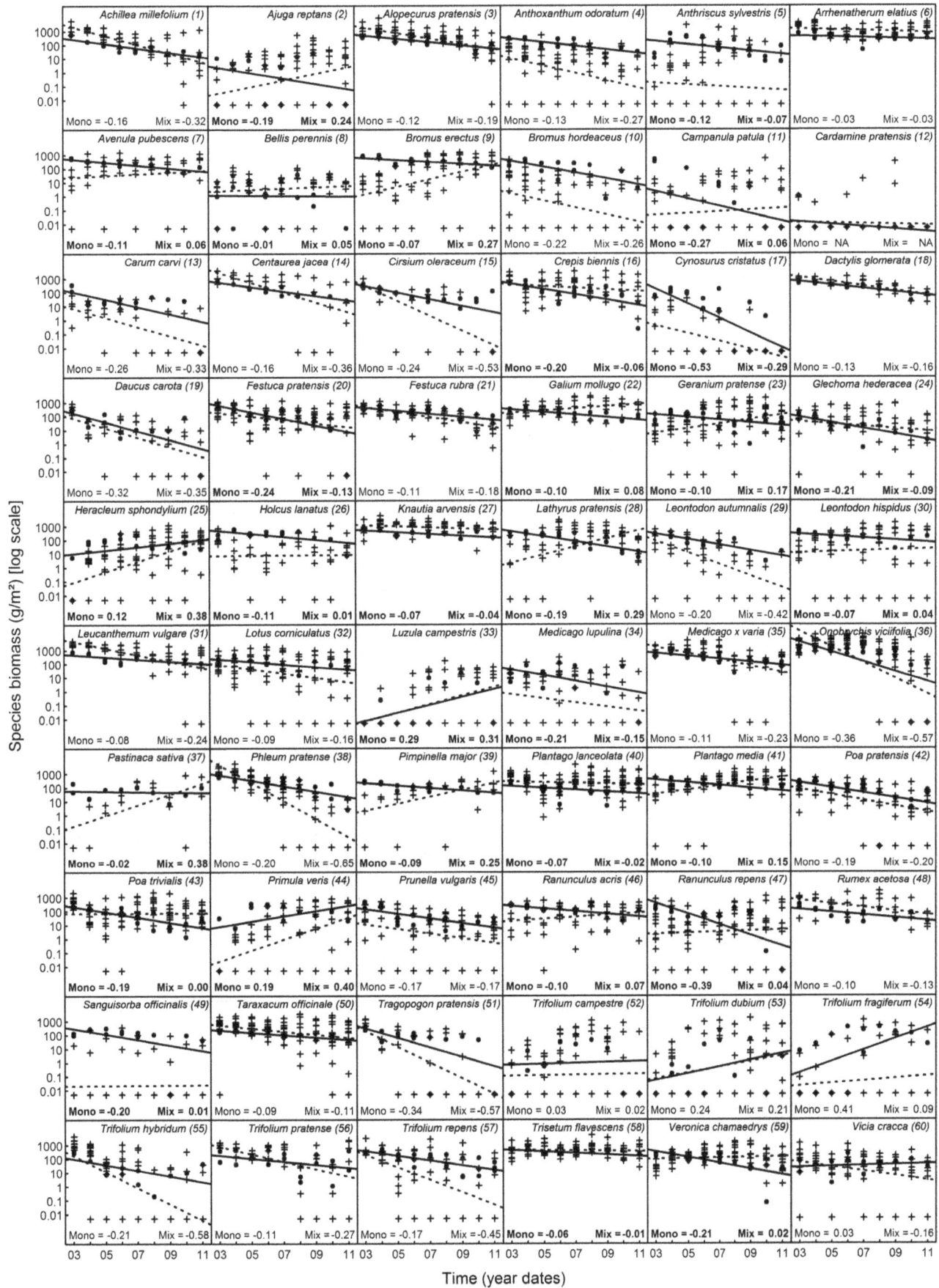

Species biomass (g/m²) [log scale]

Time (year dates)

Figure 2. Temporal trends in species specific biomass during 2003–2011. Filled circles and solid lines indicate monocultures, crosses and broken lines indicate mixtures. Bold letters indicate less positive or more negative slopes for the regression lines across the monoculture populations than for the regression lines across the mixture populations of a particular species. Species are listed in alphabetical order; the numbers are used for their identification in Fig. 3.

see Fig. 3). These metrics were calculated as follows:

$$\Delta YE_{ij} \text{ for year } x \text{ to year } (x+1) = (Ymono_i \text{ in year } x+1)$$
$$* SR_j^{-1} - (Ymono_i \text{ in year } x) * SR_j^{-1}, \text{and}$$

$$\Delta YO_{ij} \text{ for year } x \text{ to year } (x+1) = (Ymix_{ij} \text{ in year } x+1)$$
$$- (Ymix_{ij} \text{ in year } x), \text{respectively,}$$

where ΔYE_{ij} = the annual difference in the expected yield of species i (in g/m^2) in community j, ΔYO_{ij} = the annual difference in the observed yield of species i (in g/m^2) in community j, Ymono$_i$ = (the average of) the biomass of species i (in g/m^2) in its reference monoculture(s), Ymix$_{ij}$ = the biomass of species i (in g/m^2) in community j, and SR$_j$ = the number of species within j. For every annual time interval, we calculated the contributions of the individual species to changes in the net biodiversity effect of a particular plot (SC$_{ij}$) as follows

$$SC_{ij} \text{ for year } x \text{ to year } (x+1) = \Delta YO_{ij} \text{ for year } x \text{ to year } (x+1)$$
$$- \Delta YE_{ij} \text{ for year } x \text{ to year } (x+1)$$

As such, we obtained positive SC$_{ij}$-values when the annual difference in the observed yield was more positive or less negative than the annual difference in the expected yield of species i and negative SC$_{ij}$-values when the annual difference in the observed yield was less positive or more negative than the annual difference in the expected yield of species i.

The sum of SC$_{ij}$-values per plot j amounted to the annual change in the net biodiversity effect of the respective plant community.

For ΔYE_{ij}, ΔYO_{ij} and SC$_{ij}$, we calculated average values per species (per annual time interval: $\Delta \hat{Y}E_i$, $\Delta \hat{Y}O_i$ and $\hat{S}C_i$) and weighted them according to the proportion of plots on which species i was sown:

$$\Delta \hat{Y}E_i \text{ for year } x \text{ to year} (x+1) = (\sum \Delta YE_{ij} \text{ for year } x \text{ to year }$$
$$(x+1) * N_i^{-1}) * p_i,$$

$$\Delta \hat{Y}O_i \text{ for year } x \text{ to year } (x+1) = (\sum \Delta YO_{ij} \text{ for year } x \text{ to year }$$
$$(x+1) * N_i^{-1}) * p_i, \text{and}$$

$$\hat{S}C_i \text{ for year } x \text{ to year} (x+1) = (\sum SC_{ij} \text{ for year } x \text{ to year }$$
$$(x+1) * N_i^{-1}) * p_i, \text{respectively,}$$

where N$_i$ = the number of experimental plots containing species i, and p$_i$ = the proportion of plots into which species i was sown (i.e., the number of experimental plots containing species i divided by the total number of plots). Correcting the average values per species by p$_i$ was necessary to enable an analysis at the species level (instead of the plot level).

For the sake of analysing the effects of $\Delta \hat{Y}E_i$ and $\Delta \hat{Y}O_i$ on annual changes in the mean net biodiversity effect graphically, these values were square-root transformed (and the negative sign was remained if the change was negative) and two extreme data points were excluded. This allowed for a better separation of data points (see Fig. 3).

Results

Trends in Total Community Biomass

As a general trend, mean aboveground community biomass declined during our observation period in mixtures (from a

Table 1. Summary of statistical models[1] for aboveground biomass and relative biomass change rates of individual species' populations during 2003–2011, testing for changes over time and differences between monocultures and mixtures.

Aboveground biomass (productivity)						Biomass change rates (RBR)				
	DF[2]	AIC[2]	Chisq[2]	df[2]	Pr(>Chisq)[2]		DF	AIC	Chisq	df	Pr(>Chisq)
Nullmodel[3]	14	17337				Nullmodel	6	14208			
Time linear	15	17328	11.5	1	0.001***	Time linear	7	14210	0.3	1	0.565
MMC[4]	15	17331	8.3	1	0.004**	MMC	7	14207	3.4	1	0.066.
Time linear+MMC	16	17321	8.8	1	0.003**	Time linear+MMC	8	14209	3.4	1	0.064.
Time linear×MMC	17	17320	3.0	1	0.085.	Time linear×MMC	9	14209	1.3	1	0.263

[1]Linear mixed effects models fitted by the lme4-package of the statistical software R, see Methods and Tables S1 and S2 for details. Models were fitted by stepwise inclusion of variables and p-values were inferred by their hierarchical comparison.
[2]DF = model degrees of freedom, AIC = Akaike information criterion; Chisq = chi-square statistic; df = degrees of freedom required for estimating parameters, Pr(>Chisq) = associated p-value. Significance is given with *** = p<0.001;
** = p<0.01;
* = p<0.05; . = p<0.1.
[3]The Nullmodel fitted an intercept, only.
[4]MMC = Monoculture-Mixture-Contrast.

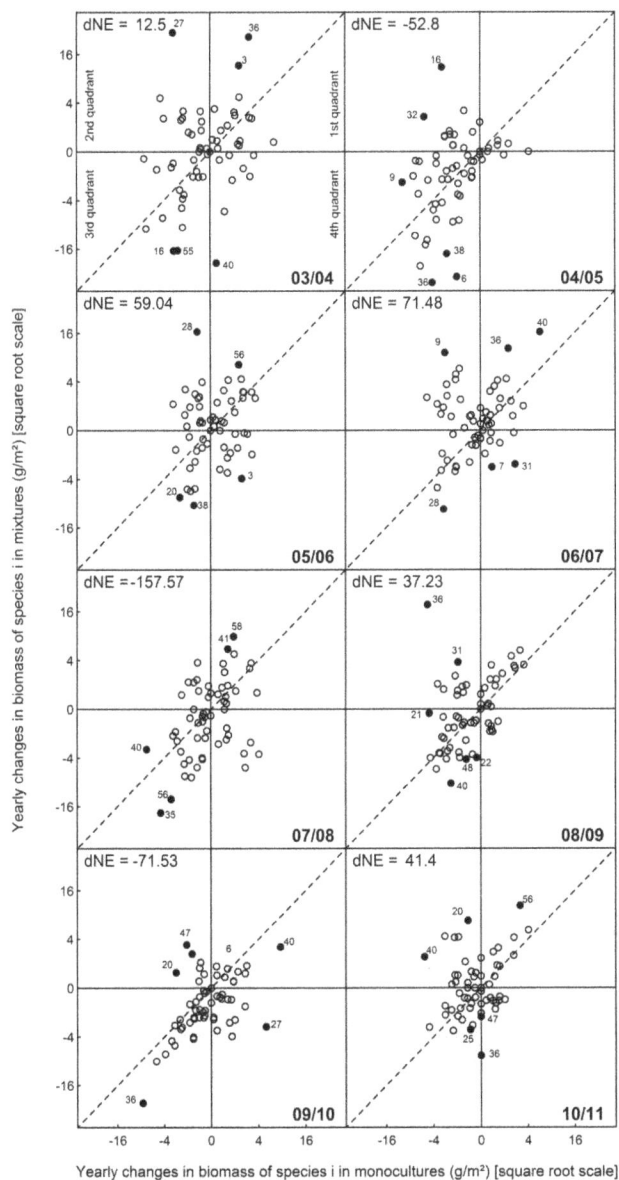

Yearly changes in biomass of species i in monocultures (g/m²) [square root scale]

Figure 3. Mean yearly changes in monoculture biomass over mean yearly changes in mixture biomass per species. The panels show annual time intervals during 2003–2011 (as indicated in the lower right of each panel). The values on the x-axis equal $\Delta \hat{Y}E_i = $ mean changes in the expected yield of a species (monoculture yield divided by species richness); the values on the y-axis equal $\Delta \hat{Y}O_i = $ mean changes in the observed yield of a species (mixture yield; see Methods). A point falling on the solid vertical line indicates that a species has not changed in its expected yield (i.e., in monoculture) during the respective time interval. A point falling on the solid horizontal line indicates that a species has not changed in its observed yield (i.e., in mixture) during the respective time interval. A point falling below the broken diagonal line contributed to a decline and a point falling above the broken diagonal line contributed to an increase in the net biodiversity effect. The perpendicular distance of a point to the diagonal equals the contribution of a particular species to the change in the net biodiversity effect (see Table 2). The small numbers next to symbols correspond to the species numbers in Fig. 2 and Table 2 and reveal the identity of the six species with the largest positive or negative contributions to changes in the net biodiversity effect. "dNE" indicates the absolute change in the net biodiversity effect (in g/m²) during the respective time interval. Note the square-root scale of the axes. The two most extreme values are not displayed to allow for a better scaling. These are the values for *O. viciifolia* during the time

intervals 2005/2006 and 2007/2008. They are given in Table 2 (non-transformed); the contribution of *O. viciifolia* to dNE was 60.9 g/m² during 2005/2006 and −106.4 g/m² during 2007/2008.

maximum of 497 g/m² in 2003 to a minimum of 141 g/m² in 2010, median values computed across all multi-species plots) as well as in monocultures (from a maximum of 262 g/m² in 2003 to a minimum of 0.11 g/m² in 2011, median values computed across all monoculture plots; see Fig. 1A). The mean net biodiversity effect ranged from a maximum of 217 g/m² in 2007 to a minimum of 64 g/m² in 2010 (median values computed across all multi-species plots) and did not increase linearly over time (Fig. 1B). Aboveground biomass as well as the net biodiversity effect were positively affected by species richness (Fig. 1A and Fig. 1B; see also [49] for the period 2003–2007).

Trends in Individual Species' Populations

At the species level, there was considerable variation in aboveground biomass production across species, communities and years (Fig. 2 and Fig. S2). On average, the individual species' populations produced slightly, but significantly, more ($10^{0.632} = 4.3$ g/m²) biomass when growing in monoculture than when growing in mixture (see Table 1 and Table S1: estimates for the fixed effect MMC; see also Fig. S2). Over time, the productivity of individual species' populations in monoculture and mixture declined with marginally significantly different slopes (monoculture slope: $10^{(-0.086+(-0.049))} = 0.7$, mixture slope: $10^{-0.086} = 0.8$, translating into a 30% or 20% decline per year in monocultures and mixtures, respectively; see Table 1 and Table S1: estimates for the fixed effects time linear and time linear×MMC; see also Fig. S2). Consistent with these results and the biomass dynamics shown in Fig. 1 and Fig. 2, the relative biomass change rates (RBR) of the individual species' populations were on average negative and marginally significantly different between monocultures ($-0.069+(-0.083) = -0.152$) and mixtures (-0.069; see Table 1 and Table S2: estimates for the intercept and MMC).

Effects of Trends in Individual Species' Populations on Community Net Biodiversity Effects

Positive annual changes in the mean net biodiversity effect did not predominantly or increasingly result from species proliferating in mixtures but deteriorating in monoculture (see Fig. 3, where the symbols representing the individual species are never predominantly or increasingly found in the second quadrant at the top left). Instead, we observed a large variation in biomass dynamics across species over nearly the entire study period (see the symbols scattering widely across all quadrants in most panels of Fig. 3; see also Fig. 2 and the large variance component attributed to species in Table S1). The strong decline among monocultures during 2004–2005 shown in Fig. 1 is clearly reflected in the second panel of Fig. 3 (the majority of symbols fall in the second and third quadrant). However, during that same time period, many species performed even worse in mixtures than in monocultures (see points below the broken diagonal in Fig. 3) which, overall, resulted in a decreasing mean net biodiversity effect. During all other time intervals, the large variation in species' behaviour shown in Fig. 3 (wide scatter of symbols, relatively evenly distributed above and below the broken diagonal during most time intervals) suggested that the overall negative trends in yields shown in Fig. 1A as well as the fluctuation in the mean net biodiversity effect shown in Fig. 1B were driven by a subset of the species, only.

No species had a continuously positive or negative impact on the net biodiversity effect across the entire study period (Table 2).

Nor did any of the species impact the net biodiversity effect consistently towards the direction of its overall mean (Table 2, exception: *Cardamine pratensis*, but its contributions to changes in the mean net biodiversity effect were zero during four out of eight time intervals due to very low abundances or its local extinction). Two species had particularly strong impacts on the net biodiversity effect: the perennial legume *Onobrychis viciifolia* (among the three most influential species during all time intervals) and the perennial small herb *Plantago lanceolata* (among the three most influential species in six out of eight time intervals). However, the direction of their impact was not persistent but switched between positive and negative. While the impact of *O. viciifolia* was mainly in the same direction as the mean changes in the net biodiversity effect, this was not true for *P. lanceolata* (compare Fig. 1B and Table 2). Thus, besides *O. viciifolia* and *P. lanceolata*, a varying set of species exerted strong effects on the mean net biodiversity effect. They either enhanced or weakened the mean trend in the net biodiversity effect during a particular annual time interval.

Discussion

As suggested by a wealth of studies on plant species richness–productivity relationships and in line with our first hypothesis, the monocultures of our experiment yielded on average less community biomass per unit area than the plant mixtures. Positive effects of species richness on aboveground community biomass and on the net biodiversity effect have previously been described for the Jena Experiment (see [50] for results on the May harvest in 2003; [6] for results on the May harvests from 2003–2009; and [49] for results on annual biomass data from 2003–2007). However, these publications either focused on the impact of different aspects of plant diversity on biomass production and did not study the role or the behaviour of individual species [49,50], or analysed the role of functional turnover for the maintenance of high biomass production [6]. Here, we focused on the differences in temporal biomass dynamics between monocultures and mixtures and evaluated whether deteriorating monocultures were a considerable driver for positive biodiversity effects persisting over time.

Building upon recent findings on negative plant–soil feedbacks [13,16,18,19,26,51,52] we expected that species in monocultures would suffer from additional growth-limiting factors compared to those in mixtures. For example, monocultures are generally regarded as being particularly prone to deplete resources unsustainably and to accumulate detrimental agents such as pathogens or toxins over time. Therefore, plant species should produce less biomass in monocultures than in mixtures and this difference should increase with time. Interestingly, we found that individual species' populations were on average slightly more productive when experiencing only intra-specific competition than when experiencing inter-specific competition. Furthermore, we did not find evidence for a particularly strong and consistent deterioration of monocultures over time. Instead, high and low biomasses were harvested regularly for monocultures as well as for mixture populations and, on average, the productivity of both declined at a rate that was only slightly, but marginally significantly, different (see Fig. 2 and Fig. S2). These findings contradicted our second and third hypotheses and suggest that mechanisms that disadvantage monocultures (such as negative feedbacks) were less common than expected or more variable over time. They may also have been partly balanced by positive feedbacks which were recently hypothesised to be more prominent in nature than recognised so far [23]. Furthermore, the mean net biodiversity effect did not increase steadily over time and was not predominately or increasingly driven by the deterioration of our

reference monocultures (hypothesis four). Instead, the mean net biodiversity effect was always the result of a wide range of species behaviours (see Fig. 1B and Fig. 3).

However, the marginally significant differences in the slopes of the productivity curves as well as in relative biomass change rates (RBR) may still reveal ecologically significant information, given the high number of species included here and the naturally large differences between them [53]; namely that the individual species' populations do develop slightly differently over time, possibly indicating the occurrence of negative feedbacks within the plant monocultures. On average, over the 9 years of the study, such negative feedbacks did not affect our monocultures much more strongly than the individual species' populations were negatively affected by inter-specific competition in the multi-species environments. The small differences in the temporal dynamics observed between monoculture and mixture populations may, however, result in stronger evidence for negative feedbacks in monocultures in the long run. Generally, the temporal dynamics in aboveground productivity were very variable across species in monoculture as well as in mixture and a more focused analysis of a particular subset of species may deliver additional insights into the role of negative feedbacks on the development of our species. For example, we found that the difference in RBR between monocultures and mixtures was more significant (monocultures having a more negative RBR than mixtures) when we restricted our analysis to those species that were actually present in our biomass samples (i.e. to species that had biomasses >0, results not shown). This corroborated our interpretation that monoculture populations were affected by factors decreasing their performance compared to those in mixtures over time, but that these factors were not yet strong enough to detect them unequivocally across a set of 60 different grassland species.

The mean net biodiversity effect was consistently positive and clearly demonstrated an average advantage of mixtures over monocultures at the community level that was much larger than the amount to which monocultures outcompeted mixtures at the population level (64–217 g/m^2 as compared with 4.3 g/m^2, see Results). This apparent discrepancy suggested that at any time, a small subset of our species profited greatly from growing within a multi-species situation. A strong overyielding of these species likely overcompensated for the lower average productivity of individual species' populations in mixture and thereby resulted in strong positive net biodiversity effects. We could only identify two species that had a strong impact on the magnitude of the net biodiversity effects during most of the years (*O. viciifolia* and *P. lanceolata*); the identity of the other influential species changed frequently over time. Such strong fluctuations in species abundance may be the result of growth rates being negatively frequency dependent [54]. Corroborating previous findings [6,45], these results suggested a substantial turnover in the species that drive the overyielding of plant mixtures in the Jena Experiment. We conclude that these driving species were few in one context, but variable across contexts, and that different mechanisms enhanced their productivity within the mixtures, including the more efficient partitioning of resources (complementarity) and other positive inter-specific interactions (such as facilitation).

Supporting Information

Figure S1 Schematic view of the field site showing the location and arrangement of the plots. Large squares symbolize the large monocultures (n = 16) and mixtures (n = 62), small squares symbolize the reference monocultures (n = 120). Small plots with grey borders were harvested until 2008. The

Table 2. Mean contributions of the individual species to mean annual changes in the net biodiversity effect ($\hat{S}C_i$) during 2003–2011.

No.	Species	$\hat{S}C_i$ [1] $_{03/04}$	$\hat{S}C_i$ $_{04/05}$	$\hat{S}C_i$ $_{05/06}$	$\hat{S}C_i$ $_{06/07}$	$\hat{S}C_i$ $_{07/08}$	$\hat{S}C_i$ $_{08/09}$	$\hat{S}C_i$ $_{09/10}$	$\hat{S}C_i$ $_{10/11}$
1	Achillea millefolium	−3.24	−7.12	−1.19	−1.5	−1.97	1.01	−0.29	0.61
2	Ajuga reptans	0.05	−0.12	0.34	−0.04	0.58	−0.18	−0.39	0.23
3	Alopecurus pratensis	**11.09**	−8.19	*−5.56*	−2.17	−4.69	−0.39	0.09	−1.24
4	Anthoxanthum odoratum	−0.51	3.4	−1.8	0.23	1.91	0.7	−0.81	0.32
5	Anthriscus sylvestris	−7.12	3.27	−1.61	6.99	−0.54	−0.81	0.1	−0.75
6	Arrhenatherum elatius	2.29	*−25.4*	1.23	3.95	−3.33	0.53	**2.61**	2.32
7	Avenula pubescens	3.27	−0.39	2.67	*−2.47*	0.83	0.76	−0.77	2.17
8	Bellis perennis	0.55	−0.55	−0.02	0.04	−0.11	0.2	−0.13	0.14
9	Bromus erectus	2.93	**9.26**	−1.8	**12.46**	−5.26	3.06	−3.53	6.8
10	Bromus hordeaceus	−2.45	3.19	0.32	−0.21	−0.31	2.43	−0.83	0.8
11	Campanula patula	7.92	−1.03	0.93	−0.07	0.04	−0.06	0.42	−0.43
12	Cardamine pratensis	0	0	0	0.07	−0.07	0.36	−0.36	0
13	Carum carvi	2.36	−0.12	0.3	−0.03	−0.36	−0.12	0.09	0.35
14	Centaurea jacea	1.49	−2.6	−0.37	0.83	−1.04	−0.47	−0.87	0.29
15	Cirsium oleraceum	−0.82	−0.02	0.23	−0.12	0.27	−0.17	−0.16	0.07
16	Crepis biennis	*−14.08*	**14.72**	−4.78	3.26	−6.26	3.07	−3.1	−1.08
17	Cynosurus cristatus	−0.7	0.23	0.76	−0.01	0.03	0	0	0
18	Dactylis glomerata	−1.88	−2.23	1.62	2.39	−3.45	−0.96	0.09	−0.54
19	Daucus carota	−2.44	−1.37	3.73	−0.37	−1.32	0.04	−1.69	1.76
20	Festuca pratensis	4.68	3.85	*−5.76*	−0.22	0.65	−3.36	**2.62**	**7.96**
21	Festuca rubra	−0.75	−1.1	−0.47	3.67	−7.8	**4.73**	−1.04	−1.12
22	Galium mollugo	1.58	2.36	−0.07	7.24	−1.67	*−3.93*	0.39	−0.78
23	Geranium pratense	−1.46	1.33	−0.09	3.22	1.82	1.04	−1.36	*−2.73*
24	Glechoma hederacea	0.91	0.62	0.19	−0.35	−0.21	1.33	−1.67	0.42
25	Heracleum sphondylium	−1.42	−0.38	−0.53	3.97	1.46	3.04	−3.47	1.88
26	Holcus lanatus	−3.86	2.59	−0.8	−0.12	0.43	1.28	0.92	0.06
27	Knautia arvensis	**26.48**	−15.63	−2.41	1.31	−4.09	−0.15	*−7.91*	2.21
28	Lathyrus pratensis	−0.27	1.43	**16.78**	*−8.06*	−2.13	1.31	−3.76	*−2.22*
29	Leontodon autumnalis	−0.33	0.51	−0.35	0.19	−1.05	0.97	0.04	−0.1
30	Leontodon hispidus	1.08	1.62	1.72	0.97	0.54	1.5	−2.7	0.38
31	Leucanthemum vulgare	3.22	−9.35	−5.42	*−4.06*	−7.08	**4.6**	−4.67	5.51
32	Lotus corniculatus	−3.43	**7.65**	2.78	0.07	−3.73	−0.07	−1.14	1.22
33	Luzula campestris	0	0	0	0.11	0.91	−1.11	0.34	−0.1
34	Medicago lupulina	−1.28	0.86	0.66	−1.21	−0.01	1.33	−1.42	1.03
35	Medicago x varia	3.06	6.45	−4.05	1.82	*−13.57*	1.14	−1.85	5.22
36	Onobrychis viciifolia	**19.6**	*−24.8*	**60.89**	9.97	*−106.35*	23.36	*−13.82*	−7.68
37	Pastinaca sativa	0.22	−0.11	0.02	0.5	−0.12	−0.24	0.75	−0.31
38	Phleum pratense	9.37	*−15.65*	*−8.77*	−1.99	−1.34	1.38	−0.68	1.53
39	Pimpinella major	0.21	0.46	0.15	0.2	−0.02	0.71	−0.55	1.47
40	Plantago lanceolata	*−20.91*	−4.96	2.12	**10.08**	4.82	*−7.75*	*−5.77*	**7.31**
41	Plantago media	1.66	6.93	3.72	2.3	**5.62**	−0.99	−1.18	0.15
42	Poa pratensis	1.11	1.27	−0.21	−0.94	1.44	−0.69	0.01	−0.12
43	Poa trivialis	−8.21	1.35	−1.1	5.56	1.97	−3.18	−1.46	−0.75
44	Primula veris	0.17	−4.24	1.24	3.35	1.74	1.46	−0.94	1.48
45	Prunella vulgaris	5.83	3.49	−2.66	−0.1	0.44	0.14	−0.38	−0.02
46	Ranunculus acris	4.09	2.64	−1.53	−1.69	0.69	0.55	1.31	−0.12
47	Ranunculus repens	−2.32	1.34	−1.76	0.53	0.11	1.01	**4.23**	−1.38
48	Rumex acetosa	0.36	−1.03	1.18	−0.97	3.21	*−3.86*	−0.28	−0.48

Table 2. Cont.

No.	Species	$\hat{S}C_i{}^1$ $_{03/04}$	$\hat{S}C_i$ $_{04/05}$	$\hat{S}C_i$ $_{05/06}$	$\hat{S}C_i$ $_{06/07}$	$\hat{S}C_i$ $_{07/08}$	$\hat{S}C_i$ $_{08/09}$	$\hat{S}C_i$ $_{09/10}$	$\hat{S}C_i$ $_{10/11}$
49	Sanguisorba officinalis	−1.26	−0.84	4.06	−1.11	−0.74	3.12	−2.61	1.2
50	Taraxacum officinale	−0.8	−1.76	−3.51	1.71	4.15	−2.05	−0.82	−1.65
51	Tragopogon pratensis	−3.85	0.74	−0.27	0.46	0.31	−0.94	0.11	0.32
52	Trifolium campestre	0.11	−0.1	0.69	0.19	3.98	−2.29	−1.19	−1.08
53	Trifolium dubium	0.24	−0.01	−0.1	−1.55	0.36	1.27	−1.15	−0.3
54	Trifolium fragiferum	−0.1	−0.48	−2.02	6.37	−7.51	4.43	−0.91	0.73
55	Trifolium hybridum	*−14.58*	−4.36	4.09	−0.23	−3.82	−0.05	0.21	−0.85
56	Trifolium pratense	−6.32	−6.99	**5.93**	5.44	*−10.83*	0.74	−1.04	**8.77**
57	Trifolium repens	−2.23	−1.81	2.01	2.92	−5.32	3.01	−2.59	1.6
58	Trisetum flavescens	3.52	7.28	−5.25	−2.25	**7.99**	−2.51	−3.63	1.54
59	Veronica chamaedrys	0.63	2.08	2.42	0.53	2.16	−1.24	−3.1	−0.24
60	Vicia cracca	−0.96	−0.98	0.52	0.42	0.07	−0.81	0.16	−0.38

[1]$\hat{S}C_i$ = mean annual contributions of the individual species to mean annual changes in the net biodiversity effect (in g/m^2). Values were scaled according to the proportion of plots on which the species were sown (see Methods). Species are listed in alphabetical order.
[2]Bold numbers indicate the three species with the highest positive contributions and numbers in bold italics indicate the three species with the most negative contributions within a particular year. These six species with the most extreme values have marked data points in Fig. 3.

different colour shading indicates the four blocks into which the field site was divided (see Methods for more details).

Figure S2 Visualisation of model estimates for aboveground biomass data of individual species' populations during 2003–2011. The data (corrected for sown diversity, 0.005 added, log10-transformed) is represented by symbols (grey: mixtures, black: monocultures); the bold lines indicate the overall intercepts and slopes as determined by the fixed part of the model (broken: mixtures, solid: monocultures). The p-value relates to the difference between the slopes of the regression lines (see the time linear×MMC interaction term in Table 1).

Table S1 Results of variance components analysis for aboveground biomass (productivity) of individual species' populations during 2003–2011.

Table S2 Results of variance components analysis for species specific relative biomass change rates (RBR) during 2003–2011.

Acknowledgments

We thank the gardeners of the Jena Experiment for their work, all student helpers for assisting in the field, J. Schumacher for statistical advice and S. Ratcliffe for proof-reading. S. Pompe provided the R-code for Fig. S2. The comments of J. Bengtsson and one anonymous referee on an earlier draft of this paper were greatly appreciated.

Author Contributions

Conceived and designed the experiments: BS CR WW. Performed the experiments: EM CR AW EDL. Analyzed the data: EM BS KN AW. Contributed reagents/materials/analysis tools: EM BS KN AW. Wrote the paper: EM.

References

1. Cardinale BJ, Matulich KL, Hooper DU, Byrnes JE, Duffy E, et al. (2011) The functional role of producer diversity in ecosystems. American Journal of Botany 98: 572–592.
2. Fargione J, Tilman D, Dybzinski R, Hille Ris Lambers J, Clark C, et al. (2007) From selection to complementarity: shifts in the causes of biodiversity-productivity relationships in a long-term biodiversity experiment. Proceedings of the Royal Society of London Series B-Biological Sciences 274: 871–876.
3. Spehn EM, Hector A, Joshi J, Scherer-Lorenzen M, Schmid B, et al. (2005) Ecosystem effects of biodiversity manipulations in European grasslands. Ecological Monographs 75: 37–63.
4. Tilman D, Reich PB, Knops JM (2006) Biodiversity and ecosystem stability in a decade-long grassland experiment. Nature 441: 629–632.
5. Van Ruijven J, Berendse F (2009) Long-term persistence of a positive plant diversity-productivity relationship in the absence of legumes. Oikos 118: 101–106.
6. Allan E, Weisser W, Weigelt A, Roscher C, Fischer M, et al. (2011) More diverse plant communities have higher functioning over time due to turnover in complementary dominant species. Proceedings of the National Academy of Sciences.
7. Reich PB, Tilman D, Isbell F, Mueller K, Hobbie SE, et al. (2012) Impacts of Biodiversity Loss Escalate Through Time as Redundancy Fades. Science 336: 589–592.
8. Burdon JJ (1987) Diseases and plant population biology: Cambridge University Press.

9. Mundt CC (2002) Use of multiline cultivars and cultivar mixtures for disease management. Annual Review of Phytopathology 40: 381–410.
10. Schenk HJ (2006) Root competition: beyond resource depletion. Journal of Ecology 94: 725–739.
11. Singh HP, Batish DR, Kohli RK (1999) Autotoxicity: Concept, organisms, and ecological significance. Critical Reviews in Plant Sciences 18: 757–772.
12. Bonanomi G, Rietkerk M, Dekker SC, Mazzoleni S (2008) Islands of fertility induce co-occurring negative and positive plant-soil feedbacks promoting coexistence. Plant Ecology 197: 207–218.
13. Petermann JS, Fergus AJF, Turnbull LA, Schmid B (2008) Janzen-Connell effects are widespread and strong enough to maintain diversity in grasslands. Ecology 89: 2399–2406.
14. Van der Putten WH, Van Dijk C, Peters BAM (1993) Plant-specific soil-borne diseases contribute to succession in foredune vegetation. Nature 362: 53–56.
15. Diez JM, Dickie I, Edwards G, Hulme PE, Sullivan JJ, et al. (2010) Negative soil feedbacks accumulate over time for non-native plant species. Ecology Letters 13: 803–809.
16. Schnitzer SA, Klironomos JN, HilleRisLambers J, Kinkel LL, Reich PB, et al. (2011) Soil microbes drive the classic plant diversity-productivity pattern. Ecology 92: 296–303.
17. Mordecai EA (2011) Pathogen impacts on plant communities: unifying theory, concepts, and empirical work. Ecological Monographs 81: 429–441.

18. Maron JL, Marler M, Klironomos JN, Cleveland CC (2011) Soil fungal pathogens and the relationship between plant diversity and productivity. Ecology Letters 14: 36–41.

19. Casper BB, Castelli JP (2007) Evaluating plant-soil feedback together with competition in a serpentine grassland. Ecology Letters 10: 394–400.

20. Van der Heijden MGA, Bardgett RD, Van Straalen NM (2008) The unseen majority: soil microbes as drivers of plant diversity and productivity in terrestrial ecosystems. Ecology Letters 11: 296–310.

21. Bever JD, Westover KM, Antonovics J (1997) Incorporating the soil community into plant population dynamics: the utility of the feedback approach. Journal of Ecology 85: 561–573.

22. Bartelt-Ryser J, Joshi J, Schmid B, Brandl H, Balser T (2005) Soil feedbacks of plant diversity on soil microbial communities and subsequent plant growth. Perspectives in Plant Ecology Evolution and Systematics 7: 27–49.

23. van der Putten WH, Bardgett RD, Bever JD, Bezemer TM, Casper BB, et al. (2013) Plant-soil feedbacks: the past, the present and future challenges. Journal of Ecology 101: 265–276.

24. Mitchell CE, Tilman D, Groth JV (2002) Effects of grassland plant species diversity, abundance, and composition on foliar fungal disease. Ecology 83: 1713–1726.

25. Mwangi PN, Schmitz M, Scherber C, Roscher C, Schumacher J, et al. (2007) Niche pre-emption increases with species richness in experimental plant communities. Journal of Ecology 95: 65–78.

26. Hendriks M, Mommer L, de Caluwe H, Smit-Tiekstra AE, van der Putten WH, et al. (2013) Independent variations of plant and soil mixtures reveal soil feedback effects on plant community overyielding. Journal of Ecology 101: 287–297.

27. Kulmatiski A, Beard KH, Heavilin J (2012) Plant-soil feedbacks provide an additional explanation for diversity-productivity relationships. Proceedings of the Royal Society B-Biological Sciences 279: 3020–3026.

28. Wagg C, Jansa J, Stadler M, Schmid B, van der Heijden MGA (2011) Mycorrhizal fungal identity and diversity relaxes plant-plant competition. Ecology 92: 1303–1313.

29. Latz E, Eisenhauer N, Rall BC, Allan E, Roscher C, et al. (2012) Plant diversity improves protection against soil-borne pathogens by fostering antagonistic bacterial communities. Journal of Ecology 100: 597–604.

30. de la Pena E, de Clercq N, Bonte D, Roiloa S, Rodriguez-Echeverria S, et al. (2010) Plant-soil feedback as a mechanism of invasion by Carpobrotus edulis. Biological Invasions 12: 3637–3648.

31. Klironomos JN (2002) Feedback with soil biota contributes to plant rarity and invasiveness in communities. Nature 417: 67–70.

32. Suding KN, Harpole WS, Fukami T, Kulmatiski A, MacDougall AS, et al. (2013) Consequences of plant-soil feedbacks in invasion. Journal of Ecology 101: 298–308.

33. Hector A, Bazeley-White E, Loreau M, Otway S, Schmid B (2002) Overyielding in grassland communities: testing the sampling effect hypothesis with replicated biodiversity experiments. Ecology Letters 5: 502–511.

34. Heisse K, Roscher C, Schumacher J, Schulze ED (2007) Establishment of grassland species in monocultures: different strategies lead to success. Oecologia 152: 435–447.

35. Dimitrakopoulos PG, Schmid B (2004) Biodiversity effects increase linearly with biotope space. Ecology Letters 7: 574–583.

36. Hille Ris Lambers J, Harpole SW, Tilman D, Knops J, Reich PB (2004) Mechanisms responsible for the positive diversity-productivity relationship in Minnesota grasslands. Ecology Letters 7: 661–668.

37. Roscher C, Schumacher J, Weisser WW, Schmid B, Schulze ED (2007) Detecting the role of individual species for overyielding in experimental grassland communities composed of potentially dominant species. Oecologia 154: 535–549.

38. Troumbis AY, Dimitrakopoulos PG, Siamantziouras ASD, Memtsas D (2000) Hidden diversity and productivity patterns in mixed Mediterranean grasslands. Oikos 90: 549–559.

39. Van Ruijven J, Berendse F (2003) Positive effects of plant species diversity on productivity in the absence of legumes. Ecology Letters 6: 170–175.

40. Tilman D, Knops J, Wedin D, Reich P, Ritchie M, et al. (1997) The influence of functional diversity and composition on ecosystem processes. Science 277: 1300–1302.

41. Marquard E, Weigelt A, Roscher C, Gubsch M, Lipowsky A, et al. (2009) Positive biodiversity-productivity relationship due to increased plant density. Journal of Ecology 97: 696–704.

42. Roscher C, Scherer-Lorenzen M, Schumacher J, Temperton VM, Buchmann N, et al. (2011) Plant resource-use characteristics as predictors for species contribution to community biomass in experimental grasslands. Perspectives in Plant Ecology Evolution and Systematics 13: 1–13.

43. Isbell FI, Polley HW, Wilsey BJ (2009) Biodiversity, productivity and the temporal stability of productivity: patterns and processes. Ecology Letters 12: 443–451.

44. Van Ruijven J, Berendse F (2007) Contrasting effects of diversity on the temporal stability of plant populations. Oikos 116: 1323–1330.

45. Roscher C, Weigelt A, Proulx R, Marquard E, Schumacher J, et al. (2011) Identifying population- and community-level mechanisms of diversity-stability relationships in experimental grasslands. Journal of Ecology 99: 1460–1469.

46. Roscher C, Schumacher J, Baade J, Wilcke W, Gleixner G, et al. (2004) The role of biodiversity for element cycling and trophic interactions: an experimental approach in a grassland community. Basic and Applied Ecology 5: 107–121.

47. Weigelt A, Marquard E, Temperton VM, Roscher C, Scherber C, et al. (2010) The Jena Experiment: six years of data from a grassland biodiversity experiment. Ecology 91: 930.

48. Loreau M, Hector A (2001) Partitioning selection and complementarity in biodiversity experiments. Nature 412: 72–76.

49. Marquard E, Weigelt A, Temperton VM, Roscher C, Schumacher J, et al. (2009) Plant species richness and functional composition drive overyielding in a 6-year grassland experiment. Ecology 90: 3290–3302.

50. Roscher C, Temperton VM, Scherer-Lorenzen M, Schmitz M, Schumacher J, et al. (2005) Overyielding in experimental grassland communities - irrespective of species pool or spatial scale. Ecology Letters 8: 419–429.

51. Bell T, Freckleton RP, Lewis OT (2006) Plant pathogens drive density-dependent seedling mortality in a tropical tree. Ecology Letters 9: 569–574.

52. Kulmatiski A, Beard KH, Stevens JR, Cobbold SM (2008) Plant-soil feedbacks: a meta-analytical review. Ecology Letters 11: 980–992.

53. Toft CA, Shea PJ (1983) Detecting community-wide patterns. Estimating power strengthens statistical inference. American Naturalist 122: 618–625.

54. Chesson P (2000) Mechanisms of maintenance of species diversity. Annual Review of Ecology and Systematics 31: 343-+.

16

Human Adaptive Behavior in Common Pool Resource Systems

Gunnar Brandt[1]*, Agostino Merico[1,2], Björn Vollan[3,4], Achim Schlüter[3,5]

1 Systems Ecology, Leibniz Center for Tropical Marine Ecology, Bremen, Germany, **2** School of Engineering and Science, Jacobs University, Bremen, Germany, **3** Institutional & Behavioural Economics, Leibniz Center for Tropical Marine Ecology, Bremen, Germany, **4** Institute of Public Finance, University of Innsbruck, Innsbruck, Austria, **5** School of Humanities and Social Sciences, Jacobs University, Bremen, Germany

Abstract

Overexploitation of common-pool resources, resulting from uncooperative harvest behavior, is a major problem in many social-ecological systems. Feedbacks between user behavior and resource productivity induce non-linear dynamics in the harvest and the resource stock that complicate the understanding and the prediction of the co-evolutionary system. With an adaptive model constrained by data from a behavioral economic experiment, we show that users' expectations of future pay-offs vary as a result of the previous harvest experience, the time-horizon, and the ability to communicate. In our model, harvest behavior is a trait that adjusts to continuously changing potential returns according to a trade-off between the users' current harvest and the discounted future productivity of the resource. Given a maximum discount factor, which quantifies the users' perception of future pay-offs, the temporal dynamics of harvest behavior and ecological resource can be predicted. Our results reveal a non-linear relation between the previous harvest and current discount rates, which is most sensitive around a reference harvest level. While higher than expected returns resulting from cooperative harvesting in the past increase the importance of future resource productivity and foster sustainability, harvests below the reference level lead to a downward spiral of increasing overexploitation and disappointing returns.

Editor: Angel Sánchez, Universidad Carlos III de Madrid, Spain

Funding: This research was fully funded by the Leibniz Center for Tropical Marine Ecology (www.zmt-bremen.de). The funders had no role in study design, data collection and analysis, decision to publish, or preparation of the manuscript.

Competing Interests: The authors have declared that no competing interests exist.

* E-mail: gunnar.brandt@zmt-bremen.de

Introduction

Many social-ecological systems (SESs) that comprise a common pool resource (CPR) face the problem of overexploitation, because it is very costly, albeit not impossible, to exclude users from subtracting resource units [1–6]. Resource appropriation in such SESs often produces benefits for the individual, while all share the costs. This gives users an obvious incentive to maximize their harvest, thus, preventing cooperation and sustainability [7]. There is, however, compelling evidence both from economic experiments [8–11] and from real systems [12–15] that under certain conditions users may overcome the egoistic temptation of maximizing individual profits.

The decision to forgo part of a possible harvest from a renewable CPR is, particularly at low resource levels, an investment into future productivity at the cost of reduced short-term returns [16]. Because such an investment into the future always comes with uncertainties and because of the human preference for proximate returns, users discount potential future pay-offs [11,17,18]. According to the standard discounted utility model, rational users integrate all current and expected future returns after discounting them at a constant rate [17]. If discounted future benefits are large enough, users are willing to forego current benefits and cooperate [8,10,19].

Users of CPRs face highly uncertain decisions for two reasons. First, many ecological systems are characterized by high intrinsic variability, which complicates predictability [20]. Second, har-

vesting itself affects the stock and eventually the productivity of the resource. By this means, the harvest behavior may also influence the weight users assign to future resource productivity giving rise to a trade-off between the harvest behavior and the expected future returns [21]. The key to sustainability in many real SES is therefore to enhance the certainty of receiving the future benefits of cooperation, a goal achieved best by creating an institutional environment that is capable of accounting for the specific characteristics of the system under consideration [21–23]. Change, whether social or ecological, may, however, overstrain also robust institutions, when it is too rapid for successful adaptation [24].

Studies of real resource-user systems often disregard these close links between harvest behavior, resource dynamics, and future certainty and focus on either ecological or social aspects. The two sub-systems are described on different levels of detail, which hinders an integrated understanding of their coupled dynamics [25,26]. Moreover, observational or experimental data from CPR systems usually cover only a short period of time or comprise many confounding factors and hence do not allow observing and understanding temporal changes. The quantitative relationship and the feedbacks between resource productivity and user behavior remain largely unknown. Consequently, regime shifts or collapses observed in overexploited ecological resources are still not fully understood [27,28].

A novel approach for understanding the combined dynamics of a CPR and the users is to reduce the number of confounding factors by studying the system under controlled laboratory conditions. In a recent study, Janssen et al. [9] presented such a computer-based laboratory experiment, in which a group of five users could harvest continuously from a renewable CPR. Each group played six consecutive rounds of $240s$ with a change of the treatment after the third round. In three of the six rounds neither communication nor punishment were possible (a treatment labeled as NCP), while in the other three rounds users could coordinate resource extraction using either communication (C), punishment (P), or a combination of both (CP).

Although the composition of each group was fixed and all properties of the game except the treatments were identical, the experimental results revealed a great variability of harvest behaviors within rounds, between rounds with the same treatment, as well as between different treatments. Users realized highest total harvests (H_{tot}) when they cooperated and allowed the CPR to grow and to produce more resource units at the beginning of a round. Communication, punishment (albeit to a lesser extend), and the total harvest realized in previous rounds influenced the harvest behavior of the users and their returns [9].

We combine here Janssen et al.'s experimental data with an adaptive model to identify the main drivers of the co-evolution of the users' harvest behavior and the CPR. Our model is based on a mechanistic trade-off between the current harvest and the discount factor of future productivity. The trade-off accounts for the effect of resource exploitation on the certainty of future returns and reflects the central decision users face while harvesting from a renewable resource with a density-dependent growth. While maximizing the current harvest reduces the certainty of future returns, because the resource stock may decline significantly or even collapse as a consequence of intensive exploitation, lower current harvests enhance the chances of higher productivities and hence higher returns in the future.

Our model simulates a renewable CPR (R, Equation 1) with a fixed number of users (C), who realize a harvest (H, Equation 9) by adopting a variable harvest strategy. We define the harvest strategy as a continuous behavioral trait (x, Equation 5) that determines the harvest rate of the users. x adapts to changes in 1) the current harvest opportunity and 2) the discounted future productivity, which we consider to be equivalent to the potential future returns. The CPR grows at a density-dependent logistic growth rate (r_r, Equation 2) and users subtract variable amounts of resource units according to a Monod-type harvest rate (r_h, Equation 3). Changes in x alter r_h via the half-saturation constant (K_S, Equation 4), but also affect the discounted future productivity of the resource (r_f, Equation 7).

More precisely, harvesting becomes less intense with increasing x, while the discount factor for future productivity (w, Equation 8) rises. The discount factor w, which is a function of the time horizon and the harvest trait, can vary between 0 and the maximum discount factor ($w^{\star}$). The parameter $w^{\star}$ sets an upper limit for the weight of future productivity and represents the maximum level of certainty that is sustained by the social system, i.e. by the rules of the game, the institutions, and the experiences of the users. It is constant on short time-scales, because rules and institutions usually change slower than the fastest processes in ecological or social systems [24], but may vary between simulations accounting for different institutional environments or different harvest experiences.

Users maximize the net present value, which is the sum of the current pay-off and all discounted future pay-offs for given maximum discount factor and CPR level. Following adaptation models of continuous traits [29–32], the temporal change of the harvest trait x is proportional to the gradient of the fitness function

F, which is the sum of current and discounted future pay-offs ($r_h + r_f$) and the costs for optional punishment (r_p, Equation 6). By adjusting x, users change their harvest behavior to increase their fitness.

In behavioral economic experiments, user behavior is measured as cooperation. Cooperation is typically expressed as a dimensionless number between 0 and 1 and determined by the user's investment relative to a potential maximum value. Following this approach, we define the average cooperation of the group (φ) by the normalized foregone harvest (Equation 10), which is the amount of resource units that the users decide not to harvest divided by the maximum possible harvest.

In our model, users are not resolved as indidivuals. By contrast, the group of users is considered as a single adaptive entity, and the state variables of the model describe the dynamics of average group properties. The model hence corresponds to the typical resolution of observational data from real SESs and does not require detailed assumptions on the behavior of each individual in the system.

While Janssen et al. [9] focused on the statistical analysis of outcomes in terms of total harvest, our aim is to find a mechanistic explanation for the dynamics of the coupled system and the observed differences between rounds. Our major assumption is that the observed variability in harvest behavior and cooperation of users is caused by a trade-off between the current harvest and the discounted future productivity, which is mainly driven by differing maximum discount factors between rounds. Therefore, we 1) constrain the proposed trade-off by experimental data [9], 2) study the influence of the social environment, which is represented by a single parameter (the maximum discount factor $w^{\star}$), on the co-evolution of the user-CPR system, and 3) assess the effect of previous experience on the users' perception of future certainty.

Results and Discussion

Trade-off between Harvest Behavior and Future Expectations

The expected future pay-offs of the CPR users were not measured directly during the experiments of Janssen et al. [9]. We instead use the relative resource productivity $P^{\star}$ as a qualitative indication of the group's expectations to constrain our model. $P^{\star}$ is defined as the cumulated resource productivity from the current time to the end of the experiment normalized to the resource level at the beginning of each round. In other words, $P^{\star}$ expresses the future productivity as a fraction of the initial resource level. Correspondingly, the inverse of the half-saturation constant Ks^{-1}, calculated from the experimental time-series of the harvest rate, is proportional to the resource affinity and used here equivalently. A high affinity value (or a low K_S) indicates aggressive harvesting already at low resource levels, while at low resource affinities users reach near maximum harvest rates only at high resource levels.

We discover a strong trade-off between $P^{\star}$ and Ks^{-1} and a high variability of these two variables in the experimental data (Figure 1C). More specifically, in the experiment, relative resource productivities significantly exceeding 1.0 only occur at low resource affinities ($K_S^{-1} < 0.005N^{-1}$). In contrast, high resource affinities ($K_S^{-1} > 0.005N^{-1}$) lead to large current harvests, while limiting the production of new resource units to values of $P^{\star} < 0.5$. Similar to known relationships in real SESs (cf. Figure 2 in [26]), this trade-off is highly non-linear and introduces a tipping point to the system that clearly separates the effects of sustainable use from overexploitation. Users, thus, face the decision of increasing either short-term benefits or the long-term resource productivity [33].

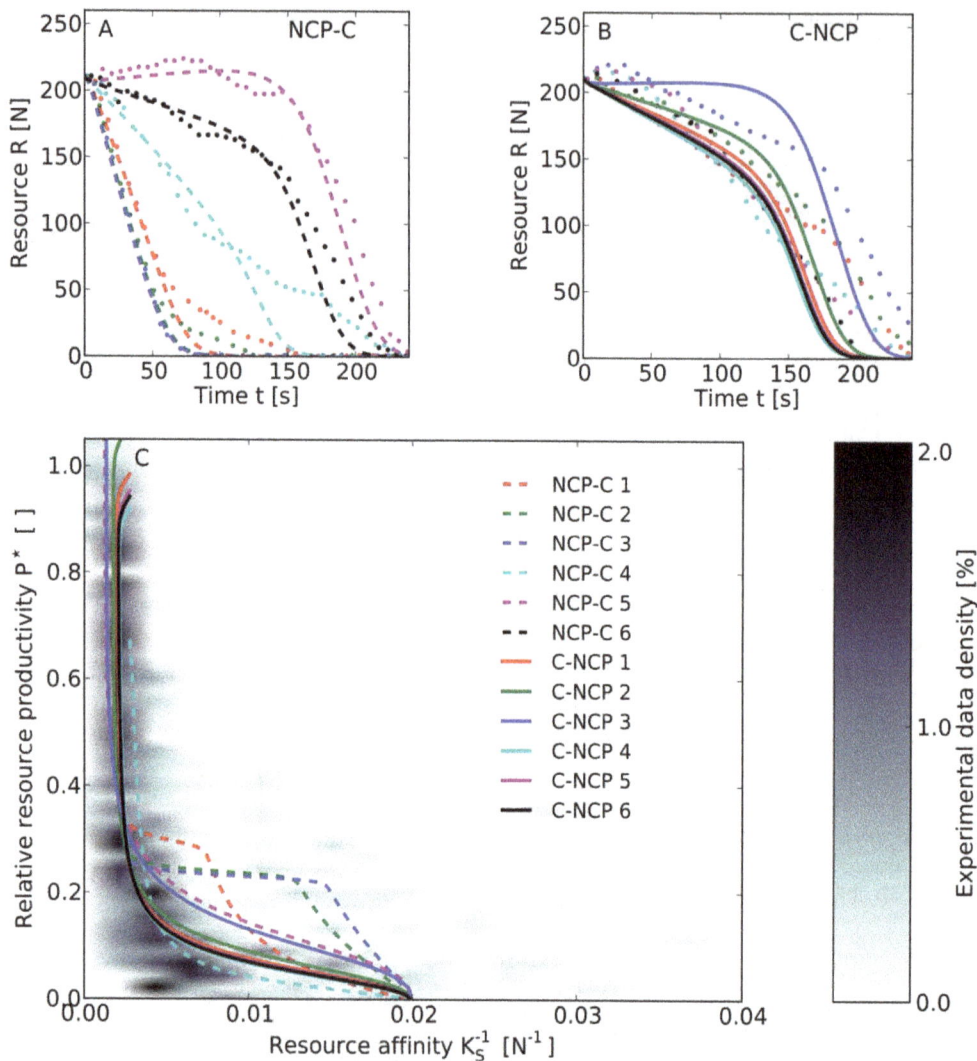

Figure 1. Trade-off between consumer's resource affinity and the productivity of a renewable resource. A–B, Resource level in six rounds of a computer-based laboratory game with five users and different treatments (dotted lines indicate experimental data [9]). While in the first three rounds of **A** neither communication nor punishment was possible (NCP-C 1–3), users could communicate in subsequent rounds (NCP-C 4–6). In the treatment C-NCP (**A**), three rounds with communication (C-NCP 1–3) were followed by three NCP rounds (C-NCP 4–6). Set-ups of model runs (dashed and solid lines in **A** and **B**, respectively) only differ in the the maximum discount factor w^* (see Equation 8, NCP-C: [4.4, 1.9, 1.4, 11.7, 32.5, 20.4], C-NCP: [18.0, 20.4, 29.3, 16.5, 17.3, 17.0]). **C,** Phase plot of the users' resource affinity, here defined as the inverse of the half-saturation constant K_S (see Equations 3 and 4), and the relative resource productivity P^*, defined as the resource productivity from the current point of time to the end of a round normalized to the initial resource level. The shaded area shows the density distribution of the experimental data from the two treatments shown in **A** and **B**. Solid lines indicate the trade-off between resource affinity and potential future harvest from the resource system in corresponding model results.

We calibrated our model to match the distribution of the experimental data and adjusted only the parameter w^* between rounds (Figure 1A and B). The model trajectories reveal the continuous change of user behavior over the course of the different rounds (Figure 1C). Starting from low affinities all model simulations end with $P^*=0$ and $Ks^{-1}=0.02\mathrm{N}^{-1}$, which represents the highest possible harvest rate and the complete exhaustion of the resource at the end of all rounds. While all rounds end similarly, they differ in the trajectories that lead to the exhaustion of the resource. When communication is possible (NCP-C 4–6, C-NCP 1–3), maximum P^* values are considerably higher than in NCP-rounds with no prior experience of communication (NCP-C 1–3). In contrast, users increase the

resource affinity in NCP-C 1–3 right from the start (cf. Figure 1A and C) and by doing so avert high resource productivities.

Effect of Different w^* on the Temporal Dynamics of the User-CPR System

The different values of the maximum discount factor w^* can be attributed to the changes in the social environment of the users, because the simulated rounds differ only by the available treatments and the history of previous round, whereas the resource characteristics and the composition of the groups of users were identical.

If the future is irrelevant to users and future returns are disregarded ($w^*=0$, Figure 2A and D), cooperation levels deteriorate within the first 30s of the simulation. In this case, the

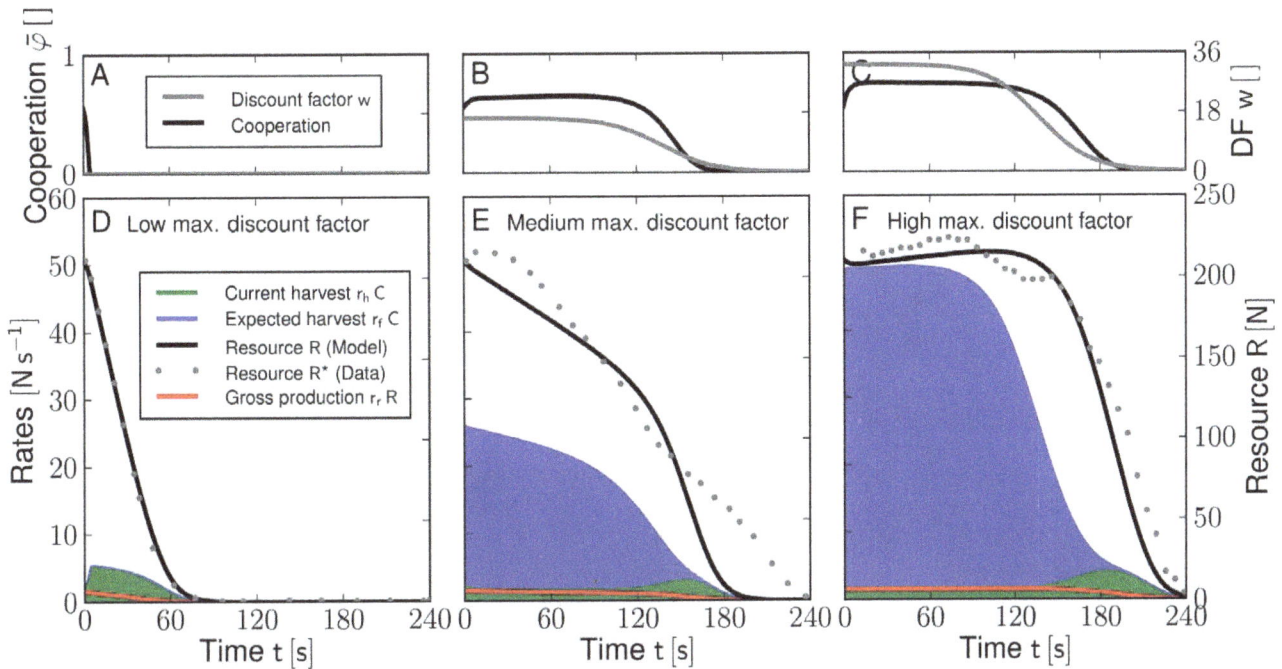

Figure 2. Time evolution of cooperation, harvest rates and a renewable resource for three different levels of future certainty.
Increasing the maximum discount factor $w^\star$ (Equation 8) lowers the current harvest $r_h C$ (Equation 3, lower panels **d–f**, green shaded area), but raises the future resource productivity $r_f C$ that is considered by the users (Equation 7, blue shaded area). Cooperation φ decreases sharply when future pay-offs are ignored (**a**, $w^\star = 0.0$) causing an immediate resource collapse (**d**, black solid line (model) and gray dots (experimental data from [9])). Larger values of $w^\star$ (**b**, $w^\star = 16.5$ and **c**, $w^\star = 32.5$) result in higher cooperation and reduce the current harvest as resource users account for a much higher proportion of future productivity (**e** and **f**). Resource collapse occurs later and the extended period of sustainable resource use leads to significantly higher total harvests (cf. Figures. S3a and S1). Red lines in panels **a–c** indicate the temporal evolution of the discount factor w.

current harvest rate significantly exceeds the growth rate of the CPR and the unsustainable use leads to a collapse within the first $100 s$ and to a poor total harvest ($H_{tot} < 300N$, Figure 3A). Increasing the importance of the future, that is increasing $w^\star$, results in higher cooperation over longer periods of time and slows down (Figure 2B and E) or even reverses overexploitation (Figure 2C and F). Towards the end of all simulations, however, w declines with the remaining time of the experiment causing an erosion of cooperation that eventually triggers the collapse of the resource (Figure 2A–C), because there is no potential future productivity to account for in a finite game.

Users continuously adjust their harvest strategy according to changing present harvest opportunities and expected future pay-offs. Therefore, cooperation and sustainable harvesting become rational when the discounted total harvest for one strategy is higher than for other strategies [34]. By treating resource users as an adaptive entity, our model unveils their great behavioral variability and the smooth transition from a sustainable to unsustainable resource use. These results support studies [35] that question stable norms of cooperation derived from "one-shot" field experiments [36–39]. The observed behavioral variability among CPR users with identical cultural background corroborates our assumption that users adapt their harvest behavior to the properties of the social and ecological environment [13,35,40].

Relation between the Maximum Discount Factor and the Total Harvest

The harvest behavior of CPR users and hence the outcome of the artificial commons vary greatly between rounds (cf. Figure S1 and S2 for the results of all rounds). A quantitative measure that

may explain those differences of the total harvest H_{tot} is the maximum discount factor $w^\star$. The total harvest indicates the success of the group's behavior and is positively related to $w^\star$ (Figure 3A), because in the model the productivity increases with R for the range of resource levels observed in the experiment ($R < 0.5K = 420.5N$). In other words, a high $w^\star$ leads to harvest rates below productivity, i.e. sustainability, at the beginning of a round and eventually to high total harvests. Given the constraints of the experiment, the highest H_{tot} of $520N$ is therefore realized at the highest value of $w^\star = 41.8$ (Figure 3A).

By only adjusting $w^\star$ to match the experimental results, we assume that differences between rounds are mainly caused by variations in the users' perception of future certainty. Allowing for variations also in the parameters α and β reduces the error between model and experimental data indicating that users also adjust the temporal dynamics of the harvest from round to round (Figure S4 and Table S1). However, the results of a systematic sensitivity analysis (Figure S5) confirm the high sensitivity of the model results to changes in $w^\star$ and corroborate the choice of $w^\star$ as the only free parameter explaining the observed differences between the rounds of the experiment.

Effect of Different Treatments on the Maximum Discount Factor

Our analysis reveals that the variations between rounds in the perception of future certainty, represented by $w^\star$ in the model, are determined by a combination of factors, including 1) treatment in the current round, 2) prior experience from rounds with the same treatment, and 3) possible exposure to a different treatment in the past.

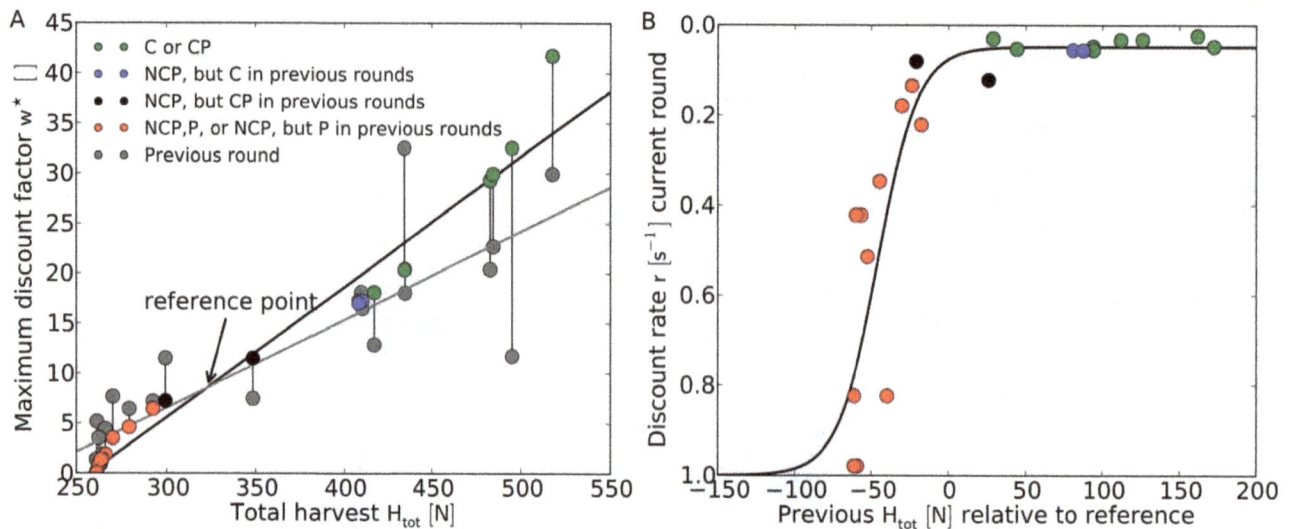

Figure 3. Outcomes of the common pool resource system. The total harvest H_{tot} is closely related to the certainty of the future, here expressed as the maximum discount factor w^* (Equation 8, colored dots). Only data from rounds that were preceded by a round with identical treatment (rounds 2, 3, 5, and 6) were included. **a,** Communication with or without punishment (CP and C, green dots) is essential to establish high w^* and to increase the H_{tot}. Punishment (P), even if experienced only in previous rounds, or the lack of communication and punishment (NCP), keep H_{tot} below the reference point (red dots). Blue and black dots indicate NCP rounds in which either communication or communication and punishment were available in previous rounds. The lines connecting two dots show the change of w^* between a current and a preceding round (gray dots) with the same treatment (at the current round's H_{tot}). The intersection of the regression lines of previous (gray line) and current (black line) discount factors reveals that the value of w^* increases from round to round if the group manages to establish a $w^* > 8.5$ (corresponding to $H_{tot} = 323$) in the preceding round, or decreases if $w^* < 8.5$. This intersection marks the sustainability threshold between positive and negative feedbacks in the system and sets the reference point for the users' expectations. **b,** H_{tot} in a preceding round determines the discount rate r, derived from the relation $r = (w^* + 1)^{-1}$. The solid, sigmoidal line indicates a least-squares fit to a logistic equation (root mean square error $RMSE = 0.155 \mathrm{s}^{-1}$).

Sustainable harvest strategies, characterized by high H_{tot} exceeding $400N$, are associated with rounds in which either communication is possible or users had experienced communication in previous rounds (Figure 3A). On the contrary, in the absence of communication or with prior experience of punishment, users harvest unsustainably throughout the simulation and, thus, realize poor harvests (below $300N$). Rounds in which the positive and negative effects of previous communication and punishment are balanced represent a transition between clearly separated strategies of successful and unsuccessful harvests (black dots in Figure 3A).

The impact of previous experience of punishment on the maximum discount factor and therefore on the total harvest becomes clear when comparing the outcomes of NCP rounds that were preceded by differing treatments. While in rounds 4 and 5 of the C-NCP treatment users manage to sustain a $w^* > 15$, the maximum discount factor drops considerably to values around 10 in corresponding rounds of the CP-NCP treatment (Figure 3A). In our model, this indirect effect on the harvest behavior is much stronger than any direct effect of punishment. As a tool to enhance the confidence of users into the future, punishment has, different from communication, no or even negative effects beyond the period of its availability [41]. Punishment alters the expectations of future returns, but communication is clearly more effective in raising the users' maximum discount factor and eventually in establishing cooperation [9,42,43]. In studies of real social-ecological systems, leadership significantly influences the successful management of the commons [12,13,44]. Supported by Janssen et al.'s observations [9], we argue that communication enables negotiation and promotes leadership in the artificial environment of this simple, computer-based CPR system.

Changes of the Maximum Discount Factor within One Treatment

A feedback between w^* and H_{tot} links the outcome of previous rounds with the same treatment to the current harvest strategy. Our results suggest that the group's total harvest will increase further in the following round with identical treatment if H_{tot} is above a certain threshold ($323 > N$, equivalent to $w^* > 8.5$). Values below this reference lead to a further deterioration of the maximum discount factor and diminish the group's total harvest in most rounds.

The reference value hence marks a sustainability threshold for the system. This feedback between rounds is similar to the mechanism proposed by Fehr & Gächter [45], who explained the decay of cooperation in a public goods game as a feedback loop of disappointed expectations that leads to lower and lower endowments of the players of the experiment.

The discount rates r, which were derived from w^* assuming exponential discounting [17], are related to the H_{tot} of the previous round by a sigmoid function (Figure 3B). We suggest that the threshold determined in Figure 3A and the discount rate correspond respectively to the reference point of the value function and to the psychological value of an outcome in prospect theory [46]. Prospect theory, which is based on gains and losses rather than on absolute outcomes, explains the discrepancies between economic rationality and observed human behavior. According to prospect theory the perceived value of an outcome does not depend linearly on its economic value. Instead, it is an asymmetric, sigmoidal function of gains and losses with respect to a reference point, the value function, which can be influenced by expectations or the current status. Furthermore, the weight humans associate to an uncertain outcome is related, but not equal, to the corresponding probability, because the human ability to objectively estimate probabilities, in particular those of rare events, is limited [46,47].

We, thus, argue that users adjust the discount rates according to the psychological value of the total harvest realized in the previous round with identical treatment. Users are highly sensitive to losses, which are caused in this system by the lack of communication or by punishment. Losses with respect to the reference value in the previous round lead to a severely impaired perception of future certainty. In contrast, small gains can be sufficient for users to adopt lower discount rates in following rounds (Figure 3B). Note, however, that we excluded the first rounds with a new treatment from our analysis, because a drastic change of the institutional environment does clearly affect future expectations and obfuscates the relationship between harvest experience and discounting as shown in Figures 3A and B.

Conclusions

We have shown that it is possible to understand main features of a CPR and the harvest dynamics of a simplified SES by reducing the social environment to its impact on the perceived future certainty of the users. Our approach extends classic models of maximum sustainable [48,49] or maximum economic [34] yield by introducing a behavioral trait that accounts for the mutual dependency of current behavior and future expectations. The social environment including the first-tier variables "users" and "governance system" of Ostrom's framework for the analysis of SESs [23] is obviously more complex than assumed here and exhibits a dynamics of its own (indicated in Figure 3A and 3B). Despite this well-recognized complexity, the harvest behavior of CPR users can be analyzed, understood, and even roughly predicted with a simple model. Our model is able to describe the co-evolution of the renewable CPR and the adaptive harvest behavior of the users following a mechanistic trade-off, a disregarded feature in classic harvest models [16]. Furthermore, by showing the influence of user experience on the perception of future certainty, we presented an approach to understand the observed variability of user behavior in apparently similar or identical situations.

We conclude that unsustainable harvest leads to reduced discounted future pay-offs and low cooperation in two ways, first, as a consequence of reduced resource productivity and, second, as a consequence of a deteriorating discount factor. Once the temporal gradient of both terms has turned negative, it is difficult for users to escape from the downward vortex of decreasing expectations and diminishing pay-offs. This feedback works also in the opposite direction towards sustainable harvest strategies, high pay-offs, and sustained cooperation among resource users. Our findings illustrate the behavioral variability of users that act rationally according to their current opportunities and their perception of future returns. By this means, our approach opens up a perspective for predicting dynamics and identifying tipping points of coupled user-resource systems.

Methods

Our adaptive model consists of three ordinary differential equations and describes the combined dynamics of the resource R, the harvest H, and the harvest trait x for a constant number of users C.

The Renewable Resource

R is changing over time t according to the difference between new production and harvest

$$\frac{dR}{dt} = r_r R - r_h C, \qquad (1)$$

where r_r is the productivity of R

$$r_r = \mu_R \left(1 - \frac{R}{K}\right), \qquad (2)$$

with μ_R and K indicating the maximum specific growth rate and the carrying capacity of the resource system, respectively. Resource growth is hence logistic with highest growth rates at $R = 0.5K$ and declining rates towards $R = 0$ as well as towards $R = K$ [16].

Adaptive Harvesting

The current harvest rate r_h is based upon Monod kinetics

$$r_h = \mu_C \left(\frac{R}{R + K_S}\right), \qquad (3)$$

with μ_C representing the maximum specific harvest rate and K_S the half-saturation constant. The harvest rate r_h is, thus, insensitive to changes in R for $R >> K_S$, but sensitive for $R < K_S$. In our model, K_S is variable and responds to changes in the harvest trait x.

$$K_S = \hat{K}_S + x^2 K_S^\star \qquad (4)$$

where $\hat{K}_S$ and $K_S^\star$ denote the minimum and the variable part of K_S, respectively. Using an adaptive modeling approach [29–32], the temporal change of x is proportional to the fitness gradient F

$$\frac{dx}{dt} = \kappa \frac{\partial F}{\partial x}, \qquad (5)$$

with $F = r_h + r_p + r_f$ and κ denoting the rate constant of the adaptive process [29]. κ hence parameterizes the speed of the adaption process, i.e. the speed of learning, of the group of users over the course of a round. While the punishment rate

$$r_p = \gamma e^{\frac{-(\hat{x}-x)^2}{2\sigma_p^2}} R K^{-1}, \qquad (6)$$

is only available in some rounds, the discounted future productivity

$$r_f = w r_r R C^{-1}, \qquad (7)$$

which stands for the future resource productivity expected by the users, is considered in all rounds. In Equation 6 γ is the specific punishment rate, $\hat{x}$ is the trait value of the punishment maximum, and σ_p is the punishment standard deviation. Consistent with the experimental results [9], $\hat{x}$ is set to intermediate values in the model so that punishment is applied mostly at intermediate levels of x. At high x users forgo a large fraction of their potential harvest, they cooperate, and punishment is therefore not necessary. By contrast, at low levels of x, which indicate egoistic harvest strategies and low importance of future pay-offs, users are not inclined to invest in a costly and uncertain measure that may support long-term sustainability. The discount factor w is a variable function of t and x

Table 1. Parameter values and variables (with initial conditions given in parenthesis).

Symbol	Name	Value	Unit
α	Shape parameter	0.055	$[s^{-1}]$
β	Shape parameter	137.5	$[s]$
C	Users	5	$[N]$
γ	Specific punishment rate	0.096	$[s^{-1}]$
H	Harvest	(0.0)	$[N]$
H_{tot}	Total harvest	(0.0)	$[N]$
μ_C	Max. specific harvest rate	1.37	$[s^{-1}]$
μ_R	Max. specific resource growth rate	0.0095	$[s^{-1}]$
K	Carrying capacity	841.0	$[N]$
K_S	Half-saturation const.	(320.0)	$[N]$
$K_S^\star$	Variable half-saturation const.	35.0	$[N]$
$\hat{K}_S$	Min. half-saturation const.	50.0	$[N]$
κ	xVariance of	1.25	$[]$
ϑ	Cooperation	(0.6)	$[]$
r_f	Future productivity	(−)	$[s^{-1}]$
r_h	Current harvest rate	(−)	$[s^{-1}]$
r_p	Punishment rate	(−)	$[s^{-1}]$
r_r	Resource growth rate	(−)	$[s^{-1}]$
R	Resource	(210.0)	$[N]$
σ_p	Punishment standard deviation	0.5	$[]$
t	Time	(−)	$[s]$
w	Discount factor	(−)	$[]$
$w^\star$	Max. discount factor	(roundspecific)	$[]$
x	Mean harvest trait	(3.0)	$[]$

$$w = w^\star (1 - e^{-x}) \left(1 - \frac{1}{1 + e^{\alpha(-t+\beta)}} \right), \qquad (8)$$

with $w^\star$, the maximum discount factor, representing the only free parameter between rounds, and α and β, two shape parameters, determining the decay of w as the time in a round elapses. While the time dependence of w is similar to the discounted utility model [17], the parameters α and β allow for a modification of the timing and the speed of decay of w with time. w acts here as a weight on future productivities and is connected to the harvest behavior of the users via the harvest trait x. We assume here that users estimate the future productivity of the resource at the current resource level. Hence, the trade-off between r_h and r_f emerges, because an increase of x raises the current harvest r_h, but erodes r_f by reducing w. The functional dependence of r_f and r_h on x is determined by the highly non-linear shape of the trade-off in the data (cf. Figure 1C) and constrained by the requirement that r_f and r_h may not be negative for any x. Integrating the harvest over time while accounting for possible costs for punishment gives the temporal evolution of the harvest H

$$\frac{dH}{dt} = (r_h - r_p)C. \qquad (9)$$

The total harvest realized over the 240s of an experimental round is then $H_{tot} = H(t=240)$.

Cooperation

Cooperation is a diagnostic variable in our model. It is defined by the non-realized current harvest normalized to the maximum possible harvest $r_h(x=0)$

$$\varphi = 1 - \frac{r_h(x)}{r_h(x=0)}. \qquad (10)$$

In other words, not harvesting anything results in $\varphi = 1$, whereas maximizing the current harvest rate by adopting $x=0$ leads to $\varphi = 0$. Note that considering users as an adaptive entity implies that the properties x and also φ are mean properties of the group. Unlike in similar evolutionary dynamics models, the dynamics of the trait distribution is not determined by the reproductive fitness of individuals bearing a certain trait, because we assume that the change of a behavioral trait does not require sexual reproduction [50]. In our model, the fixed group of resource users is able to quickly adapt the harvest strategy according to the state of the resource and the maximum discount factor, an assumption that is corroborated by the observed variability of harvest rates in the laboratory experiments [9]. Consistent with the published results of the experimental study, the model is not spatially explicit, because the dynamics of the spatial averages of the resource and the group's harvest in the homogeneous system can be adequately described by zero-dimensional approach.

Simulations

The parameter-set of the model was manually calibrated to fit the temporal evolution of the resource and the total harvest observed in the 36 rounds of the laboratory experiment (cf. Table 1 and Figure S2). The data [9] are averages of five or six replicates for each round. All simulations were conducted with identical initial conditions and parameter values except for the maximum discount factor $w^\star$. Changes of $w^\star$ account for all variability in the model results we show in the main text. The different treatments and the learning of the users are, thus, reduced to their impact on the expectations of future pay-off, which is represented by the maximum discount factor $w^\star$ in the model. Additional results with three variable parameters are presented in the Supporting Information (Figures S4 and S5, Text S1, and Table S1).

Supporting Information

Figure S1 Total harvest for different treatments. Comparison of total harvest H_{tot} for combinations of different treatments, namely communication (C), costly punishment (P), communication and costly punishment (CP), neither communication nor punishment (NCP). Respectively, bars and dots with error bars denote mean values and standard deviations of experimental results obtained from the laboratory study of Janssen et al. [9].

Figure S2 Temporal dynamics of the resource. Times-series of resource levels for six experiments consisting of six rounds each. Treatments are communication (C), costly punishment (P), communication and costly punishment (CP), neither communication nor punishment (NCP) and change after three rounds. Solid lines indicate model results and dotted lines indicate the experimental results [9]. Only the maximum discount factor $w^\star$ was varied between rounds to fit the experimental results, all other parameter values are reported in Table 1.

Figure S3 Model-data comparison. Comparison of all experimental data from [9] shown in Figure S2 with corresponding model data. A linear regression yields $r^2 = 0.972$.

Figure S4 Temporal dynamics of the resource. Times-series of resource levels for six experiments consisting of six rounds each. Treatments are communication (C), costly punishment (P), communication and costly punishment (CP), neither communication nor punishment (NCP) and change after three rounds. Solid lines indicate model results and dotted lines indicate the experimental results obtained from the laboratory study of Janssen et al. [9]. Only the parameters $w^\star$, α, and β vary between rounds to fit the experimental results, all other parameter values are reported in Table 1.

Figure S5 Sensitivity analysis. Sensitivity of the root mean square (RMS) error between simulated and experimental data to changes in the three parameters $w^\star$ (A), α (B), and β (C), all other parameter values are reported in Table 1. The ranges of variation

were $w^\star = [0 - 70]$, $\alpha = [0.00 - 0.04\text{s}^{-1}]$, and $\beta = [10 - 200\text{s}]$. Each of the 36 experimental resource time-series was compared to the results of 112000 model runs with unique combinations of the three variable parameters to find the optimal parameter values (cf. Figure S4 for the best results). The panels A–C show how the RMS error increases from the optimum when only one of the three parameters is varied while the other two are held constant at their optimum value.

Table S1 Root mean square error (RMSE) between experimental and simulated data for models with one ($w^\star$) and three free parameters ($w^\star$, α, and β).

Text S1 Additional model results.

Acknowledgments

We are grateful to two anonymous reviewers for their thoughtful and constructive comments, which helped to improve the manuscript considerably.

Author Contributions

Conceived and designed the experiments: GB AM. Performed the experiments: GB. Analyzed the data: GB AM. Contributed reagents/materials/analysis tools: GB. Wrote the paper: GB AM BV AS.

References

1. Pauly D, Christensen V, Dalsgaard J, Froese R, Torres Jr F (1998) Fishing down marine food webs. Science 279: 860–863.
2. Pauly D, Christensen V, Guenette S, Pitcher TJ, Sumaila UR, et al. (2002) Towards sustainability in world fisheries. Nature 418: 689–695.
3. Jackson JB, Kirby MX, Berger WH, Bjorndal KA, Botsford LW, et al. (2001) Historical overfishing and the recent collapse of coastal ecosystems. Science 293: 629–37.
4. Achard F, Eva H, Stibig H, Mayaux P, Gallego J (2002) Determination of deforestation rates of the world's humid tropical forests. Science 297: 999–1002.
5. Pandolfi J, Jackson J, Baron N, Bradbury R (2005) Are US coral reefs on the slippery slope to slime? Science 307: 1725–1726.
6. Wada Y, van Beek LPH, van Kempen CM, Reckman JWTM, Vasak S, et al. (2010) Global depletion of groundwater resources. Geophys Res Lett 37: 1–5.
7. Levin SA (2010) Crossing scales, crossing disciplines: collective motion and collective action in the Global Commons. Philos Trans R Soc London, Ser B 365: 13–8.
8. Dal Bó P, Fréchette G (2011) The evolution of cooperation in in_netely repeated games: experimental evidence. Amer Econ Rev 101: 411–429.
9. Janssen MA, Holahan R, Lee A, Ostrom E (2010) Lab experiments for the study of social-ecological systems. Science 328: 613–617.
10. Dal Bó P (2005) Experimental cooperation under the shadow of the future: evidence from infitely repeated games. Amer Econ Rev 95: 1591–1604.
11. Gintis H (2000) Beyond Homo economicus: evidence from experimental economics. Ecol Econ 35: 311–322.
12. Gutiérrez NL, Hilborn R, Defeo O (2011) Leadership, social capital and incentives promote successful fieries. Nature 470: 386–9.
13. Rustagi D, Engel S, Kosfeld M (2010) Conditional Cooperation and Costly Monitoring Explain Success in Forest Commons Management. Science 330: 961–965.
14. Dolsak N, Ostrom E, editors (2003) The Commons in the New Millennium: Challenges and Adaptations. Cambridge: The MIT Press.
15. Ostrom E (1990) Governing the Commons. Cambridge: Cambridge University Press.
16. Clark C (1976) Mathematical Bioeconomics. New York: Wiley.
17. Frederick S, Loewenstein G, O'Donoghue T (2002) Time discounting and time preference: a critical review. J Econ Lit XL: 351–401.
18. Holt Ca, Laury SK (2002) Risk aversion and incentive effects. Amer Econ Rev 92: 1644–1655.
19. Axelrod R (1984) The Evolution of Cooperation. New York: Basic Books.
20. Scheéér M, Carpenter S, Foley JA, Folke C, Walker B (2001) Catastrophic shifts in ecosystems. Nature 413: 591–6.
21. Kortenkamp KV, Moore CF (2006) Time, uncertainty, and individual differences in decisions to cooperate in resource dilemmas. Pers Soc Psychol B 32: 603–615.
22. Ostrom E (2009) A general framework for analyzing sustainability of social-ecological systems. Science 325: 419–422.
23. Ostrom E (2007) A diagnostic approach for going beyond panaceas. Proc Nat Acad Sci USA 104: 15181–15187.
24. Dietz T, Ostrom E, Stern PC (2003) The struggle to govern the commons. Science 302: 1907–12.
25. An L (2011) Modeling human decisions in coupled human and natural systems: Review of agentbased models. Ecol Modell 229: 25–36.
26. Liu J, Dietz T, Carpenter SR, Alberti M, Folke C, et al. (2007) Complexity of coupled human and natural systems. Science 317: 1513–6.
27. Pinsky ML, Jensen OP, Ricard D, Palumbi SR (2011) Unexpected patterns of fisheries collapse in the world's oceans. Proc Nat Acad Sci USA 108: 8317–8322.
28. Lees K, Pitois S, Scott C, Frid C, Mackinson S (2006) Characterizing regime shifts in the marine environment. Fish Fish 7: 104–127.
29. Abrams Pa, Matsuda H, Harada Y (1993) Evolutionarily unstable fitness maxima and stable fitness minima of continuous traits. Evol Ecol 7: 465–487.
30. Wirtz KW, Eckhardt B (1996) Effective variables in ecosystem models with an application to phytoplankton succession. Ecol Modell 92: 33–53.
31. Norberg J, Swaney DP, Dushoff J, Lin J, Casagrandi R, et al. (2001) Phenotypic diversity and ecosystem functioning in changing environments: a theoretical framework. Proc Nat Acad Sci USA 98: 11376–11381.
32. Merico A, Bruggeman J, Wirtz KW (2009) A trait-based approach for downscaling complexity in plankton ecosystem models. Ecol Modell 220: 3001–3010.
33. Janssen MA, Anderies JM (2007) Robustness trade-offs in social-ecological systems. Int J Comm 1: 43–66.
34. Grafton RQ, Kompas T, Hilborn RW (2007) Economics of overexploitation revisited. Science 318: 1601.
35. Lamba S, Mace R (2011) Demography and ecology drive variation in cooperation across human populations. Proc Nat Acad Sci USA 2011: 1–5.
36. Gächter S, Herrmann B (2009) Reciprocity, culture and human cooperation: previous insights and a new cross-cultural experiment. Philos Trans R Soc London, Ser B 364: 791–806.
37. Herrmann B, Thöni C, Gächter S (2008) Antisocial punishment across societies. Science 319: 1362–1367.
38. Henrich J (2004) Cultural group selection, coevolutionary processes and large-scale cooperation. J Econ Behav Organ 53: 3–35.
39. Henrich J, Boyd R, Bowles S, Camerer C, Fehr E, et al. (2001) In search of Homo economicus: behavioral experiments in 15 small-scale societies. Amer Econ Rev 91: 73–78.

40. Hayo B, Vollan B (2012) Group interaction, heterogeneity, rules, and co-operative behaviour: evidence from a common-pool resource experiment in South Africa and Namibia. J Econ Behav Organ 81: 9–28.

41. Gneezy URI, Rustichini A, Glaeser E, Levine D, Nyhus E, et al. (2000) A fine is a price. J Legal Stud XXIX: 1–17.

42. Rand DG, Dreber A, Ellingsen T, Fudenberg D, Nowak MA (2009) Positive interactions promote public cooperation. Science 325: 1272–5.

43. Dreber A, Rand DG, Fudenberg D, Nowak MA (2008) Winners don't punish. Nature 452: 348–51.

44. Kenward RE, Whittingham MJ, Arampatzis S, Manos BD, Hahn T, et al. (2011) Identifying governance strategies that effectively support ecosystem services, resource sustainability, and biodiversity. Proc Nat Acad Sci USA 108: 5308–12.

45. Fehr E, Fischbacher U (2003) The nature of human altruism. Nature 425: 785–791.

46. Kahneman D, Tversky A (1979) Prospect theory: an analysis of decision under risk. Econometrica 47: 263–292.

47. Tversky A, Kahneman D (1974) Judgment under uncertainty: heuristics and biases. Science 185: 1124–1131.

48. Lande R, Engen S, Saether B (1994) Optimal harvesting, economic discounting and extinction risk in uctuating populations. Nature 372: 88–90.

49. Clark C (1973) The economics of overexploitation. Science 181: 630–634.

50. Levin SA (2006) Learning to live in a global commons: socioeconomic challenges for a sustainable environment. Ecol Res 21: 328–333.

17

Caribbean-Wide, Long-Term Study of Seagrass Beds Reveals Local Variations, Shifts in Community Structure and Occasional Collapse

Brigitta I. van Tussenbroek[1]*, Jorge Cortés[2], Rachel Collin[3], Ana C. Fonseca[2], Peter M. H. Gayle[4], Hector M. Guzmán[3], Gabriel E. Jácome[3], Rahanna Juman[5], Karen H. Koltes[6], Hazel A. Oxenford[7], Alberto Rodríguez-Ramirez[8¤a], Jimena Samper-Villarreal[2], Struan R. Smith[9¤b], John J. Tschirky[10], Ernesto Weil[11]

1 Instituto de Ciencias del Mar y Limnología, Universidad Nacional Autónoma de México, Cancún, Mexico, 2 Centro de Investigación en Ciencias del Mar y Limnología (CIMAR), Universidad de Costa Rica, San Pedro, Costa Rica, 3 Smithsonian Tropical Research Institute, Panama, Republic of Panama, 4 Discovery Bay Marine Laboratory, Discovery Bay, Jamaica, 5 Institute of Marine Affairs, Trinidad, Trinidad and Tobago, 6 Office of Insular Affairs, Department of the Interior, Washington DC, United States of America, 7 CERMES, University of the West Indies, Barbados, West Indies, 8 Instituto de Investigaciones Marinas y Costeras (INVEMAR), Santa Marta, Colombia, 9 Bermuda Biological Station for Research, St. George, Bermuda, 10 Garrett Park, Maryland, United States of America, 11 Department of Marine Sciences, University of Puerto Rico, Mayaguez, Puerto Rico, United States of America

Abstract

The CARICOMP monitoring network gathered standardized data from 52 seagrass sampling stations at 22 sites (mostly *Thalassia testudinum*-dominated beds in reef systems) across the Wider Caribbean twice a year over the period 1993 to 2007 (and in some cases up to 2012). Wide variations in community total biomass (285 to >2000 g dry m^{-2}) and annual foliar productivity of the dominant seagrass *T. testudinum* (<200 and >2000 g dry m^{-2}) were found among sites. Solar-cycle related intra-annual variations in *T. testudinum* leaf productivity were detected at latitudes > 16°N. Hurricanes had little to no long-term effects on these well-developed seagrass communities, except for 1 station, where the vegetation was lost by burial below ~1 m sand. At two sites (5 stations), the seagrass beds collapsed due to excessive grazing by turtles or sea-urchins (the latter in combination with human impact and storms). The low-cost methods of this regional-scale monitoring program were sufficient to detect long-term shifts in the communities, and fifteen (43%) out of 35 long-term monitoring stations (at 17 sites) showed trends in seagrass communities consistent with expected changes under environmental deterioration.

Editor: Judi Hewitt, University of Waikato (National Institute of Water and Atmospheric Research), New Zealand

Funding: CARICOMP received support from the John D. and Catherine T. MacArthur Foundation, UNESCO Environment and Development in Coastal Regions and in Small Islands (CSI), US National Science Foundation-Division of International Programs and Division of Ocean Sciences, CARICOMP data Management Centre, Centre for Marine Sciences, University of West Indies, Jamaica, and the directors and administrators of the participating institutions. The funders had no role in study design, data collection and analysis, decision to publish, or preparation of the manuscript.

Competing Interests: The authors have declared that no competing interests exist.

* E-mail: vantuss@cmarl.unam.mx

¤a Current address: The University of Queensland, Brisbane, Australia,
¤b Current address: Bermuda Aquarium Museum and Zoo, Bermuda,

Introduction

Seagrass beds are among the most extensive shallow marine coastal habitats worldwide [1]. Their ecosystem services include sustaining diverse faunal communities [1], supporting fisheries [2], providing coastal protection through stabilization of sediments [3], cycling of nutrients [4] and carbon sequestration [5,6]. In the Caribbean, seagrasses are associated with marine/brackish protected bays and estuaries or reef systems (reef lagoons between the coastlines and the coral reefs). In reef systems, seagrass communities fulfil the above-mentioned services, and additionally provide important ecological linkages with the adjacent coral reefs and/or mangroves. Seagrass communities support the existence of coral reefs through the export of organic materials [7] and provide grazing grounds and/or nurseries for coral reef fishes and other

reef fauna [8–10]. In addition, associated calcareous macro-algae and epiphytes (algae and invertebrates) on seagrass leaves are major providers of calcium carbonate sediments [11–13].

In places where funding resources are particularly limited such as developing countries in the Caribbean, bio-indicators can provide warning of changes in the biological condition of a coastal system and are thereby valuable to natural resource managers [14]. Despite such relevance, comprehensive spatio-temporal analyses of bio-indicators in tropical countries are scarce because of the lack of long-term monitoring data. Seagrasses are widely distributed, rooted in the substrate and respond to changes in the environment in terms of morphology and population characteristics; thus they can serve as biological indicators (or bio-indicators) for assessing changes in the status of coastal systems [15]. Retreat of seagrass beds to shallower areas, with a consequent reduction in coverage or biomass has been used as an indicator of decreasing water clarity [16,17]. Seagrasses respond to nutrient enrichment physiologically by increasing N or P content [18,19], changing morphology such as leaf width [20,21] and changing biomass distribution between above- and below-ground plant parts [22,23]. Changes in water quality (clarity, salinity or nutrients) result in changes in species composition [22–24] and density of the foliar shoots of the seagrasses [25].

This study presents the results of a long-term (1993–2007, with some continuing to the present) Caribbean-wide seagrass monitoring initiative: the Caribbean Coastal Marine Productivity (CARICOMP) program. The program was established in 1992 to study land-sea interaction processes and to monitor changes through time in the productivity and structure of the three principal tropical coastal ecosystems: mangroves, seagrasses and coral reefs, with the ultimate goal of providing scientific information for management of coastal resources [26]. The CARICOMP program has generated a Caribbean-wide dataset using a simple, low-cost but standardized sampling protocol, consistent among sites over time. While the Caribbean region corresponds to the "Tropical Atlantic" seagrass bioregion [27] which has relatively high species diversity (10 species, [28]), most CARICOMP seagrass study areas were shallow reef lagoons dominated by two species (*Thalassia testudinum* and *Syringodium filiforme*, [29]). The present work aims to document changes within seagrass communities at unprecedented spatial and temporal (more than a decade) scales. We demonstrate that the low-cost standardized CARICOMP sampling protocol (consistent among sites and over time) can provide, in addition to the responses of seagrass communities to season and climate, evidence of deterioration of the environment along Caribbean coastlines.

Materials and Methods

Ethics Statement

Sample permits were issued by Florida Keys National Marine Sanctuary for USA-Florida Keys (site 2), SAGARPA for Mexico-Puerto Morelos (site 5, since 2003), Ministerio del Ambiente de Costa Rica for Costa Rica (site 21, since 2000); specific permissions were not required for any other site or date, and the field studies did not involve endangered or protected species.

The sampling protocols and organization of the monitoring network are described in CARICOMP [26,29]. Seagrass monitoring was conducted at 22 sites with 52 sampling stations (Fig 1) from 1993 until 2012, although many concluded the monitoring program before 2007 (Table S1). At each site, generally two *Thalassia testudinum*-dominated stations were selected by the participants; one station representing the most developed *T. testudinum* bed ("high productivity") and the second an average or typical bed. However, some sites had only one station, whereas others had up to six stations (Table S1). Bi-annual sampling intervals were specified in the protocol, but participation in the program was voluntary, and sampling frequencies and periods varied among sites (Table S1).

Growth and productivity of the seagrass *T. testudinum* were determined in 4–6 haphazardly placed quadrats (10×20 cm) per station. The leaves were marked just above the colorless basal sheath (or at the level of the quadrats) by punching one or two holes with a syringe needle and left to grow for 7–14 days. After this time, the foliar shoots were counted and the leaves were cut at the levels of the previous basal marks. Alternatively, the leaves were marked again at the base, and the entire foliar shoots (including sheaths) were retrieved from the sediments, and the foliar shoots were counted in the laboratory. In the laboratory, leaf

tissues were separated into new growth (newly emerged leaves and leaf sections below the mark of old leaves) and old fractions. The epiphytes were removed by rinsing the blades in a 10% acid solution and/or scraping with a razorblade. The leaf fractions were dried and dry weight was determined. The dry weights of new growth fractions represented the production (g m^{-2} d^{-1}) and the combined weight of both fractions corresponded with leaf biomass (g m^{-2}). Based on expected changes in growth due to the solar cycle, sampling was planned twice a year: once in the high-growth season (March through August) and once during the low-growth season (September through February). Annual productivity rates were determined by averaging the daily productivity rates (per m^{-2}) of all samples collected during the low- and high growth seasons for each year and multiplying by 365.

Biomass of the seagrass community was determined by taking two to four core samples at each station with a PVC or steel corer 15–20 cm in diameter (depending on site). The seagrasses (*T. testudinum* and "other grasses", mostly *Syringodium filiforme*) were separated into above- and below-ground fractions. Above-ground fractions were also separated and analyzed for the rooted calcareous and fleshy algae (below-ground parts were excluded from the analysis). The fractions were cleaned and dried before the weight was determined. The calcareous algae were decalcified in a 10% acid solution before being dried and weighed to determine their somatic weight. Annual biomass (per m^{-2}) was determined as the means of all samples collected during a year.

Possible spatial patterns for mean daily productivity rates of *T. testudinum* related to latitude or Physical Environments of the Caribbean Sea (PECS) defined by Chollet et al. [30] were explored using a Random Forest analysis [31]. The standard setting for the analysis defined by R v. 2.15.3 (creation of 500 randomly selected decision trees) and overall mean values (Table S2) for the stations were used for this analysis. The same analyses were applied to total (above-and below-ground) biomass of the seagrass community (data in table S3). Posteriorly, the terms latitude, depth and Secchi reading (table S2) were combined (latitude*depth*Secchi reading) to discern whether a combination of these terms could explain *T. testudinum* productivity or total community biomass.

Intra-annual variation in productivity of *T. testudinum* was determined for twenty-four stations at eleven sites that had ten or more sampling events. General mean productivity was computed for each of the twenty-four stations as the average of all measurements of productivity at that station. Intra-annual variation (ΔP) was determined by calculating the deviation from the general mean productivity for each sampling event (per station), expressed as the percentage of the general mean productivity. ΔP was plotted separately for high- and low- growth season at each station. For each degree of latitude, a One-sample t-test was applied to test whether ΔP differed from zero. At Long Key, Florida (Site 2), and Bon Accord Lagoon, Tobago (Site 18), sampling was conducted more than twice a year. Correlations between mean monthly Sea Surface Temperatures, (SST), hours of daylight and growth rates per shoot (g dry shoot^{-1}d^{-1}) were determined for each station at these two sites.

An interim report of the CARICOMP program identified increased terrestrial run-off (sewage, fertilizers and/or sediments) as the major and most prevalent anthropogenic influence in the monitoring region [29]. Consequences of increased terrestrial run-off into coastal waters are mainly increasing nutrients loads and/or decreasing water clarity. The following indicators of long-term changes in community and seagrass parameters were used as indicators of potential changes in coastal conditions: 1) Total (above-, and below ground) community biomass (seagrasses and

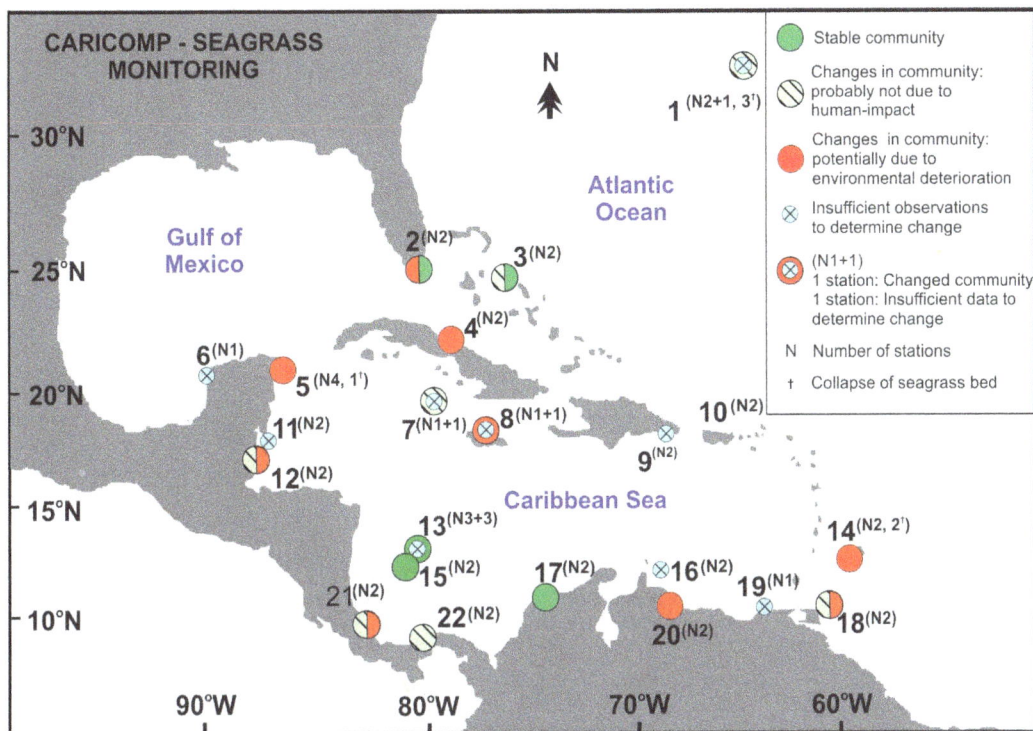

Figure 1. Map of CARICOMP seagrass sites, ordered according to latitude. 1. Bermuda, 2. USA-Long Key, 3. Bahamas-San Salvador, 4. Cuba-Cayo Coco, 5. Mexico-Puerto Morelos, 6. Mexico-Celestun, 7. Cayman Islands-Grand Cayman, 8. Jamaica-Discovery Bay, 9. Dominican Republic-Parque Nacional Este, 10. Puerto Rico-La Parguera, 11. Belize-Turneffe Island, 12. Belize-Twin Cays/Carrie Bow Cay, 13. Colombia-Isla Providencia, 14. Barbados-St. Lawrence, 15. Colombia-Isla San Andres, 16. Curaçao-Spaanse Water, 17. Colombia-Chengue Bay, 18. Tobago-Bon Accord Lagoon, 19. Venezuela-Isla de Margarita, 20. Venezuela-Morrocoy, 21. Costa Rica-Cahuita, 22. Panama-Isla de Colon.

algae), 2) Relative dominance (above-ground biomass seagrass/above-ground community biomass) of faster-growing seagrasses (classified as "other seagrasses" in the CARICOMP protocol, but mainly consisting of *S. filiforme*), 3) Relative dominance of faster growing fleshy algae, 4) Above-ground biomass relative to total biomass of the seagrasses (the most abundant seagrass *T. testudinum* was used for this assessment), 5) Productivity of *T. testudinum* and 6) Foliar shoot density of *T. testudinum*. Responses to changes in the environment depend on local settings [14], but combinations of consistent trends in two or more of the above-mentioned parameters may indicate environmental degradation. Both decrease in water clarity and increase in nutrients were expected to increase the dominance of faster growing seagrasses (parameter 2) or algae (parameter 3), and/or increase the relative investment in above-ground biomass (parameter 4). Total community biomass (parameter 1) and productivity of *T. testudinum* (parameter 5) are expected to decrease at increasing turbidity and to increase at increasing nutrient input into oligotrophic or mesotrophic systems. Foliar shoot density of *T. testudinum* (parameter 6) is expected to decrease with increasing turbidity. Significant slopes of linear regressions for each of these parameters versus year were considered to indicate a consistent pattern of change over the sampling period. These long-term trends were determined only when sampling covered at least five years (although intermittent at some stations) sampling effort with at least six sampling events, and included 35 stations at 17 sites.

Results

Annual productivity of leaves of *Thalassia testudinum* varied by an order of magnitude among the stations: lowest productivity was registered at Long Key, Florida Keys, USA (site 2-station 4, $<$ 200 g dry m^{-2} y^{-1}) whereas highest leaf productivity ($>$ 2000 g dry m^{-2} y^{-1}) was attained at Puerto Rico (site 10-stations 21 and 22) and Tobago (site 18-station 45, Fig. 2, Table S2). Community above-ground biomass varied over 20-fold among stations (Table S3), from 16 g dry m^{-2} at a mono-specific *T. testudinum* bed at Bermuda to 325 g dry m^{-2} at a coastal fringe in Puerto Morelos, Mexico, where 59% of biomass was accounted for by the large fleshy algae *Avrainvillea* spp. (Fig. 3, Table S3). The highest total (above-, and below ground) biomass of *T. testudinum* was registered at Twin Cays, Belize (1960 g dry m^{-2}, Fig 3). Inter-annual variations in both productivity and total biomass were considerable at all sampling stations (Figs. 2, 3). Of the 52 stations included in the analysis of biomass, only five ($<$10%) were monospecific beds of *T. testudinum* (without other seagrasses or macro-algae apart from epiphytes, Fig. 3, Table S3), indicating that the seagrass vegetative communities in the tropical Atlantic reef systems are typically multi-species associations.

No clear classification trees could be constructed relative to latitude for *T. testudinum* leaf productivity and total (above-and below-ground) community (seagrass and macro-algae) biomass. The Random Forests only explained 32.5% (productivity) or 57.9% (biomass) of the variance in the data. When depth and Secchi readings were added as terms, the fits of the models increased for productivity (52.8% of variance explained) and for

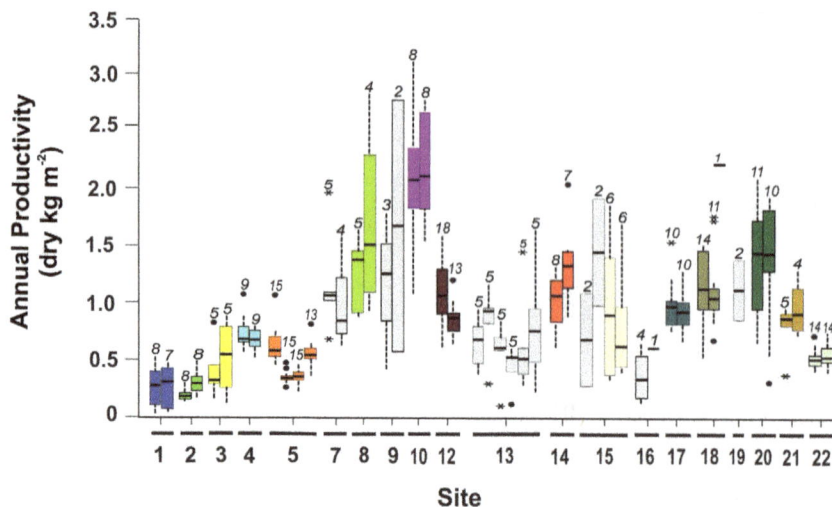

Figure 2. Annual leaf productivity of *Thalassia testudinum* per sampling station. The stations are grouped per site (underlined, 1–6 stations per site), and stations only sampled during one season are excluded. The boxes and vertical bars represent inter-annual variation. The horizontal lines correspond with the median values, 50% of the cases are within the box limits and the vertical bars indicate the smallest or largest values that are not outliers, • represent values more than 1.5 box lengths from lower/upper box limit, and * represent values more than 3 box-lengths from lower/upper box limit. The digits above the bars indicate N (the number of sampling years). Grey bars represent stations that were not included in the long-term analysis.

biomass (59.8% of variance explained). The CARICOMP seagrass sites were in 10 of the 16 physicochemical provinces in the Caribbean defined by Chollett et al. [30], and the models constructed for productivity and PECS did not result in a precise fit either for productivity (34.8% of the variance was explained) or total community biomass (35.1% of the variance).

Intra-annual variation in leaf productivity was evident at latitude 16°48′ (Belize) and higher (Fig. 4, Table S4). At Long Key, Florida (site 2) correlations among mean monthly shoot growth rates of *T. testudinum*, mean monthly SST and median hours of daylight were significant (Figure S1, Table S5). None of these correlations was significant at the more southern Bon Accord Lagoon in Tobago (site 18).

The seagrass communities at the majority (25) of the 35 stations included in the analysis for longer-term trends in the community showed changes in at least one of the six selected parameters (Table 1, Table S6). At six stations the seagrass beds collapsed: in Bermuda (3 stations) the decline was due to excessive grazing by sea turtles; in Barbados (2 stations) poor water quality followed by a population explosion of sea urchins and subsequent storms were responsible; and in Mexico, a coastal bed (1 station) was buried by sediments during a hurricane. Most monitoring stations (46 out of 52) were exposed at least once to a major meteorological event (hurricane or tropical storm, Table S1) during the study period, but apart from the above-mentioned collapse of communities in Mexico and Barbados (where the storms were not the main cause of collapse), minor impacts of storms were registered only at Belize-station 26 and Venezuela-stations 47 and 48 (Table 1). At 15 (43%) out of 35 studied stations (Table 1, Fig. 5), changes in the seagrass beds were consistent with hypothesized change scenarios of increased turbidity (Site 4-Stations 8 & 9 and Site 21-Station 49) or increased nutrient input (Site 2-Station 5; Site 5-Stations 10 thru13; Site 8-Station17; Site 10-Station 21; Site 12-Station 25; Site 14-Stations 33 & 34 Site 20-Stations 47 & 48). Most stations that showed shifts in community structure consistent with environmental degradation were reported to have received little or only moderate human-induced impacts at the onset of the study (Fig. 6).

Discussion

The CARICOMP monitoring program shows wide variation in seagrass productivity and biomass across the Caribbean, reflecting the different environmental settings among the sampling sites, although most were associated with coral reef systems. This study included a broad spectrum of seagrass community types dominated by *Thalassia testudinum*, from highly productive almost mono-specific beds to multi-species communities with several seagrass species and benthic macro-algae. The physicochemical provinces (PECS) defined by Chollett et al. [30] could not reliably predict the mean leaf productivity of *T. testudinum* or total biomass of the community. The 16 PECS were defined based on sea surface temperature, water clarity (from satellite images), salinity, wind-driven exposure and exposure to hurricanes. The criteria for the classification into these 16 provinces (PECS) likely did not include all relevant parameters that determine seagrass development. For example, Zieman et al. [32] suggested that in the Caribbean higher standing crops may be expected at sites with relief and considerable rainfall that supply nutrients for the development of larger plants, such as Jamaica, Puerto Rico, Belize, Venezuela (Morrocoy) and Panama. Latitude determines water temperatures and light regimes, and it was a better predictor for community biomass (but not productivity of *T. testudinum*) than the PECS. Combining latitude with local depth and mean Secchi reading (indicator for water transparency) resulted in more precise predictions; but less than 60% of the variances in mean leaf productivity of *T. testudinum* or total community biomass were explained by these combined predictors, suggesting that other factors also influence the leaf dynamics of *T. testudinum* and status of the seagrass communities.

Intra-annual changes in the growth of *T. testudinum* were registered at latitude 16°48′N (Site 12-Belize) and higher, and they appeared to be mainly driven by the seasonal solar-cycle. Also, at the northerly Florida Keys site, the initiation of the 'high-growth season' shows a lag of 1–2 months in comparison with more southerly sites (Figure S1), most likely in response to relatively low

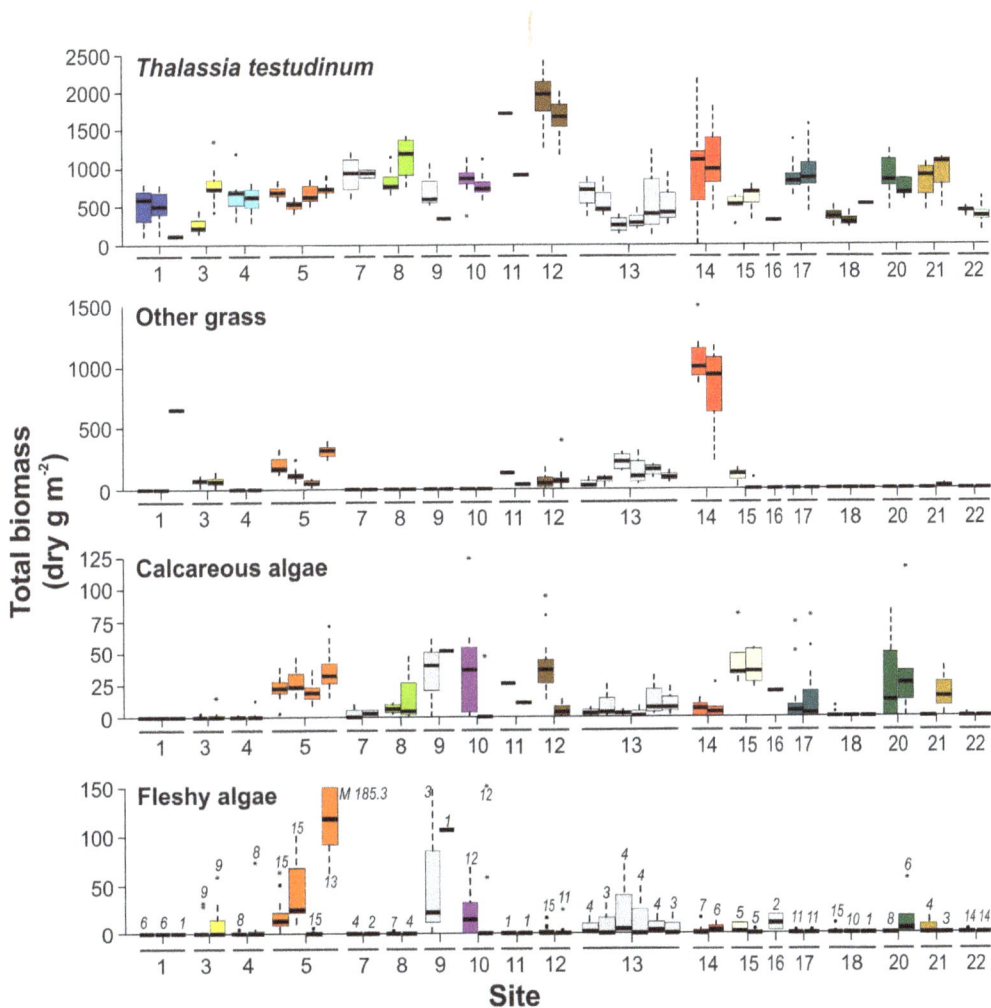

Figure 3. Total (above- and below-ground) biomass of the principal components of the community per sampling station grouped per site. Other grass: species of seagrass other than *Thalassia testudinum*, mostly *Syringodium filiforme*. Somatic (decalcified) above-ground weight of the calcareous algae is considered. The boxes and bars represent inter-annual variation, and stations with only one sampling event are excluded. The digits above the bars in the bottom graph indicate N (the number of sampling years). M median of fleshy algae at site 5-station 13. See legend of Fig. 2 for further explanation.

temperatures with corresponding reduced growth rates of *T. testudinum* until May (Table S5).

Most seagrass beds in this study have been exposed to hurricanes or major storms during the monitoring period (Table S1, Figure S2), but they were strikingly resilient. The seagrasses at a few stations were negatively affected (see Table 1), but recovery was rapid (within 1 to several years). Eradication only occurred at one station in a narrow (20–50 m wide) coastal fringe in Puerto Morelos, Mexico, during Hurricane Wilma (2005) by burial below ~1 m of sand [33]. However, recovery of this seagrass bed is now in progress (Van Tussenbroek, unpublished data). The vegetation at most stations consisted of *T. testudinum* dominated beds in sub-tidal reef-systems, where *T. testudinum* is a large and persistent seagrass that invests much of its biomass in below-ground tissue, which aids in firm anchorage of the plants [34] and stabilization of the sediments. Seagrass beds in the tropical or sub-tropical Atlantic that are not associated with reefs are often dominated by less robust seagrass species, such as those in the Gulf of Mexico, may be much more susceptible to damage or destruction by hurricanes or storms [Heck et al., 1996 (in [35]), [36–38], although there are exceptions [35].

Forty-three percent of the seagrass communities at the 35 long-term study stations at nine sites show changes that potentially indicate degradation of the environment between 1993 and 2007 (2012 for some stations, Table 1, Fig. 5). These changes over a relatively short time-span (6–18 years) across many sites is a worrying trend, particularly because most of these sites were only moderately disturbed by humans at the outset of the study (Fig. 6). Only two originally undisturbed sites, Colombia-Isla Providencia (site 13, Stations 29–31), and Colombia-Chengue Bay (site 17, stations 41 and 42) remained in 'pristine' condition up to the end of the monitoring period (2007 and 2005 respectively, Table 1). Several sites, such as Bahamas-San Salvador (site 3), Colombia-Isla San Andres (site 15), Tobago-Bon Accord lagoon (site 18), Panama-Isla de Colon (site 22), have been impacted by human development for decades or more than a century [39], but we did not detect indications of further degradation during the study period.

The consequences of these changes in seagrass communities across the Caribbean are difficult to assess at this point because baseline information concerning the structure, processes and drivers of Caribbean seagrass beds is deficient. However, it is likely that the ecosystem services offered by the seagrass communities

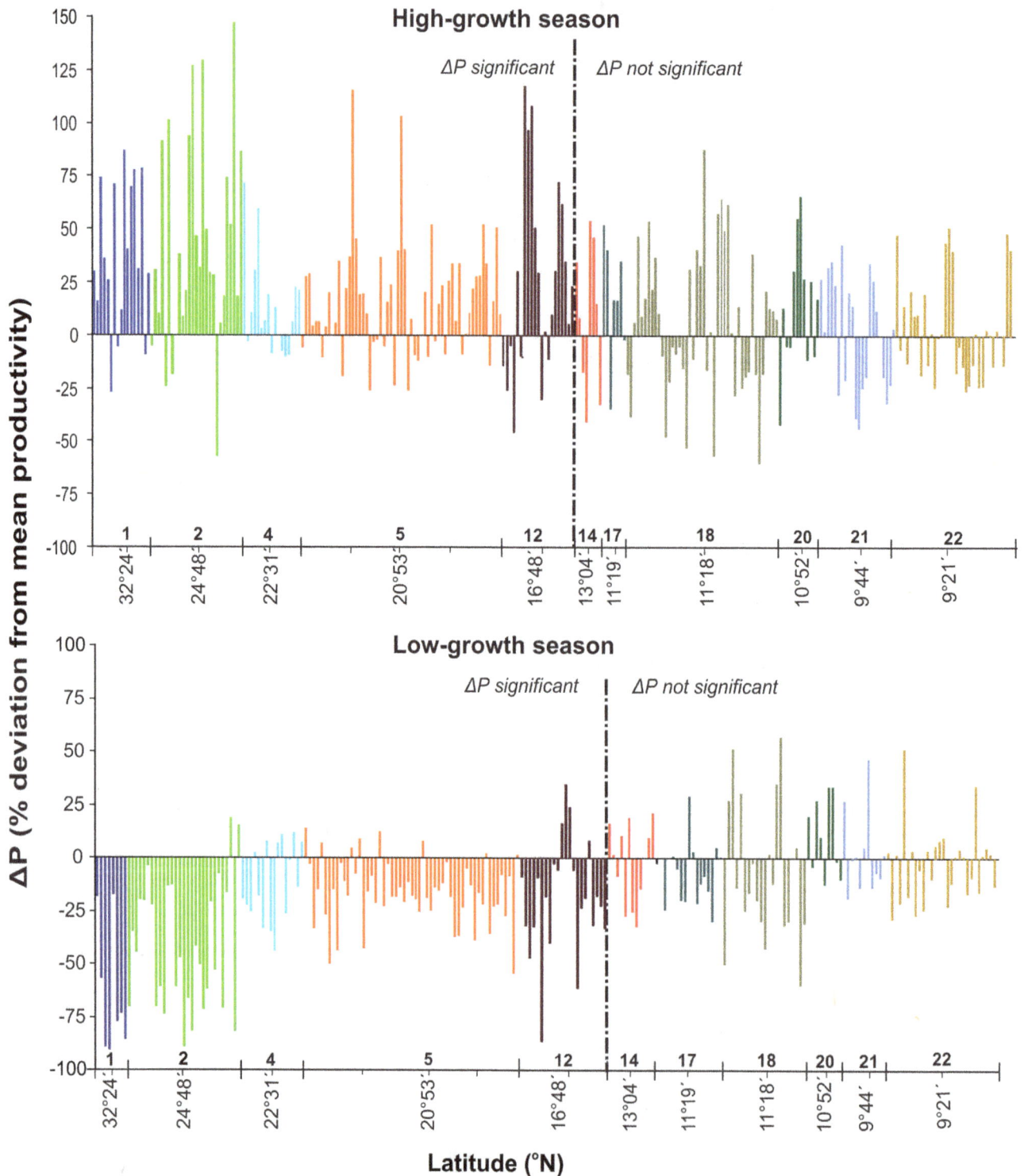

Figure 4. Deviations from general mean leaf productivity (ΔP) of *Thalassia testudinum* **per station during High growth season (May–September at site 1 and 2, March –August at all other sites) and Low growth season (October-April at sites 1 and 2, September-February at all other sites).** See Table S4 for significance differences of ΔP. Only stations with at least 10 sampling events were included. Numbers above the X-axis indicate site number, and the minor ticks indicate the different sampling stations at those sites.

will be compromised by changes in their productivity and composition of the seagrass community. For example, a community shift from *T. testudinum* to faster-growing seagrass and/or algal species (observed at 7 stations, Fig. 5) will result in a change in the overall structure of the seagrass canopy, and possibly a change in associated fauna. Sediment retention is likely to be compromised as seagrass communities shift away from broad-bladed and deeply-rooted *T. testudinum*. Faster-growing seagrass species have a less

Table 1. Long-term trends at CARICOMP seagrass stations, including observations on disturbances.

Site	Station	Country/Territory (sampling period: yr-yr)	Biomass	Rel. ab. - Other seagrass	Rel. Ab. - Fleshy algae	% Above/Total Biomass	Productivity	Shoot density	Condition at beginning of monitoring	
1	1†††	Bermuda	D	-	-	D	D	I	PRIST	The pristine seagrass beds have showed catastrophic declines since 1997, and they were extirpated in 2001, likely caused by excessive grazing by the green turtle *Chelonia mydas* [1,2].
	2†††	(94-02)	D	-	-	D	D	D		
2	4	USA	-	-	-	D	D	D	INT	Florida Keys suffer from eutrophication through groundwater contamination from septic tanks[3,4]; only the inshore station 5 showed indications of disturbance.
	5*	(96-03)	I^N	-	-	-	I^N	I		
3	6	Bahamas	n	n	-	n	-	-	EUTR	San Salvador had plantations, but has become a tourist destination since 1970s [5].
	7	(/94-06)	n	D	-	n	-	-		
4	8*	Cuba	D^T	-	-	I^{NT}	n	D^T	PRIST	Cayo Coco was considered as pristine, but increasing sedimentation (decreased light) could have forced changes in seagrass beds[5].
	9*	(94-02)	D^T	-	-	I^{NT}	n	D^T		
5	10*	Mexico	n	I^{NT}	I^{NT}	I^{NT}	I^N	I	PRIST	Increased eutrophication through ground-water discharge likely forced changes in the seagrass beds at Puerto Morelos reef lagoon[6,7,8]. Vegetation at the coastal fringe (station 13) was buried during hurricane Wilma (2005)[7].
	11*	(93-09)	I^N	n	I^{NT}	I^{NT}	n	I		
	12*		n	I^{NT}	-	I^{NT}	I^N	n		
	13*†††		I^N	n	n	I^{NT}	n	n		
7	15	Cayman I.	-	-	-	-	I^N	D^T	INT	The high number of visitors to the Grand Cayman compromise carrying capacity and cause environmental degradation, but not in protected CARICOMP areas[9].
		(97-03)								
8	17*	Jamaica	I^N	-	-	I^{NT}	I^N	-	DIST	Discovery Bay has been affected by terrestrial runoff from agricultural developments and possibly by proliferation of urban developments without a central sewage system[9].
		(93-99)								

Table 1. Cont.

Site	Station	Country/Territory (sampling period: yr-yr)	Biomass	Rel. ab. - Other seagrass	Rel. Ab. - Fleshy algae	% Above/Total Biomass	Productivity	Shoot density	Condition at beginning of monitoring	
10	21*	Puerto Rico	I^{N}	-	I^{NT}	I^{NT}	n	n	INT	Residential and tourism developments at and near La Parguera have increased terrigenous sediment load, and raw- and secondary sewage effluents from a treatment plant[9] possibly changed the seagrass community at station 21.
	22	(94-06)	n	-	n	I^{NT}	n	n		
12	25*	Belize	n	I^{NT}	n	I^{NT}	I^{N}	D^{T}	PRIST	Both sites have been subjected to loss of water clarity due to increased input of sediments and nutrients from coastal development and agriculture. *T. testudinum* shoot density has declined at Site 25(Twin Cays) possibly due to increased sedimentation[10]. Station 26 (Carrie Bow Cay) was scoured by Hurricane Mitch (1998) but recovering[11].
	26	(93-12) (97-12)	n	n	n	I^{NT}	n	I		
13	29	Isla	n	n	-	n	n	n	PRIST	Isla Providencia (Columbia) was for a long time sparsely inhabited; recently tourism has increased, but the changes in seagrass beds were not consistent with environmental degradation. Hurricane Beta passed close by, but the seagrasses received no impact[12].
	30	Providencia	n	D	-	n	n	n		
	31	(00-07)	n	n	-	n	D^{T}	D^{T}		
14	33*†††	Barbados	n	I^{NT}	-	I^{NT}	n	D^{T}	DIST	St. Lawrence has been affected by anthropogenic activities since 1880s from sugar cane cultivation, residential developments and eutrophication. Combined effects of increased sedimentation due to frequent flushing of a new sewage pipe system, excessive sea urchin grazing and storms caused collapse of the seagrass beds[5, 13].
	34*†††	(93-01)	n	I^{NT}	-	I^{NT}	n	-		
15	37	San Andres	n	n	-	n	-	I	DIST	San Andres Island (Colombia) is the most densely populated oceanic island in the Caribbean. Past population increase with poorly planned development resulted in degradation of coastal ecosystems[5] The seagrass community did not show changes.
	38	(99-07)	n	n	-	n	-	n		

Table 1. Cont.

Site	Station	Country/Territory (sampling period: yr-yr)	Biomass	Rel. ab. - Other seagrass	Rel. Ab. - Fleshy algae	% Above/Total Biomass	Productivity	Shoot density	Condition at beginning of monitoring	
17	41	Colombia	n	-	-	n	n	n	PRIST	Seagrass beds, mangroves and coral reefs at Chengue Bay are healthy and stable[14]. Abundant *Halimeda opuntia* before 1996 disappeared without obvious cause[5].
18	42	(94-05)	n	-	-	n	n	n		
	43	Tobago	n	-	-	n	n	n	DIST	Bon Accord Lagoon has received impacts from tourism and sewage for ~50 y[9]. But apart from a decrease in productivity at station 45, the seagrass conditions were stable.
	44	(92-07)	n	-	-	n	D^T	n		
20	47*	Venezuela	I^N	-	-	I^{NT}	I^N	I	DIST	Morrocoy Park has been subjected to increasing land-based constructions, mangrove demolition and sewage effluents since 1970s[9]. Heavy rainfall in 1999 caused loss of seagrass leaves, but recovery followed within months[5].
	48*	(93-06)	I^N	-	-	I^{NT}	I^N	I		
21	49*	Costa Rica	D^T	-	-	n	D^T	D^T	PRIST	The Limón earthquake in 1991 affected the seagrass beds at Cahuita which fully recovered after 1 y[5]. Seagrass beds at Station 49 may have deteriorated due to increased turbidity by sewage load, sediments and fertilizers (citrus and banana farming)[15, 16].
	50	(99-05)	-	-	-	-	D^T	n		
22	51	Panama	I^N	-	-	-	n	n	$DIST^{16}$	The narrow *Thalassia testudinum* bed fringing a mangrove swamp showed increased biomass over the years, which may be attributed to natural causes[17].
	52	(99-06)	I^N	-	-	-	n	I		

Biomass: total above-ground biomass of the community; Rel abund: relative abundance (biomass) of faster growing seagrass and algal species; Other seagrass: seagrass species other than *Thalassia testudinum* (mostly *Syringodium filiforme*); % Above/Total Biomass: percentage of above-ground of total biomass of *T. testudinum* (*S. filiforme* for site 14, because *T. testudinum* was absent in later years at station 33); Productivity: productivity of leaves of *T. testudinum*. ??? Collapse of seagrass bed, * seagrass beds showed changes that potentially indicate with human-induced environmental deterioration. Trends: I increase, D decrease, n without change, - not determined, N expected change due to increasing nutrient load, T expected change due to increasing turbidity, NT expected change due to either increasing turbidity or nutrient load, changes without symbol were not consistent with expectations of water quality deterioration (See text for further explanation). Conditions at the beginning of monitoring: PRIST (relatively) pristine (undisturbed by humans); INT Moderate disturbance; DIST Disturbed (eutrophication, terrestrial runoff, or overfishing, from [26], [29], [47]). See Table S6 for information on regression lines. Source: 1. [48], 2. [49], 3. [50], 4. [51], 5. [29], 6. [52], 7. [33], 8. [53], 9. [38], 10. [44], 11. Pers. Obs. K. Koltes, 12.Pers.Obs. H.A Oxenford, 13. [54], 14. [55], 15. [56], 16. [57], 17. [58].

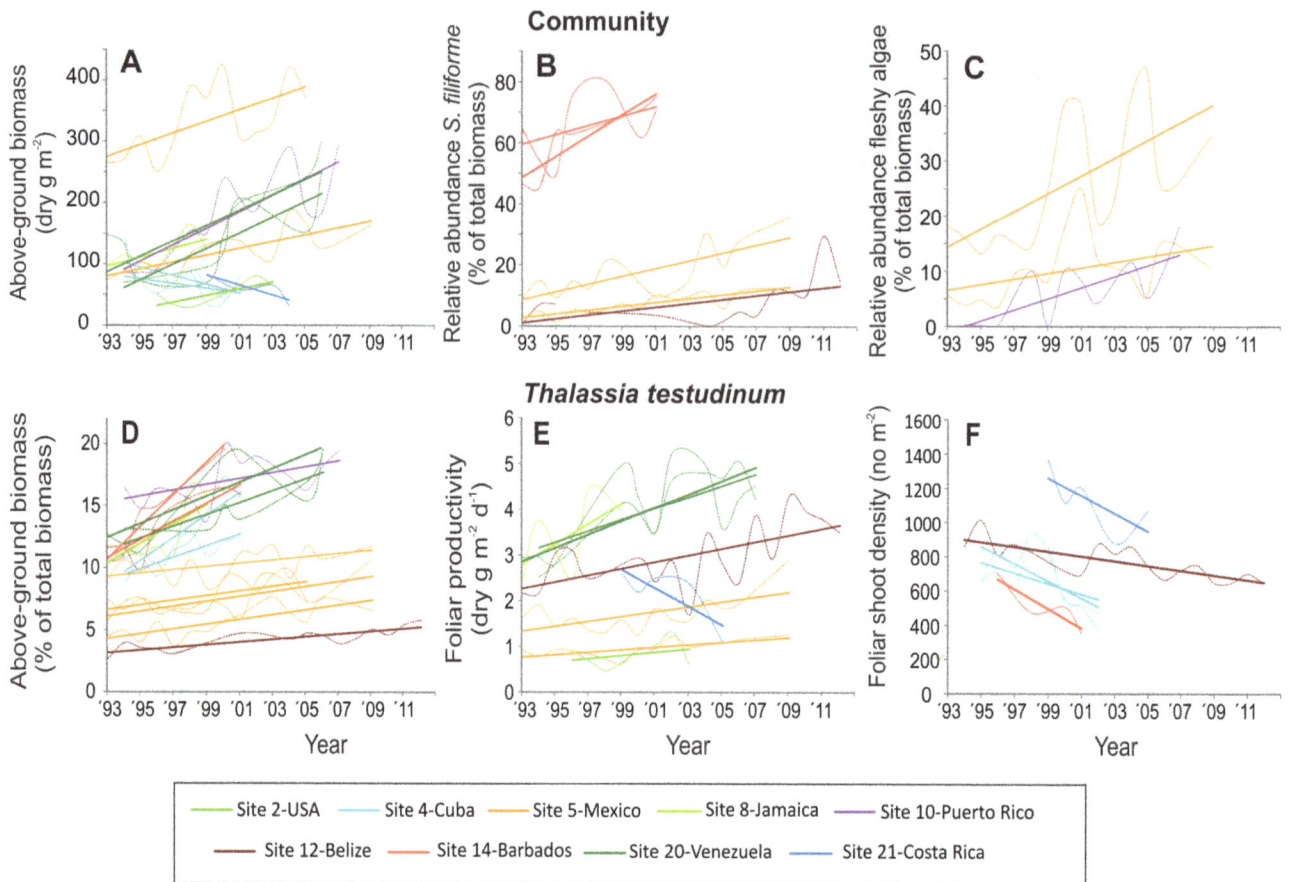

Figure 5. Significant long-term trends in seagrass attributes and community parameters at CARICOMP monitoring stations across the nine sites that showed changes consistent with deterioration of the environmental conditions. The broken smoothed lines connect annual average values and serve to illustrate the inter-annual variability in the data. Data from all samples per year (N = 4-9, Table S3) were used to determine the regression lines (Table S6). D. For Site14, the relationship was determined for the more persistent *Syringodium filiforme*.

developed below-ground rhizome-root system [40] which could also have negative consequences for below-ground carbon sequestration [41], or resistance of seagrass beds to hurricanes. Impacts of hurricanes are most deleterious in already disturbed beds [42], and the faster growing *S. filiforme* is much more susceptible to dislodgement than *T. testudinum* [34,43]. The fate of the seagrass beds at Barbados-St. Lawrence lagoon (site 14) is a good example of how long-term (chronic) anthropogenic stress can act synergistically with acute extreme disturbance events [overgrazing by an exceptionally strong recruitment of sea urchins, Hurricane Ivan (2004) and Tropical Storm Emily (2005)] to cause collapse of an ecosystem. Even after 7 years, this lagoon has shown only minimal recovery, with just a few impoverished *T. testudinum* plants in areas of coral rubble and a very sparse vegetation of *Halodule wrightii* appearing in the sand areas (H. Oxenford, unpublished data).

Interpretation of the long-term shifts in the seagrass communities is not unequivocal, because responses of individual communities depend on local conditions and the state of the community when monitoring began. For example, productivity and biomass were expected to decrease with decreasing water clarity, a relationship reported for Cuba (site 4) and Costa Rica (site 20-station 49). At Belize (site 12), conflicting trends at the two stations (25 and 26) resulted from differing environmental contexts and initial states of the two seagrass meadows. At station 25, a relatively low energy site inside the lagoon, increased productivity

is more likely a response to nutrient enrichment associated with the well-documented increases in turbidity [44] than to the declining light levels at this shallow (1 m) depth. Further evidence for this is that shoot density of *T. testudinum* declined by 50% from 1993 to 2012 (Fig. 5F) while the relative abundance of *S. filiforme* increased (Fig. 5B), both typical responses to declining light levels and/or nutrient enrichment. By contrast, station 26, established in 1997 in a higher-energy zone adjacent to a cut in the barrier reef-line, was scoured by Hurricane Mitch in 1998. Trends at this station largely reflect recovery of the seagrass meadow over the first 8-10 years of monitoring (Table 1).

The area of seagrass sampled by the CARICOMP protocol (0.08–0.12 m^2 for foliar productivity and 0.04–0.09 m^2 for biomass) is smaller than that of more recently established monitoring programs such as Seagrass Watch [45], Seagrass Monitoring in the Florida Keys National Marine Sanctuary [46], or Seagrass Net [25], that employ cover estimates of the vegetation in 5–11 quadrats (0.25–1.0 m^2) along one to three 50m-long transects (supplemented with small samples of leaves, complete plants, sediments or seeds). Small sample size assumes a relatively homogeneous distribution of the species (or species groups). Long-living seagrasses (such as *T. testudinum*) in continuous beds, without obvious environmental gradients, may fulfil this assumption; however, a larger scale-sampling scheme (such as that of Seagrass Watch or Seagrass Net) may be necessary for less-abundant and more irregular distributed species such as the more ephemeral

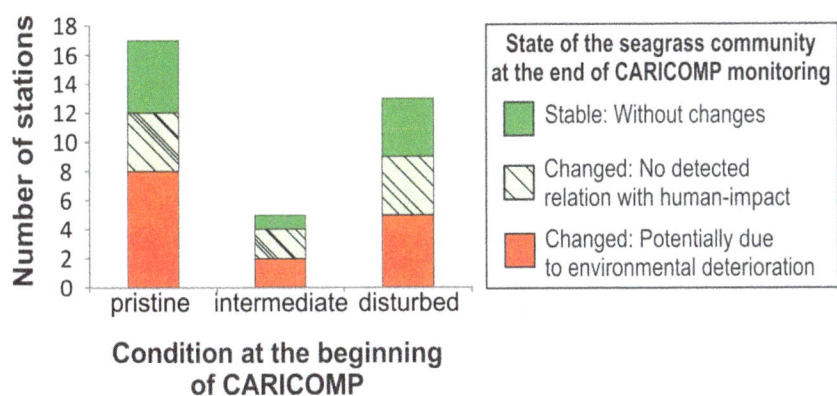

Figure 6. State of the long-term monitoring stations at the beginning (1993) and end (2007–2012) of the CARICOMP program. Pristine: Relatively undisturbed stations at the start of the program. Intermediate: Stations moderately disturbed by human-impact at the beginning of the program. Disturbed: Stations which had undergone chronic human-induced impacts before the initiation of the monitoring program.

seagrasses (e.g *Halodule* spp. or *Halophila* spp.) or rooted macro-algae. The CARICOMP measures of foliar productivity and total plant biomass are destructive and therefore cannot be employed on a large scale (and also because sample processing is labor-intensive), but these measures are less subjective than cover estimates. In addition, they may be more sensitive to small shifts in plant abundance or biomass distribution and detect responses to environmental change sooner than monitoring programs based on less precise estimates of abundance such as vegetation cover. However, cover estimates are useful to assess loss of seagrasses over large areas under regimes of relatively severe (human-induced) stress (e.g.[25]). Thus, optimal design of a monitoring protocol depends on its goals, site- and plant characteristics and logistics. Regardless of the protocol, it is paramount for determining trends in seagrass communities that consistent observations be made over long periods (5-10 y at least). The CARICOMP program depended on voluntary participation and local resources, which resulted in large differences in sampling frequency and periods among sites. Only 17 of the 22 sites obtained sufficient data for analysis of long term trends in the seagrass communities; and even among these sites the sampling period (5–15 y, Table 1) and frequency (24–124 biomass samples, 30–202 productivity samples, Table S6) varied considerably. We may therefore, have missed possible changes at stations which were sampled infrequently or for shorter periods (5 or 6 y).

CARICOMP was a pioneering monitoring program, and this study has shown that the simple and low-cost methods used were sufficient to discern long-term trends in the seagrass communities. We suggest that changes across various parameters are consistent with deterioration of the coastal environment, thereby indicating sites that would benefit from further studies and management efforts. We recognize that drivers of change were poorly covered in this program, and suggest that future monitoring of the long-term trends should include relevant environmental measures, such as nutrient availability (C, N, P contents in plant tissues) and sediment (anoxia, organic matter) conditions, in addition to (pulse fluctuations in) water transparency, temperature and salinity. Environmental degradation often involves multiple interacting stressors, and long-term monitoring programs such as CARI-COMP can only determine causal factors of change when ecological data are viewed together with supplemental data on environmental conditions.

Supporting Information

Figure S1 Plots of the mean monthly shoot growth rates. Vertical bars represent 95% confidence limits. Horizontal bar above X-axis represents periods of High (black) and Low (grey) growth season. See legend Table S5 for further information. Note differences in ordinate scales for growth. +: Mean monthly SST.

Figure S2 Passage of hurricanes and storms. Tracks of named storms or hurricanes that passed within 1 degree (60 nm or 111 km) of any CARICOMP seagrass site (indicated by the numbered circles) during the observation periods reported in Table S1. The size of the event when passing the site was not considered, thus the atmospheric and hydrological impacts at the locations may have varied from weak to severe, and other events of impact (e.g surge or excessive rains) may not be included. Servicio Académico de Monitoreo Meteorológico y Oceanográfico, Unidad Académica de Sistemas Arrecifales, Instituto de Ciencias del Mar y Limnología, Universidad Nacional Autónoma de México).

Table S1 CARICOMP seagrass monitoring sites. General information on the sites and stations (ordered from North to South), together with sampling periods (as mm/yy) for *Thalassia testudinum* leaf productivity (Table S2) and community biomass (Table S3). Hurricanes/Storms: year of passage ('yy) and max. strength when passing the affected location (T Tropical Storm, H hurricane) in parenthesis (see Figure S2). nd not determined. British OT: British Overseas Territory.

Table S2 *Thalassia testudinum* leaf dynamics. Depths (below MTL) at the stations and average values (± SE) for parameters of *T. testudinum* leaf productivity, biomass and density. N: number of samples (10×20 cm, see S1 for period). N for foliar shoot density is less at some sites because this measure was introduced later in the protocol. Secchi: mean Secchi readings from 1993-1995 (from CARICOMP, 1998).*Celestun is in the Gulf of Mexico, na: not available, nd: not determined.

Table S3 Community biomass. Average values (± SE) of the biomass of the community by vegetation group. N: number of (core) samples (see S1 for period). Core diam: Diameter of the core samples. Total: biomass of above- and belowground live tissues. AB: above-ground biomass. Biomass of calcareous algae expressed as somatic (decalcified) weight, ~85% of the calcified dry weight is $CaCO_3$. Below-ground tissues of the algae were not considered. Biomass cores were not taken at Site 2 (Florida). "Other grass": mostly *Syringodium filiforme* but includes *Halodule wrightii* at Station 7. * Celestun is in the Gulf of Mexico. na: not applicable.

Table S4 Intra-annual variability in *Thalassia testudinum* leaf productivity. Results of One-sample t-test for significant differences of ΔP (deviations from general mean leaf productivity) during High- and Low-growth season at different latitudes. H0: Average $\Delta P = 0$, $\alpha = 0.05$.

Table S5 Correlations between temperature, light and *Thalassia testudinum* leaf growth. Correlations between mean monthly SST (Sea Surface temperature, °C), H daylight (Hours of daylight) and shoot growth rates of *Thalassia testudinum* at USA-Florida Keys (Site 2, 1996–2003), Mexico-Puerto Morelos (Site 5, 1990–1991: from Van Tussenbroek BI [1995] *Thalassia testudinum* leaf dynamics in a Mexican Caribbean reef lagoon. Mar Biol 122: 33–40) and Tobago-Bon Accord Lagoon (Site 18, 1997–2007), N number of months, ns: not significant. Hours daylight were obtained from http://astro.unl.edu/classaction/animations/coordsmotion/daylighthoursexplorer.html. Mean monthly SST were from: NOAA Coral Reef Watch, Coral Bleaching Virtual Stations (http://www.osdpd.noaa.gov/ml/ocean/cb/virtual_stations.html)-*Sombrero Reef, Florida* (Site 2), Rodríguez-Martínez RE, Ruíz-Rentería F, Van Tussenbroek BI, Barba-Santos G, Escalante-Mancera E et al. [2010] State and environmental tendencies of the Puerto Morelos CARICOMP site, Mexico. Rev Biol Trop 58: 23–43 (Site 5), and R. Juman, J. Gomez [unpublished data] (Site 18).

Table S6 Regression lines of trends. Results for the linear regressions of selected parameters *vs* year to indicate possible trends Regressions were computed for those stations and parameters when the sampling covered at least 5 y with at least six sampling events and at least 50% of the samples had values >0. *: not determined, ns: not significant, negative t indicated a

negative slope. The significance level (α) is 0.05, but a Bonferroni correction is applied to this level of significance, because the parameters are derived from the same cores (Total above-ground biomass. Relative abundance of faster-growing seagrass, Relative abundance of faster growing fleshy algae, % Above-ground/total biomass for *Thalassia testudinum*) or quadrats (Productivity, Foliar shoot density of *T. testudinum*). The results of the regressions of foliar weight per shoot of *T. testudinum* is also given, to facilitate interpretations of the results, although this was not a parameter for potential degradation of the coastal environment.

Acknowledgments

The following persons or institutions also aided in seagrass sampling: Fourqurean Laboratory at Florida International University, Donald T. Gerace Research Center, Daily Zuñiga and Lídice Clero (Centro de Investigaciones de Ecosistemas Costeros de Cayo Coco, Cuba), M. Guadalupe Barba Santos (UNAM, Mexico), Jorge A. Herrera-Silveira (CINVESTAV, Mexico), Phillippe G. Bush (Dept. of Environment, Grand Cayman), Francisco X. Geraldes (CIBIMA, Dominican Republic), Jorge R. García (University of Puerto Rico, Puerto Rico), Eden Garcia (University College of Belize Marine Research Centre, Belize), June M. Mow Robinson (CORALINA, Colombia), Jaime Garzón-Ferreira (IN-VEMAR, Colombia), Renata Goodridge (CERMES, Barbados), Carmabi Foundation, Curaçao, Ramón Varela (EDIMAR, Venezuela), Aldo Cróquer (Universidad Simón Bolivar, Venezuela), Daisy Perez (INTEC-MAR, Venezuela), Penny Barnes and Arcadio Castillo (Smithsonian Tropical Research Institute, Panama). The CARICOMP seagrass team is grateful to J. Zieman who developed the sampling protocol and to John C. Ogden and members of the CARICOMP Steering Committee. This is contribution number 938 of the Caribbean Coral Reef Ecosystems Program (CCRE), Smithsonian Institution.

Author Contributions

Performed the experiments: BIT JC RC ACF PMHG HMG GEJ RJ KHK HAO ARR JSV SRS JJT EW. Analyzed the data: BIT. Contributed reagents/materials/analysis tools: BIT JC RC ACF PMHG HMG GEJ RJ KHK HAO ARR JSV SRS JJT EW. Wrote the paper: BIT JC RC ACF PMHG HMG GEJ RJ KHK HAO ARR JSV SRS JJT EW.

References

1. Green EP, Short FT (2003) Editors, World Atlas of Seagrasses. Prepared by the UNEP World Conservation Monitoring Centre. University of California Press, Berkeley, USA. 298 p.
2. Gillanders BM (2006) Seagrasses, fish and fisheries. In: Larkum AWD, Orth RJ, Duarte CM, editors. Seagrasses: Biology, Ecology and Conservation, Springer, Dordrecht, The Netherlands. pp. 503–536.
3. Madsen JD, Chambers PA, James WF, Koch EW, Westlake DF (2001) The interaction between water movement, sediment dynamics and submersed macrophytes. Hydrobiologia 444: 71–84.
4. Romero J, Lee K-S, Pérez M, Mateo MA, Alcoverro T (2006) Nutrient dynamics in seagrass ecosystems. In: Larkum AWD, Orth RJ, Duarte CM, editors, Seagrasses: Biology, Ecology and Conservation, Springer, Dordrecht, The Netherlands. pp. 227–254.
5. Duarte CM, Middelburg JJ, Caraco N (2005) Major role of marine vegetation on the oceanic carbon cycle. Biogeosciences 2: 1–8.
6. Fourqurean JW, Duarte CM, Kennedy, Marbà N, Holmer M, et al. (2012) Seagrass ecosystems as a globally significant carbon stock. Nature Geosc 5: 505–509.
7. Heck KL Jr, Carruthers TJB, Duarte CM, Hughes AR, Kendrick G, Orth RJ, Williams SW (2008) Trophic transfers from seagrass meadows subsidize diverse marine and terrestrial consumers. Ecosystems 11: 1198–1210.
8. Nagelkerken I, Kleijnen S, Klop T, Van den Brand RACJ, Cocheret de la Morinière E, et al. (2001) Dependence of Caribbean reef fishes on mangroves and seagrass beds as nursery habitats: a comparison of fish faunas between bays with and without mangroves/seagrass beds. Mar Ecol Prog Ser 214: 225–235.
9. Unsworth RKF, Salinas de León P, Garrard SL, Jompa J, Smith DJ, et al. (2008) High connectivity of Indo-Pacific seagrass fish assemblages with mangrove and coral reef habitats. Mar Ecol Prog Ser 353: 213–224.
10. Verweij MC, Nagelkerken I, De Graaff D, Peeters M, Bakker EJ, et al. (2006) Structure, food and shade attract juvenile coral reef fish to mangrove and seagrass habitats: a field experiment. Mar Ecol Prog Ser 306: 257–268.
11. Wefer G (1980) Carbonate production by algae *Halimeda, Penicillus* and *Padina*. Nature 285: 323–324.
12. Nelsen JE, Ginsburg RN (1985) Calcium carbonate production by epibionts on Thalassia testudinum in Florida Bay. J Sedim Petrol 56:622–628
13. Van Tussenbroek B I, Van Dijk J-K (2007) Spatial and temporal variability in biomass and production of psammophytic *Halimeda incrassata*, in a Caribbean reef lagoon. J Phycol 43: 69–77.
14. Linton DM, Warner GF (2003) Biological indicators in the Caribbean coastal zone and their role in integrated coastal management. Ocean Coast Manage 46: 261–276.
15. Orth RJ, Carruthers TJB, Dennison WC, Duarte CM, Fourqurean JW, et al. (2006) A global crisis for seagrass ecosystems. Bioscience 56: 987–996.
16. Dennison WC, Orth RJ, Moore KA, Stevenson JC, Carter V (1993) Assessing water quality with submersed aquatic vegetation. Bioscience 43: 86–94.
17. Livingston RJ, McGlynn SE, Niu X (1998) Factors controlling seagrass growth in a gulf coastal system: Water and sediment quality and light. Aq Bot 60: 135–159.
18. Duarte CM (1990) Seagrass nutrient content. Mar Ecol Prog Ser 67: 201–207.
19. Fourqurean JW, Zieman JC, Powel GVN (1992) Phosphorus limitation of primary production in Florida Bay: Evidence from C:P:N ratios of the dominant seagrass *Thalassia testudinum*. Limnol Oceanogr 37: 162–171.
20. Short FT (1987) Effects of sediment nutrients on seagrasses: Literature review and mesocosm experiment. Aq Bot 27: 41–67.
21. Udy JW, Dennison WC (1997) Growth and physiological responses of three seagrass species to elevated sediment nutrients in Moreton Bay, Australia. J Exp Mar Biol Ecol 217: 253–277.

22. Armitage AR, Frankovich TA, Fourqurean JW (2011) Long-term effects of adding nutrients to an oligotrophic coastal environment. Ecosystems 14: 430–444.

23. Duarte CM (1991) Seagrass depth limits. Aq Bot 40: 363–377.

24. Fourqurean JW, Boyer JN, Durako MJ, Hefty LN, Bradley J, et al. (2003) Forecasting responses of seagrass distributions to changing water quality using monitoring data. Ecol Appl 13: 474–489.

25. Short FT, Koch EW, Creed JC, Magalhães KM, Fernandez E, et al.(2006) SeagrassNet monitoring across the Americas: case studies of seagrass decline. Mar Ecol 27: 277–289.

26. CARICOMP (1994) CARICOMP Methods Manual-Level I. Manual of Methods for Mapping and Monitoring of Physical and Biological Parameters in the Coastal Zone of the Caribbean. CARICOMP Data Management Centre, University of the West Indies, Mona, Kingston, Jamaica (available at http://www.ima.gov.tt/home/images/stories/caricomp_manual_2001.pdf)

27. Short F, Caruthers T, Dennison W, Waycott M (2007) Global seagrass distribution and diversity: A bioregional model. J Exp Mar Biol Ecol 350: 3–20.

28. Short FT, Polidoro B, Livingstone SR, Carpenter KE, Bandeira S, et al. (2011) Extinction Risk Assessment of the World's Seagrass Species. Biol Conserv 144: 1961–1971.

29. CARICOMP (2004) Caribbean Coastal Marine Productivity Program: 1993-2003, Linton D, Fisher T, editors. CARICOMP Program, 88 p.

30. Chollett I, Mumby PJ, Müller-Karger FE, Hu C (2012) Physical environments of the Caribbean Sea. Limnol Oceanogr 57: 1233–1244.

31. Liaw A (2012) CRAN-R randomforest package: Breiman and Cutler's random forests for classification and regression. http://cran.r-project.org/web/packages/randomForest.

32. Zieman J, Penchaszadeh P, Ramirez JR, Perez D, Bone D, et al. (1997) Variation in ecological parameters of *Thalassia testudinum* across the CARICOMP network. Proceedings of the 8th International Coral Reef Symposium 1, Panama, 663–668.

33. Rodríguez-Martínez RE, Ruíz-Rentería F, Van Tussenbroek BI, Barba-Santos G, Escalante-Mancera E, et al. (2010) State and environmental tendencies of the Puerto Morelos CARICOMP site, Mexico. Rev Biol Trop 58 (Suppl. 3): 23–43.

34. Cruz-Palacios V, Van Tussenbroek BI (2005) Simulation of hurricane-like disturbances on a Caribbean seagrass bed. J Exp Mar Biol Ecol 324: 44–60.

35. Byron D, Heck KL Jr (2006) Hurricane effects on seagrass along Alabama's Gulf Coast. Estuar Coasts 29: 939–942.

36. Montcrieff CA, Randall TA, Calswell JD, McCall RK, Blackburn BR, et al. (1999) Short-term effects of hurricane Georges on seagrass populations in the north Chandeleur Island: Patterns as a function of sampling scale. Gulf Res Reports 11: 74–75. 36.

37. Fonseca MS, Kenworthy WJ, Whitfield PE (2000) Temporal dynamics of seagrass landscapes: a preliminary comparison of chronic and extreme disturbance events. Biol Mar Medit 7:373–376.

38. Davis SE, Cablet JE, Childers DL (2004) Importance of storm events in controlling ecosystem structure and function in a Florida Gulf Coast Estuary. J Coast Res 20: 1198–1208.

39. CARICOMP (1998) CARICOMP - Caribbean coral reef, seagrass and mangrove sites, Kjerve B, editor. Coastal region and small islands papers 3 (UNESCO, Paris), XIV + 347 p.

40. Marbà N, Duarte CM (2003) Scaling of ramet size and spacing in seagrasses: implication for stand development. Aq Bot 77: 87–98.

41. McLeod E., Chmura GL, Bouillon S, Salm R, Björk M, et al. (2011) A blueprint for blue carbon: toward an improved understanding of the role of vegetated coastal habitats in sequestering CO_2. Frontiers Ecol Environ9:552–560.

42. Whitfield PE, Kenworthy WJ, Hammerstrom KK, Fonseca MS (2002) The role of a hurricane in expansion of disturbances initiated by motor vessels on subtropical seagrass banks. J Coast Res 37: 68–99.

43. Van Tussenbroek BI, Barba Santos MG, Van Dijk JK, Sanabria-Alcaraz SNM, Téllez-Calderón ML (2008) Selective elimination of rooted plants from a tropical seagrass bed in a back-reef lagoon: a hypothesis tested by hurricane Wilma (2005). J Coast Res 24: 278–281.

44. Koltes KH, Opishinski T (2009) Patterns of water quality and movement in the vicinity of Carrie Bow Cay, Belize. Smiths Contrib Mar Sc 38: 379–390.

45. McKenzie LJ, Long L, Coles RG, Roder CA (2000) Seagrass-Watch: Community based monitoring of seagrass resources. Biol Mar Medit 7: 393–396.

46. Fourqurean JW, Willsie A, Rose C D, Rutten LM (2001) Spatial and temporal pattern in seagrass community composition and productivity in south Florida. Mar Biol 138: 341–354.

47. Jordán-Dahlgren E, Cortés J, Smith SR, Parker C, Oxenford HA, et al. (1997) Physiography and setting of CARICOMP sites: a pattern analysis. Proc 8th Int Coral Reef Symp 1, Panama. pp. 647–650.

48. Murdoch TJT, Glasspool AF, Outerbridge M, Ward J, Manuel S, et al. (2007) Large-scale decline in offshore seagrass meadows in Bermuda. Mar Ecol Prog Ser 339: 123–130.

49. Fourqurean JW, Manuel S, Coates, Kenworthy WJ, Smith SR (2010) Effects of excluding sea turtle herbivores from a seagrass bed: Overgrazing may have led to loss of seagrass meadows in Bermuda. Mar Ecol Prog Ser 419: 223–232.

50. Szmant AM, Forrester A (1996) Water column and sediment nitrogen and phosphorus distribution patterns in the Florida Keys, USA. Coral Reefs 15: 21–41.

51. Fourqurean JW, Zieman JC (2002) Nutrient content of the seagrass *Thalassia testudinum* reveals regional patterns of relative ability of nitrogen and phosphorus in the Florida Keys USA. Biogeochemistry 61: 229–245.

52. Carruthers TJB, Van Tussenbroek BI, Dennison WC (2005) Influence of submarine springs and wastewater on nutrient dynamics of Caribbean seagrass meadows. Estuar Coast Shelf Sci 64: 191–199

53. Van Tussenbroek BI (2011) Dynamics of seagrasses and associated algae in coral reef lagoons. Hidrobiológica 21: 293–310.

54. Rodríguez-Ramírez A, Reyes-Nivia MC (2008) Evaluación rápida de los efectos del huracán Beta en la Isla Providencia (Caribe Colombiano). Boletín de Investigaciones Marinas y Costeras INVEMAR 37 (www.oceandocs.net/handle/1834/3481)

55. Rodriguez-Ramírez A, Garzón-Ferreira J, Batista-Morales A, Gil DL, Gómez-López DI, et al. (2010) Temporal patterns in coral reef, seagrass and mangrove communities from Chengue bay CARICOMP site (Colombia): 1993-2008. Rev Biol Trop 58 (Suppl. 3): 45–62.

56. Fonseca AC, Nielsen V, Cortés J (2007) Monitoreo de pastos marinos en Perezoso, sitio CARICOMP en Cahuita, Costa Rica. Rev Biol Trop 55: 55–66.

57. Cortés J, Fonseca AC, Nivia-Ruíz J, Samper-Villareal J, Salas E (2010) Monitoring coral reefs, seagrasses and mangroves in Costa Rica (CARICOMP). Rev Biol Trop 58 (Suppl. 3): 1–22.

58. D'Croz L, Del Rosario JB, Góndola P (2005) The effect of fresh water runoff on the distribution of dissolved inorganic nutrients and plankton in the Bocas del Toro Archipelago, Caribbean Panama. Caribb J Sc 41: 414–429.

18

Effects of Warming and Clipping on Ecosystem Carbon Fluxes across Two Hydrologically Contrasting Years in an Alpine Meadow of the Qinghai-Tibet Plateau

Fei Peng*, Quangang You, Manhou Xu, Jian Guo, Tao Wang, Xian Xue*

Key Laboratory of Desert and Desertification, Chinese Academy of Sciences, Cold and Arid Regions Environmental and Engineering Research Institute, Chinese Academy of Sciences, Lanzhou, China

Abstract

Responses of ecosystem carbon (C) fluxes to human disturbance and climatic warming will affect terrestrial ecosystem C storage and feedback to climate change. We conducted a manipulative experiment to investigate the effects of warming and clipping on soil respiration (Rs), ecosystem respiration (ER), net ecosystem exchange (NEE) and gross ecosystem production (GEP) in an alpine meadow in a permafrost region during two hydrologically contrasting years (2012, with 29.9% higher precipitation than the long-term mean, and 2013, with 18.9% lower precipitation than the long-tem mean). Our results showed that GEP was higher than ER, leading to a net C sink (measured by NEE) over the two growing seasons. Warming significantly stimulated ecosystem C fluxes in 2012 but did not significantly affect these fluxes in 2013. On average, the warming-induced increase in GEP ($1.49 \ \mu$ mol m^{-2}s^{-1}) was higher than in ER ($0.80 \ \mu$ mol m^{-2}s^{-1}), resulting in an increase in NEE ($0.70 \ \mu$ mol m^{-2}s^{-1}). Clipping and its interaction with warming had no significant effects on C fluxes, whereas clipping significantly reduced aboveground biomass (AGB) by 51.5 g m^{-2} in 2013. These results suggest the response of C fluxes to warming and clipping depends on hydrological variations. In the wet year, the warming treatment caused a reduction in water, but increases in soil temperature and AGB contributed to the positive response of ecosystem C fluxes to warming. In the dry year, the reduction in soil moisture, caused by warming, and the reduction in AGB, caused by clipping, were compensated by higher soil temperatures in warmed plots. Our findings highlight the importance of changes in soil moisture in mediating the responses of ecosystem C fluxes to climate warming in an alpine meadow ecosystem.

Editor: Ben Bond-Lamberty, DOE Pacific Northwest National Laboratory, United States of America

Funding: Financial support came from the Foundation for Excellent Youth Scholars of CAREERI, CAS (351191001), National Natural Science Foundation of China (41301210, 41201195 and 41301211), and Chinese Academy of Sciences (Hundred Talents Program). The funders had no role in study design, data collection and analysis, decision to publish, or preparation of the manuscript.

Competing Interests: The authors have declared that no competing interests exist.

* Email: pengguy02@yahoo.com (FP); xianxue@lzb.ac.cn (XX)

Introduction

Global mean temperature has increased by 0.76°C since the year 1850 and is predicted to rise an additional 1.8–4°C by the end of the 21st century [1]. Elevated global temperature can substantially impact the global carbon (C) budget, resulting in positive or negative feedbacks to global climate change [2,3]. The balance between C fixed by photosynthesis and C emitted to the atmosphere through plant and heterotrophic respiration determines the rate of terrestrial C storage [4].

Studies have shown that global warming could stimulate both ecosystem C uptake and emission across various terrestrial biomes [5]. However, the response of net C balance to warming is highly variable because of different temperature and soil moisture sensitivities in the processes that control C uptake and emission [6]. It is generally assumed that the terrestrial ecosystem might act as a net C source under a global warming scenario because the processes controlling ecosystem C emission are more sensitive to higher temperatures than the processes controlling C uptake [3,7,8]. However, some evidence indicates that warming could increase net C uptake, and global C models project enhanced terrestrial CO_2 uptake in response to warming through the middle of this century [9,10]. Current and completed experimental studies that have investigated warming effects have focused mostly on net primary productivity (NPP), biomass and soil respiration [5,11], from which the change in the C balance change was estimated. However, responses of gross primary production (GPP) and ecosystem respiration (ER), the major components net ecosystem exchange (NEE), to warming in field experiments [12] have received less attention in the alpine area [13].

Mowing (clipping) or grazing in grasslands, which account for 20% of the land use of the global terrestrial ice-free surface, may have substantial effects on ecosystem C fluxes, especially on a short-term basis [14]. Clipping would result in rapid changes in nutrient cycling [15], vegetation cover, plant community composition [16], and soil microclimate [17]. Collectively, these processes appear to stimulate the rate of ecosystem C cycling, however, their impacts on the net C balance are inconsistent [18,19].

Carbon stored in permafrost at high latitudes and in mountain areas is one of the major components of the terrestrial C pool. It is

Table 1. Results (F-values) of a three-way ANOVA on the effects of warming (W), clipping (C), measuring month (M), and their interactions on soil respiration (Rs), ecosystem respiration (ER), net ecosystem exchange (NEE) and gross ecosystem production (GEP).

	M	W	C	M×W	M×C	W×C	M×C×W
Rs	88.3**	3.9^	0.1	0.1	0.3	4.2*	1.1
ER	21.8**	8.3**	0.2	0.4	0.1	0.1	0.4
NEE	43.0**	4.8**	0.8	1.2	0.4	0.1	0.1
GEP	44.3**	9.5**	0.0	0.7	0.1	0.1	0.1
Rs/ER	1.5	0.3	0.1	2.5**	1.8	4.0**	0.1
ER/GEP	13.3**	0.4	0.5	2.1^	0.9	2.1	0.8
AGB	117.0**	3.8^	22.2**	1.2	2.3^	0.3	2.1^
RB	1.0	1.3	2.0	0.1	0.1	3.0^	0.2
AGB/RB	26.6**	10.3**	19.6**	6.6**	8.1**	13.0**	8.5**

Significance: ^, $P<0.1$; *, $P<0.05$; **, $P<0.01$.

estimated that soils in the permafrost regions store as much as 1672 Pg C (1 Pg = 10^{15} g), which is equivalent to double the atmospheric C pool [20,21]. Ecosystems in permafrost regions are C sinks because microbial decomposition of soil organic matter is inhibited under low annual mean temperature, and there is limited availability of organic C in frozen soil [22–24]. Altered growing season length, and changes in plant growth, ecosystem energy exchange and land use, together with the thawing of permafrost under a changing climate are projected to enhance the capability of ecosystem C uptake [21]. However, these altered dynamics do not appear to be able to compensate for the C released from thawing permafrost, resulting in ecosystems in permafrost regions acting as positive feedback to global change [21].

The Qinghai-Tibet Plateau (QTP) is experiencing a "much greater than average" increase in surface temperature, based on data observed at meteorological stations [25] and predictions from coupled climate-carbon models [1]. Grassland in the QTP is the largest vegetation unit of the Eurasian continent and covers an area of approximately 2.5 million km^2 [26]. Grazing is the most prevalent land use practice in the grassland. Results from eddy covariance measurements showed that the alpine meadow in the QTP is a weak C sink with annual variations [23,24]. Several studies have examined responses of ER and aboveground biomass to warming and clipping [17,27,28], in which the alpine meadow is thought to be a net C sink based on C balance calculations [28]. However, no field experiment has been conducted in the permafrost region of the QTP to measure the response of NEE, which provides a direct measure of the C balance. We conducted a two-year warming and clipping experiment to investigate how NEE and its components (GPP and ER) respond to warming and clipping, and how the associated changes in soil moisture, soil temperature, above- and belowground biomass affect the responses of ecosystem C fluxes in the permafrost region of the QTP.

Materials and Methods

Experimental site

The study site is situated in the region of the Yangtze River source, inland of the QTP near the Beilu River research station (34°49′N, 92°56′E, no specific permissions were required for activities in this location) at an altitude of 4635 m. This area has a typical alpine climate: mean annual temperature is −3.8°C and monthly air temperature ranges from −27.9°C in January to 19.2°C in July. Mean annual precipitation is 290.9 mm, of which over 95% falls during the warm growing season (May to October). Mean annual potential evaporation is 1316.9 mm, mean annual relative humidity is 57%, and mean annual wind velocity is 4.1 m s^{-1} [29]. The study site is a winter-grazed range, dominated by alpine meadow vegetation: *Kobresia capillifolia*, *K. pygmaea*, and *Carex moorcroftii*, with a mean plant height of 5 cm. Plant roots occur mainly within the 0–20 cm soil layer, and average soil organic C is 1.5%. The soil development is weak, and the soil belongs to alpine meadow soil (Chinese soil taxonomy), or is classified as a Cryosol according to World Reference Base, with a Mattic Epipedon at a depth of approximately 0–10 cm, and an organic-rich layer at a depth of 20–30 cm [30]. The parent soil material is of fluvioglacial origin and is composed of 99% sand. The Mattic Epipedon lowers the saturated soil water content, but increases soil water storage, and plant roots are dense and compressed within this layer. Permafrost thickness observed near the experimental site is 60–200 m and the depth of the active layer is 2.0–3.2 m [29,31]. However, because of climatic warming, the thickness of the active layer has been increasing at a rate of 3.1 cm y^{-1} since 1995 [32]. The experimental field is on a mountain slope

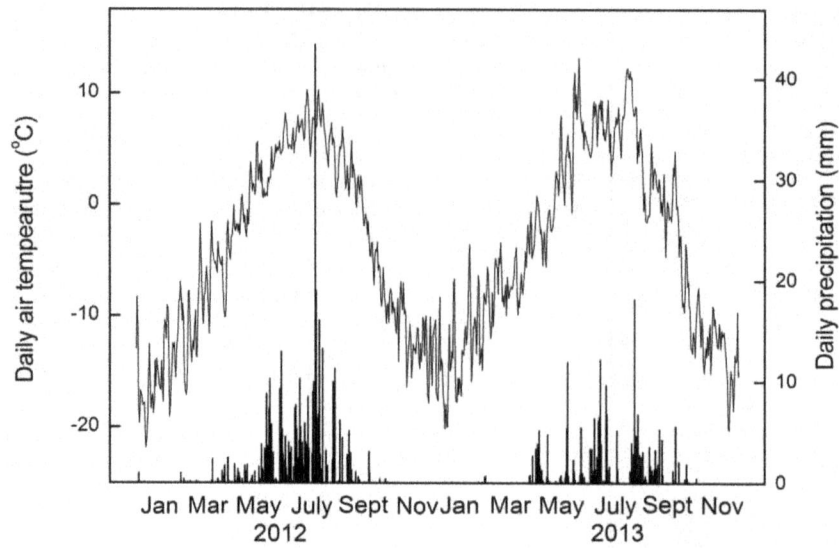

Figure 1. Daily precipitation (columns) and daily mean air temperature (line) in 2012 and 2013. Data are from the micro-meteorological station adjacent (approx. 100 m) to the experimental plots.

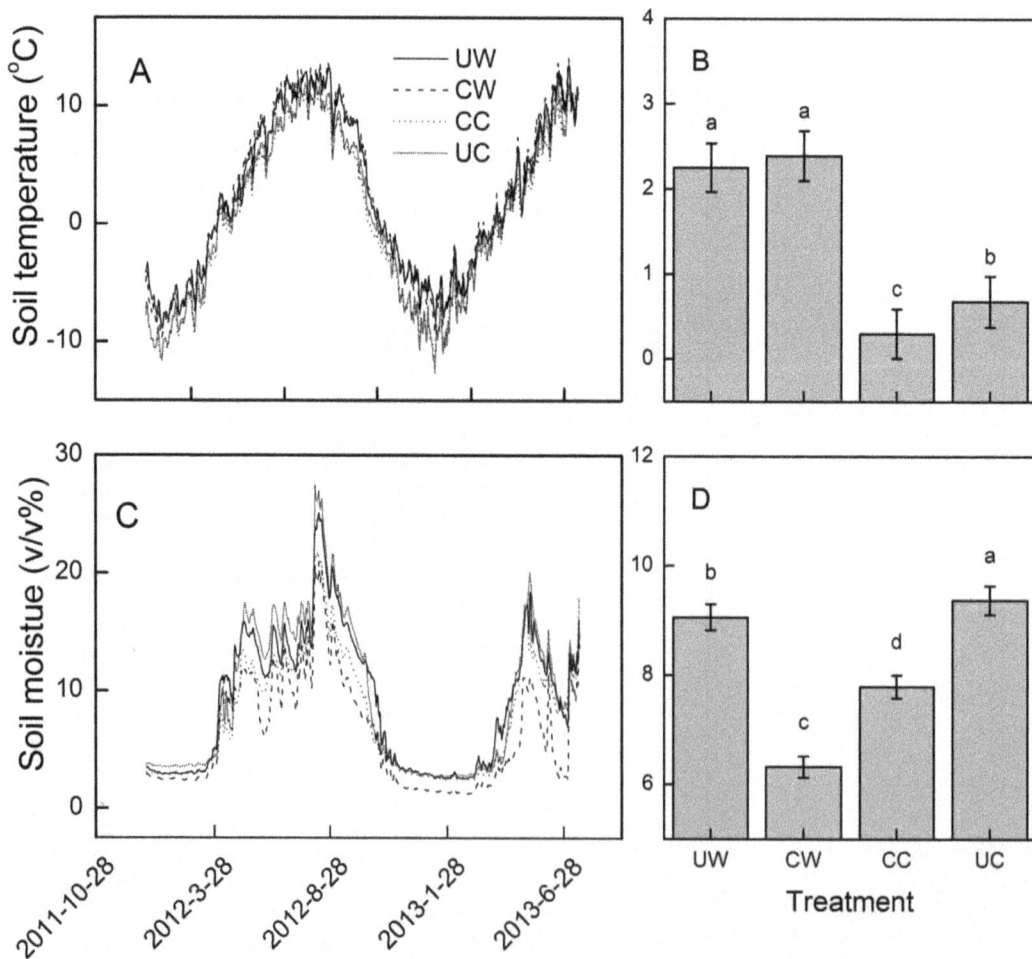

Figure 2. Daily soil temperature measured at a depth of 5 cm and volumetric soil moisture measured at a depth of 0–10 cm (A, C), and effects of warming and clipping on average (2012–2013) soil temperature (B) and soil moisture (d). UW, unclipped warming; CW, clipped warming; CC, clipped control; UC, unclipped control.

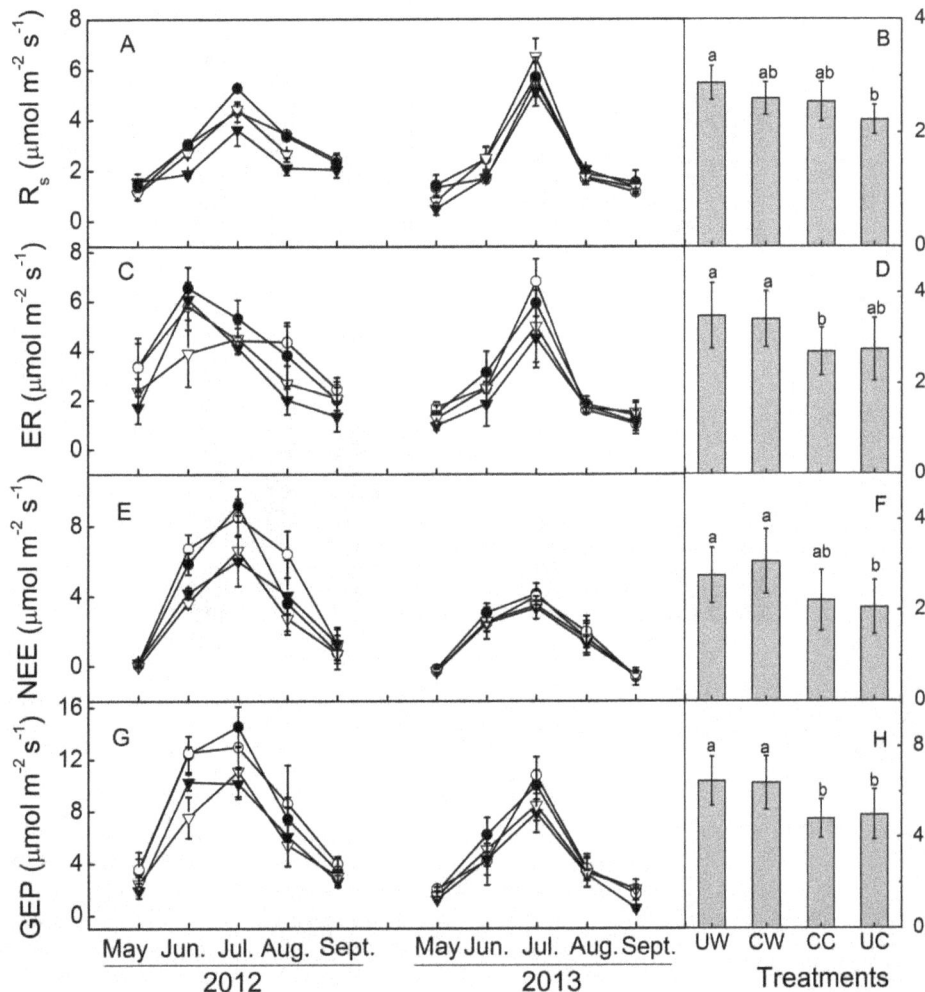

Figure 3. Mean monthly and overall means of soil respiration (Rs, A, B), ecosystem respiration (ER, C, D), net ecosystem exchange (NEE, E, F) and gross ecosystem production (GEP, G, H). Clipping was conducted in September 2011 and 2012. UW, unclipped warming; CW, clipped warming; CC, clipped control; UC, unclipped control.

with a mean incline of 5°. Detailed information about the soil properties is presented in Table 1.

Experimental design and measurement protocols

Experimental design. A two factorial experimental design (warming and clipping) was used with five replicates in each of the four treatments, i.e. unclipped control (UC), clipped control (CC), unclipped warming (UW) and clipped warming (CW). In total, 20 plots $(2 \times 2$ m$)$ were used in a complete randomized block distribution in the field. Plots were selected for homogeneity of topography, soil texture, aboveground biomass, and species composition. In each warmed plot, one 165 cm$\times$15 cm infrared heater (MR-2420, Kalglo Electronics Inc., Bethlehem, PA, USA) was suspended in the middle of each plot at a height of 1.5 m above the ground with a radiation output of 150 watts m^{-2}. The heating has been operating continuously since July 1st 2010. To simulate the shading effects of the heaters, one "dummy" heater,

made of a metal sheet with the same shape and size as the heaters, was also installed in the control plot.

Plants in the clipped plots were clipped at the soil surface on an annual basis, usually in last September. The rotational grazing system is that two ranches for each family, one is for the summer grazing and another is for the winter grazing. The rotational use of ranches is implemented usually in September. We consulted to the owner and they ensured that grassland of our study site is a winter grazing ranch. In our study, the clipping treatment was conducted in last September. One sheep needs 1.46×10^6 g grass per year [33] and the carrying capacity for alpine meadow is 1.39 head of sheep per hectare. Based on those data, about 203 g m^{-2} grass per year would be grazed for the alpine meadow. The aboveground biomass in September 2013 was about 320 g m^{-2}. In the clipping treatment, biomass cut was <320 g m^{-2}. Therefore we believe that our clipping treatment provided a reasonable simulation of local grazing practices.

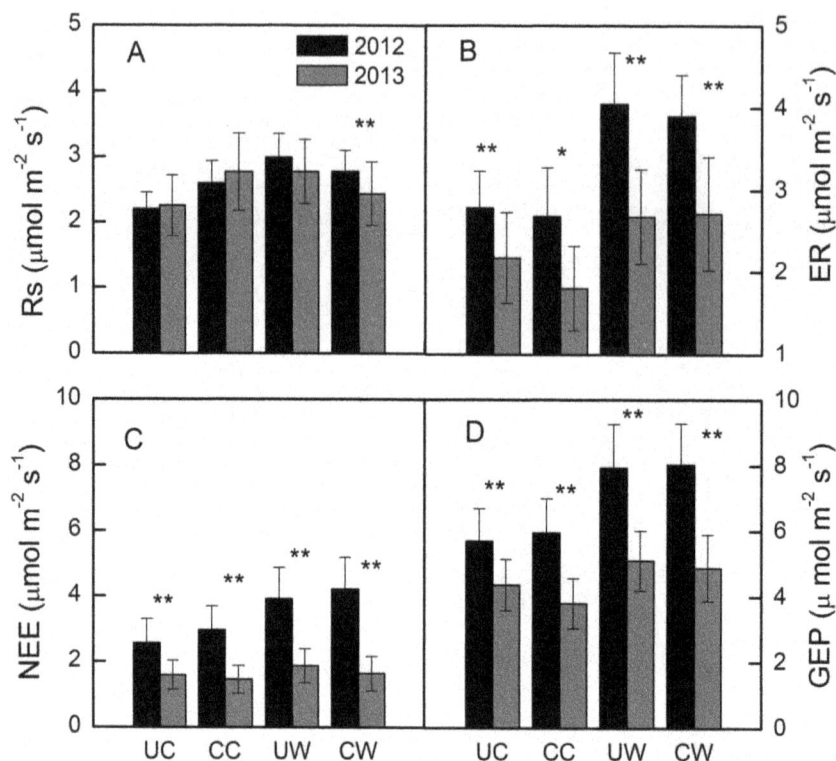

Figure 4. Mean annual soil respiration (Rs), ecosystem respiration (ER), net ecosystem exchange (NEE) and gross ecosystem production (GEP) under unclipped control (UC), clipped control (CC), unclipped warming (UW) and clipped warming (WC) treatments in 2012 and 2013. Symbols above the bars represent significant differences at $p < 0.05$ (*) and $p < 0.01$ (**).

Measurement protocol. Air temperature, water vapor pressure and relative humidity were monitored automatically at a height of 20 cm above the soil surface in the center of each plot using a Model HMP45C probe (Campbell Scientific Inc., Bethlehem, PA, USA). Nine thermistors were installed to monitor soil temperatures at depths of 5, 15, 30, 60, 100, 150, 200, 250 and 300 cm. All the probes were connected to a CR1000 datalogger (Campbell Scientific Inc.). Data recorded every 10 min were averaged and reported as daily values. Pavelka et al. (2007) stated that for grassland ecosystems, surface soil temperature is the most suitable depth for measuring soil temperature because of the optimized regression coefficient between surface soil temperature and soil respiration [34]. Therefore, we used soil temperature measured at a depth of 5 cm in the following analyses.

An EnviroSmart sensor (Sentek Pty Ltd., Stepney, Australia), which used frequency domain reflection, was used to monitor volumetric soil moisture at depths of 0–10, 10–20, 20–40, 40–60 and 60–100 cm. These soil moisture data were also recorded using a CR1000 datalogger. When analyzing the relationships between C fluxes and soil moisture, we used the daily average soil moisture data that were collected when ecosystem C flux measurements were conducted.

Soil respiration (Rs) was measured by using Licor-6400-09 (Lincoln, NE, USA) on PVC collars 5 cm in height and 10.5 cm in diameter, which were permanently inserted 2–3 cm into the soil in the center of each plot. Small living plants were cut at the soil surface at least one day before measurements to eliminate the effect of respiration from aboveground biomass [35]. ER and NEE were measured with a transparent chamber (0.5×0.5×0.5 m) attached to an infrared gas analyzer (IRGA, Licor-6400, Lincoln, NE, USA). The transparent chamber is a custom-designed chamber made of Polytetrafluoroethene (4 mm in thickness) with light transmittance about 99%. During measurements, a foam gasket was placed the chamber and the soil surface to minimize leaks. One small fan ran continuously to mix the air inside the chamber during measurements. Nine consecutive recordings of CO_2 concentration were taken in each plot at 10 s intervals during a 90 s period. Following the measurement of NEE, the chamber was vented for several minutes and covered with an opaque cloth for measuring ER, as the opaque cloth eliminated light (and hence photosynthesis). CO_2 flux rates were determined from the time-course of the CO_2 concentrations used to calculate NEE and ER. The method used was similar to that reported by Steduto et al. (2002) [36] and Niu et al. (2008) [37]. Gross ecosystem productivity (GEP) was the calculated as the sum of NEE and ER. Rs, NEE and ER were measured in each plot on a monthly basis from May to September in 2012 and 2013.

Aboveground biomass (AGB) was obtained from a step-wise linear regression with AGB as the dependent variable, and coverage and plant height as independent variables. 100 small plots (30 cm×30 cm) were included in the regression analysis (AGB = 22.76×plant height + 308.26×coverage − 121.80, $R^2 = 0.74$, $P < 0.01$). Coverage of each experimental plot was measured using a 10 cm×10 cm frame in four diagonally divided subplots replicated eight times. Plant height was measured 40 times by a ruler and averaged for each experimental plot. A biomass index was used as the ratio of the derived biomass on any given date to the maximum biomass during the entire study period [38]. Root biomass (RB) was obtained from soil samples that were air-dried for one week and passed through a 2- mm diameter sieve to remove large particles. Roots were separated from the soil by washing, and a 0.25-mm diameter sieve was used to retrieve fine

Table 2. Results (F-values) of a three-way ANOVA on the effects of warming (W), clipping (C), measuring month (M), and their interactions on soil respiration (Rs), ecosystem respiration (ER), net ecosystem exchange (NEE) and gross ecosystem production (GEP) in contrasting years.

	M	W	C	M×W	M×C	W×C	M×C×W
2012							
Rs	54.2**	11.6**	0.8	2.2^	0.3	5.1*	1.4
ER	13.9**	9.4**	0.1	0.6	0.4	0.01	1.2
NEE	46.9**	8.6**	2.6	2.0	1.3	0.0	0.04
GEP	40.3**	12.2**	0.2	0.9	0.6	0.1	0.2
ER/GEP	16.6**	0.05	0.2	0.6	0.1	0.8	0.8
AGB	33.9**	2.1	2.0	0.6	0.4	0.5	0.2
RB	7.2**	0.7	1.0	0.5	0.09	2.8	0.2
AGB/RB	2.4^	1.2	2.2	0.5	0.2	1.6	0.2
2013							
Rs	78.3**	0.1	0.02	0.8	0.6	3.3^	0.6
ER	31.0**	2.6	0.2	1.1	0.2	0.2	0.3
NEE	51.2**	1.1	0.3	0.3	0.1	0.01	0.1
GEP	44.2**	3.2^	0.9	0.7	0.4	0.2	0.3
ER/GEP	22.1**	9.0**	11.0**	13.9**	8.3**	8.8**	10.9**
AGB	89.9**	0.06	20.1**	1.3	0.8	0.5	1.1
RB	2.9*	0.7	1.2	0.2	0.3	0.8	0.2
AGB/RB	20.7**	1.5	6.2*	1.5	2.3^	2.6	2.7*

Significance: ^, $P<0.1$; *, $P<0.05$; **, $P<0.01$.

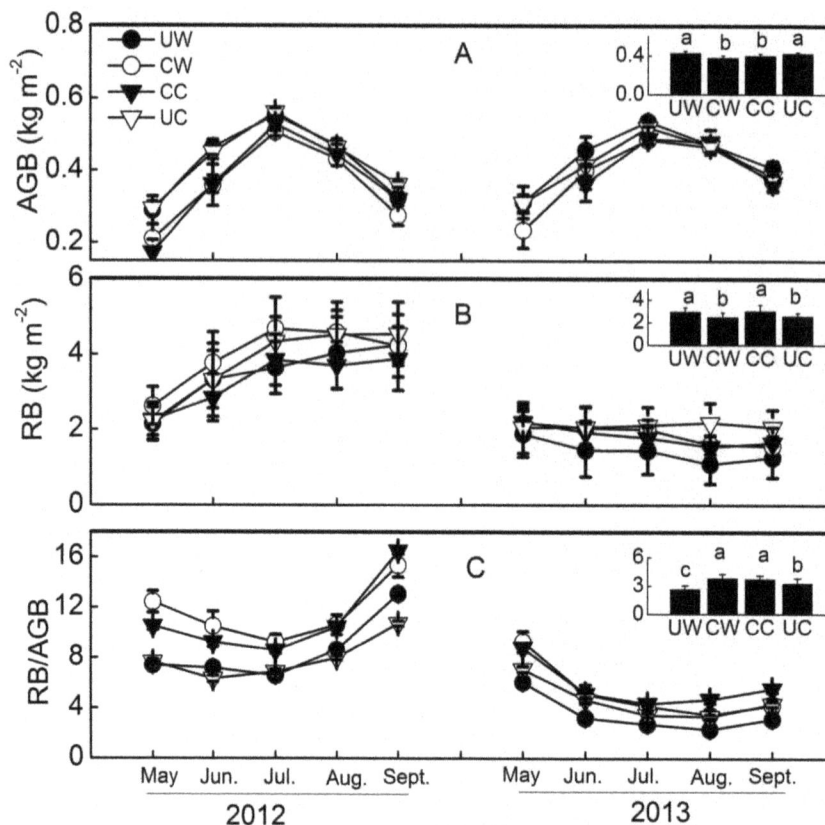

Figure 5. Temporal variations and overall means (inserted panels) of aboveground biomass (AGB, A), root biomass (RB, B) and the ratio of RB to AGB (RB/AGB, C). Clipping was conducted in September 2011 and 2012. See Figures 2 and 3 for notes and abbreviations.

roots. Living roots were separated from dead roots by their color and consistency [39]. Separated roots were dried at 75°C for 48 h.

Data analysis

Temperature and soil moisture data used in analyses were from January 1st 2012 to July 18th 2013 because a power failure prevented data from being collected from July 19th 2013 onwards. The effect of the warming and clipping treatments on soil temperature (5 cm), soil moisture (0–10 cm), Rs, ER, NEE, GEP, AGB and RB were determined with a three-way analysis of variance (ANOVA) using SPSS Version 18.0. (SPSS, Inc., Chicago, IL, USA).

Relationships between C fluxes and soil microclimate (soil temperature and soil moisture) were examined using daily soil microclimate data that were collected when ecosystem C fluxes

were measured. Linear regression analyses were used to examine the relationships of C fluxes with abiotic (soil moisture and soil temperature) and biotic factors (monthly AGB and RB).

Results

Microclimate

In comparison to the long-term average (1981–2008) mean annual air temperature (MAT, $-5.1°C$), higher MAT values were recorded in 2012 and 2013 ($-3.5°C$ and $-3.8°C$, respectively). Annual precipitation in 2012 (420.1 mm, Fig. 1) was higher than the long-term mean annual precipitation (294.5 mm), but it was lower than the long-term mean in 2013 (238.8 mm) (Fig. 1).

Experimental warming significantly elevated annual mean soil temperature (Figs. 2A and 2B, $P<0.05$). In unwarmed plots, the average daily soil temperature at 5 cm depth was 0.65°C and

Table 3. Fitted quadratic models of the relationships between ecosystem respiration (ER), net ecosystem exchange (NEE), gross ecosystem production (GEP) and soil moisture (θ, v/v%, 10 cm). Max. F, θ represents the value of θ when ER, NEE and GEP are at their maximum.

	ER/μmol m^{-2}s^{-1}	NEE/μmol m^{-2}s^{-1}	GEP/μmol m^{-2}s^{-1}
Fitted model	$ER = -0.058\theta^2 + 1.71\theta - 7.72$	$NEE = -0.076\theta^2 + 2.37\theta - 12.53$	$GEP = -0.139\theta^2 + 4.21\theta - 21.02$
R^2	0.31	0.45	0.36
p	0.007	<0.001	<0.001
Max. F, θ/%	14.7%	15.6%	15.1%

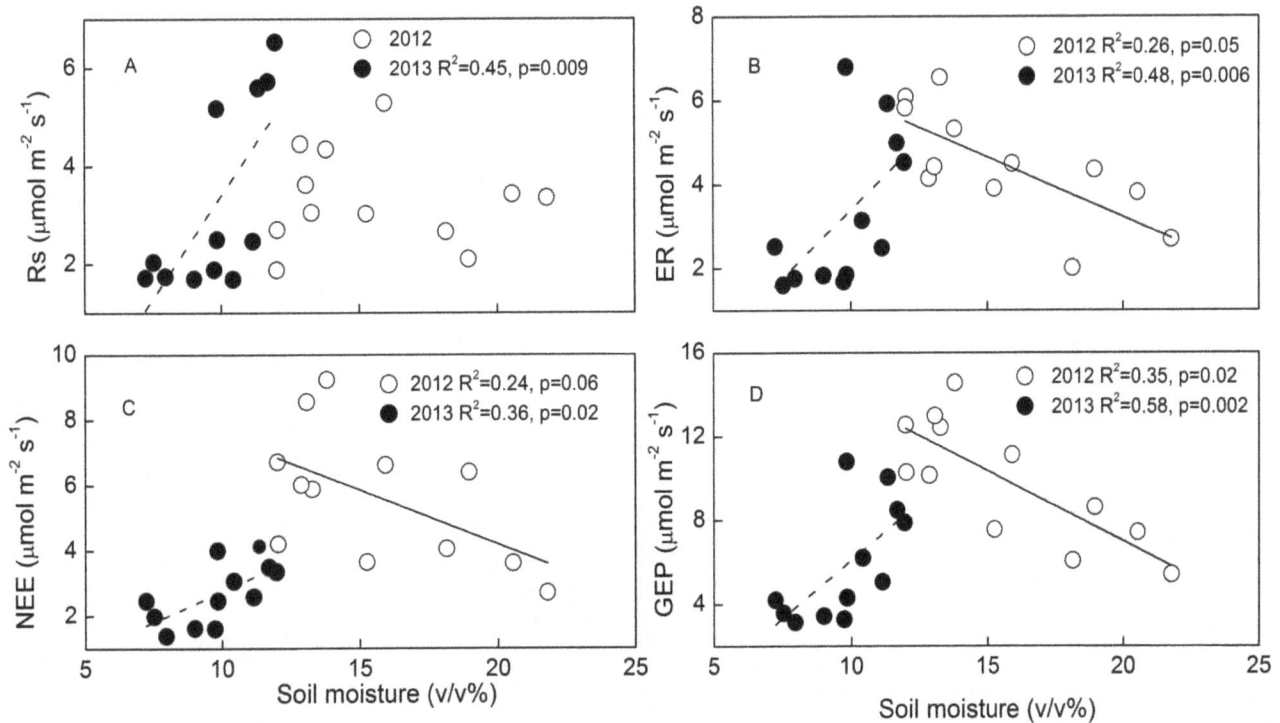

Figure 6. The relationships between soil moisture (0–10 cm) and ecosystem C fluxes: soil respiration (Rs), ecosystem respiration (ER), net ecosystem exchange (NEE) and gross ecosystem productivity (GEP) in 2012 (hollow circles) and 2013 (solid circles), respectively. Soil moisture and ecosystem C fluxes data were the average for all plots in each month. Data for both years were collected from June to August.

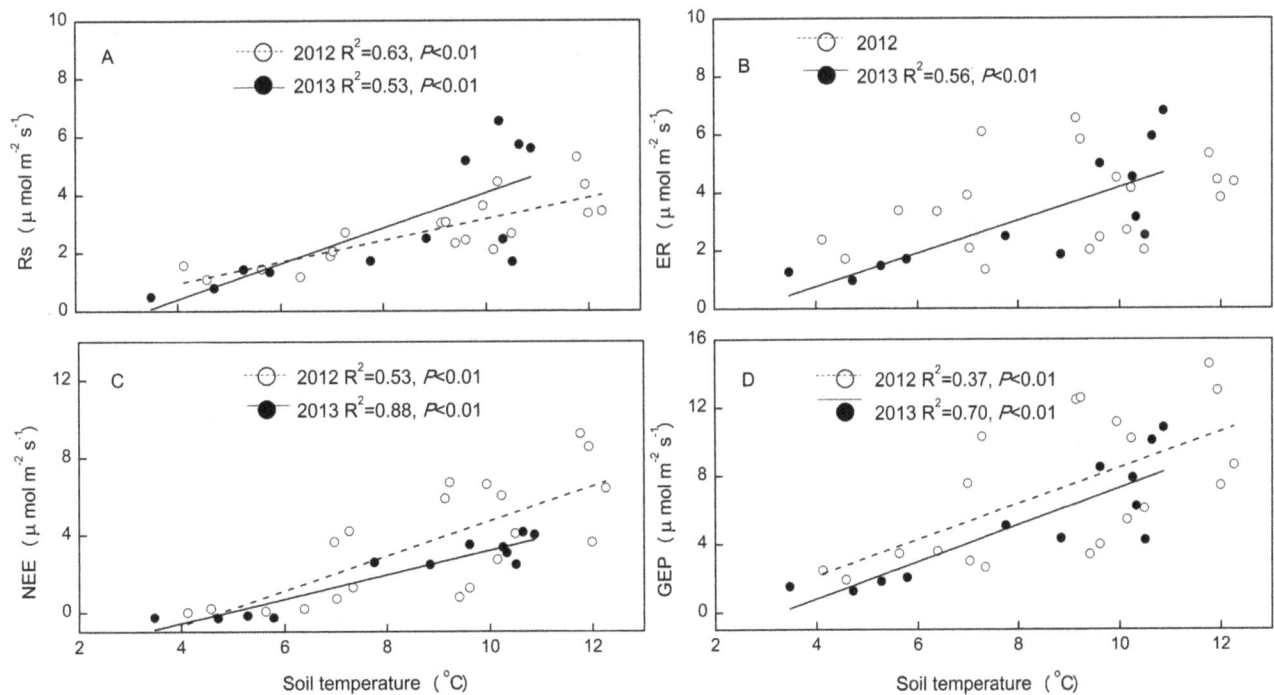

Figure 7. The relationships between soil temperature (5 cm) and ecosystem C fluxes: soil respiration (Rs), ecosystem respiration (ER), net ecosystem exchange (NEE) and gross ecosystem productivity (GEP). Soil temperature and ecosystem C fluxes data were the average of all plots in each month.

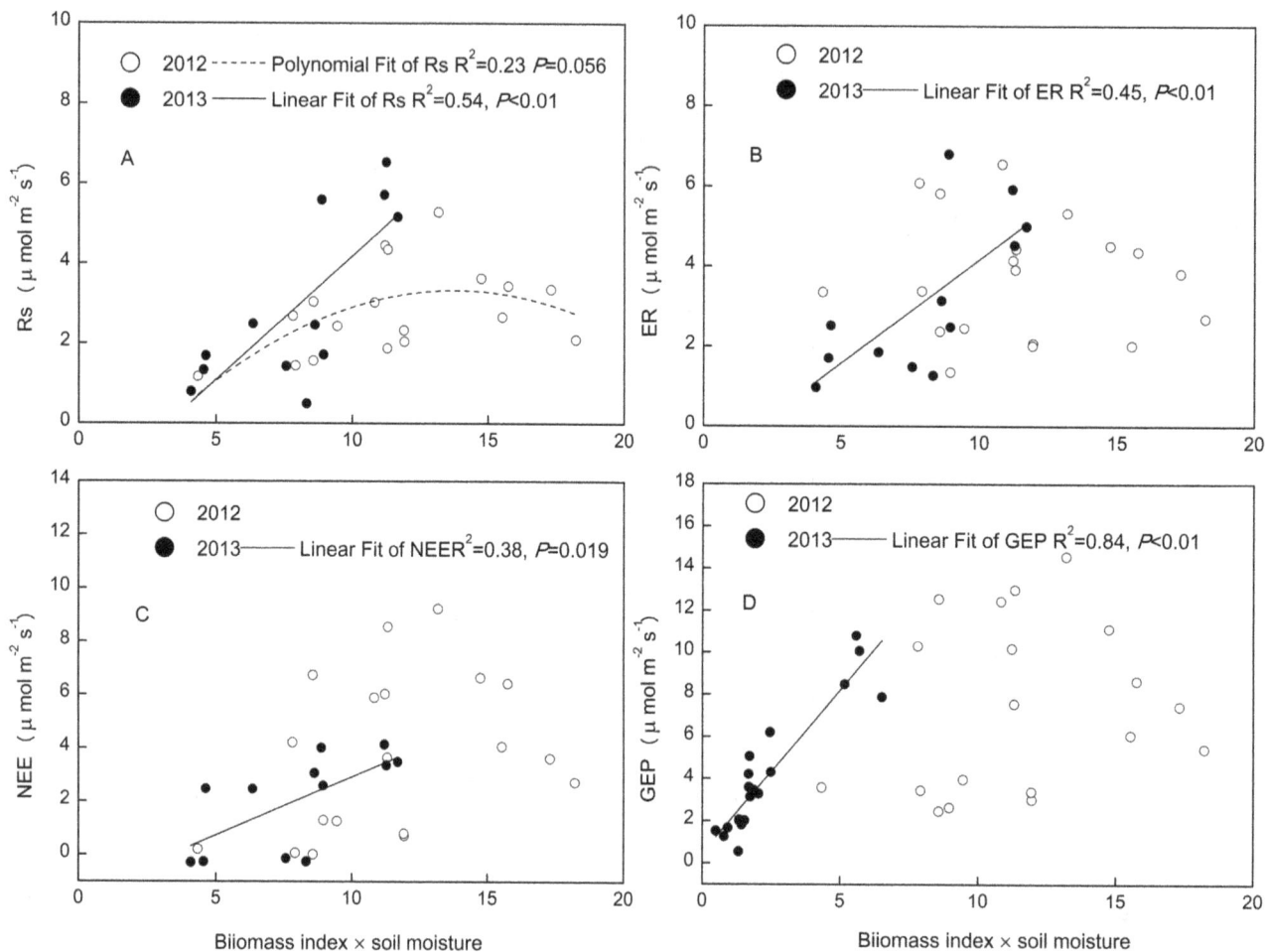

Figure 8. The relationships between changes in ecosystem C fluxes: soil respiration (Rs), ecosystem respiration (ER), net ecosystem exchange (NEE) and gross ecosystem productivity (GEP), and the changes in the product of abveground biomass index and soil moisture during 2012 and 2013.

0.14°C in 2012 and 2013, respectively. Warming significantly increased the soil temperature by 1.96°C (2012) and 1.59°C (2013) (Fig. 2B). The average daily soil temperature in unclipped plots was 1.69°C and 0.99°C in 2012 and 2013, respectively, but was unaffected by clipping. Volumetric soil moisture measured over 0–10 cm fluctuated greatly over the study period (Fig. 2C). The average daily soil moisture in unwarmed plots was 9.31 (v/v%) and 7.22 (v/v%), and warming significantly reduced soil moisture by 7.4% and 17.4% (P<0.05) in 2012 and 2013, respectively (Fig. 2D). The average daily soil moisture in unclipped plots was 10.08 (v/v%) and 7.61 (v/v%), and clipping significantly decreased it by 28.3% and 36.5% in 2012 and 2013, respectively (Fig. 2D).

Warming and clipping effects on C fluxes

The temporal dynamics of Rs, ER, NEE, and GEP followed the seasonal patterns of air and soil temperature in both years, which peaked in mid-growing season (Figs. 1 and 3). Substantial inter-annual variations in ecosystem C fluxes were observed in this study (Fig. 3). The annual average ER, NEE and GEP were all significantly higher in 2012 than in 2013 (Fig. 4) in all treatments, but higher Rs in 2012 was only observed in the CW treatment (Fig. 4A). On average, NEE, ER and GEP were 47%, 22% and 34% higher, respectively, in 2012 than in 2013.

Warming significantly increased NEE ($P = 0.03$), whereas no significant effects of clipping ($P = 0.37$) or its interaction with warming ($P = 0.83$) were detected (Table 1). When analyzed separately by year using a three-way ANOVA, warming only significantly increased NEE in 2012, by 28.5% (Table 2). Warming induced an enhanced growing season mean NEE in 2012, which was lower in clipped (17%) than in unclipped plots (30%). Measuring date had a significant effect on NEE, but the interaction of measuring date with warming or clipping had no effect on NEE in either year (Table 2).

Similar to NEE, average GEP was significantly increased by warming ($P = 0.003$) but not by clipping ($P = 0.97$) or by their interaction ($P = 0.87$, Table 1). When analyzed separately by year using a three-way ANOVA, warming significantly increased average GEP (Table 2) by 2.13 μ mol m^{-2}s^{-1} in 2012 and marginally enhanced it by 0.82 μ mol m^{-2}s^{-1} in 2013. The increased GEP caused by warming was lower in clipped (23%) than in unclipped plots (28.3%) in 2012, but GEP was higher in clipped (22.6%) than in unclipped plots (13.4%) in 2013.

Warming also significantly increased average Rs ($P = 0.052$) and ER ($P = 0.005$), but clipping had no significant effect on average Rs ($P = 0.73$) or ER ($P = 0.66$, Table 1). Similar to NEE, when analyzed separately by year using a three-way ANOVA, the effect of warming on ER and Rs was only significant in 2012 (Table 2),

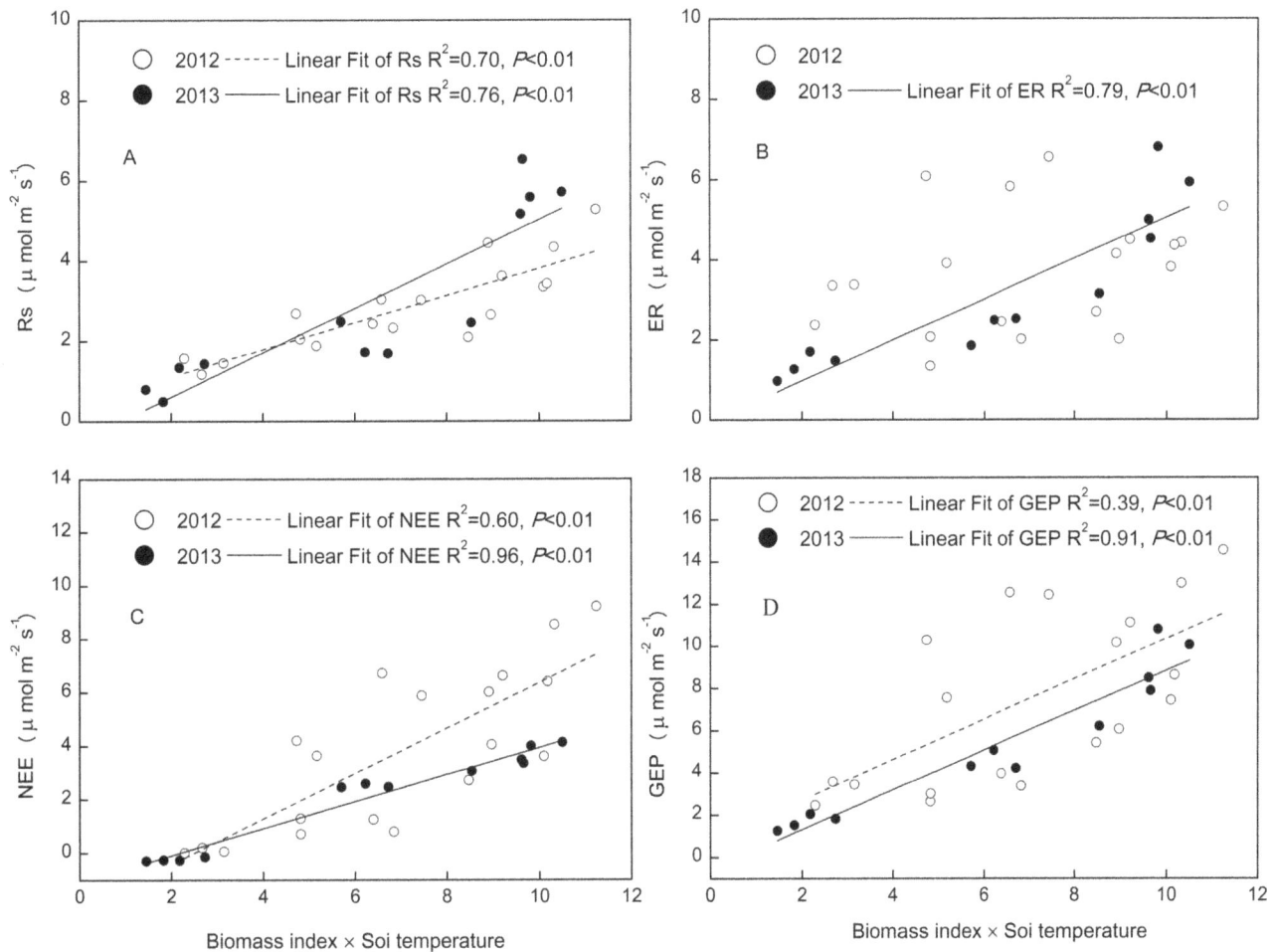

Figure 9. The relationships between changes in ecosystem C fluxes: soil respiration (Rs), ecosystem respiration (ER), net ecosystem exchange (NEE) and gross ecosystem productivity (GEP), and the changes in the product of aboveground biomass index and soil temperature during 2012 and 2013.

which increased by 25.8% and 17%, respectively. The interaction between warming and clipping had a significant effect on Rs but not on ER (Table 1). The average increase in Rs caused by warming was lower in clipped (5.9%) than in unclipped plots (27.4%).

There was no significant effect of warming or clipping on the ER to GEP ratio (ER/GEP, $P = 0.51$ and $P = 0.46$ for 2012 and 2013, respectively, Table 1). When analyzed separately by year using a three-way ANOVA, ER/GEP was significantly affected by warming, clipping, measurement date and their interactions in 2013 (Table 2).

Warming and clipping effects on biomass

Similar to the inter-annual variation of ecosystem C fluxes, RB was significantly lower in 2013 (by a value of 50.2%) than in 2012, whereas there was no difference in AGB over the two growing seasons (Fig. 5).

Warming marginally increased AGB ($p = 0.053$) and clipping significantly reduced AGB ($p<0.001$), but there was no significant effect on RB ($p = 0.26$ and $p = 0.16$ for warming and clipping, respectively, Table 1). When analyzed separately by year using a three-way ANOVA, warming had no significant effect on AGB or RB in either year, whereas clipping significantly reduced AGB in

2013 (Table 2). The reduction in AGB by clipping was higher in unwarmed (14.4%) plots than in warmed plots (10.3%) in 2013.

Impacts of biotic and abiotic factors on ecosystem C fluxes

Over the two growing seasons, there was no clear relationship between ER, NEE, GEP and soil moisture, but there was a quadratic relationship between these variables and soil moisture when May and September data were excluded (Table 3). The optimal soil moisture for ER, NEE and GEP was about 15% (Table 3). When plotted separately, ER, NEE, and GEP decreased linearly with increasing soil moisture in 2012, and increased linearly with increasing soil moisture in 2013 (Fig. 6).

The temperature response curves for Rs, ER, NEE and GEP in 2012 were quite similar to those recorded in 2013. However, the slope of relationship between Rs and soil temperature was higher in 2012 than in 2013, but that between NEE and soil temperature was smaller in 2012 than in 2013 (Fig. 7).

The statistical interaction term of above-ground biomass index and soil moisture showed polynominal relationship only with Rs in 2012 (Fig. 8a), whereas it linearly correlated with all the ecosystem C fluxes in 2013 (Fig. 8). The interaction term of above-ground biomass and soil temperature explained more variation in ecosystem C fluxes than did the interaction term of above-ground

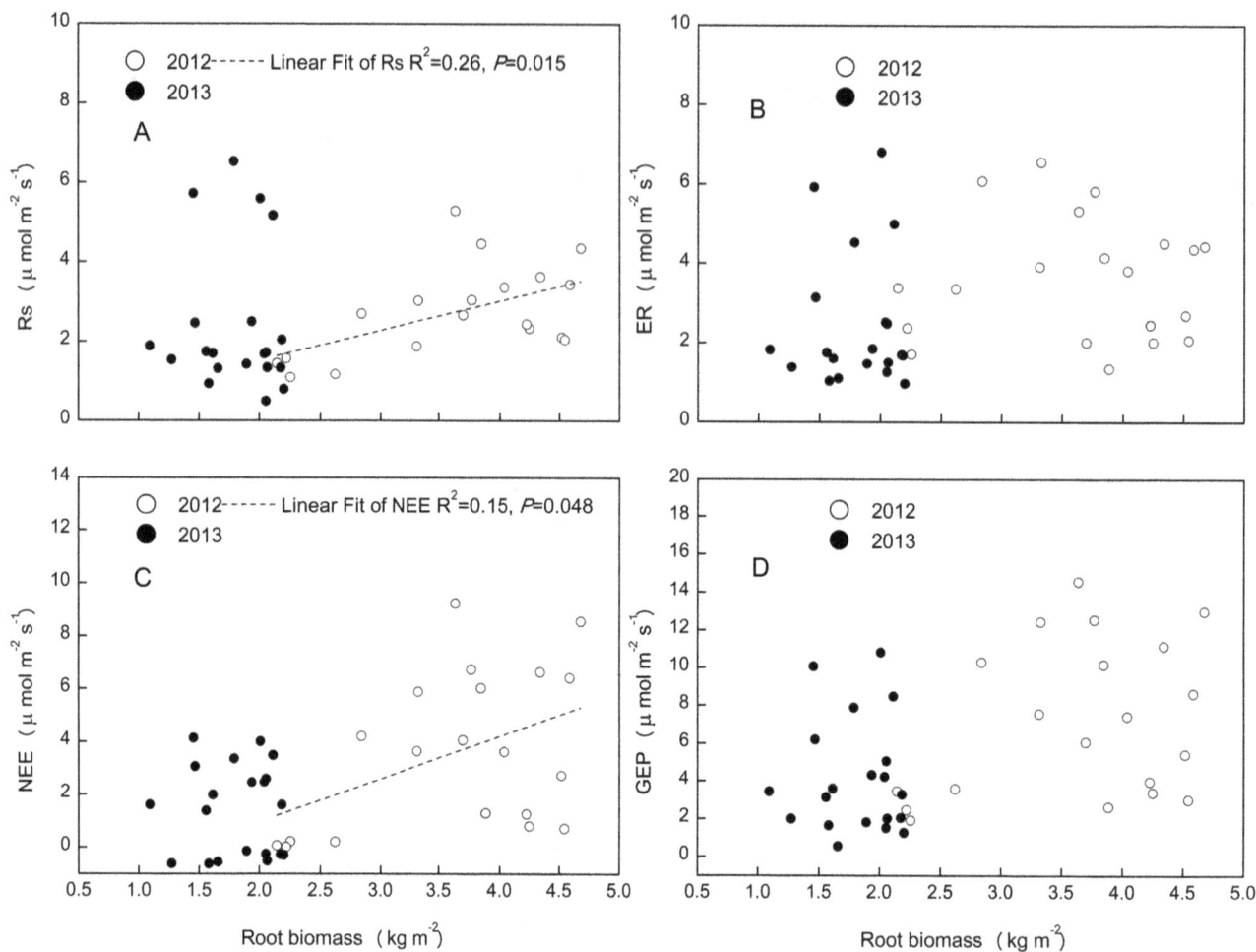

Figure 10. Relationship between root biomass and ecosystem C fluxes: soil respiration (Rs), ecosystem respiration (ER), net ecosystem exchange (NEE) and gross ecosystem productivity (GEP).

biomass and soil moisture (Figs. 8 and 9). The fitting slope between Rs and the interaction of above-ground biomass and soil temperature was higher in 2013 than in 2012 (Fig. 9a), but that between NEE and the interaction of above-ground biomass and soil temperature was smaller in 2013 than in 2012 (Fig. 9c). Root biomass only had effect on ecosystem C fluxes in wet year (Fig. 10).

Discussion

C fluxes and their inter-annual variation

In this alpine meadow ecosystem studied in the QTP, a higher uptake of C (GEP) than release (ER) resulted in a net C sink (Figs. 3–4). This result is similar to that reported in seasonally frozen areas in the QTP [23,24], and in some arctic ecosystems [7].

Higher precipitation and associated higher soil moisture in 2012 than in 2013, and the similar MAT between the two years suggest that drought reduced annual ER, NEE and GEP in 2013. Our results are in agreement with those from temperate grassland ecosystems [37,40,41]. The quadratic relationship between ecosystem C fluxes and soil moisture (Table 3) at the temporal scale support the above findings, as ecosystem C fluxes were positively related to soil moisture in 2013 (Fig. 6). There was greater fluctuation in GEP than ER: GEP was 34% lower in 2013

than in 2012, compared to ER, which was 22% lower in 2013 than in 2012. These results are consistent with those from boreal and temperate forests [42] and temperate grasslands [40], which indicate that GEP is more sensitive to inter-annual climatic variation than ER in alpine meadow ecosystems. Differences in the magnitudes of the inter-annual variation in GEP and ER could be explained by differences in the slopes between these variables when plotted against soil moisture (Fig. 6). The greater dependence of GEP than ER on soil moisture across both years suggests that GEP is more sensitive to changes in soil moisture than ER. Despite the significant inter-annual difference in ER, NEE and GEP, Rs did not differ significantly between the two years. This indicates that annual variations in Rs may be controlled by other factors, such as soil temperature, as Rs had the highest slope when plotted against this variable (Fig. 7).

Drought typically reduces aboveground biomass in grasslands [43,44]. However, in our study, no significant reduction in AGB was observed in 2013. One reason could be the various reactions of different species to drought [45,46], and this compensation may hold community AGB constant. The 50.2% decrease in RB in 2013 could be attributed to the reduction in soil moisture because RB in alpine ecosystems is positively correlated with annual precipitation [47]. The relative reduction rate of RB in our study (50.2% reduction in RB and 2.1 m^3 m^{-3} reduction in soil moisture) was higher than in a temperate grassland ecosystem

(23% reduction in RB and 5.2 m^3 m^{-3} reduction in soil moisture) [48]. The divergent responses of RB in the surface and deep soil layers [46,49] could cancel each other out and therefore lead to a lower relative decrease in RB across the whole soil profile.

However, only one wet and dry year was included in this study. The wet year was followed directly by a dry year, which might lowers the effect of drought because soil drought might lag the meteorological drought.

Main effects of warming

Experimental warming stimulated GEP more than ER (1.49 μ mol $m^{-2}s^{-1}$ vs. 0.79 μ mol $m^{-2}s^{-1}$), leading to an increase in NEE in the warming treatment of this alpine meadow ecosystem in a permafrost area of the QTP. Warming effects on ecosystem C exchange are likely modulated by soil water regimes [7,37]. For example, Oberbauer (2007) reported that higher soil moisture in wet tundra limited increases in ER relative to increases in GEP under warming conditions, indicating the dependence of the warming effect on hydrological conditions. In the current alpine meadow ecosystem, differences in the responses of NEE to warming between 2012 and 2013 (Table 2) differed from results from a temperate steppe, in which NEE demonstrated no change under a warming treatment over two hydrologically contrasting years [40]. Soil moisture showed positive impacts on C fluxes in 2013 and negative impacts on fluxes in 2012 (Fig. 6). ABG and RB were positively correlated with ecosystem C fluxes (Figs. 8, 9). As there were no significant effects of warming on AGB or RB in either year (Tables 1, 2), the significant increase in NEE in warmed plots in 2012 could be attributed to the higher stimulation of GEP (2.09 μ mol $m^{-2}s^{-1}$) than ER (1.07 μ mol $m^{-2}s^{-1}$).

The insensitivity of NEE to warming in 2013 could be attributed to the effect of the soil moisture deficit on GEP and ER (Table 2, Fig. 6). The positive responses of GEP and ER to warming (Table 1) are consistent with those in a tundra ecosystem [7], but differ from those in a subalpine meadow ecosystem, where soil moisture stress induced by warming reduced ER [27,40]. ER is composed of Rs and respiration of AGB. Therefore, the significant increase in ER in 2012 could be attributed mainly to the stimulation in Rs (Table 2), as AGB was insensitive to warming (Table 2). This indicates that the response of ER to warming was determined by Rs even though AGB respiration is the major component of ER in alpine meadow ecosystems [50]. Rs is composed of root respiration and microbial decomposition of soil organic matter [51]. There was no significant change in RB at a depth of 0–10 cm in the warming treatment (Tables 1, 2), which suggests that the response of ecosystem C emission to warming is determined by soil organic matter decomposition. The non-significant response of ER to warming in 2013 likely resulted from lower soil moisture (less than 15%, the optimal soil moisture for ecosystem C fluxes, Table 3), and the warming-induced reduction in soil moisture (Fig. 2). This is because the negative effects of drought and warming induced soil water stress on Rs and ER, which could override the positive effect of warming on these variables, which has been shown for a Montane meadow [52] and a subalpine meadow ecosystem in the QTP [27]. The marginal increase in GEP in 2013 (Table 2) likely resulted from a change in the species composition, which was observed in an open top chamber warming experiment nearby our study site, where coverage of grass and sedges decreased but that of forbs increased with warming [46]. The increased forbs biomass could ameliorate the negative impact of warming-induced soil moisture stress and the effect of lower AGB on GEP [40].

Although experimental warming tends to have a positive effect on plant productivity across ecosystems, experiments in grasslands indicate that clear increases in plant productivity in response to warming are relatively rare [53]. We did not detect a significant change in RB. We attributed this to the fact that we sampled the RB at a depth of 0–10 cm, which was constrained by a reduction in soil water, whereas the RB at a depth of 10–50 cm was significantly stimulated by warming. In contrast to the response of forbs, sedges and grass [28] may have cancelled each other out, leading to the non-significant change in AGB in this alpine meadow ecosystem. AGB and RB were positively correlated with ecosystem C fluxes (Figs. 8, 9). There was no significant change in AGB or RB with warming, whereas ecosystem C fluxes were significantly stimulated by warming (Tables 1, 2). It is possible that this resulted from the large seasonal variation in AGB and RB compared to the relatively smaller warming-induced changes in these biomass pools.

Main effects of clipping

There was no significant effect of clipping on C fluxes, which contrasted with results from other studies where increases in GEP, ER, and NEE have been reported for a temperature steppe [54] and tallgrass prairie [13], and decreases in GEP, ER, and NEE have been reported for a Swiss grassland [55]. The negative impact of clipping on ecosystem C fluxes is attributed to the grass being cut in the middle of the growing season, which may reduce the green leaf area and thus C fluxes [55]. Positive effects of clipping on C fluxes may result primarily from improved light conditions with the removal of standing litter [54] and compensatory growth from clipping [56]. In the current study, we clipped the plants in late September once they had started to senesce, and this could be one reason for the non-significant effect of clipping on GEP, NEE and ER. In addition, soil temperature has been found to influence CO_2 exchange in alpine meadow ecosystems [24], and we did not detect a significant effect of clipping on soil temperature (Fig. 2). Besides temperature, biomass also affects C fluxes in an alpine meadow ecosystem [24], as was observed in our study (Fig. 8). The significant decrease in AGB in 2013 under the clipping treatment with a non-significant change in C fluxes, suggests that soil temperature is the major factor controlling the response of ecosystem C fluxes to clipping.

Conclusion

Ecosystem C fluxes responded positively to elevated temperature, with a higher relative increase in GEP than in ER, leading to a net C gain in this alpine meadow ecosystem. Clipping and its interaction with warming had no significant effect on ecosystem C fluxes because clipping did not significantly affect soil temperature. In addition, this study was conducted during two hydrologically contrasting years (wet in 2012 and dry in 2013), which provided a unique opportunity to understand how drought affects ecosystem C fluxes and their response to warming and clipping in an alpine meadow ecosystem. In the dry year, positive effects of warming on ecosystem C fluxes were cancelled by lower soil moisture. However, we caution that our study encompassed only a single wet and dry year, and thus our inferences of drought need to be supported by future research. Our findings will improve our understanding of the response of ecosystem C fluxes to the combined effects of climate change factors and human activities in an alpine meadow ecosystem in the permafrost region of the QTP.

Acknowledgments

Authors thank Prof. Yongzhi, Liu, Hanbo Yun, Guilong Wu, and Yuanwu Yang for their help in setting up the field experiment.

Author Contributions

Conceived and designed the experiments: FP XX. Performed the experiments: FP QY MX JG. Analyzed the data: FP. Contributed to the writing of the manuscript: FP TW XX.

References

1. IPCC (2007) Climate change 2007: The physical Science Basis Contributin of Working group I to the Fourth Assessment Report of the Intergovernmental Panel on Climate Change. In: S S., D Qin, M Manning, Z Chen, M Marquis, K. B Averyt, M Tignor and H. L Miller, editors. Cambridge, United Kingdom/ New York, NY USA: Cambridge University Press. 749–766.
2. Luo YQ, Wan SQ, Hui DF, Wallance LL (2001) Acclimatization of soil respiration to warming in a tall grass prairie. Nature 413: 622–625.
3. Melillo JM, Steudler PA, Abler JD (2002) Soil warming and carbon-cycle feedbacks to the climate system. Science 298: 2173–2175.
4. Friedlingstein P, Cox P, Betts R, Bopp L, von Bloh W, et al. (2006) Climate-Carbon Cycle Feedback Analysis: Results from the C4MIP Model Intercomparison. J. Climate 19: 3337–3353.
5. Rustad LE, Campbell JL, Marion GM, Norby RJ, Mitchell MJ, et al. (2001) A meta-analysis of the response of soil respiration, net nitrogen mineralization, and aboveground plant growth to experimental ecosystem warming. Oecologia 126: 543–562.
6. Peñuelas J, Gordon C, Llorens L, Nielsen T, Tietema A, et al. (2004) Nonintrusive Field Experiments Show Different Plant Responses to Warming and Drought Among Sites, Seasons, and Species in a North-South European Gradient. Ecosystems 7: 598–612.
7. Oberbauer SF, Tweedie CE, Welker JM, Fahnestock JT, Henry GHR, et al. (2007) Tundra CO2 fluxes in response to experimental warming across latitudinal and moisture gradients. Ecol. Monogr. 77: 221–238.
8. Kirschbaum MF (1995) The temperature dependence of soil organic mater decomposition, and the effect of global warming on soil organic C storage. Soil Biolo. Biochem. 27: 753–760.
9. Cramer W, Bondeau A, Woodward FI, Prentice IC, Betts RA, et al. (2001) Global response of terrestrial ecosystem structure and function to CO2 and climate change: results from six dynamic global vegetation models. Global Change Biolo. 7: 357–373.
10. Canadell JG, Le Quéré C, Raupach MR, Field CB, Buitenhuis ET, et al. (2007) Contributions to accelerating atmospheric CO2 growth from economic activity, carbon intensity, and efficiency of natural sinks. PNAS 104: 18866–18870.
11. Wu ZT, Dijkstra P, Koch GW, PeÑUelas J, Hungate BA (2011) Responses of terrestrial ecosystems to temperature and precipitation change: a meta-analysis of experimental manipulation. Global Change Biolo. 17: 927–942.
12. Lu M, Zhou XH, Yang Q, Li H, Luo YQ, et al. (2013) Responses of ecosystem carbon cycle to experimental warming: a meta-analysis. Ecology.
13. Niu SL, Sherry RA, Zhou XH, Luo YQ (2013) Ecosystem carbon fluxes in responses to warming and clipping in a tallgrass prairie. Ecosystems.
14. Bahn M, Knapp M, Garajova Z, Pfahringer N, Cernusca A (2006) Root respiration in temperate mountain grasslands differing in land use. Global Change Biolo. 12: 995–1006.
15. Ross DJ, Tate KR, Scott NA, Feltham CW (1999) Land-use change: effects on soil carbon, nitrogen and phosphorus pools and fluxes in three adjacent ecosystems. Soil Biolo. Biochem. 31: 803–813.
16. Klein J, Harte J, Zhao X (2004) Experimental warming causes large and rapid species loss, dampened by simulated grazing, on the Tibetan Plateau. Ecol. Lett. 7: 1170–1179.
17. Luo CY, Xu GP, Chao ZG, Wang SP, Lin XW, et al. (2010) Effect of warming and grazing on litter mass loss and temperature sensitivity of litter and dung mass loss on the Tibetan plateau. Global Change Biol. 16: 1606–1617.
18. Derner JD, Boutton TW, Briske DD (2006) Grazing and ecosystem carbon storage in the North American Great Plains. Plant Soil 280: 77–90.
19. Niu SL, Sherry RA, Zhou XH, Wan SQ, Luo YQ (2010) Nitrogen regulation of the climate-carbon feedback: evidence from a long-term global change experiment. Ecology 91: 3261–3273.
20. Tarnocai C, Canadell JG, Schuur EAG, Kuhry P, Mazhitova G, et al. (2009) Soil organic carbon pools in the northern circumpolar permafrost region. Global Biogeochem. Cy. 23: GB2023.
21. Schuur EAG, Bockheim J, Canadell JG (2008) Vulnerability of permafrsot carbon to climate change:implication for the global carbon cycle. BioScience 58: 701–714.
22. Harden JW, Sundquist ET, Stallard RF, Mark RK (1992) Dynamics of soil carbon during deglaciation of the Laurentide ice sheet. Science 258: 1921–1924.
23. Kato T, Tang Y, Gu S, Cui X, Hirota M, et al. (2004) Carbon dioxide exchange between the atmosphere and an alpine meadow ecosystem on the Qinghai-Tibetan Plateau, China. Agr. Forest Meteorol. 124: 121–134.
24. Kato T, Tang Y, Gu S, Hirota M, Du M, et al. (2006) Temperature and biomass influences on interannual changes in CO2 exchange in an alpine meadow on the Qinghai-Tibetan Plateau. Global Change Biol. 12: 1285–1298.
25. Liu XD, Chen BD (2000) Climatic warming in the Tibetan Plateau during recent decades. Inter. J. Climatol. 20: 1729–1742.
26. Zheng D, Zhang QS, Wu SH (2000) Mountain Geoecology and sustainable development of the Tibetan Plateau. Dordrecht, Netherlands: Kluwer Academic Publishers.
27. Lin XW, Zhang ZH, Wang SP, Hu YG, Xu GP, et al. (2011) Response of ecosystem respiration to warming and grazing during the growing seasons in the alpine meadow on the Tibetan plateau. Agr. Forest Meteorol. 151: 792–802.
28. Li N, Wang GX, Yang Y, Gao YH, Liu GS (2011) Plant production, and carbon and nitrogen source pools, are strongly intensified by experimental warming in alpine ecosystems in the Qinghai-Tibet Plateau. Soil Biol. Biochem. 43: 942–953.
29. Lu Z, Wu Q, Yu S, Zhang L (2006) Heat and water difference of active layers beneath different surface conditions near Beiluhe in Qinghai-Xizang Plateau. J. Glaciol. Geogryol. 28: 642–647.
30. Wang G, Wang Y, Li Y, Cheng H (2007) Influences of alpine ecosystem responses to climatic change on soil properties on the Qinghai-Tibet Plateau, China. Catena 70: 506–514.
31. Pang Q, Cheng G, Li S, Zhang W (2009) Active layer thickness calculation over the Qinghai-Tibet Plateau. Cold Regions Science and Technology 57: 23–28.
32. Wu QB, Liu YZ (2004) Ground temperature monitoring and its recent change in Qinghai-Tibet Plateau. Cold Reg. Sci. Technol. 38: 85–92.
33. Yang ZL, Yang GH (2000) Potential productivity and livestock carrying capacity of high-frigid grassland in China. Resources Sci. 22: 72–77.
34. Pavelka M, Acosta M, Marek MV, Kutsch W, Janous D (2007) Dependence of the Q10 values on the depth of soil temperature measuring point. Plant Soil 292: 171–179.
35. Zhou XH, Wan SQ, Luo YQ (2007) Source components and interannual variability of soil CO2 efflux under experimental warming and clipping in a grassland ecosystem. Global Change Biol. 13: 761–775.
36. Steduto P, Çetinkökü Ö, Albrizio R, Kanber R (2002) Automated closed-system canopy-chamber for continuous field-crop monitoring of CO2 and H2O fluxes. Agr. Forest Meteorol.111: 171–186.
37. Niu S, Wu M, Han Y, Xia J, Li L, et al. (2008) Water-mediated responses of ecosystem carbon fluxes to climatic change in a temperate steppe. New Phytol. 177: 209–219.
38. Lawrence BF, Bruce GJ (2005) Interacting effects of temperature, soil moisture and plant biomass production on ecosystem respiration in a northern temperate grassland. Agr. Forest Meteorol. 130: 237–253.
39. Yang Y, Fang J, Ji C, Han W (2009) Above- and belowground biomass allocation in Tibetan grasslands. J. Veg. Sci. 20: 177–184.
40. Xia JY, Niu SL, Wan SQ (2009) Response of ecosystem carbon exchange to warming and nitrogen addition during two hydrologically contrasting growing seasons in a temperate steppe. Global Change Biol. 15: 1544–1556.
41. Williams M, Law BE, Anthoni PM, Unsworth MH (2001) Use of a simulation model and ecosystem flux data to examine carbon-water interactions in ponderosa pine. Tree Physiol. 21: 287–298.
42. Barr AG, Griffis TJ, Black TA, Lee X, Staebler RM, et al. (2002) Comparing the carbon budgets of boreal and temperate deciduous forest stands. Can. J. Forest Res. 32: 813–822.
43. Kahmen A, Perner J, Buchmann N (2005) Diversity-dependent productivity in semi-natural grasslands following climate perturbations. Funct. Ecol. 19: 594–601.
44. Gilgen AK, Buchman N (2009) Responses of tempeartue grasslands at different altitudes to simulated summer drought differed but scaled with annual precipitation. Biogeosciences 6: 2525–2539.
45. Sebastià M-T (2007) Plant guilds drive biomass response to global warming and water availability in subalpine grassland. J. Appl. Ecol. 44: 158–167.
46. Li N, Wang GX, Yang Y, Gao YH, Liu LA, et al. (2011) Short-term effects of temperature enhancement on community structure and biomass of alpine meadow in the Qinghai-Tibet Plateau. Acta Ecol. Sinica 31: 895–905.
47. Li XJ, Zhang XZ, Wu JS, Shen ZX, Zhang YJ, et al. (2011) Root biomass distribution in alpine ecosystem of the northern Tibet Plateau. Environ. Earth Sci. 64: 1911–1919.
48. De Boeck HJ, Lemmens CMHM, Gielen B, Bossuyt H, Malchair S, et al. (2007) Combined effects of climat warming and plant diversity loss on above- and below-ground productivity. Environ. Exp.l Bot. 60: 95–104.
49. Xu MH, Peng F, Xue X, You QG, Guo J (2014) All-year warming and autumnal clipping lead to the downward movement of the root biomass, carbon and total nitrogen in the soil of an alpine meadow. Environ. and Exp. Bot.: in press.
50. Zhang PC, Tang YH, Hirota M, Yamamoto A, Mariko S (2009) Use of regression method to partition sources of ecosystem respiration in an alpine meadow. Soil Biol. Biochem. 41: 663–670.
51. Hanson PJ, Edwards NT, Garten CT, Andrews JA (2000) Separating root and soil microbial contributions to soil respiratin: a review of methods and observations. Biogeochem. 48: 115–146.

52. Saleska S, Harte K, Torn M (1999) The effect of experimental ecosystem warming on CO_2 fluxes in a montane meadow. Global Change Biol. 5: 125–141.

53. Dukes JS, Chiariello NR, Cleland EE, Moore LA, Shaw MR, et al. (2005) Responses of Grassland Production to Single and Multiple Global Environmental Changes. PLoS Biol 3: e319.

54. Niu SL, Wu M, Han Y, Xia J, Zhang Z, et al. (2010) Nitrogen effects on net ecosystem carbon exchange in a temperate steppe. Global Change Biol. 16: 144–155.

55. Rogiers N, Eugster W, Furger M, Siegwolf R (2005) Effect of land management on ecosystem carbon fluxes at a subalpine grassland site in the Swiss Alps. Theor. Appl. Climatol. 80: 187–203.

56. Zhao W, Chen S-P, Lin G-H (2008) Compensatory growth responses to clipping defoliation in Leymus chinensis (Poaceae) under nutrient addition and water deficiency conditions. Plant Ecol. 196: 85–99.

19

Ancient Clam Gardens Increased Shellfish Production: Adaptive Strategies from the Past Can Inform Food Security Today

Amy S. Groesbeck[1]*, Kirsten Rowell[2], Dana Lepofsky[3], Anne K. Salomon[1]*

1 School of Resource and Environmental Management, Simon Fraser University, Burnaby, British Columbia, Canada, **2** Department of Biology, University of Washington, Seattle, Washington, United States of America, **3** Department of Archaeology, Simon Fraser University, Burnaby, British Columbia, Canada

Abstract

Maintaining food production while sustaining productive ecosystems is among the central challenges of our time, yet, it has been for millennia. Ancient clam gardens, intertidal rock-walled terraces constructed by humans during the late Holocene, are thought to have improved the growing conditions for clams. We tested this hypothesis by comparing the beach slope, intertidal height, and biomass and density of bivalves at replicate clam garden and non-walled clam beaches in British Columbia, Canada. We also quantified the variation in growth and survival rates of littleneck clams (*Leukoma staminea*) we experimentally transplanted across these two beach types. We found that clam gardens had significantly shallower slopes than non-walled beaches and greater densities of *L. staminea* and *Saxidomus giganteus*, particularly at smaller size classes. Overall, clam gardens contained 4 times as many butter clams and over twice as many littleneck clams relative to non-walled beaches. As predicted, this relationship varied as a function of intertidal height, whereby clam density and biomass tended to be greater in clam gardens compared to non-walled beaches at relatively higher intertidal heights. Transplanted juvenile *L. staminea* grew 1.7 times faster and smaller size classes were more likely to survive in clam gardens than non-walled beaches, specifically at the top and bottom of beaches. Consequently, we provide strong evidence that ancient clam gardens likely increased clam productivity by altering the slope of soft-sediment beaches, expanding optimal intertidal clam habitat, thereby enhancing growing conditions for clams. These results reveal how ancient shellfish aquaculture practices may have supported food security strategies in the past and provide insight into tools for the conservation, management, and governance of intertidal seascapes today.

Editor: Simon Thrush, University of Auckland, New Zealand

Funding: This work was funded by a National Geographic Research and Exploration Grant (8636-09) to Dana Lepofsky, a Natural Sciences and Engineering Research Council of Canada Discovery Grant (385921-2009) to Anne K. Salomon, a Social Sciences and Humanities Research Council Insight Grant (2011-0833) to Dana Lepofsky, and a David and Lucile Packard Foundation Grant (2008-32497) to Kirsten Rowell. The funders had no role in study design, data collection and analysis, decision to publish, or preparation of the manuscript.

Competing Interests: The authors have declared that no competing interests exist.

* E-mail: amysue72@gmail.com (ASG); anne.salomon@sfu.ca (AKS)

Introduction

Sustaining global food production presents one of the greatest environmental and humanitarian challenges of the 21st century. Given current global population and consumption trajectories, the world's food production must double by 2040 [1], [2] and its footprint must shrink substantially to reduce the degradation of land, water, biodiversity, and climate. Consequently, society will need to develop clever ways to meet demands on terrestrial and marine resources and spaces efficiently, while maintaining ecosystem productivity and resilience. Fortunately, evidence from the past often offers solutions to contemporary quandaries [3], [4]. Here, we provide empirical evidence of an ancient form of mariculture that magnified shellfish production in a limited space, providing practical insights into sustainable marine management techniques which may inform local food security strategies of today.

Humans have been altering, exploiting, and managing marine and terrestrial ecosystems for millennia [5], [6], [7], [8]. Throughout history, human hunting and fishing in coastal ecosystems has caused declines in key species [9], reduced prey size [10], [11], triggered trophic cascades [12], [13], and facilitated ecosystem regime shifts [11]. In other cases, the archaeological record indicates long term sustained yields, with little indication of resource depression [14]. Recent archaeological evidence and oral historical knowledge suggests that First Peoples around the world actively managed and enhanced nearshore ecosystems to maintain and increase productivity [15], [16], [17], [8].

Several ancient environmental engineering and resource management strategies have been documented among coastal indigenous peoples. Enforced harvest size restrictions of forest resources in Fiji [18]; construction and tending of root gardens [16]; transplantation and cultivation of berries to increase yields [19]; and prescribed burns to clear land and magnify plant production [20] along the Northwest Coast provide examples of intentional management of terrestrial coastal resources. Globally, marine examples include stone ponds for fish aquaculture in Polynesia [21]; complex wooden fish weirs in Brittany [22]; and temporal and age-specific harvest restrictions to conserve reef fish in Oceania [15]. Along the northwest coast of North America,

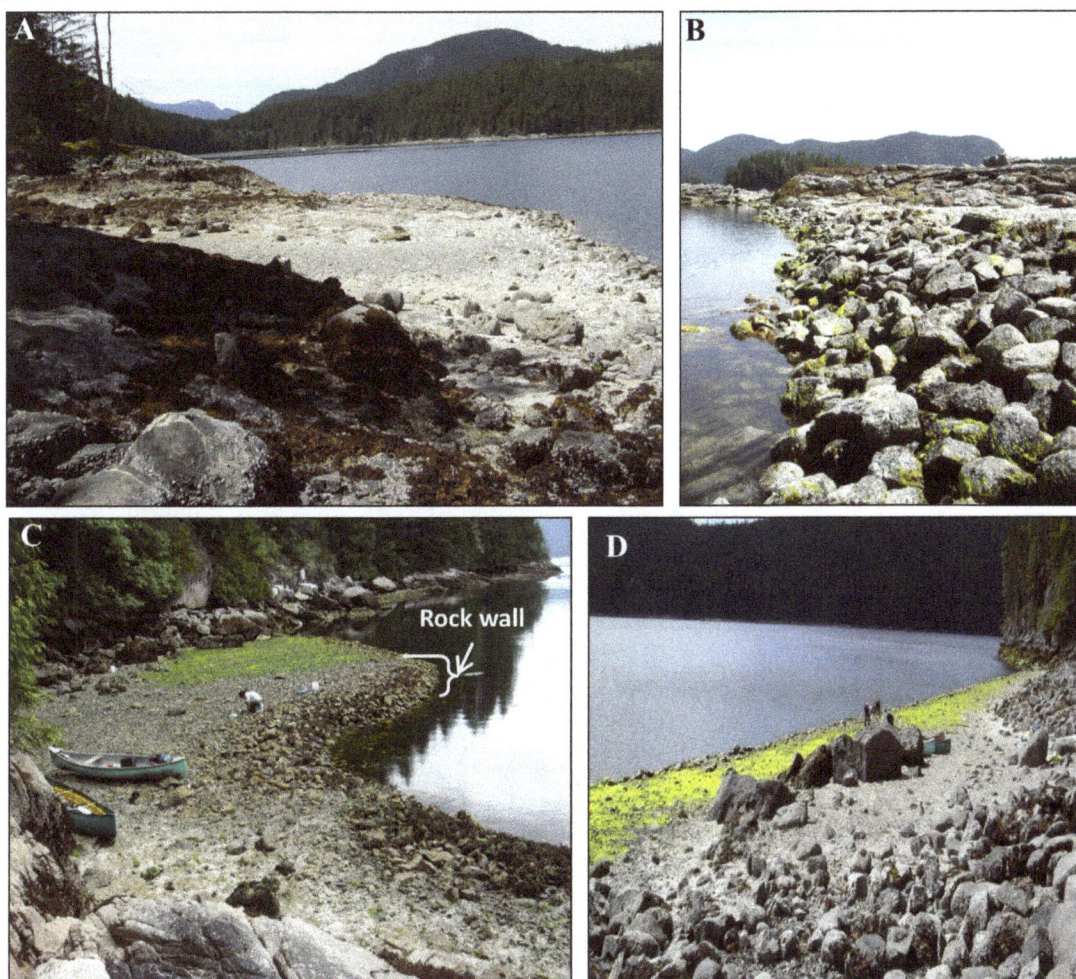

Figure 1. Clam Garden Images. A) Ancient clam gardens on Quadra Island, BC, Canada, are intertidal beach terraces built by humans by constructing B) a rock wall at low tide typically between 0.7–1.3 m above chart datum. C, D) Quadra Island clam gardens range in size and shape but generally create shallow sloping intertidal terraces encompassing tidal heights of 0.9–1.5 m above chart datum.

marine examples include intertidal stone fish traps at the mouths of salmon bearing streams and intertidal wooden fish weirs [23], [24] in conjunction with size selective fishing practices to enhance salmon productivity [25] and the reduction of predatory sea otters to increase shellfish abundance [7]. These records of direct environmental manipulations, tending, and stewardship practices suggest complex systems of resource management which increased local food security [5], [26], [27].

Researchers have recently turned their attention to ancient clam gardens along the Northwest coast highlighting another example of ancient marine enhancement and management. These human-engineered intertidal terraces, thought to have been constructed in the late Holocene, have been recorded from Alaska through British Columbia (BC) and into Washington State [28]. Made by building rock walls in the low intertidal of soft-sediment clam beaches, clam gardens are thought to stabilize sediments at a specific tidal height, presumably to enhance shellfish productivity [28], [29], [30], [17]. Clam gardens walls exist at the mouths and along the edges of embayments, parallel to coastlines, and vary in shape, length, width, and intertidal height (Fig 1A-D). Although the age of these ancient features is currently unresolved, the

immense shell middens associated with clam garden walls suggest the significance of shellfish as a staple food source for Northwest Coast First Nations for at least 5000 years [31]. The combination of the widespread occurrence of clam gardens on the northwest coast, their associated shell middens, and traditional knowledge of clam garden tending passed down in song, story, and practice [32], [16], underscores the importance of these features and suggests that they were constructed to increase clam yields. Knowing how and the extent to which clam gardens boost clam yields may offer insights into contemporary investments in food security for coastal communities.

In this study, we quantified the productivity of ancient clam gardens on Quadra Island, BC with surveys and an *in situ* transplant experiment. We measured how bivalve communities and beach morphology differs between clam gardens and non-walled beaches, and what environmental factors contribute to these differences. Specifically, we ask, do clam gardens have higher clam densities, biomass, and growth rates compared to non-walled beaches? And if so, what physical characteristics best explain these differences?

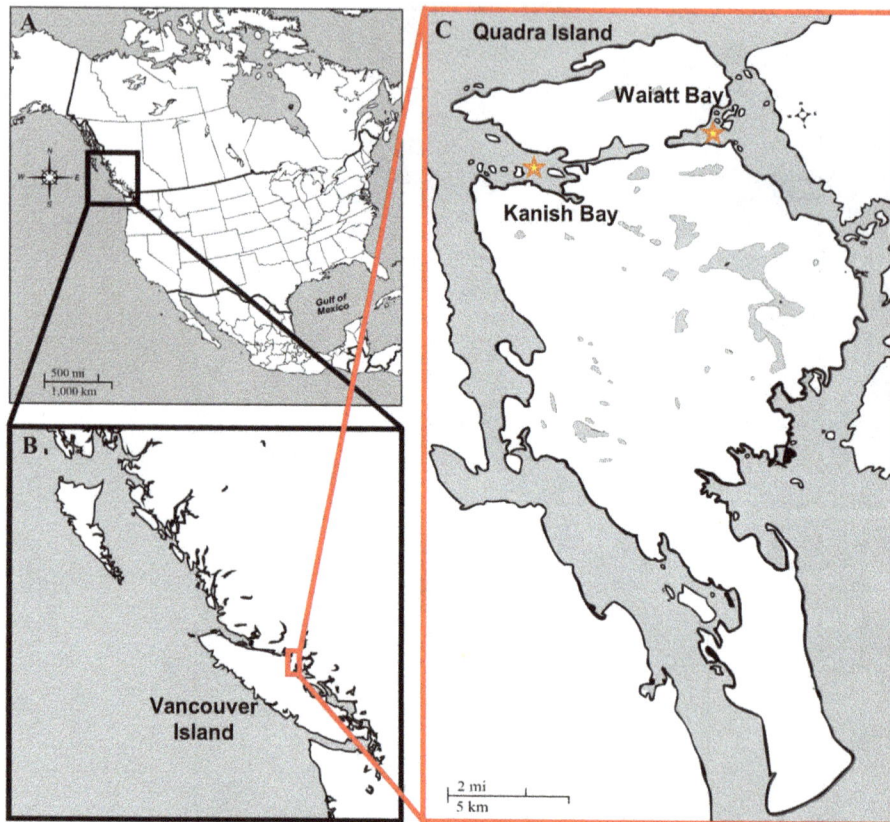

Figure 2. Map of Study Area. This research was conducted on A) the west coast of British Columbia, Canada, in the Inside Passage between B) Vancouver Island and the mainland on the northern end of C) Quadra Island, in Kanish Bay (West, starred) and Waiatt Bay (East, starred).

Methods

Study Area

We conducted our research on northern Quadra Island in British Columbia (BC), Canada, where an exceptionally high density of clam gardens have been documented [30] in Kanish (n = 45 clam gardens) and Waiatt Bays (n = 49 clam gardens) (Fig 2). Quadra Island has an abundance of archaeological sites found throughout the landscape, with shell middens representing both permanent settlements and short term camps. Today, the northern part of the island falls within the traditional territories of the Northern Coast Salish and the Southern Kwakwaka'wakw (now Laich-kwil-tach) First Nations. Some of the descendants of the ancient settlements live in nearby Indian Reserves and town centers. Presently, Kanish and Waiatt Bays are only sparsely settled and bordered by second growth forests and active wood lots. The bays are popular recreation areas and anchorages, encompassing two provincial parks and an active scallop farm. Clam digging, once a mainstay of the dense human population in the bays, is now only conducted recreationally and sporadically.

The soft sediment, low wave energy, intertidal shores of Kanish and Waiatt Bays foster bivalve communities of *Leukoma staminea* (native littleneck) and *Saxidomus giganteus* (butter clams), both ecologically, economically, and culturally important clam species. Other common bivalves include the native *Macoma spp.* (macoma clams), *Clinocardium nuttallii* (heart cockles), *Tresus nuttallii* and *Tresus capax* (horse clams), and the non-native *Venerupis philippinarum* (Japanese littlenecks), and *Mya arenaria* (eastern softshell clams).

Field Surveys

We located 5 non-walled clam beaches in each bay that we deemed appropriate controls for our study. We then chose at least 5 clam gardens in each bay as treatments for comparison. The total number of non-walled beaches (n = 10) and clam gardens (n = 11) surveyed for physical and biological comparison was constrained by the number of low tide days available during the spring and summer field season of 2011.

We characterized beach slope at 11 clam gardens and 10 non-walled beaches to quantify the physical differences in intertidal clam habitat between clam gardens and non-walled clam beaches. At each site, we established a vertical transect, perpendicular to shore, with 15 randomly stratified stations. The tidal height of each station was quantified using a total station or laser level in meters above Canada chart datum, lowest low water large tide (LLWLT). Transects began at the highest intertidal height at which clams were found in test pits. In clam gardens, the bottom tidal height of each transect was anchored by the landward edge of the human-made rock wall. At non-walled clam beaches, the bottom tidal height of each transect was anchored at ~1.0 m above chart datum, the mean tidal height of the landward edge of the clam garden rock-walls.

To test for differences in bivalve composition, density, size, and biomass between the 11 clam gardens and 10 non-walled clam beaches, we dug sample units (25×25×30 cm = 0.018 m³) at the 15 tidal stations along the vertical transect (Fig 3A). Live clams

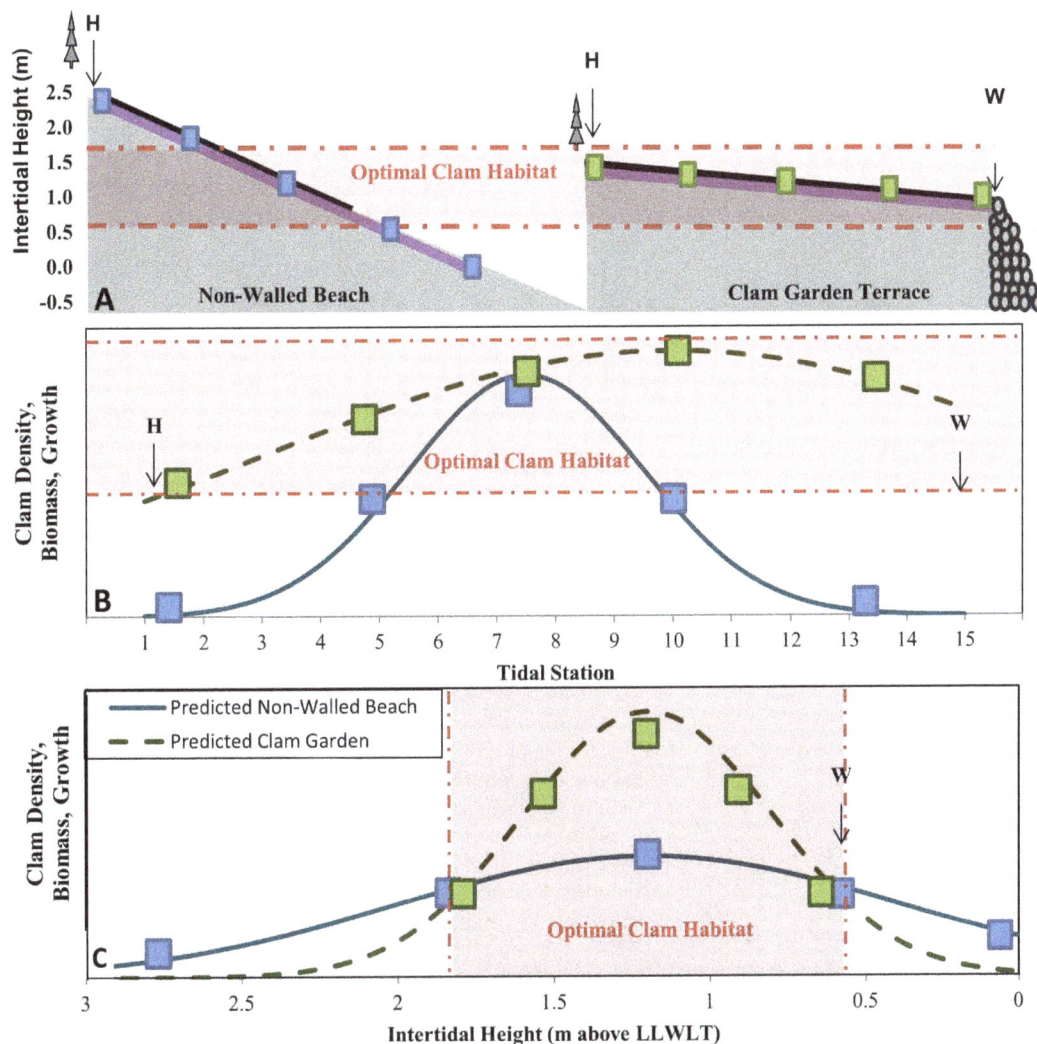

Figure 3. Study Design and Predictions. A) We surveyed clams across a vertical transect (black line) from the top of clam habitat (H) to ~1.0 m in non-walled beaches (left) and to the edge of the rock wall (W) in clam gardens (right). We then transplanted clams in mesh bags at 5 evenly stratified tidal stations (blue and green colored squares) across a vertical transect (purple line) from the top of clam habitat (H) to ~0 m in 5 Non-Walled Beaches (left) and to the edge of the rock wall (W) in 6 clam gardens (right). B) Hypothesis 1: predicted clam productivity as a function of tidal station. Tidal station 1 = top of clam habitat, tidal station 15 = top of clam garden wall in clam gardens or ~1.0 m tidal height in non-walled beaches. C) Hypothesis 2: predicted clam productivity as a function of intertidal height.

were identified to species, and their wet weight, maximum longitudinal valve length, and width were measured.

Clam Transplant Experiment

To test if clam gardens increase the growth rates of *L. staminea*, we conducted a transplant experiment across six clam gardens and five non-walled clam beaches during the growing season from May to October 2011 [33]. Clams 11–34 mm in length were collected from Waiatt and Kanish Bay, labelled with two uniquely numbered vinyl tags, measured to the nearest 0.1 mm and weighed to the nearest 1.0 gram. Fifteen individuals, representing the range of sizes collected, were placed inside a 34 cm×24 cm Vexar mesh bag. Five bags of *L. staminea* were evenly spaced at five tidal stations along a single vertical transect, perpendicular to shore, from the top of clam habitat to ~0 m intertidal height at non-walled beaches and at the edge of rock wall in clam gardens

(Fig 3A). Each bag was buried approximately 10 cm below the surface, based on our natural history observations for clams of this size and recommendations from a contemporary aquaculture facility. Each transplant bag was then secured with a flagged and labelled rebar stake. Transplanted clams were left *in situ* for 160 days. Upon retrieving the transplanted clams, max lengths, widths, and weights were recorded. We noted all losses, mortalities, and evidence of predation.

We chose *L. staminea* for our experiment because they are a biologically and culturally significant species in the area, and they possess distinct annuli, which have been verified by the annual temperature-driven $\delta^{18}O$ signature observed between annuli (K Rowell, pers. comm.). For the same reasons *L. staminea* will also be used in a companion archeological study comparing ancient clam growth rates through deep time, both before and after clam garden construction.

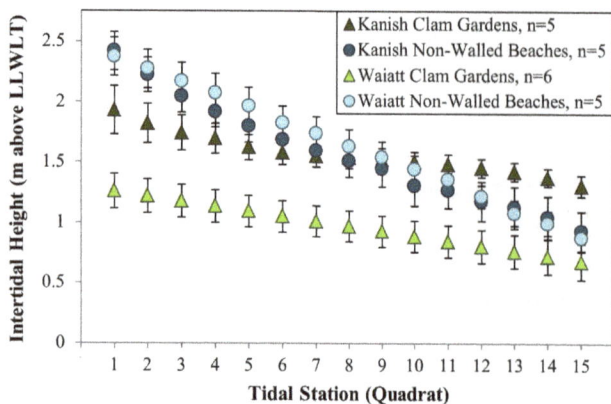

Figure 4. Site Descriptions: Intertidal Height by Tidal Station.
Intertidal height (mean +/–SE) and relative slope ($\triangle y/\triangle x$) from top of clam habitat (tidal station = 1) to top of rock wall feature in clam gardens and ~1.0 m above LLWLT in non-walled beaches (tidal station = 15).

Predictions

By constructing a rock-walled terrace in the intertidal, we predicted that clam gardens expand optimal clam habitat by altering the slope of the beach and thus increasing clam habitat at the intertidal height at which clams grow and survive best (Fig 3A). In response to increased habitat and enhanced conditions we predicted that clam density, biomass, growth, and survival would increase at the first and last tidal stations (Fig 3B), i.e., the extreme high and low of intertidal heights (Fig 3C), in clam gardens compared to non-walled beaches. We also predicted that Gaussian models of optimal clam productivity as a function of intertidal height would peak at the same intertidal height in both clam gardens and non-walled beaches, but that the magnitude of productivity would be greater (i.e. due to increased water retention, differences in sediment composition, or both) and the variance smaller (i.e. due to the reduction in beach slope) in clam gardens relative to non-walled beaches (Fig 3C).

We selected clam garden sites based on the presence of a complete rock wall spanning an embayment. Non-walled control beaches were beaches that lacked a rock wall but encompassed clam habitat, specifically intertidal soft sediment and the full array of clam species observed in the area. When comparing non-walled beaches and clam gardens, we assume that contemporary non-walled beaches are representative of the pre-construction state of clam garden sites.

Data Analysis

Physical Site Characteristics. To test for an effect of *beach type* (i.e., clam garden [n = 11] vs. non-walled beach [n = 10]) and an effect of *bay* (Kanish vs. Waiatt Bay) on beach slope, we used a general linear model (GLM). We used the same strategy to examine differences in heights of clam garden walls and mean heights of garden terraces between bays. Models of Slope, Wall Top Height, and Terrace Height were fit with a Restricted Maximum Likelihood (REML), a Gaussian error distribution, and identity link function using the lme function in the nlme package [34].

Field Surveys. To test for differences in clam density and biomass between clam gardens (n = 11) and non-walled clam beaches (n = 10), we constructed general linear mixed effects

models (GLMMs) where *beach type* was treated as a fixed effect and *site* was treated as a random effect. These models were constructed for the three most dominant species (*L. staminea*, *S. giganteus*, and *Macoma* spp.) and total clams. To test for differences in *L. staminea* density among different size classes, we ran the same models described above on five size classes of clam binned into categories based on 12 mm increments. Bin size was determined by the size range of the surveyed clams and number of clams in each bin that would yield sample sizes large enough for sufficient statistical power. Differences in clam density and biomass between clam gardens (n = 11) and non-walled clam beaches (n = 10) as a function of tidal station in both Kanish and Waiatt Bay were assessed using the same GLMMs as described above, with the additional fixed effects of *bay* and the interaction of *beach type*tidal station*. Beach type, bay, and type*tidal station* were specifically chosen as treatments to be tested, beach *type* to detect a clam garden effect, *bay* to detect an effect of oceanographic context, and *type*tidal station* to detect our predicted across-beach effect of tidal station in clam gardens and non-walled beaches (Fig 3B). Clam biomass models were fit with a REML, a Gaussian error distribution, and identity link function using the lme function in the nlme package [34]. Clam density models were fit with Laplace Approximation, a Poisson error distribution, and log link function using the lmer function and lme4 package [35]. All GLM and GLMM modelling was conducted in R [36].

Optimal Clam Habitat Models. To assess if and how clam garden engineering altered intertidal height and optimal growing conditions for clams, we modeled the relationship between intertidal height and a) density and biomass of surveyed *L. staminea* and b) survivorship and growth of transplanted *L. staminea* in clam gardens and non-walled beaches in both bays, by fitting Gaussian models (Eq.1) to each metric of clam productivity (*y*) as a function of intertidal height where:

$$y = \alpha * e^{-0.5(\frac{x-\mu}{\sigma})^2} \qquad (\text{Eq.1})$$

α (curve height) describes the magnitude of clam productivity (biomass, density, or growth rate), μ (curve mean) is the intertidal height at which productivity is greatest, and σ (curve width) describes the standard deviation in clam productivity. We then compared fitted model parameters across clam gardens and non-walled beaches in both bays based on our predictions (Fig 3).

Experimental Transplants. We tested for differences in survival and growth rates of *L. staminea* transplanted in clam gardens (n = 6) and non-walled clam beaches (n = 5) in Waiatt Bay, using generalized GLMMs where *beach type* was a fixed effect and *site* was a random effect. To test for differences in *L. staminea* growth and survival across tidal stations within clam gardens and non-walled beaches, we constructed the same GLMMs as above, with *beach type*, *bay* and *beach type*tidal station* as fixed effects. Growth rates models were fit with REML, a Gaussian error distribution, and identity link function using the lme function in the nlme package [34]. Survivorship models were fit with a Laplace Approximation, a binomial error distribution, and logit link function using the lmer function and lme4 package [35]. Survivorship and growth were compared among different juvenile clam size classes using the same models described above on three size classes of transplanted *L. staminea* binned by 5mm increments. Bin size was determined by the size of the transplanted clams and number of clams in each bin that would yield sample sizes large enough for sufficient statistical power.

Ethics Statement. All field work was conducted with the following permits: a Contaminated Shellfish Collection for

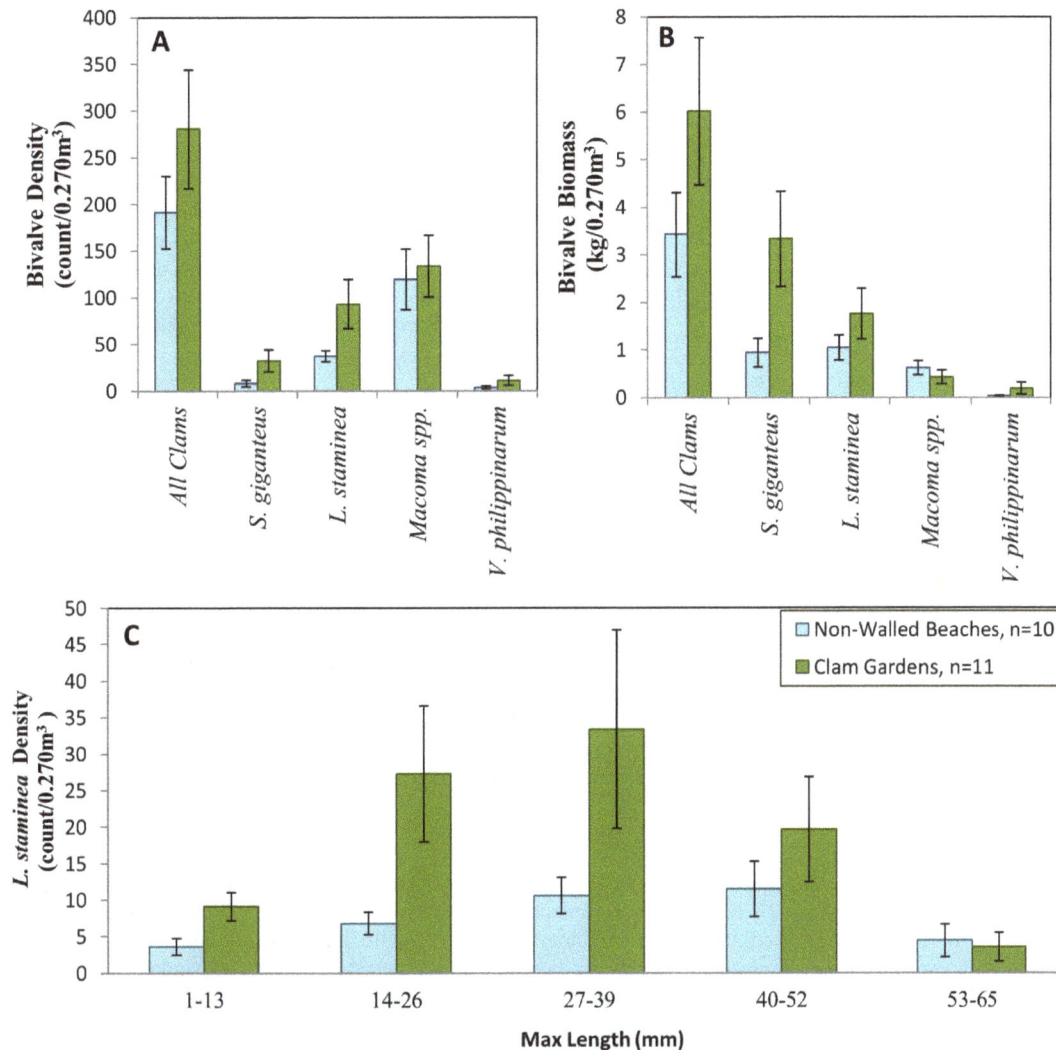

Figure 5. Survey: Bivalve Biomass and Density. B) Density (count/0.270 m3 +/–SE) and C) Biomass (kg/0.270 m3 +/–SE) of four most abundant bivalve species; A) *L. staminea* density (count/0.270 m3 +/–SE) of 5 size classes in Clam Gardens and Non-Walled Beaches.

Scientific Purpose Permit from Department of Fisheries and Oceans Canada, a Park Use Permit from BC Parks, an Animal Care Protocol Exemption from Canadian Council on Animal Care for working with invertebrates, and permission from Laichwiltach Treaty Society. The individual pictured in this manuscript (Figure 1C and Striking Image) has given written informed consent (as outlined in PLOS consent form) to publish their likeness. Our study did not involve endangered or protected species. All necessary permits were obtained for the described study, which complied with all relevant regulations.

Results

Physical Characteristics of Beach Types

On Quadra Island, BC, clam garden terrace heights varied between bays, but their slopes were consistently shallower than unaltered beaches (Fig 4, S2A, Table S1, $F_{(1,19)} = 6.914$, $p = 0.017$). On average, mean intertidal heights of clam garden terraces in Waiatt Bay were significantly lower than those in Kanish Bay (Fig 4, S1, S2A, Table S1, $F_{(1,9)} = 15.848$, $p = 0.003$). In Waiatt Bay,

clam garden terraces were located on average at 0.97 m (+/– 0.31 SE) above chart datum, while the tops of rock wall features averaged 0.68 m (+/– 0.36 SE) in intertidal height (Fig 4, S1). In contrast, Kanish clam garden terraces were located on average at 1.57 m (+/–0.21 SE) above chart datum and the tops of rock wall features averaged 1.3 m (+/–0.19 SE) in intertidal height (Table S2). Non-walled beach slopes and mean intertidal heights did not differ between bays (Fig 4, S2A,B).

We observed greater variation in the intertidal height of clam gardens in Waiatt Bay. There, clam garden terraces were located between 0.53–1.45 m above chart datum, with four of the six clam gardens having mean terrace heights between 0.78–1.16 m, and two outliers having mean terrace heights of 0.53 m and 1.45 m respectively (Fig 4, S1).

Field Surveys

L. staminea and *S. giganteus* dominated the subsistence species bivalve community in clam gardens and non-walled beaches, both in density and biomass, (Fig 5A,B). We detected significantly higher

Table 1. Clam garden effect on density and biomass.

Response variable	Fixed Effect		Random Effect	
	Beach Type		Site (Beach Type)	
Density	z	p	Variance	StdDev
L. staminea (All)	**-2.24**	**0.03***	0.60	0.78
S. giganteus	**-2.25**	**.03***	2.38	1.54
V. philippinarum	-0.69	0.49	4.66	2.16
Macoma spp.	0.05	0.96	1.51	1.23
TOTAL clam	-1.01	0.32	0.53	0.73
L. staminea (1–13 mm)	**-2.49**	**0.01***	0.61	0.78
L. staminea (14–26 mm)	**-2.76**	**<0.01***	0.77	0.87
L. staminea (27–39 mm)	**-2.11**	**0.04***	0.73	0.86
L. staminea (40–52 mm)	-1.06	0.29	1.12	1.06
L. staminea (53–65 mm)	0.18	0.86	3.70	1.92
Biomass	t	p	Residual	
L. staminea	-1.16	0.26	0.10	
S. giganteus	-1.77	0.09	0.24	
V. philippinarum	-1.23	0.24	0.03	
Macoma spp.	1.41	0.17	0.04	
TOTAL clam	-0.20	0.85	0.33	

The effect of clam gardens (Beach Type) on density and biomass (per survey transect, 0.027 m³) of *L. staminea* (littleneck clam), *S. giganteus* (butter clam), *V. philippinarum* (Japanese littleneck clam), *Macoma* spp (macoma clams) and total clams. * designates significant p-values (p≤0.05).

densities of *L. staminea*, in clam gardens (93 +/-26SE #/0.270 m³) than in non-walled beaches (37 +/-6 SE #/0.270 m³, $p = 0.03$, Fig 5A, Table 1, S2). Differences were more pronounced at smaller size classes (Fig 5C, Table 1, S2). Densities of *S. giganteus* were also significantly greater in clam gardens (32 +/-12 SE #/0.270 m³) compared to non-walled beaches (8 +/-4 SE #/0.270 m³)(Fig 5A, Table 1,S2), and these clams tended to be larger in clam gardens, yielding on average higher biomass (3.3 +/-1.0SE kg/0.270 ³) compared to non-walled beaches (0.94 +/-0.3SE kg/0.270 m³)(Fig 5B, Table 1, S2). The density and biomass of other documented bivalve species did not differ as a function of beach type (Table 1). Surveyed clam biomass of all species combined was nearly double within clam gardens (6.02 +/-1.55SE kg/0.270 m³) compared to non-walled beaches (3.43 +/- 0.88SE kg/0.270 m³ ³) (Fig 5B), however due to the variation among sites within beach type, overall bivalve biomass did not significantly differ between beach types (Fig 5B, Table 1, S2).

By examining clam density and biomass as a function of tidal station and beach type in both bays, we found that *L. staminea* and *S. giganteus* densities and biomass were significantly greater in clam gardens than non-walled beaches and as predicted (Fig 3B), this relationship varied as a function of tidal station (Fig 6). This is reflected by the significant interaction terms in Table 2. Specifically, clam densities and biomass tended to be higher at the first 6–7 tidal stations. The effect of tidal station position was highly significant for total clam densities and densities of *L. staminea* (all sizes, 14–26 mm, 27–39 mm) (Fig 6, Table 2), *S. giganteus*, and *V. philippinarum* (Table 2). The density and biomass of the invasive *V. philippinarum* varied significantly as a function of tidal station and was significantly higher in Kanish Bay (Table 2).

Effect of Beach Engineering on Optimal Clam Habitat

As predicted (Fig 3c), the magnitude of *L. staminea* productivity (*a*) in terms of density, biomass, and growth, was higher in clam gardens than non-walled beaches, and the standard deviation (σ) was lower (Fig 7A-D, Table 3). Contrary to expectations, the intertidal height at which little neck clams reach their maximum density, biomass, and growth (μ) was consistently higher in clam gardens than non-walled beaches in Kanish Bay (Fig 7, S4, Table 3), suggesting that optimal habitat shifted ~0.5 m higher up the beach in clam gardens in Kanish. In Waiatt, maximum *L. staminea* productivity (μ) did not differ substantially between beach types, and biomass did not conform to a Gaussian relationship within non-walled beaches.

Experimental Transplants

Transplanted *L. staminea* grew significantly faster in clam gardens than non-walled beaches, and this effect varied as a function of tidal station (Fig 8, S3, Table 4,S3). Clams grew proportionally faster at tidal station extremes (the first and last tidal station) in clam gardens compared to non-walled beaches. In line with our expectations, the overall magnitude of growth rates as a function of tidal height was higher in clam gardens than non-walled beaches (Fig7E, S4, Table 3). Size appears to be a major predictor of survivorship - small size classes of *L. staminea* (11−16 and 17−22 mm) were more likely to survive in clam gardens than non-walled beaches, although clam garden habitat did not appear to effect survivorship when all size classes were pooled. This size dependent effect varied as a function of tidal station (Table 4).

Discussion

Strong evidence from both our surveys and experimental transplants, suggests that ancient clam gardens increased clam productivity. By altering the slope of soft-sediment beaches (Fig 4), these human-made, intertidal terraces expanded the optimal intertidal habitat and enhanced growing conditions for clams. Specifically, we detected significantly greater densities of *S. giganteus* and *L. staminea* in clam gardens compared to non-walled beaches, particularly among smaller size classes of pre-reproductive clams (Fig 5). Overall, clam gardens contained 4 times as many butter clams (*S. giganteus*) and over twice as many little neck clams (*L. staminea*) relative to non-walled beaches (Fig S2). As predicted, the magnitude of this relationship varied as a function of intertidal height, whereby clam density and biomass was enhanced in clam gardens compared to non-walled beaches at the top and bottom of the beach, the areas where clam gardens extend optimal clam habitat (Fig 6,8). The pattern of increased clam productivity by clam gardens appears to be driven by the modification of intertidal height (Fig 7) and is supported by our experimental results indicating higher *L. staminea* growth rates within clam gardens (Fig 7E,8). Even though clam gardens on Quadra Island have not been actively tended or managed for decades, we detected significant signals of enhanced shellfish production across these engineered beaches simply due to their modified slopes. Furthermore, elevated clam densities, biomass, and growth rates at equivalent intertidal heights in clam gardens compared to non-walled beaches suggests that additional mechanisms, in addition to tidal height modification, appear to be magnifying secondary production in these clam terraces.

Mechanisms Enhancing Productivity in Clam Gardens

In addition to altering beach slope and thereby extending the area of intertidal habitat at the optimum tidal height for clam survival and growth, clam gardens terraces may have enhanced

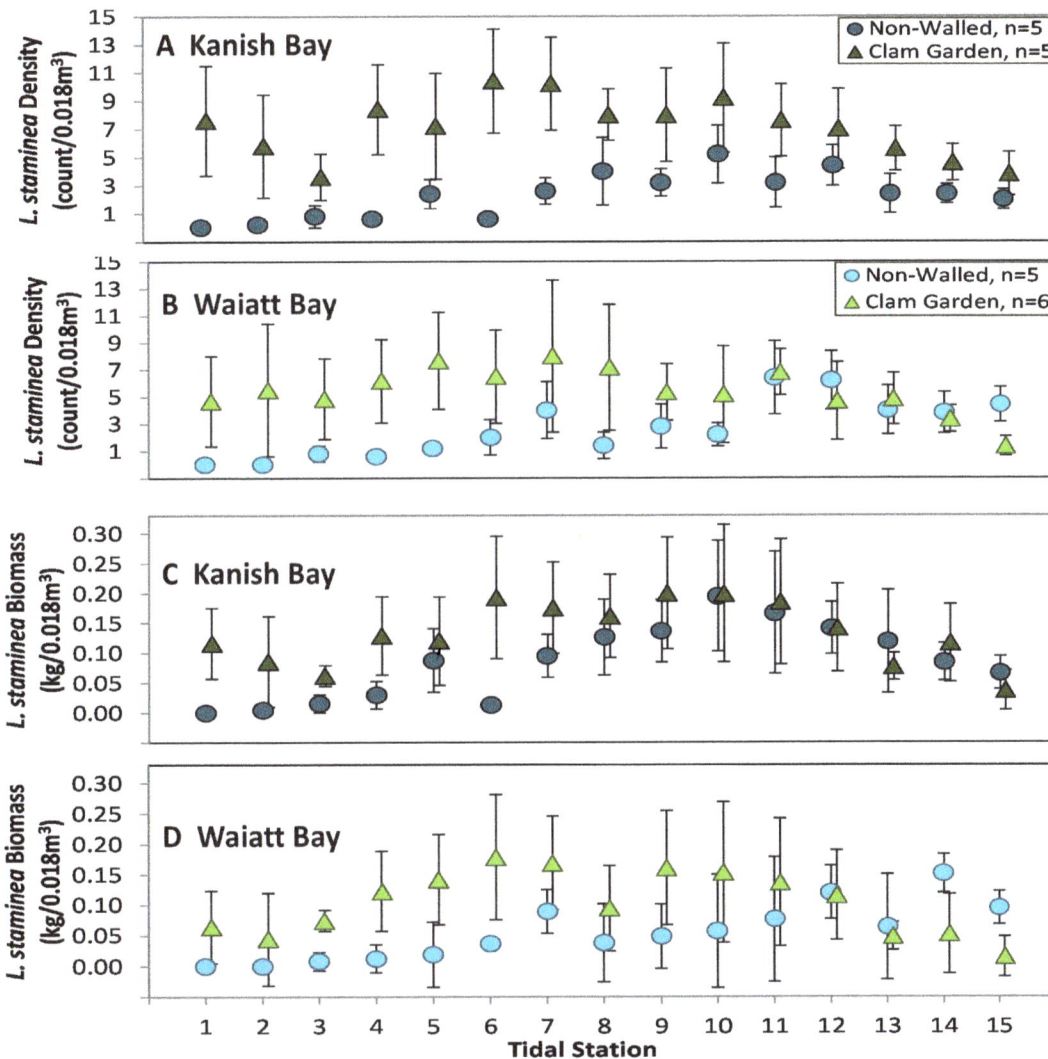

Figure 6. Survey: *L. staminea* **Biomass and Density by Tidal Station.** Surveyed A&B) density (count/0.018m3 +/– SE) and C&D) biomass (kg/0.018m3 +/– SE) and of *L. staminea* as a function of bay (Kanish or Waiatt), site type (Clam Garden or Non-Walled Beach), and tidal station (1 = top of clam habitat, 15 = top of clam garden wall or ~0.75 m intertidal height in non-walled beaches).

clam productivity in multiple ways. For example, over the diurnal tidal exchange, we observed increased water retention over clam gardens relative to natural sloping non-walled beaches. Water retention may increase the opportunity and success of larval clam recruitment and survivorship in a clam garden. In fact, a Hul'qumi'num First Nation reported that clam gardens were a way to "trap the seeds and keep them here" [27]. Low water velocities create optimal conditions for larval settlement and recruitment [37]. Water retained over shallow sloping clam gardens could increase in temperature in the late spring and summer months thereby increasing phytoplankton growth rates [38] and fuelling secondary production[39], [40]. Furthermore, increased water temperatures in temperate intertidal systems are known to enhance bivalve growth rates [41], [42] and trigger bivalve spawning events [33].

By intentionally modifying substrate, reducing density dependence, and excluding competitors and predators in clam garden terraces, indigenous people may have enhanced clam productivity in clam gardens even further. On Quadra Island, we observed that clam garden substrate tended to be higher in gravel and shell hash content compared to non-walled beach substrate, which tended to have more fine sediments. Similar observations have been reported elsewhere [23], [28] and *L. staminea*, are commonly found on natural beaches in course sand or fine gravel mixed with mud, stones, or shells [43], [44]. According to a Heiltsuk First Nation knowledge holder; "We put gravel in the garden to increase the number of clams" (D Wilson, pers. comm.). Increasing the gravel content in clam gardens may have created larger interstitial spaces in the substrate which is likely important for porewater flow and reduced fine silts and clays that are known to smother newly settled *L. staminea* larvae [43]. The act of aerating beach sediments by rolling rocks, or "turning over beaches" is also commonly reported [32], [5], [27] and aims to reduce anoxic conditions that can reduce productivity. In addition, Hul'qumi'num First Nation knowledge holders report returning crushed and whole clam shells to clam gardens as a management practice [27] and adult bivalve shell has been shown to offer an important settling cue for shellfish like oysters and clams [45], [46]. In fact, it

Table 2. Effects of clam garden treatment and oceanographic context.

Response Variable	Fixed Effect								Random Effect	
	Beach Type		Bay		Tidal Station		Beach Type x Tidal Station		Site	
Density	z	p	z	p	z	p	z	p	Var	SD
L. staminea (All)	**-5.78**	**<0.01***	-0.21	0.83	**-3.47**	**<0.01***	**10.34**	**<0.01***	0.60	0.77
S. giganteus	**-4.72**	**<0.01***	-0.52	0.96	**-4.29**	**<0.01***	**7.14**	**<0.01***	2.38	1.54
V. philippinarum	-0.32	0.75	-0.40	0.69	**-3.47**	**<0.01***	-1.16	0.25	4.61	2.15
Macoma	**-2.69**	**0.01***	-0.87	0.39	0.12	0.91	**15.97**	**<0.01***	1.46	1.21
TOTAL clam	**-4.50**	**<0.01***	-0.55	0.58	**-4.03**	**<0.01***	**18.21**	**<0.01***	0.53	0.73
L. staminea (1–13 mm)	**-3.02**	**<0.01***	0.59	0.56	0.90	0.37	**1.93**	**0.05***	0.62	0.79
L. staminea (14–26 mm)	**-5.63**	**<0.01***	0.32	0.75	**-3.62**	**<0.01***	**6.17**	**<0.01***	0.77	0.88
L. staminea (27–39 mm)	**-4.96**	**<0.01***	1.48	0.14	**-3.18**	**<0.01***	**6.13**	**<0.01***	0.67	0.82
L. staminea (40–52 mm)	**-2.74**	**<0.01***	-1.72	0.09	-0.06	0.95	**3.57**	**<0.01***	0.92	0.96
L. staminea (53–65 mm)	-1.19	0.23	0.94	0.35	0.77	0.44	**2.81**	**<0.01***	3.59	1.90
Biomass	t	p	t	p	t	p	t	p	Residual	
L. staminea	**-2.77**	**0.01***	-0.71	0.49	-0.59	**0.55**	**3.92**	**<0.01***	0.10	
S. giganteus	**-2.59**	**0.02***	1.05	0.31	-0.89	**0.37**	**2.21**	**0.03***	0.24	
V. philippinarum	**-2.16**	**0.04***	-1.45	0.17	-3.03	**<0.01***	1.83	0.07	0.03	
Macoma spp.	-0.92	0.37	-1.56	0.14	-0.62	0.53	4.14	0.00	0.04	
TOTAL clam	-0.83	0.42	-0.56	0.58	0.60	0.55	1.36	0.17	0.33	

The effects of clam gardens (Beach Type), oceanographic context (Waiatt Bay vs. Kanish Bay), and Tidal Station on the biomass and density of surveyed *L. staminea*, *S. giganteus*, *V. philippinarum*, *Macoma spp*, and total clams (per survey transect, 0.027 m^3). * designates significant p-values (p≤0.05).

has been demonstrated that gravel and shell increase clam settlement and survival [47]. Finally, Turner [5] reports that clam gardens were "thinned," reducing densities of large adult clams via harvest, giving smaller clams the space and resources to grow, and thus increasing overall yields [48]. We also hypothesize that predators such as sea stars (*Pycnopodia helianthoides*, *Pisaster brevispinus*.), large crabs (*Metacarcinus magister*) and mammalian coastal predators (river otters, sea otters), may have been intentionally excluded from these gardens to decrease both direct predator mortality and negative non-lethal predator effects on clam productivity [49], [50].

In addition to these ecological factors, several key social factors, including systems of tenure and control, may have equally enhanced and maintained the productivity of clam gardens. Indigenous peoples of the northwest coast had territorial governance systems and complex protocols that delineated access rights to the land and sea [51], [5], [52]. Clam gardens, like seaweed picking areas, root gardens, and fish traps, would have been embedded in these traditional systems of marine governance and tenure. Among the Heiltsuk First Nation, families owned and tended productive clam beaches (D Wilson, pers. comm.). Building and maintaining clam gardens were intentional acts, clearly showing cultural investment. Territorial access rights, via family-based proprietorship, established a governance system over common pool fisheries resources that likely conferred resilience to societies on the Northwest coast for millennia [52], [8]. Similarly, empirical evidence from contemporary fisheries management highlights the importance of designating access rights to enhance resource sustainability [53], [54].

Assumptions

Documenting ancient resource management within contemporary landscapes presents a challenge for various reasons, one being the identification of adequate controls. Comparisons of areas with and without archaeological features are complicated by the uncertainty in why suitable land and seascapes were not modified. In this study, we assumed naturally shallow sloping non-walled clam beaches are appropriate clam garden controls; however a variety of alternative and non-mutually exclusive hypotheses could explain why some beaches were left unmodified. For example, non-walled beaches could have been of poor quality habitat, used for other purposes, owned by other title holders or may have been at an inappropriate tidal height due to past sea level change.

To further examine the potential productivity of clam gardens and what constitutes a true control, several experiments and surveys could be performed. We recommend sampling clam garden beach sediment cores to quantify pre-modification beach characteristics. Indeed, building experimental clam gardens or deconstructing them, while quantifying ecological and physical responses before and after the modification at both treatment and control sites would be the ultimate way to test the effects of clam gardens on clam productivity. The experimental modification of substrate (i.e. adding shell hash), elimination of clam predators (i.e. *Pycnopodia helianthoides* and *Pisastser brevispinus*), and reduction of density dependence (i.e. reducing the density of juvenile clams) could also help tease apart other detailed mechanisms driving increased productivity in clam gardens.

Enhancing Food Security Confers Resilience to Social Ecological Systems

An increased appreciation of the coupling between ecosystems and human well-being has triggered a paradigm shift in the

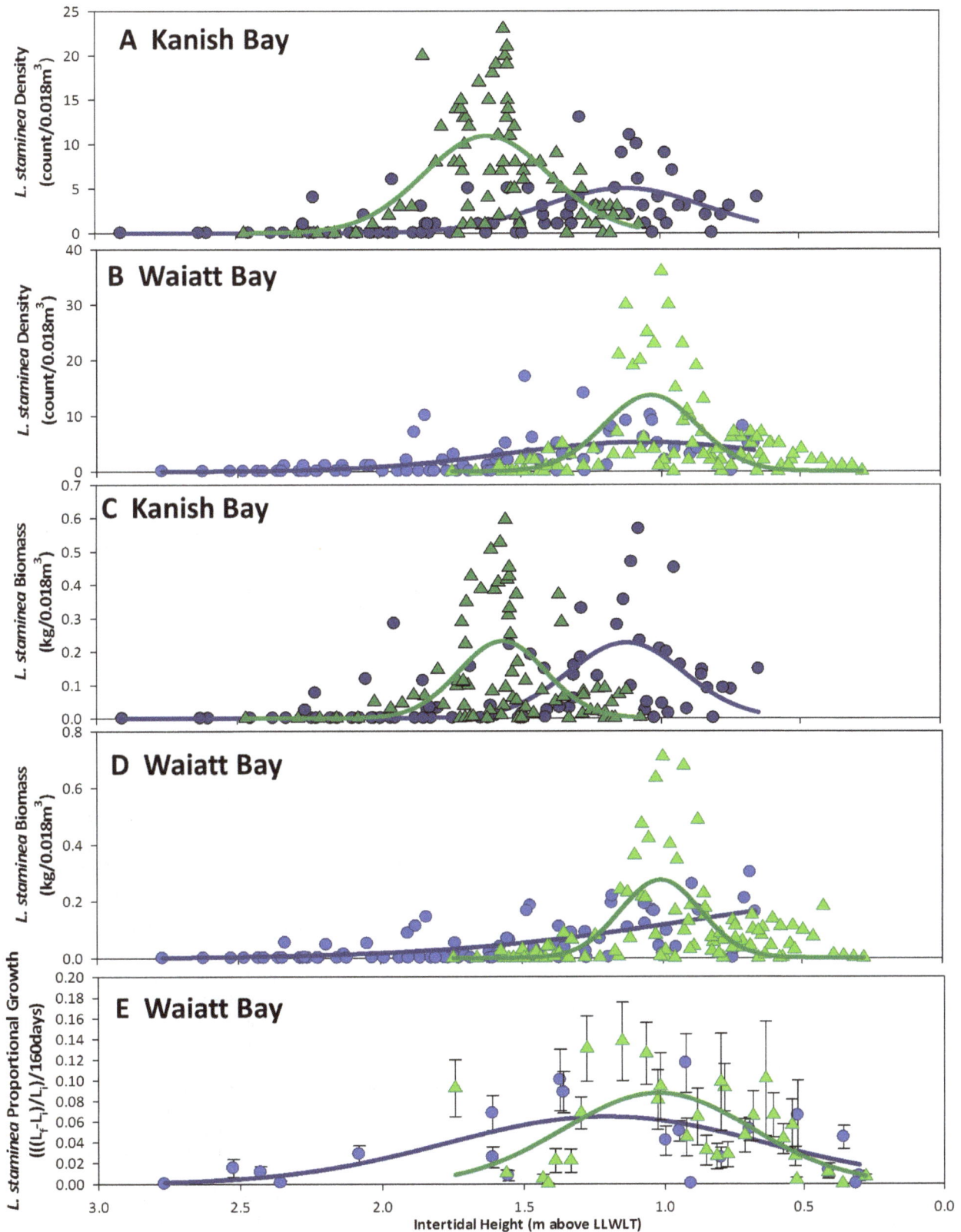

Figure 7. Survey & Experiment: *L. staminea* **Biomass, Density and Growth by Intertidal Height vs. Model Predictions.** Actual and predicted A&B) density (count/0.018 m3 +/− SE) C&D) biomass (kg/0.018 m3 +/− SE) and E) growth (mean +/−SE) of surveyed (A-D) and transplanted (E) *L. staminea* as a function of intertidal height (m above LLWLT) in clam gardens (green triangles) and non-walled beaches (blue circles) in Kanish and Waiatt bays, British Columbia, Canada.

Table 3. Gaussian models: effects of clam garden treatment and intertidal height.

Predictive Gaussian Curves, 3 parameter: $y = a*\exp(-0.5*((x-\mu)/\sigma)^2)$

Method	Bay	Type	Response	a (height)	μ (mean)	σ (variance)	R²
Survey	Kanish	NW	*L.s.* Density	5.000	1.126	0.286	0.327
Survey	Kanish	CG	*L.s.* Density	10.913	1.613	0.222	0.380
Survey	Waiatt	NW	*L.s.* Density	5.219	1.065	0.487	0.291
Survey	Waiatt	CG	*L.s.* Density	13.640	1.038	0.167	0.323
Survey	Kanish	NW	*L.s.* Biomass	0.226	1.125	0.207	0.315
Survey	Kanish	CG	*L.s.* Biomass	0.231	1.564	0.155	0.258
Survey	Waiatt	NW	*L.s.* Biomass	0.216	0.015	0.882	0.398
Survey	Waiatt	CG	*L.s.* Biomass	0.274	1.009	0.144	0.322
Transplant	Waiatt	NW	*L.s.* Growth	0.065	1.209	0.566	0.362
Transplant	Waiatt	CG	*L.s.* Growth	0.088	1.027	0.331	0.283

Parameters for the modeled responses of biomass (kg/0.018 m³ +/− SE), density (count/0.018 m³ +/− SE), and growth (mean +/−SE) of surveyed and transplanted *L. staminea* (*L.s.*) as a function of intertidal height. Each response was predicted by modeling a Gaussian curve to the data, $y = a*\exp(-0.5*((x-\mu)/\sigma)^2)$ (Eq. 1), where y = response, x = intertidal height, a = height, μ = mean, and σ = standard deviation.

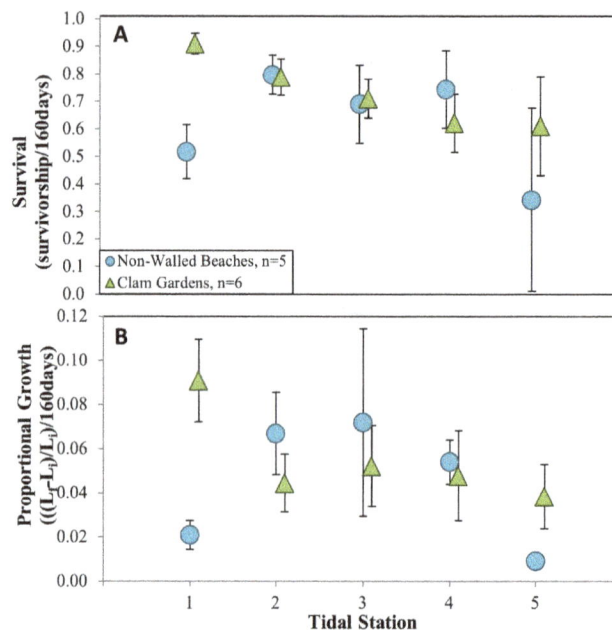

Figure 8. Experiment: *L. staminea* Growth and Survival by Tidal Station. A) Survival and B) growth rates (+/−SE) of transplanted *L. staminea* as a function of tidal station. Tidal station 1 was anchored at top of clam habitat and tidal station 5 was located at the top of the rock wall feature in clam gardens, and at ~0 m below LLW in non-walled beaches.

applied ecological sciences towards a focus on understanding the dynamics of coupled social-ecological systems (SES), linked systems of people and nature [55], [56], [23]. In many marine systems, current management approaches have demonstrably failed to halt or reverse fisheries declines [57], in part due to the inadequate recognition of the strong links between social and ecological processes [58], [59]. Ancient clam gardens and their governance by coastal communities are an example of an adaptive strategy that likely enhanced regional food security and thus conferred resilience to these coupled human-coastal ocean ecosystems.

Our observations on the variation in this ancient form of mariculture also highlight key aspects of resilient social-ecological-systems. The general uniformity in tidal height of clam garden walls on Quadra -- which likely date to different time periods and were owned and managed by different social groups -- reflects knowledge that was shared inter-generationally and across communities. As for variation, we suggest that the two clam gardens that fell outside the optimal tidal height range could be representations of engineering errors and learning, or perhaps, were built during times of differing sea levels. Alternatively, these features may have been built to target other clam species or may have had purposes other than shellfish harvest.

Food security is not only a contemporary issue. It has motivated ingenuity and development of civilizations throughout time. Investigations of how ancient clam gardens work will provide information on possible solutions to local food security and economic resiliency of coastal communities. Based on our clam surveys on Quadra Island, densities of *L. staminea* and *S. giganteus* within clam gardens are elevated on average by 151% and 300%, respectively, within clam gardens (Table S2). Clam garden biomass of *L. staminea* and *S. giganteus* were elevated on average by 68% and 253%, respectively (Table S2). Estimates made from our experiments indicate that clam gardens within optimal clam habitat can enhance growth rates of *L. staminea* on average by 89%, meaning that clams reach harvestable size at a faster rate (Table S2). If we had chosen to transplant *S. giganteus*, we predict

that we would have detected higher growth rates in clam gardens compared to non-walled beaches, but isolated to the lower half of clam gardens, within *S. giganteus* optimal habitat. The archaeological record is clear; abundant shellfish have supported large populations of people on the Northwest Coast through history [60], [8]. This new evidence helps emphasize the value of incorporating traditional management techniques into future strategies towards sustainable solutions, contributing to local food security efforts globally.

Informing Contemporary and Future Marine Management

Finding solutions to meet ecologically sound food production for the growing demands is a global effort, even though successful remedies may be locally adapted. Local food production is essential to community food security and autonomy [61]. Autonomous economies have been found to, out of necessity, recognize ecological limits, and protect biological, cultural and social diversity [62]. Some of today's benthic shellfish aquaculture practices have been shown to alter the community composition of nearshore systems [63], change sediment characteristics [64], and facilitate the introduction of invasive species [65], [66]. Ecosystem impacts of modern harvest techniques that do not prioritize conservation of ecosystem biodiversity as well as productivity can undermine nearshore ecosystem resilience [67]. These are very real concerns for coastal First Nations. Documenting these traditional practices and their ecological and societal benefits will help First Nations during a pivotal time, as First Nations continue to assert their rights to access traditional lands and resources and secure sustainable food production into the future.

Table 4. Experimental effects of clam garden treatment and tidal station.

Response Variable	Fixed Effect								
	Beach Type			Tidal Station			Beach Type x Tidal Station		
Growth Rates	df	F	p	df	F	p	df	F	p
Growth rate (All sizes)	1,9	8.79	**0.02***	1,666	22.05	**<0.01***	1,666	8.62	**<0.01***
Growth rate (11−16 mm)	1,9	1.57	0.242	1,87	3.90	**0.05***	1,87	0.38	0.54
Growth rate (17−22 mm)	1,9	<0.01	0.980	1,197	0.13	0.720	1,197	0.08	0.78
Growth rate (23−28 mm)	1,9	0.23	0.646	1,125	1.30	0.256	1,125	0.01	0.92
Survival	SE	z	p	SE	z	p	SE	z	p
Survivorship (All sizes)	1.53	−0.98	0.33	0.31	−1.28	0.20	0.47	0.77	0.44
Survivorship (11−16 mm)	0.95	−1.92	**0.05***	0.14	−2.66	**<0.01***	0.27	1.19	0.23
Survivorship (17−22 mm)	0.81	−2.50	**0.01***	0.14	−4.15	**<0.01***	0.20	2.73	**<0.01***
Survivorship (23−28 mm)	1.18	−0.01	0.99	−0.48	0.20	**0.02***	0.32	0.07	0.95

The effects of clam gardens (Beach Type) and tidal station on the growth and survivorship of transplanted *L. staminea*. * designates significant p-values (p≤0.05).

Supporting Information

Figure S1 Mean clam garden terrace intertidal height. Mean Intertidal Height (+/− min and max terrace height) of eleven clam gardens in Waiatt Bay and Kanish Bay, British Columbia, Canada. n = 6. Dashed lines represent optimal tidal height for *L. staminea* in Kanish Bay (darker line, 0.7–1.9 m) and Waiatt Bay (lighter line, 0.6–1.6 m) as determined by our survey data of *L. staminea* density experimental growth rates of non-walled beaches (Fig. S4a,b).

Figure S2 Site characteristics: Mean intertidal height and slope. A) Mean intertidal height (m above LLWLT +/− SE) and B) slope ($\triangle$y/$\triangle$x +/−SE) across survey transects spanning from the top of clam habitat to top of clam garden wall within clam gardens, and to ~0.75 m intertidal height within non-walled beaches in Waiatt and Kanish Bay, British Columbia, Canada.

Figure S3 Effect of clam garden treatment on survival and growth. A,C) Survival (+/−SE) and B,D) growth (+/−SE) of transplanted *L. staminea* (n = 15 individuals/outplant bag) over 160 days in clam gardens (n = 6) and non-walled beaches (n = 5). Note: D) includes gardens WB33,36,39,42 and excludes WB10 and WB31.

Figure S4 Mean proportional growth of transplanted *L. staminea*. A) Proportional growth (mean of site means +/−SE) of transplanted *L. staminea* over 160 days in Clam Gardens (n = 6 sites, n_{site} = 5) and Non-Walled Beaches (n = 5, n_{site} = 5) ($F_{(4,9)}$ = 1.576, p = 0.241). and B) growth excluding gardens terraces at optimal tidal height extremes WB10 and WB31 ($F_{(4,7)}$ = 11.947, p = 0.011*).

Table S1 Site Characteristics. Means and standard errors of all measured site characteristics by Site Type and Bay.

Table S2 Summary Table: Effect of clam garden treatment on all measured responses. Means and standard errors for all measured response variables of bivalve productivity by site type.

Table S3 GLMMs Summary. The effects of clam gardens (Beach Type) on experimentally transplanted *L. staminea* survivorship and growth. Analysis of GLMMs with Beach Type as a fixed effect (i.e. clam garden vs. non-walled beach) and Site as a random effect. * designates significant p-values (p≤0.05).

Acknowledgments

This project was conducted within the traditional territories of the Northern Coast Salish and Southern Kwakwaka'wakw First Nations. We are grateful to the Laich-kwil-tach Treaty Society for their support and guidance. Project and field assistance from C. Adams, K. Baglot, J. Benner, M. Caldwell, T. Clark, D. Cullon, B. Davis, J. Earnshaw, H. Graham, R. Groesbeck, C. Gruman, H. Howard, J. Isabella, L. Johannesen, P. Johanson, S. Johnson, E. Jordan, S. Jossul, N. Lee, G. Lertzman-Lepofsky, F. Munro, H. Munro, E. Nixon, S. Oakes, D. Parker, M. Puckett, A. Schmitt, G. Singh, C. Springer, T. Storr, T. Storr, J. Tewksbury, L. Wilson and M. Wunsch made this project possible. We are thankful to X. Basurto, B. Dumbauld, and S. Thrush for their insightful suggestions and edits to the manuscript.

Author Contributions

Conceived and designed the experiments: AKS KR DL. Performed the experiments: ASG AKS KR. Analyzed the data: ASG AKS. Contributed reagents/materials/analysis tools: ASG AKS KR DL. Wrote the paper: ASG AKS KR DL.

References

1. Foley JA, Ramankutty N, Brauman KA, Cassidy ES, Gerber JS, et al. (2011) Solutions for a Cultivated Planet. Nature 478: 337–342.

2. Food and Agriculture Organization of the United Nations (2009) The State of Food Insecurity in the World: Economic Crises - Impacts and Lessons Learned. Rome: FAO. 62 p.

3. Rick TC, Erlandson J (2008) Human Impacts on Ancient Marine Ecosystems: A Global Perspective. Berkeley: University of California Press. 332 p.

4. Stimpson CM (2012) Conservation Biology and Applied Zooarchaeology. Tucson (AZ): The University of Arizona Press. 264 p.

5. Turner NJ (2005) The Earth's Blanket. Madeira Park: Douglas & McIntyre. 306 p.

6. Lepofsky D, Lertzman K (2008) Documenting Ancient Plant Management in the Northwest of North America. Botany 86: 2: 129–145.

7. Erlandson JM, Rick TC, Braje TJ, Steinberg A, Vellanoweth RL (2008) Human Impacts on Ancient Shellfish: a 10,000 Year Record from San Miguel Island, California. J Archaeol Sci 35: 8: 2144–2152.

8. Lepofsky D, Caldwell ME (2013) Indigenous Marine Resource Management on the Northwest Coast of North America. Ecological Processes 2: 1: 12.

9. Jackson JBC, Kirby MX, Berger WH, Bjorndal KA, Botsford LW, et al. (2001) Historical Overfishing and the Recent Collapse of Coastal Ecosystems. Science 293: 5530: 629–637.

10. Braje TJ, Kennett DJ, Erlandson JM, Culleton BJ (2007) Human Impacts on Nearshore Shellfish Taxa: A 7,000 Year Record from Santa Rosa Island, California. Am Antiq 72:4: 735–756.

11. Erlandson JM, Rick TC (2010) Archaeology Meets Marine Ecology: The Antiquity of Maritime Cultures and Human Impacts on Marine Fisheries and Ecosystems. Ann Rev Mar Sci 2: 1: 231–251.

12. Simenstad CA, Estes JA, Kenyon KW (1978) Aleuts, Sea Otters, and Alternate Stable-State Communities. Science 200: 4340: 403–411.

13. Steneck RS (2012) Apex Predators and Trophic Cascades in Large Marine Ecosystems: Learning from Serendipity. Proc Natl Acad Sci 109: 21: 7953–7954.

14. Campbell SK, Butler VL (2010) Archaeological Evidence for Resilience of Pacific Northwest Salmon Populations and the Socio-ecological System over the Last ~7,500 Years. Ecol Soc 15: 1-17.

15. Johannes RE (1978) Traditional Marine Conservation Methods in Oceania and Their Demise. Annu Rev Ecol Syst 9: 1: 349–364.

16. Turner NJ, Deur D (2005) Keeping It Living Traditions of Plant Use and Cultivation on the Northwest Coast of North America. Vancouver: University of British Columbia Press. 404 p.

17. Caldwell ME, Lepofsky D, Combes G, Washington M, Welch JR, et al. (2012) A Bird's Eye View of Northern Coast Salish Intertidal Resource Management Features, Southern British Columbia, Canada. Journal of Island and Coastal Archaeology 7: 2: 219–233.

18. Thompson L (1949) The Relations of Men, Animals, and Plants in an Island Community (fiji). Am Anthropol 51: 2: 253–267.

19. Turner NJ, Turner KL (2007) Traditional Food Systems, Erosion and Renewal in Northwestern North America. Indian Journal of Traditional Knowledge 6: 1: 57–68.

20. Lepofsky D, Lertzman K, Hallett D, Mathewes R (2005) Climate Change and Culture Change on the Southern Coast of British Columbia 2400-1200 Cal. B.P.: An Hypothesis. Am Antiq 70: 2: 267–293.

21. Costa-Pierce BA (1987) Aquaculture in Ancient Hawaii. J BioSci. 37: 5: 320–331.

22. Langouët L, Daire M-Y (2009) Ancient Maritime Fish-Traps of Brittany (France): A Reappraisal of the Relationship Between Human and Coastal Environment During the Holocene. Journal of Maritime Archaeology 4: 2: 131–148.

23. Haggan N, Turner NJ, Carpenter J, Jones J, Mackie Q, et al. (2006) 12,000+ Years of Change: Linking Traditional and Modern Ecosystem Science in the Pacific Northwest. UBC Fisheries Centre: University of British Columbia Press. 30 p.

24. White (Xanius) E (2006) Heiltsuk Stone Fish Traps: Products of My Ancestors' Labour. Master's Thesis, Department of Archaeology. Burnaby: Simon Fraser University. 152 p.

25. Brown F, Brown YK (2009) Staying the Course, Staying Alive Coastal First Nations Fundamental Truths: Biodiversity, Stewardship and Sustainability. Victoria: Biodiversity British Columbia. 82 p.

26. Cannon A, Burchell M (2009) Clam Growth-stage Profiles as a Measure of Harvest Intensity and Resource Management on the Central Coast of British Columbia. J Archaeol Sci 36: 4: 1050–1060.

27. Parks Canada (2011) Gulf Island National Park Reserve - Hul'qumi'num Treaty Group Shellfish Traditional Knowledge Research Project Report. Sidney: Parks Canada. 26 p.

28. Harper JR, Haggerty J, Morris MC (1995) Final report, Broughton Archipelago clam terrace survey. BC Ministry of Government Services, Victoria: Land Use Coordination Office.

29. Williams J (2006) Clam Gardens: Aboriginal Mariculture on Canada's West Coast. Vancouver: New Star Books. 127 p.

30. Harper JR (2007) Clam Garden Field Inventory, CORI Project 06-02. Sydney: Coastal & Ocean Resources Inc. 61 p.

31. Cannon A, Burchell M, Bathurst R (2008) Trends and Strategies in Shellfish Gathering on the Pacific Northwest Coast of North America. In: Antcsak A, Cipriani R, editors. Early Human Impact on Megamolluscs. Oxford: Archeopress. pp. 7–22.

32. Szimanski A, director (2005) Ancient Sea Gardens: Mystery of the Pacific Northwest. Woods DJ, Woods D, producers Toronto: aquaCULTURE Pictures, Inc. Film.

33. Quayle DB (1943) Sex, Gonad Development and Seasonal Gonad Changes in Paphia Staminea. Can J Fish Aquat Sci 6: 2: 140–151.

34. Pinheiro J, Bates D, DebRoy S, Sarkar D, R Development Core Team (2012) Nlme: Linear and Nonlinear Mixed Effects Models (version 3.1-106). R Package.

35. Bates D, Maechler M, Bolker B (2012) Lme4: Linear Mixed-effects Models Using S4 Classes (version 0.999999-0). R Package.

36. R Core Team (2012) R: A Language and Environment for Statistical Computing. Vienna: R Foundation for Statistical Computing.

37. Roegner GC (2000) Transport of Molluscan Larvae through a Shallow Estuary. J Plankton Res 22: 9: 1779–1800.

38. Eppley RW (1972) Temperature and Phytoplankton Growth in the Sea. Fish Bull 70: 4: 1063–1085.

39. Phibbs FD (1971) Temperature, Salinity and Clam Larvae. Proc Nat Shellfish Assoc. 61:12. (Abstr.)

40. Quayle DB, Bourne N (1972) Bulletin 179: The Clam Fisheries of British Columbia. Ottowa: Fisheries Research Board of Canada.70p.

41. Shaw WN (1986) Species Profiles: Life Histories and Environmental Requirements of Coastal Fishes and Invertebrates (Pacific Southwest) -- Common Littleneck Clam. U.S Fish and Wildlife Biological Report 82(11.46). Washington: U.S. Army Corps of Engineers. 11 p.

42. Menge BA, Chan F, Lubchenco J (2008) Response of a Rocky Intertidal Ecosystem Engineer and Community Dominant to Climate Change. Ecol Lett 11: 2: 151–162.

43. Fraser CM, Smith GM (1928) Notes on the Ecology of the Little Neck Clam, Paphia Staminea Conrad. Trans R Soc Can 22: 5: 249–269.

44. Fitch JE (1953) Fish Bulletin 90: Common Marine Bivalves of California. Sacramento: California Department of Fish and Game.102 p.

45. Tamburri MN, Zimmer-Faust RK, Tamplin ML (1992) Natural Sources and Properties of Chemical Inducers Mediating Settlement of Oyster Larvae: A Re-examination. The Biol Bull 183: 2: 327–338.

46. Butman CA, Grassle JP, Webb CM (1988) Substrate Choices Made by Marine Larvae Settling in Still Water and in a Flume Flow. Nature 333: 6175: 71–773.

47. Thompson D (1995) Substrate additive studies for the development of hardshell clam habitat in waters of Puget Sound in Washington State: An analysis of effects on recruitment, growth, and survival of the Manila clam, *Tapes philippinarum*, and on the species diversity and abundance of existing benthic organisms. Estuaries Coasts 18:1: 91-107.

48. Peterson CH, Beal BF (1989) Bivalve Growth and Higher Order Interactions: Importance of Density, Site, and Time. Ecology 70: 5: 1390–1404.

49. Nakaoka M (2000) Nonlethal Effects of Predators on Prey Populations: Predator-mediated Change in Bivalve Growth. Ecology 81: 4: 1031–1045.

50. Smee DL, Weissburg MJ (2006) Hard Clams (Mercenaria Mercenaria) Evaluate Predation Risk Using Chemical Signals from Predators and Injured Conspecifics. J Chem Ecol 32: 3: 605–619.

51. Trosper R (2002) Northwest Coast Indigenous Institutions That Supported Resilience and Sustainability. Ecol Econ 41: 2: 329–344.

52. Trosper R (2009) Resilience, Reciprocity and Ecological Economics: Northwest Coast Sustainability. New York: Routledge. 188 p.

53. Ostrom E (2009) A General Framework for Analyzing Sustainability of Social-Ecological Systems. Science 325: 5939: 419–422.

54. Pinkerton E, Silver J (2011) Cadastralizing or Coordinating the Clam Commons: Can Competing Community and Government Visions of Wild and Farmed Fisheries Be Reconciled? Marine Policy 35: 1: 63–72.

55. Berkes F, Folke C, Colding J (2000) Linking Social and Ecological Systems: Management Practices and Social Mechanisms for Building Resilience. Massachusetts: Cambridge University Press. 480 p.

56. Folke C (2006) Resilience: The emergence of a perspective for social–ecological systems analyses. Glob Environ Change 16: 3: 253−267.

57. Pauly D, Christensen V, Guénette S, Pitcher TJ, Sumaila UR, et al. (2002) Towards Sustainability in World Fisheries. Nature 418: 6898: 689–695.

58. Gelcich S, Hughes TP, Olsson P, Folke C, Defeo O, et al. (2010) Navigating Transformations in Governance of Chilean Marine Coastal Resources. Proc Natl Acad Sci 107: 39: 16794–16799.

59. Horan RD, Fenichel EP, Drury KLS, Lodge DM (2011) Managing Ecological Thresholds in Coupled Environmental–human Systems. Proc Natl Acad Sci 108: 18: 7333–7338.

60. Moss ML, Cannon A, editors (2011) The Archaeology of North Pacific Fisheries. Fairbanks: University of Alaska Press. 326 p.

61. Menezes F (2001) Food Sovereignty: A Vital Requirement for Food Security in the Context of Globalization. Development 44: 4: 29–33.

62. Starr A, Adams J (2003) Anti-globalization: The Global Fight for Local Autonomy. New Polit Sci 25: 1: 19–42.

63. Whiteley J, Bendell-Young L (2007) Ecological Implications of Intertidal Mariculture: Observed Differences in Bivalve Community Structure Between Farm and Reference Sites. J Appl Ecol 44: 3: 495–505.

64. Wagner E, Dumbauld BR, Hacker SD, Trimble AC, Wisehart LM, et al. (2012) Density-dependent effects of an introduced oyster, *Crassostrea gigas*, on a native intertidal seagrass, *Zostera marina*. Mar Ecol Prog Ser 468: 149−160.

65. Williams SL, Smith JE (2007) A Global Review of the Distribution, Taxonomy, and Impacts of Introduced Seaweeds. Annu Rev Ecol Evol Syst 38:1: 327−359.

66. Miller KA, Aguilar-Rosa LE, Pedroche FF (2011) A review of non-native seaweeds from California, USA and Baja California, Mexico. Hidrobiologica 21:3: 365−379.

67. Gunderson LH, Pritchard L (2002) Resilience and the Behavior of Large-Scale Systems. Washington: Island Press. 316 p.

20

Oxidative Stress Is a Mediator for Increased Lipid Accumulation in a Newly Isolated *Dunaliella salina* Strain

Kaan Yilancioglu[1], Murat Cokol[1,2], Inanc Pastirmaci[1], Batu Erman[1], Selim Cetiner[1]*

1 Faculty of Engineering and Natural Sciences, Sabanci University, Orhanlı, Istanbul, Turkey, **2** Sabanci University Nanotechnology Research and Application Center, Orhanlı, Istanbul, Turkey

Abstract

Green algae offer sustainable, clean and eco-friendly energy resource. However, production efficiency needs to be improved. Increasing cellular lipid levels by nitrogen depletion is one of the most studied strategies. Despite this, the underlying physiological and biochemical mechanisms of this response have not been well defined. Algae species adapted to hypersaline conditions can be cultivated in salty waters which are not useful for agriculture or consumption. Due to their inherent extreme cultivation conditions, use of hypersaline algae species is better suited for avoiding culture contamination issues. In this study, we identified a new halophilic *Dunaliella salina* strain by using 18S ribosomal RNA gene sequencing. We found that growth and biomass productivities of this strain were directly related to nitrogen levels, as the highest biomass concentration under 0.05 mM or 5 mM nitrogen regimes were 495 mg/l and 1409 mg/l, respectively. We also confirmed that nitrogen limitation increased cellular lipid content up to 35% under 0.05 mM nitrogen concentration. In order to gain insight into the mechanisms of this phenomenon, we applied fluorometric, flow cytometric and spectrophotometric methods to measure oxidative stress and enzymatic defence mechanisms. Under nitrogen depleted cultivation conditions, we observed increased lipid peroxidation by measuring an important oxidative stress marker, malondialdehyde and enhanced activation of catalase, ascorbate peroxidase and superoxide dismutase antioxidant enzymes. These observations indicated that oxidative stress is accompanied by increased lipid content in the green alga. In addition, we also showed that at optimum cultivation conditions, inducing oxidative stress by application of exogenous H_2O_2 leads to increased cellular lipid content up to 44% when compared with non-treated control groups. Our results support that oxidative stress and lipid overproduction are linked. Importantly, these results also suggest that oxidative stress mediates lipid accumulation. Understanding such relationships may provide guidance for efficient production of algal biodiesels.

Editor: Douglas Andrew Campbell, Mount Allison University, Canada

Funding: This work was financially supported by research fund of Sabanci University. The funders had no role in study design, data collection and analysis, decision to publish, or preparation of the manuscript.

Competing Interests: The authors have declared that no competing interests exist.

* E-mail: cetiner@sabanciuniv.edu

Introduction

The idea of using biofuels has gained prominence, since they provide a cleaner alternative to the currently used fossil fuels. It has recently been estimated that utilization of biofuels will result in a 30% decrease in CO_2 emissions in the United States. Biofuels can be derived from different kinds of resources including microalgae, animal fats, soybeans, corns and other oil crops. While none of these options currently has the efficiency to produce the required amounts of biofuel [1], microalgae are considered the most promising venue of biofuel production due to their ease of cultivation, sustainability, and compliance in altering their lipid content resulting in higher biofuel production.

High lipid accumulation and biomass productivity are the two manifestly desired phenotypes in algae for biodiesel production. However, various studies conducted under nutrient depleted conditions have demonstrated that biomass productivity and lipid accumulation are negatively related [2]. These studies have established that stress conditions, which by definition reduce the biomass production, increase lipid content of algae. This problem was addressed by using a two-stage reactor where algal species such as *Oocysti sp.* and *amphora sp.* are grown in optimal conditions

for maximum biomass, followed by stress conditions for maximum lipid accumulation [3]. Within this context, nitrogen depletion can be still considered as a strategy for increasing lipid accumulation since it has been still defined as one of the best lipid accumulator stress condition in algae to date. However the mechanistic insights of this phenomenon are still needed.

Nitrogen deprivation as a stress condition is known to maximize the lipid content up to 90% [4]. However, underlying mechanisms have not been well described in terms of its physiological and molecular aspects. Despite the fact that oxygen itself is not harmful for cells, the presence of reactive oxygen species (ROS) may lead to oxidative damage to the cellular environment, ultimately leading to toxicity resulting from excessive reactive oxygen stress [5]. Redox reactions of the reactive forms of oxygen including hydrogen peroxide (H_2O_2), superoxide (O_2^-) or hydroxyl (OH^-) radicals with cellular lipids, proteins, and DNA result in oxidative stress [6]. A previous study showed that nitrogen depletion results in the co-occurrence of ROS species and lipid accumulation in diatoms [7]. Association of increased reactive oxygen species levels and cellular lipid accumulation under different environmental stress conditions was also shown in green microalgae [8] ROS is known to be an important factor in cellular response and it is well

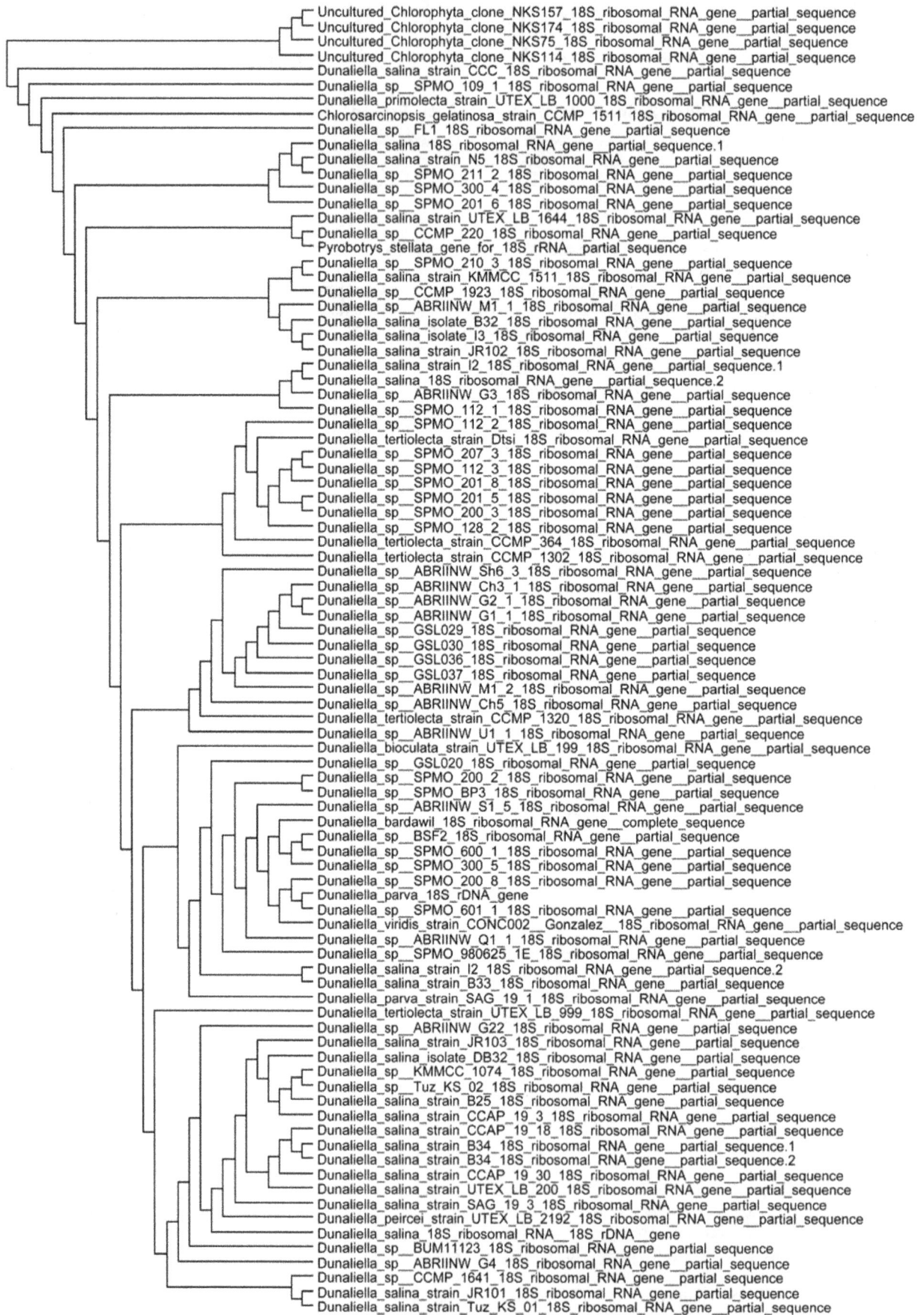

Uncultured_Chlorophyta_clone_NKS157_18S_ribosomal_RNA_gene__partial_sequence
Uncultured_Chlorophyta_clone_NKS174_18S_ribosomal_RNA_gene__partial_sequence
Uncultured_Chlorophyta_clone_NKS75_18S_ribosomal_RNA_gene__partial_sequence
Uncultured_Chlorophyta_clone_NKS114_18S_ribosomal_RNA_gene__partial_sequence
Dunaliella_salina_strain_CCC_18S_ribosomal_RNA_gene__partial_sequence
Dunaliella_sp__SPMO_109_1_18S_ribosomal_RNA_gene__partial_sequence
Dunaliella_primolecta_strain_UTEX_LB_1000_18S_ribosomal_RNA_gene__partial_sequence
Chlorosarcinopsis_gelatinosa_strain_CCMP_1511_18S_ribosomal_RNA_gene__partial_sequence
Dunaliella_sp_FL1_18S_ribosomal_RNA_gene__partial_sequence
Dunaliella_salina_18S_ribosomal_RNA_gene__partial_sequence.1
Dunaliella_salina_strain_N5_18S_ribosomal_RNA_gene__partial_sequence
Dunaliella_sp__SPMO_211_2_18S_ribosomal_RNA_gene__partial_sequence
Dunaliella_sp__SPMO_300_4_18S_ribosomal_RNA_gene__partial_sequence
Dunaliella_sp__SPMO_201_6_18S_ribosomal_RNA_gene__partial_sequence
Dunaliella_salina_strain_UTEX_LB_1644_18S_ribosomal_RNA_gene__partial_sequence
Dunaliella_sp__CCMP_220_18S_ribosomal_RNA_gene__partial_sequence
Pyrobotrys_stellata_gene_for_18S_rRNA__partial_sequence
Dunaliella_sp__SPMO_210_3_18S_ribosomal_RNA_gene__partial_sequence
Dunaliella_salina_strain_KMMCC_1511_18S_ribosomal_RNA_gene__partial_sequence
Dunaliella_sp__CCMP_1923_18S_ribosomal_RNA_gene__partial_sequence
Dunaliella_sp__ABRIINW_M1_1_18S_ribosomal_RNA_gene__partial_sequence
Dunaliella_salina_isolate_B32_18S_ribosomal_RNA_gene__partial_sequence
Dunaliella_salina_isolate_I3_18S_ribosomal_RNA_gene__partial_sequence
Dunaliella_salina_strain_JR102_18S_ribosomal_RNA_gene__partial_sequence
Dunaliella_salina_strain_I2_18S_ribosomal_RNA_gene__partial_sequence.1
Dunaliella_salina_18S_ribosomal_RNA_gene__partial_sequence.2
Dunaliella_sp__ABRIINW_G3_18S_ribosomal_RNA_gene__partial_sequence
Dunaliella_sp__SPMO_112_1_18S_ribosomal_RNA_gene__partial_sequence
Dunaliella_sp__SPMO_112_2_18S_ribosomal_RNA_gene__partial_sequence
Dunaliella_tertiolecta_strain_Dtsi_18S_ribosomal_RNA_gene__partial_sequence
Dunaliella_sp__SPMO_207_3_18S_ribosomal_RNA_gene__partial_sequence
Dunaliella_sp__SPMO_112_3_18S_ribosomal_RNA_gene__partial_sequence
Dunaliella_sp__SPMO_201_8_18S_ribosomal_RNA_gene__partial_sequence
Dunaliella_sp__SPMO_201_5_18S_ribosomal_RNA_gene__partial_sequence
Dunaliella_sp__SPMO_200_3_18S_ribosomal_RNA_gene__partial_sequence
Dunaliella_sp__SPMO_128_2_18S_ribosomal_RNA_gene__partial_sequence
Dunaliella_tertiolecta_strain_CCMP_364_18S_ribosomal_RNA_gene__partial_sequence
Dunaliella_tertiolecta_strain_CCMP_1302_18S_ribosomal_RNA_gene__partial_sequence
Dunaliella_sp__ABRIINW_Sh6_3_18S_ribosomal_RNA_gene__partial_sequence
Dunaliella_sp__ABRIINW_Ch3_1_18S_ribosomal_RNA_gene__partial_sequence
Dunaliella_sp__ABRIINW_G2_1_18S_ribosomal_RNA_gene__partial_sequence
Dunaliella_sp__ABRIINW_G1_1_18S_ribosomal_RNA_gene__partial_sequence
Dunaliella_sp__GSL029_18S_ribosomal_RNA_gene__partial_sequence
Dunaliella_sp__GSL030_18S_ribosomal_RNA_gene__partial_sequence
Dunaliella_sp__GSL036_18S_ribosomal_RNA_gene__partial_sequence
Dunaliella_sp__GSL037_18S_ribosomal_RNA_gene__partial_sequence
Dunaliella_sp__ABRIINW_M1_2_18S_ribosomal_RNA_gene__partial_sequence
Dunaliella_sp__ABRIINW_Ch5_18S_ribosomal_RNA_gene__partial_sequence
Dunaliella_tertiolecta_strain_CCMP_1320_18S_ribosomal_RNA_gene__partial_sequence
Dunaliella_sp__ABRIINW_U1_1_18S_ribosomal_RNA_gene__partial_sequence
Dunaliella_bioculata_strain_UTEX_LB_199_18S_ribosomal_RNA_gene__partial_sequence
Dunaliella_sp__GSL020_18S_ribosomal_RNA_gene__partial_sequence
Dunaliella_sp__SPMO_200_2_18S_ribosomal_RNA_gene__partial_sequence
Dunaliella_sp__SPMO_BP3_18S_ribosomal_RNA_gene__partial_sequence
Dunaliella_sp__ABRIINW_S1_5_18S_ribosomal_RNA_gene__partial_sequence
Dunaliella_bardawil_18S_ribosomal_RNA_gene__complete_sequence
Dunaliella_sp__BSF2_18S_ribosomal_RNA_gene__partial_sequence
Dunaliella_sp__SPMO_600_1_18S_ribosomal_RNA_gene__partial_sequence
Dunaliella_sp__SPMO_300_5_18S_ribosomal_RNA_gene__partial_sequence
Dunaliella_sp__SPMO_200_8_18S_ribosomal_RNA_gene__partial_sequence
Dunaliella_parva_18S_rDNA_gene
Dunaliella_sp__SPMO_601_1_18S_ribosomal_RNA_gene__partial_sequence
Dunaliella_viridis_strain_CONC002__Gonzalez__18S_ribosomal_RNA_gene__partial_sequence
Dunaliella_sp__ABRIINW_Q1_1_18S_ribosomal_RNA_gene__partial_sequence
Dunaliella_sp__SPMO_980625_1E_18S_ribosomal_RNA_gene__partial_sequence
Dunaliella_salina_strain_I2_18S_ribosomal_RNA_gene__partial_sequence.2
Dunaliella_salina_strain_B33_18S_ribosomal_RNA_gene__partial_sequence
Dunaliella_parva_strain_SAG_19_1_18S_ribosomal_RNA_gene__partial_sequence
Dunaliella_tertiolecta_strain_UTEX_LB_999_18S_ribosomal_RNA_gene__partial_sequence
Dunaliella_sp__ABRIINW_G22_18S_ribosomal_RNA_gene__partial_sequence
Dunaliella_salina_strain_JR103_18S_ribosomal_RNA_gene__partial_sequence
Dunaliella_salina_isolate_DB32_18S_ribosomal_RNA_gene__partial_sequence
Dunaliella_sp__KMMCC_1074_18S_ribosomal_RNA_gene__partial_sequence
Dunaliella_sp__Tuz_KS_02_18S_ribosomal_RNA_gene__partial_sequence
Dunaliella_salina_strain_B25_18S_ribosomal_RNA_gene__partial_sequence
Dunaliella_salina_strain_CCAP_19_3_18S_ribosomal_RNA_gene__partial_sequence
Dunaliella_salina_strain_CCAP_19_18_18S_ribosomal_RNA_gene__partial_sequence
Dunaliella_salina_strain_B34_18S_ribosomal_RNA_gene__partial_sequence.1
Dunaliella_salina_strain_B34_18S_ribosomal_RNA_gene__partial_sequence.2
Dunaliella_salina_strain_CCAP_19_30_18S_ribosomal_RNA_gene__partial_sequence
Dunaliella_salina_strain_UTEX_LB_200_18S_ribosomal_RNA_gene__partial_sequence
Dunaliella_salina_strain_SAG_19_3_18S_ribosomal_RNA_gene__partial_sequence
Dunaliella_peircei_strain_UTEX_LB_2192_18S_ribosomal_RNA_gene__partial_sequence
Dunaliella_salina_18S_ribosomal_RNA__18S_rDNA__gene
Dunaliella_sp__BUM11123_18S_ribosomal_RNA_gene__partial_sequence
Dunaliella_sp__ABRIINW_G4_18S_ribosomal_RNA_gene__partial_sequence
Dunaliella_sp__CCMP_1641_18S_ribosomal_RNA_gene__partial_sequence
Dunaliella_salina_strain_JR101_18S_ribosomal_RNA_gene__partial_sequence
Dunaliella_salina_strain_Tuz_KS_01_18S_ribosomal_RNA_gene__partial_sequence

Figure 1. Phylogenetic analysis of *Dunaliella salina* **strain** *Tuz_KS_01* **(GeneBank accession no. JX880083).** Dendrogram was generated using the neighbor-joining analysis based on 18S rDNA gene sequences. The phylogenetic tree shows the position of *Dunaliella salina* strain *Tuz_KS_01* (GeneBank accession no. **JX880083**) relative to other species and strains of *Dunaliella* deposited in NCBI GeneBank.

established that ROS increases when microalgae are exposed to various stresses. However, a mechanistic understanding of the connection between ROS increase and increased lipid accumulation in algae species requires further investigation [9].

Nitrogen depleted conditions trigger reactive oxygen species accumulation, increased cellular lipid content and protein production impairment. However, the temporal order and the causal links between these events are yet to be explored. Here, we aimed at finding the relationship between oxidative stress and increased cellular lipid content under nitrogen depleted conditions in a hypersaline green alga in order to have a better understanding of this phenomenon.

Dunaliella genus [10] is one of the microalgae genus that has been considered for lipid production. *Dunaliella* species are particularly attractive due to their strong resistance characteristics to various unfavourable environmental conditions such as high salinity. Obtaining such strong algal species for lipid production under conditions that are otherwise not useful is an important economical consideration in terms of biodiesel production. In addition, cultivation of algae species in freshwaters may not be feasible due to the limited supply and population expansion. Understanding the mechanisms behind increased lipid accumulation in halophilic *Dunaliella* in response to different stress conditions, especially nitrogen depletion, is crucial to enable key manipulations at the genetic, biochemical and physiological level for decreasing biodiesel production costs.

Materials and Methods

Organism and Culture Conditions

Dunaliella tertiolecta (*D.t.* #LB999) and *Chlamydomonas reinhardtii* (*C.r.* #90) were obtained from UTEX, Collection Culture of Algae, USA and cultivated artificial sea water medium and soil extract medium as instructed by the culture collection protocols respectively. The alga used in this study was isolated from the hypersaline lake "Tuz", which is located in Middle Anatolia, Turkey with the research permission of Republic of Turkey Ministry of Food, Agriculture and Livestock (Permit issue: B.12.0.TAG.404.03.10.03.03–1607). Field studies did not involve endangered or protected species. Collection of water samples was done and isolation location was recorded using a GPS device as 39°4'23.97"K - 33°24'33.11"E at the southern east part of Sereflikochisar province. 10 ml water samples were collected and enriched with the same volume of Bold's Basal Medium (BBM) modified by addition of 5% NaCl. BBM was pH 7.4 and consisted of 5 mM NaNO$_3$ along with CaCl$_2$·2H$_2$O 0.17 mM, MgSO$_4$·7H$_2$O 0.3 mM, K$_2$HPO$_4$ 0.43 mM, KH$_2$PO$_4$ 1.29 mM, Na$_2$EDTA·2H$_2$O 2 mM, FeCl$_3$·6H$_2$O 0.36 mM, MnCl$_2$·4H$_2$O 0.21 mM, ZnCl$_2$ 0.037 mM, CoCl$_2$·6H$_2$O 0.0084 mM, Na$_2$MoO$_4$·2H$_2$O 0.017 mM, Vitamin B$_{12}$ 0.1 mM and 5 mM NaCO$_3$ was supplied as carbon source.

Water samples were plated on petri dishes with modified BBM 5% NaCl and 1% bacteriological agar, while same water samples were also subjected to dilutions with fresh modified BBM 5% NaCl in 48-well plates (1:2, 1:4, 1:8, 1:16, 1:32, 1:64) to obtain monocultures. After 2 and 4 weeks cultivation periods of 48-well plate liquid cultures and petri dishes, respectively, clones were isolated. These clones were transferred to fresh mediums in 100 ml canonical flasks in a final volume of 25 ml for obtaining cell stocks

at 25°C under continuous shaking and photon irradiance of 80 rpm and 150 μEm^{-2}s^{-1}.

All experiments were carried out under same cultivation conditions in 250 ml batch cultures but different nitrogen concentrations, 0.05, 0.5 and 5 mM NaNO$_3$ were used for low, medium and high nitrogen concentrations, respectively. Concentrations ranging between 200 uM and 8 mM were used for H$_2$O$_2$ experiments.

Isolation and Purification of DNA and Amplification of 18S rRNA Encoding Gene

DNA isolation was done by using DNeasy Plant Mini Kit (Qiagen) as instructed by the manufacturer. Quantification of the genomic DNA obtained and assessment of its purity was done on a Nanodrop Spectrophotometer ND-1000 (Thermo Scientific) and on 1% agarose gel elecrophoresis. MA1 [5'-CGGGATCCG-TAGTCATATGCTTGTCTC-3'] and MA2 [5-GGAATT-CCTTCTGCAGGTTCACC-3'] were designed from 18S rDNA genes and were previously reported by Olmos et al [11]. PCR reactions were carried out in a total volume of 50 μl containing 50 ng of chromosomal DNA in dH$_2$O and 200 ng MA1 and MA2 conserved primers. The amplification was carried out using 30 cycles in a MJ Mini Personal Thermal Cycler (BioRad), with a T$_m$ of 52°C for all reactions. One cycle consisted of 1 minute at 95°C, 1 minute at 52°C and 2 minutes at 72°C.

Sequencing and Phylogenetic Analysis

MA1–MA2 PCR products were utilized to carry out sequencing reactions after purification with a QIAquick PCR purification kit (Qiagen). The sequencing reactions were run by MCLAB (San Francisco, CA), employing primers MA1–MA2 in both reverse and forward directions. DNA sequences were imported to BLAST for identification and to search for phylogenetic relationship correlations between other 18S rDNA gene sequences of *Dunaliella* species/strains deposited in NCBI Gene Bank. Dendrogram data generated by BLAST was converted into newick format and submitted to Phyfi [12] for generating phylogenetic tree.

Growth Analysis

Specific growth rate and biomass productivity was calculated according to the equation; $K' = Ln(N2/N1)/(t2 - t1)$ where $N1$ and $N2$, biomass at time1 ($t1$) and time2 ($t2$) respectively ($t2 > t1$) [13]. Divisions per day and the generation (doubling) time were calculated according to the equations below:

$$Div \times day^{-1} = K'/Ln2$$

$$Gen't = 1/Div \times day^{-1}$$

Extraction and Measurements of Lipid Contents and Fluorescence Microscopy

Lipid was extracted according to Bligh and Dyer wet extraction method. Briefly, to a 15 ml glass vial containing 100 mg dried algal biomass, 2 ml methanol and 1 ml chloroform were added

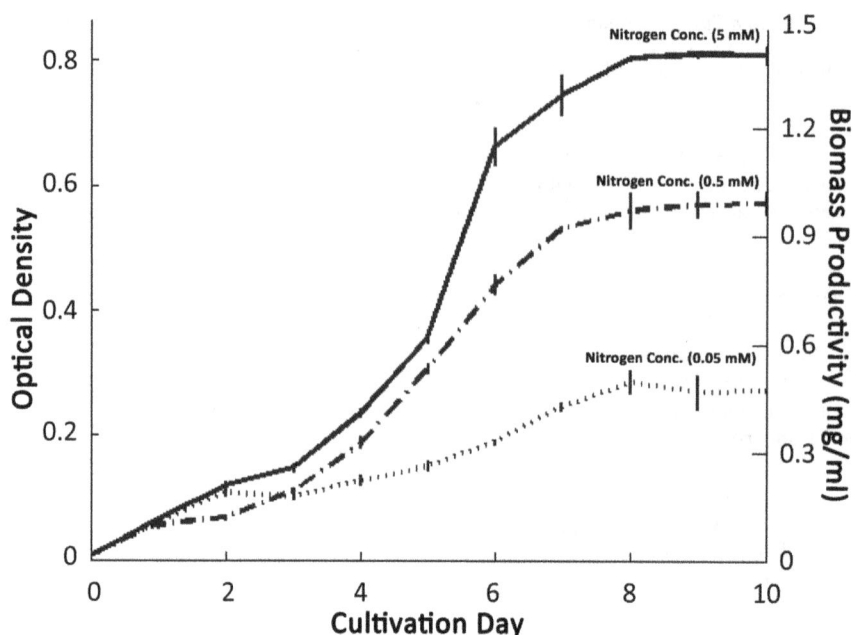

Figure 2. Effects of different nitrogen concentrations on the growth and biomass productivity of *Dunaliella salina strain Tuz_KS_01* *(*GeneBank accession no. JX880083). 0.05 mM, 0.5 mM and 5 mM NaNO$_3$ are referred as low, medium and high nitrogen concentrations respectively. Shown OD$_{600}$ optical density and biomass values are the means of three replicates. The error bars correspond to ±1 SD of triplicate optical density measurements.

and kept for 24 h at 25°C. The mixture was then vortexed and sonicated for 10 minutes. One milliliter of chloroform was again added, and the mixture shaken vigorously for 1 min. Subsequently, 1.8 ml of distilled water was added and the mixture vortexed again for 2 min. The aqueous and organic phases were separated by centrifugation for 15 min at 2,000 rpm. The lower (organic) phase was transferred into a previously weighed clean vial (V1). Evaporation occurred in a thermo-block at 95°C, and the residue was further dried at 104°C for 30 min. The weight of the vial was again recorded (V2). Lipid content was calculated by subtracting V1 from V2, and expressed as dcw %. The correspondence between Nile Red fluorescence intensity and % lipid content was determined by plotting relative fluorescence units against % cellular lipid content obtained from triplicate samples. For fluorescence microscopy analysis of Nile Red, cells were stained with 5 µl 0.5 mg/mL Nile Red (Sigma, USA) stock solution after fixing cells with 5% paraformaldehyde and imaged by epifluorescence microscopy with a Leica DMR microscope (Leica Microsystems).

Spectrofluorometric Microplate Analysis for Determination of Lipid and Reactive Oxygen Species Accumulation

Nile red, 9-diethylamino-5H-benzo[alpha]phenoxazine-5-one was first reported that is an excellent vital stain for the detection of intracellular lipid droplets by fluorescence microscopy and flow cytoflourometry [14] and it has been widely used for measuring and comparing cellular lipid content in various organisms in numerous studies [15–19]. A stock solution of Nile Red (NR) (Sigma, 72485) was prepared by adding 5 mg of NR to 10 ml of acetone. The solution was kept in a dark colored bottle and stored in the dark at −20°C. 1 ml of algal cells from a culture of 250 ml glass erlenmeyer flasks containing 100 ml growth media with

different nitrogen concentrations were transferred to 1.5 ml eppendorf tubes for 5 min centrifugation at 5,000 rpm, washed twice with fresh medium, and measured in a spectrophotometer at 600 nm. Each sample was adjusted to an OD$_{600}$ of 0.3 in a 1 ml final volume by dilution with fresh medium. 5 µl of Nile Red solution was added to each tube and mixed well, followed by 20 min incubation in the dark. Finally, cellular neutral lipids were quantified using a 96-well microplate spectrofluorometry (SpectraMAX GEMINI XS) with an excitation wavelength of 485 nm and an emission wavelength of 612 nm.

Dichloro-dihydro-fluorescein diacetate (DCFH-DA) is the most widely used fluorometric probe for detecting intracellular oxidative stress. This probe is cell-permeable and is hydrolyzed intracellularly to the DCFH carboxylate anion. Two-electron oxidation of DCFH results in the formation of a dichlorofluorescein (DCF) as a fluorescent product. The amount of this fluorescent product is highly correlated with the cellular oxidation/oxidative stress level. Investigators have routinely used DCFH-DA to measure intracellular generation of reactive oxidants in cells in response to intra- or extracellular activation with oxidative stimulus [20]. Determination of ROS production related to oxidative stress was done by using DCFH-DA (Sigma, USA). Vital staining for determination of cell survival under H$_2$O$_2$ treatment was done by using fluorescein diacetate (FDA) (Sigma, USA). FDA is incorporated into live cells and it is converted into fluorescein by cellular hydrolysis [21]. 0.5 mg/ml stock solutions for both stains were prepared in acetone and the same protocol was used for FDA (Sigma, USA) and DCFH-DA (Sigma, USA) staining as described above for Nile Red staining. Data were recorded as relative fluorescence units [22] for all spectrofluorometric staining experiments.

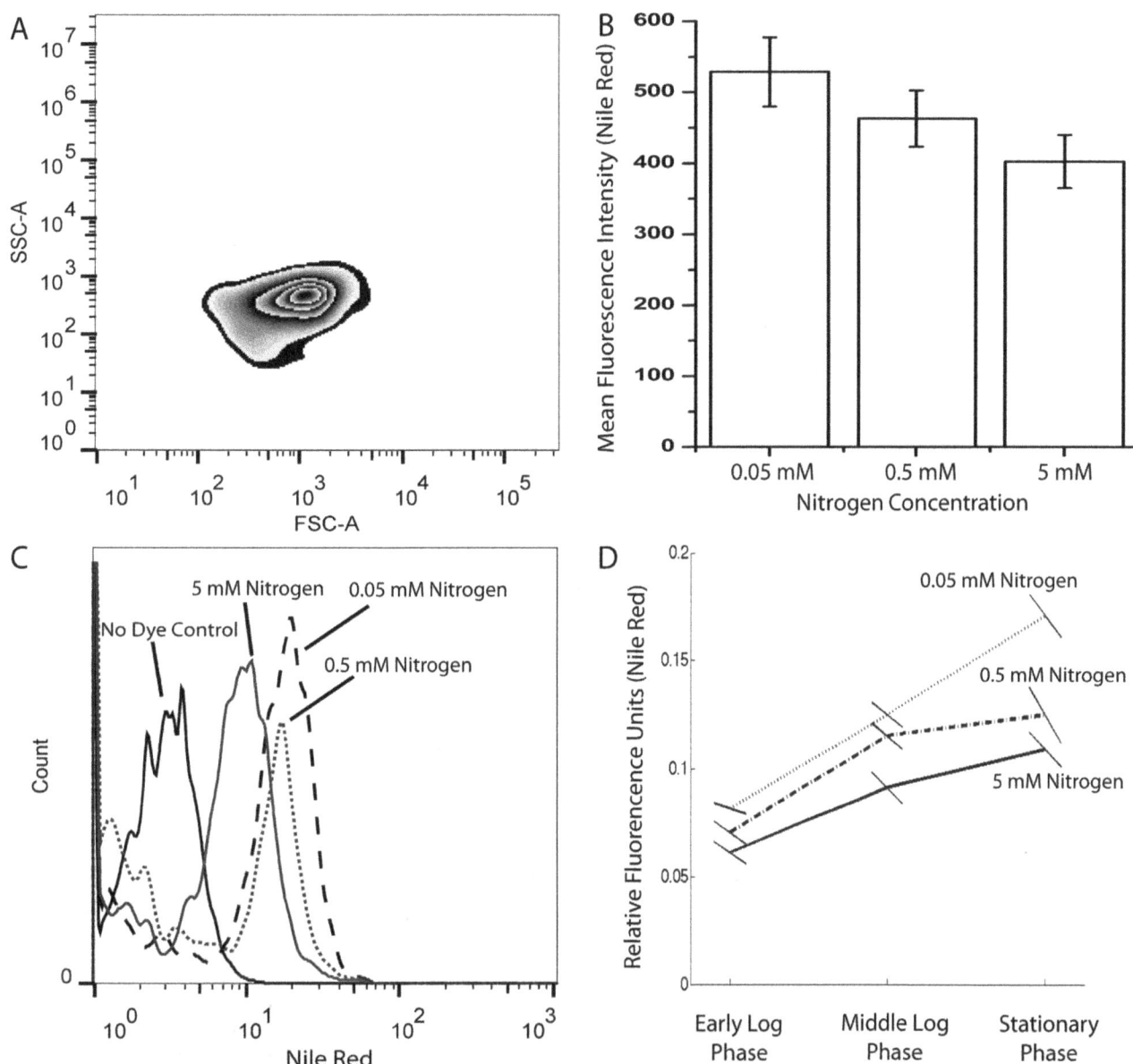

Figure 3. Lipid content analysis of *Dunaliella salina* strain *Tuz_KS_01* under different nitrogen conditions. A) Zebra-plot of SSC (Side Scatter) and FSC (Forward Scatter) expressing cellular granulation and cellular size, respectively, under a representative nitrogen depletion condition (0.05 mM). **B)** A representative histogram of flow-cytometric analysis of lipid contents under different nitrogen concentrations **C)** Mean fluorescent intensities (MFI) of flow cytometric analysis. **D)** Fluorometric microplate Nile Red analysis of early logarithmic, late logarithmic and stationary growth phases. Data represent the mean values of triplicates. Standard error for each triplicate is shown as tilted lines for clarity, where the minimum and maximum y values of each line corresponds to ±1 SE.

Flow Cytometric Analysis for Determination of Lipid and Reactive Oxygen Species Accumulation

5 µl of Nile Red (Sigma, USA) from stock solution (0.5 mg/mL) was added to 1 ml of a cell suspension at an OD_{600} of 0.3 after washing cells twice with fresh medium. This mixture was gently vortexed and incubated for 20 minutes at room temperature in dark. Nile Red uptake was determined using a BD-FACS Canto flow cytometer (Becton Dickinson Instruments) equipped with a 488 nm argon laser. Upon excitation by a 488 nm argon laser, NR exhibits intense yellow-gold fluorescence when dissolved in neutral lipids. The optical system used in the FACS Canto collects yellow and orange light (560–640 nm, corresponding to neutral

lipids). Approximately 10,000 cells were analysed using a log amplification of the fluorescent signal. Non-stained cells were used as an autofluorescence control. Nile Red fluorescence was measured using a 488 nm laser and a 556 LP+585/42 band pass filter set on a FACS Canto Flow Cytometer. Data were recorded as mean fluorescence intensity (MFI).

DCFH-DA (Sigma, USA) from stock solution (0.5 mg/ml) was used as described above for flow-cytometric microplate Nile Red staining method in order to analyze cellular oxidative stress status. Briefly, 1 ml of algal cells in different nitrogen concentrations were transferred into 1.5 ml eppendorf tubes, centrifuged 5 min at 5,000 rpm, washed twice with fresh medium, and measured in a spectrophotometer at 600 nm. Each sample was adjusted to an

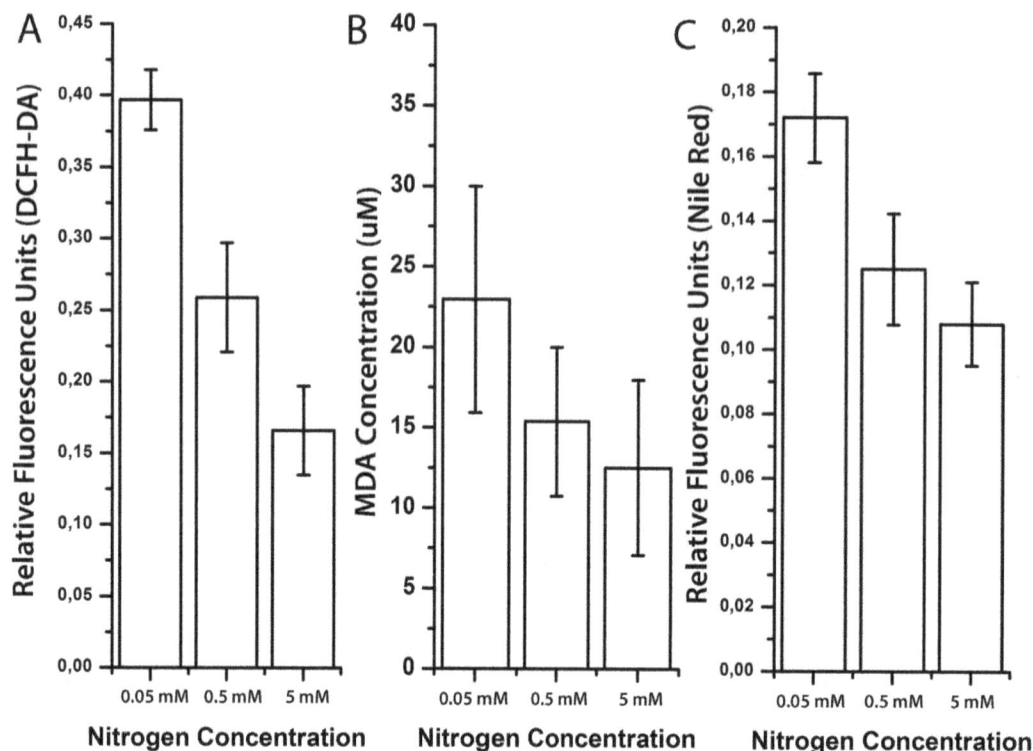

Figure 4. Oxidative stress and lipid accumulation under different nitrogen concentrations. A) Fluorometric microplate DCFH-DA analysis under different nitrogen concentrations. **B)** TBARS analysis for lipid peroxidation under different nitrogen concentrations. **C)** Fluorometric microplate Nile-Red analysis under different nitrogen concentrations. Data represent the mean values of triplicates ±1 SE.

OD_{600} of 0.3 in a 1 ml final volume by dilution with fresh medium. 5 µl of dye solution was added to each tube and mixed well, followed by 20 min incubation in the dark and analyzed. Nonfluorescent DCFH-DA is taken up and converted into diclorodihydrofluorescein (DCF) by the action of cellular esterases. Fluorescence from oxidized DCF was measured using a 488 nm laser and a 556/Long Pass filter set on a BD-FACS Canto Flow Cytometer. Data were expressed as mean fluorescence intensity (MFI). The data of both staining methods were evaluated using Flowjo Ver. 7.6.1 (Tree Star, Inc.).

Protein, Chlorophyll, Carotenoid, TBARS Analyses and Enzymatic Assays

Total protein isolation was done according to Barbarino et al. [23]. Protein content was determined following the method described by Bradford [24]. Cellular chlorophyll and carotenoid isolations were done using the methanol extraction method. Calculation of chlorophyll and carotenoid contents were carried out using the formula for methanol extraction described by Wellburn [25]. Lipid peroxidation analysis upon different nitrogen concentration regimes was analysed by using thiobarbituric acid reactive substances (TBARS) method described by Sabatini et al [26]. For superoxide dismutase (SOD) analysis 50 mg biomass was homogenized in 2 ml 0.5 M phosphate buffer (pH 7.5) and centrifuged at 13,000 rpm for 10 min at 4°C. SOD activity was determined in the supernatant by inhibition of nitroblue tetrazolium (NBT) using a reaction mixture of 1.5 ml Na_2CO_3 (1 M), 200 µl methionine (200 mM), 100 µl NBT (2.25 mM), 100 µl EDTA (3 mM), 100 µl riboflavin (60 ?M) and 1.5 ml phosphate buffer (pH 7.8, 0.1 M). The absorbance was recorded at 560 nm. One unit of SOD per gram of protein was defined as the amount

causing 50% inhibition of photochemical reduction of NBT [27]. For catalase (CAT) analysis, 50 mg biomass was homogenized in 2 ml phosphate buffer (0.5 M, pH 7.5), centrifuged at 12,000 rpm at 4°C for 30 min and supernatant was taken for CAT activity. A reaction mixture containing 1.6 ml phosphate buffer (pH 7.3), 100 µl EDTA (3 mM), 200 µl H_2O_2 (0.3%) and 100 µl supernatant (containing enzyme extract) was taken in a cuvette and CAT activity in supernatant was determined by monitoring the disappearance of H_2O_2, by measuring a decrease in absorbance at 240 nm against a blank of same reaction mixture without 0.3% H_2O_2 [28]. For analysis of ascorbate peroxidase (APX), 50 mg biomass was homogenized in 2 ml phosphate buffer (0.5 M, pH 7.5) and centrifuged at 12,000 rpm at 4°C for 30 min. The supernatant was taken for APX activity. A reaction mixture containing 1 ml phosphate buffer (pH 7.3), 100 µl EDTA (3 mM), 1 ml ascorbate (5 mM), 200 µl H_2O_2 (0.3%) and 100 µl supernatant (containing enzyme extract) was prepared in a cuvette. The reaction was followed for 3 min at a wavelength of 290 nm against a blank of same reaction mixture without 0.3% H_2O_2 [29].

Results and Discussion

Isolation and Identification of the New *Dunaliella salina* Strain

We isolated a new hypersaline microalga from the biggest salt lake of Turkey located in the Middle Anatolian region in July 2011. The microalga, which is unicellular, biflagellate and tolerant up to 20% NaCl was identified as *Dunaliella salina* based on the morphological characterization prior to molecular identification. We observed that cell color was green under normal conditions and the color changed to red under stress conditions. Morpho-

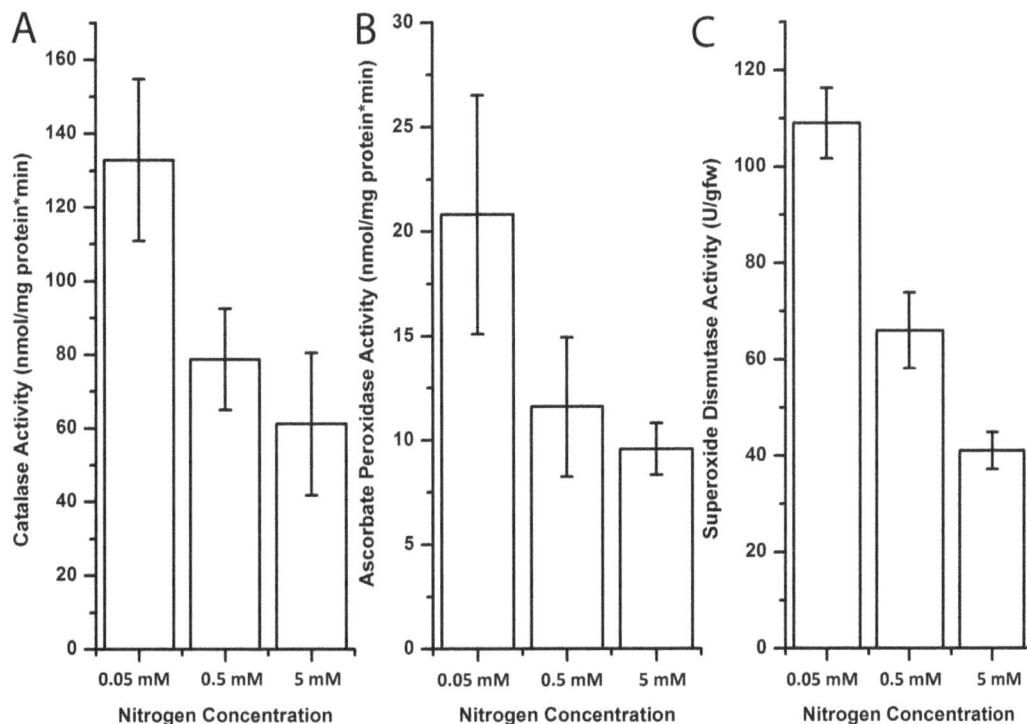

Figure 5. Antioxidant enzyme activities under different nitrogen concentrations. A–C) Spectrophotometric enzymatic assays for catalase (CAT), ascorbate peroxidase (APX) and superoxide dismutase (SOD). Data demonstrate the mean values of triplicates ±1 SE.

logically, ovoid, spherical and cylindrical cell shapes were observed. Stigma was not clearly visible or diffuse. Large pyrenoid with distict amylosphere was observed and refractile granules are

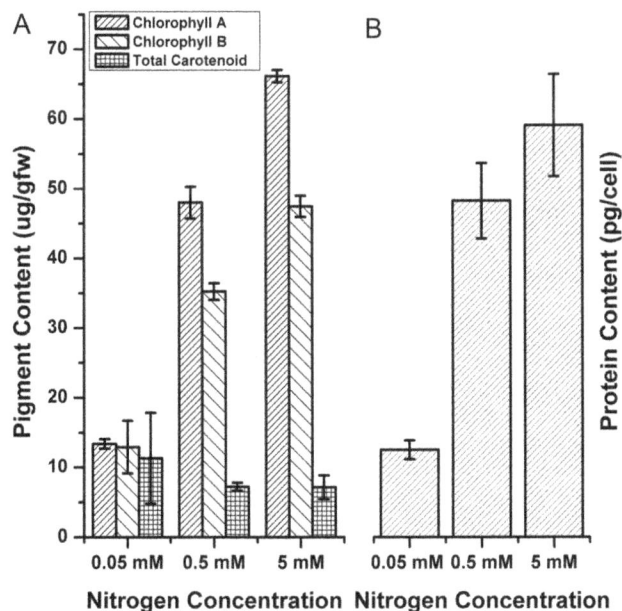

Figure 6. Pigment and protein contents of *Dunaliella salina* **strain under different nitrogen concentrations. A)** Chlorophyll A, B and total carotenoid contents under different nitrogen concentrations. **B)** Protein contents of *Dunaliella salina* strain Tuz_KS_01 cells cultivated under different nitrogen concentrations Data demonstrate the mean values of triplicates ±1 SE.

absent. Cellular size was measured as ~20 μm. These observed characteristics were consistent with previous morphological characteristics of *Dunaliella salina*.

This strain was deposited in −196°C ultra freeze conditions at Sabanci University algae culture collection as *Dunaliella salina strain Tuz_KS_01*. Molecular identification was based on 18S rDNA gene sequence amplification and sequencing. A single 18S rDNA PCR amplicon was about ~1800 bp in size, in accordance with Olmos et al. [11]. The partial sequence of the 18S rDNA encoding gene was submitted to National Center for Biotechnology Information (NCBI) GeneBank database as *Dunaliella salina strain Tuz_KS_01* (GeneBank accession no. **JX880083**). According to Basic Local Aligment Search Tool (BLAST) analysis, the isolated sequence had very high percentage of identity with other deposited 18S rDNA sequences of *Dunaliella* species shown in the phylogenetic tree plotted in Figure 1. Molecular identification showed a high percentage identity to *Dunaliella salina* strains (Identity 93%, Query Coverage 70%).

Growth Analysis of the New Dunaliella Strain Under Different Nitrogen Concentrations

The newly isolated halotolerant *Dunaliella salina* strain was cultivated under different nitrogen concentrations for growth analysis. 0.05 mM, 0.5 mM and 5 mM NaNO$_3$ concentrations were considered as low, medium and high nitrogen concentrations. Specific growth rates and/or average biomass productivities (ABP) were found to be 370 mg/dayL, 430 mg/dayL, 520 mg/dayL for low, medium and high nitrogen groups respectively. The highest biomass concentration under low nitrogen regimes was 495 mg/l, compared with 994 mg/l and 1409 mg/l for medium and high nitrogen regimes shown in Figure 2. Other kinetic growth data including doubling time/generation time and division per day were also calculated. Doubling times were calculated as

1.84, 1.59, and 1.32 whereas division of cells per day was calculated as 0.54, 0.62, and 0.75 for low, medium and high nitrogen concentration groups respectively.

Biomass productivity is one of the most important parameters for the feasibility of utilizing algal oil for biodiesel production. Hence, numerous studies have been conducted in various algal species. Tang et al. (2011) demonstrated that *Dunaliella tertiolecta* had a highest biomass concentration of ~400 mg/l [30]. In another study conducted by Ho et al. (2010) the optimal biomass productivity of freshwater alga *Scenedesmus obliquus* were found to be 292.50 mg/dayL [31]. Griffiths et al. (2012) studied 11 different algal species including freshwater, marine and halotolerant species and demonstrated their biomass productivities [32]. Compared to the reported biomass productivities in previous studies, both the highest biomass concentration and average biomass productivity of isolated *Dunaliella salina strain Tuz_KS_01* showed very good potential, eventhough small batch cultivation process was utilized instead of using more sophisticated and efficient cultivation systems.

Lipid Accumulation Analysis of the New *Dunaliella salina* Strain Under Different Nitrogen Concentrations

Experimental groups were subjected to flow cytometric Nile red analysis for measuring lipid accumulation at single cell level in stationary growth phase. Cell populations were chosen according to cellular size and granulation as quantified by forward and side scatter values of populations, respectively which is demonstrated in Figure 3A. Figure 3B shows a representative flow cytometric analysis of lipid contents under different nitrogen concentrations, indicating high lipid content under low nitrogen condition. Figure 3C shows the mean and standard error for three biological replicates, clearly exhibiting a negative relationship between lipid accumulation and nitrogen levels. Therefore, cultivation under low nitrogen conditions stimulates lipid accumulation in *Dunaliella salina strain Tuz_KS_01*, in agreement with previous studies [33,34].

In addition, we measured lipid accumulation of the green alga under different nitrogen levels using spectrofluorophotometer [35] at early-logarithmic, late-logarithmic and stationary growth phases. While the cellular lipid accumulation in early and late logarithmic phases did not show a drastic change, we observed a marked increase of lipid content in the stationary phase, which was significantly higher under low nitrogen cultivation conditions shown in Figure 3D.

Nitrogen depletion is well known to result in increased lipid accumulation in algal species [36], eventhough its association with other related factors has not been well defined. To demonstrate the oxidative stress under nitrogen limited cultivation conditions, fluorometric DCFH-DA and related TBARS lipid peroxidation assays were utilized. According to these results, increased ROS production (Figure 4A) and lipid peroxidation (Figure 4B) were both observed under nitrogen limited conditions, in association with increased lipid accumulation especially at stationary phase shown in Figure 4C.

Antioxidant Enzyme Activities, Pigment Composition and Protein Analyses Under Different Nitrogen Concentrations

ROS accumulation is prevented by a powerful intrinsic antioxidant system in photosynthetic organisms, involving enzymes such as superoxide dismutase (SOD), catalase (CAT), ascorbate peroxidase (APX) [37]. Next, we analyzed these three oxidative stress indicator antioxidant enzymes to demonstrate the

effect of nitrogen depletion on the intracellular oxidative stress status as also supplementary to DCFH-DA measurements.

Intracellular SOD converts O_2^- to O_2 and H_2O_2, acting as the first line of defence against oxidative stress [38]. We observed that SOD activity of cells increased under low nitrogen concentrations, especially at stationary growth phase. This data suggests that superoxides may be elevated under nitrogen depleted conditions, necessitating increased SOD activity (Figure 5A). CAT is a heme-containing enzyme that catalyzes the conversion of H_2O_2 into oxygen and water [37]. APX is involved in the ascorbate-glutathione cycle occuring in chloroplasts, cytoplasm, mitochondria and perixisomes [39]. We observed elevated levels of CAT and APX activity under nitrogen depleted conditions (Figure 5B, 5C). These data strongly indicate that oxidative stress is induced under nitrogen depletion in *Dunaliella salina strain Tuz_KS_01*.

Next, we analyzed the chlorophyll, carotenoid and protein content change in response to nitrogen depletion in *Dunaliella salina strain Tuz_KS_01*. Chlorophyll content is an indicator of photosynthetic efficiency, rate and nitrogen [40] status. Carotenoids can perform an essential role in photoprotection by quenching the triplet chlorophyll and scavenging singlet oxygen and other reactive oxygen species [41]. We observed that chlorophyll A and B levels sharply declined in low nitrogen conditions, as expected (Figure 6A). In contrast, we observed no decrease in total carotenoid contents under nitrogen depletion conditions. This observation suggests that cells increase caroteoid content to overcome with oxidative stress induced by singlet oxygens, similar to SOD activity increase discussed above (Figure 6A).

Upon nitrogen starvation, algal cells reduce the synthesis of protein and nucleic acid synthesis and carbon flow is directed from protein synthesis to fatty acid and carbonhydrate synthesis [42]. We found that neutral lipids increase while the protein content of algal cells was reduced under nitrogen depletion conditions (Figure 6B). This indicates that carbon flow direction is to fatty acid synthesis under nitrogen limitation at stationary phase, in accordance with the literature [43].

Reactive Oxygen Species (ROS) Production Induces Lipid Accumulation

Our analysis thus far hints that lipid accumulation might be partially triggered by ROS accumulation and oxidative stress under nitrogen depleted condition (Figure 4). In order to test this hypothesis, we used H_2O_2, a well-known oxidative stress inducer [44]. We treated cells at exponential growth phase in high nitrogen containing media (5 mM) with different concentrations of H_2O_2 and we measured ROS, via using flowcytometric DCFH-DA method, a robust fluorescent assay [45]. In addition, cell viability was measured by using FDA fluorescent assay while using Nile red staining method for determination of lipid contents under different H_2O_2 concentrations. Although cellular size and granularity under H_2O_2 treated conditions were similar to previous experimental conditions (Figure 7A and 3A), cell survival under higher H_2O_2 concentrations were observed significantly reduced due to the oxidative stress induction (Figure 7B–7C). Although reduction of cell survival is to ~20% at 4 mM H_2O_2 concentration, at this high concentration of H_2O_2, new isolate *Dunaliella salina strain Tuz_KS_01* showed highly tolerant characteristics to oxidative stress (Figure 7D).

Lipid accumulation of algae is known to increase by various factors including temperature, excessive light, and pH, which may also related to oxidative stress induced by ROS accumulation [33,46]. We used fluorometric Nile red analysis for analysis of lipid accumulation under different H_2O_2 treatments. We observed increased lipid accumulation with increasing concentrations of

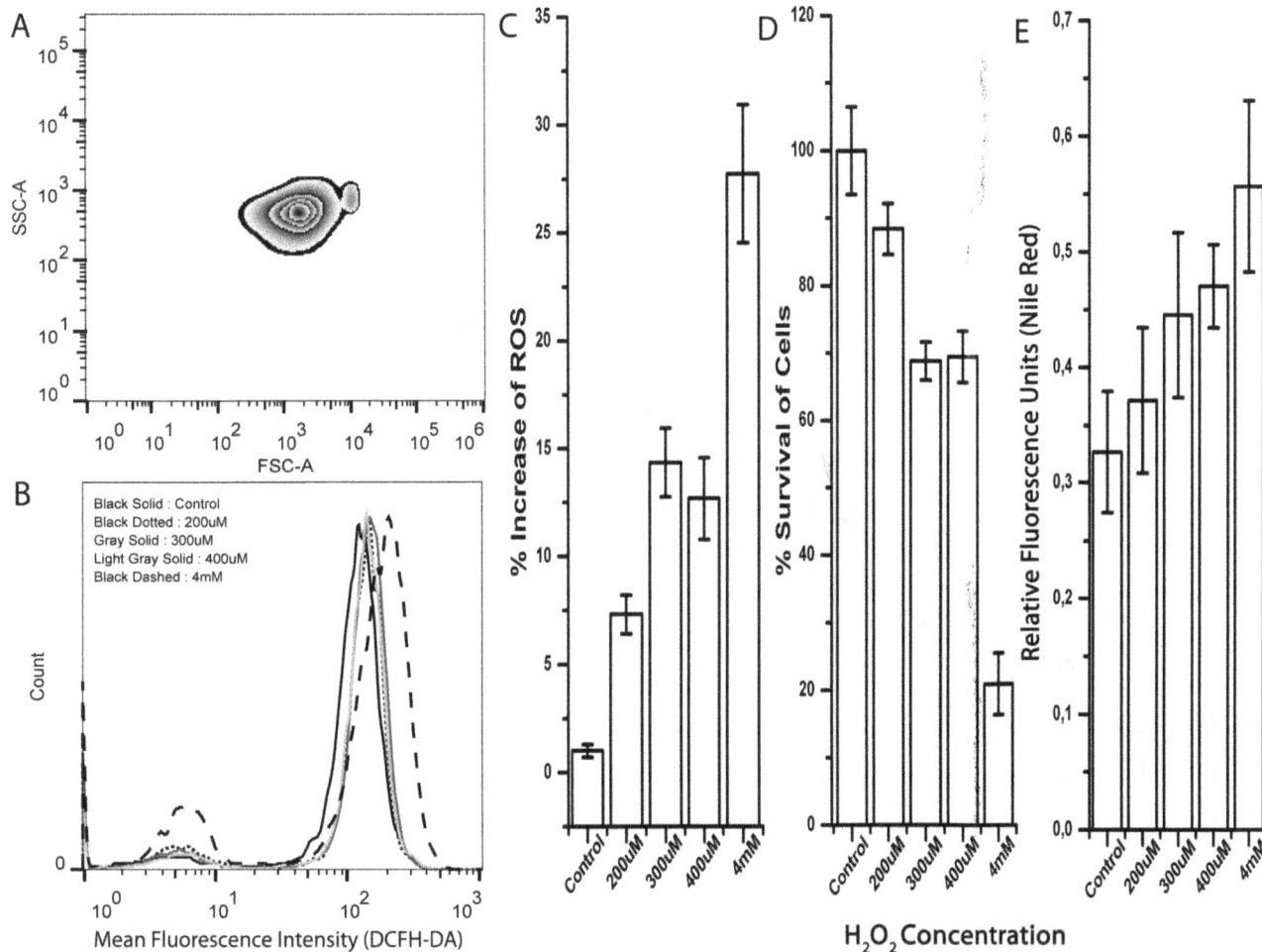

Figure 7. Effect of H₂O₂, a known oxidative stress inducer, on *Dunaliella salina strain Tuz_KS_01.* **A)** Zebra-plot of SSC (Side Scatter) and FSC (Forward Scatter) expressing cellular granulation and cellular size under a representative H₂O₂ condition (4 mM). **B)** Histogram of flow-cytometric analysis of ROS accumulation under different H₂O₂ concentrations. **C)** Percentage increase of ROS production based on flow-cytometric DCFH-DA analysis. **D)** Fluorometric microplate fluorescent diacetate (FDA) survival analysis cultivated under different H₂O₂ concentrations. **E)** Fluorometric microplate Nile-Red analysis under different H₂O₂ concentrations. Data represent the mean values of triplicates ±1 SE.

H₂O₂ (Figure 7E). This result supports that the lipid accumulation observed under nitrogen depletion conditions is at least partly mediated by oxidative stress. Importantly, this result also suggests a method for obtaining high lipid from green alga, namely by growing cells in high nitrogen (optimal) conditions followed by oxidative stress induction at stationary phase. Such a method may be economically more feasible than two stage reactors described above.

We next wished to visually observe the effect of nitrogen depletion and H₂O₂ treatment on lipid accumulation. After cultivation, we collected and stained cells with Nile red and observed with fluorescence microscopy. We used three different conditions: 1) control, 2) 0.05 mM nitrogen and 3) 4 mM H₂O₂. Microscopy results were in strong agreement with our previous results, as lipid accumulation (increased size and number of cytoplasmic lipid droplets) after nitrogen depletion or H₂O₂ treatment was higher than the control (Figure 8A–8B–8C).

Moreover, in order to validate the lipid increasing effect of nitrogen limitation and H₂O₂ induced oxidative stress, solvent extraction gravimetric lipid analysis in nitrogen depleted (0.05 mM) and H₂O₂ containing media (4 mM) were done. We

found that, nitrogen limitation led to increased lipid accumulation up to 35%, and H₂O₂ induced oxidative stress led to increased lipid accumulation up to 44% as shown in Figure 8D. We also measured the lipid content by Nile red fluorescence analysis for each of these conditions. Consistent with previous studies, we found that gravimetric and fluorometric measurements were correlated ($r^2 = 0.82$) (Figure 8E) [17]. These data suggest that exogenously induced oxidative stress triggered by the application of H₂O₂ resulted to increased lipid accumulation. Therefore, as previously shown by the fluorometric analyses above, oxidative stress and increased lipid accumulation association is coupled under nitrogen limitation conditions. Induction of lipid accumulation by applying exogenous oxidative stress inducers such as H₂O₂ may assist more effective lipid production compared with biomass production lowering nitrogen starvation strategy.

ROS production resulting from various stress factors is known to affect nearly all cellular processes by impairing the structural stability of functional macromolecules including DNA, proteins and structural lipids. Since the green algae life-cycle is reliant on its photosynthetic activity and cellular integrity, it is crucial to protect against oxidative stress. Otherwise cells can not tolerate oxidative

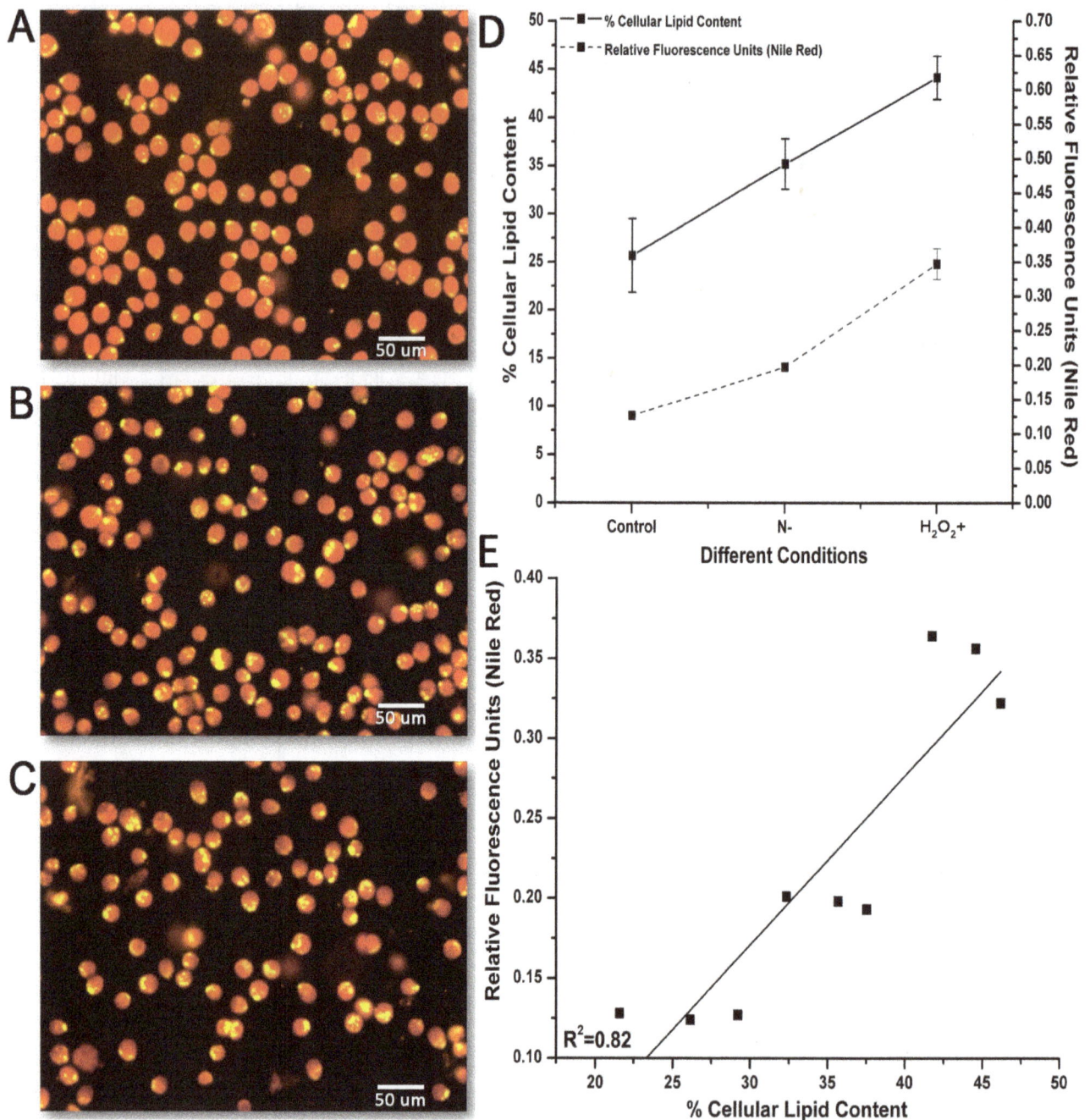

Figure 8. Fluorescence microphotographs of *Dunaliella salina strain Tuz_KS_01* **stained with Nile-Red fluorescence dye and screened under 400X magnification. A)** Control group, 5 mM nitrogen concentration cultivation condition. **B)** Nitrogen limitation group, 0.05 mM nitrogen condentration cultivation condition **C)** Oxidative stress group, 4 mM H_2O_2 cultivation condition. Gravimetric lipid content analysis and Nile-Red fluorescence measurement correlation **D)** Gravimetric and fluorometric Nile-Red lipid content analysis of *Dunaliella salina strain Tuz_KS_01* under 0.05 mM nitrogen and 4 mM H_2O_2 cultivation conditions. Data represent the mean values of triplicates ±1 SE **E)** Correlation plot of gravimetric and fluorometric Nile-Red lipid content analysis ($r^2 = 0.82$).

damage and eventually die [47–49]. Even though there is limited explanation about the correlation between ROS production and lipid accumulation in algae cultivated under nitrogen depleted conditions. Our results indicate that nitrogen depletion results in ROS accumulation and lipid peroxidation especially at stationary growth phase. This relationship might also be related to survival response of the alga against excessive oxidative stress conditions [7].

Although a few recent studies have suggested an association of ROS levels and cellular lipid accumulation [7,8], underlying mechanistic principles are not clear. ROS is well demonstrated to modify cellular responses against different stressors in corresponding signal transduction pathways. ROS levels increase in microalgae cells when exposed to the different environmental stresses [9]. In a recent study, nitrogen depletion was shown to correlate with increased ROS accumulation and increased cellular

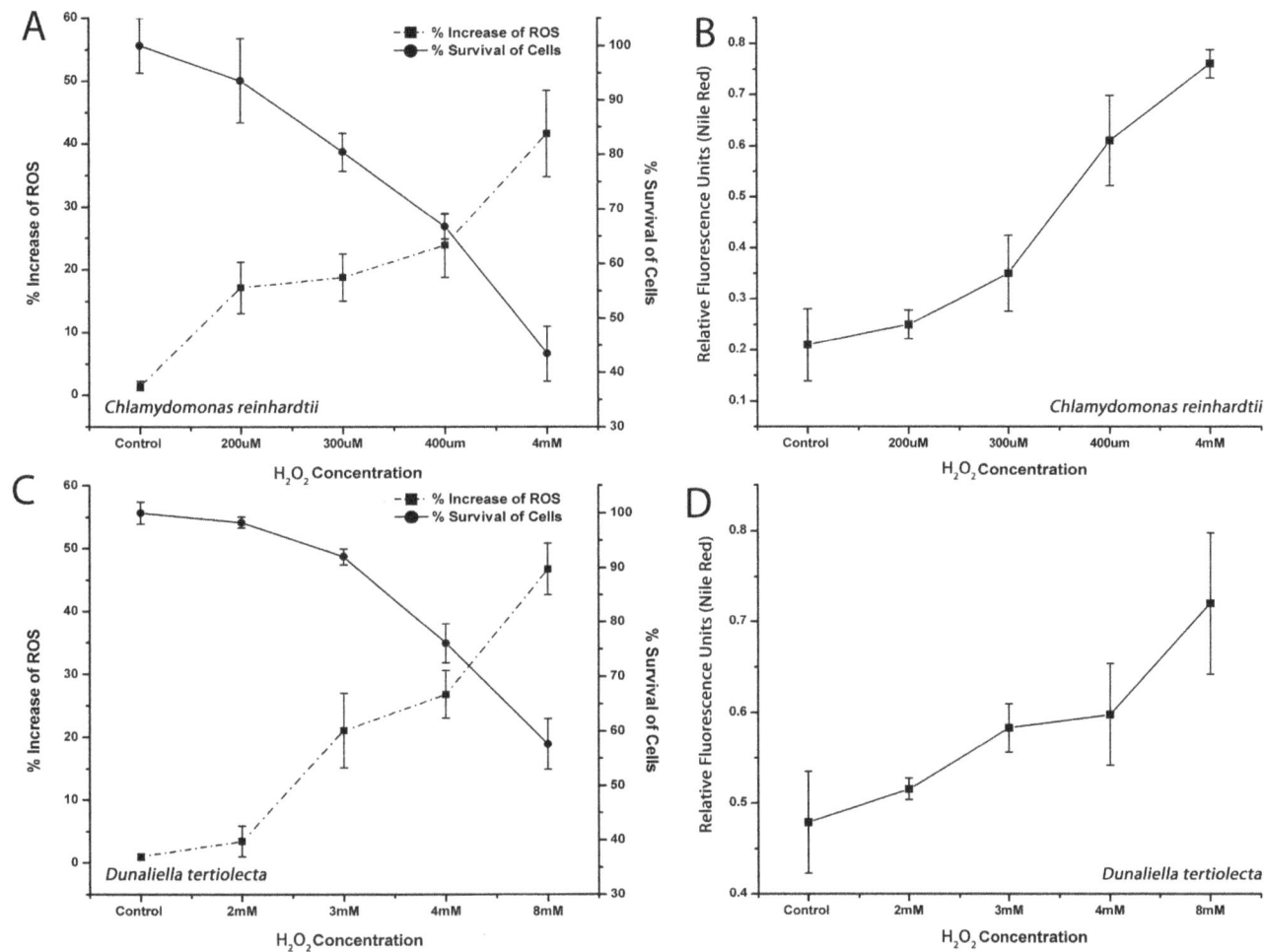

Figure 9. Effect of H₂O₂ induced oxidative stress on cellular lipid accumulation of different algae species. A) Fluorometric DCFH-DA and FDA analyses showing percantage change of ROS production and percentage change of cell survivability of *Chlamydomonas reinhardtii* cells cultivated under different H₂O₂ concentrations. **B)** Fluorometric Nile red analysis showing the effect of different H₂O₂ concentrations on cellular lipid accumulation of *Chlamydomonas reinhardtii* cells **C)** Fluorometric DCFH-DA and FDA analyses showing percantage change of ROS production and percentage change of cell survivability of *Dunaliella tertiolecta* cells cultivated under different H₂O₂ concentrations. **D)** Fluorometric Nile Red analysis showing the effect of different H₂O₂ concentrations on cellular lipid accumulation of *Dunaliella tertiolecta* cells. Data represent the mean values of triplicates ±1 SE.

lipid accumulation in a freshwater algae. This study showed that increased MDA concentration is an indicator of membrane peroxidation which implied increased ROS levels. It was suggested that H_2O_2 induced exogenous oxidative stress was an effective factor for neutral lipid induction in *C. sorokiniana C3* [50].

Microalgae can modify its photosynthetic system under stress, resulting in a decrease in the gene expression of various proteins forming up the photosystem complexes I and II [51]. Such adjustments are thought to occur for minimizing oxidative stress via decreasing photosynthesis rate [52]. Nitrogen deprivation is closely associated with the degradation of ribulose-1,5-bisphosphate carboxylase oxygenase to recycle nitrogen [53]. Degradation of this protein may result in alterations in photosynthesis rate, which is consistent with the observed decrease in chlorophyll content under nitrogen depleted condition [54]. As a result of decreased photosynthesis rate, overall anabolic reaction flux is severely constrained. In this context, algae cells may favor storage of energetic molecules, such as lipids, instead of consumption.

Effects of H₂O₂ Induced Oxidative Stress on Lipid Accumulation in *Dunaliella tertiolecta* and *Chlamydomonas reinhardtii*

Green algae, *Dunaliella tertiolecta* and *Chlamydomonas reinhardtii* were chosen for further investigation and confirmation of the H_2O_2 induced oxidative stress effects on cellular lipid accumulation. We cultivated cells in high nitrogen containing media (5 mM) with different concentrations of H_2O_2 as described for *Dunaliella salina* strain *Tuz_KS_01*. We measured ROS by using the DCFH-DA method, cell viability by using the FDA fluorescent assay, and lipid content by using the fluorometric Nile Red method, as described above. *Dunaliella tertiolecta* cells were found to be more tolerant to H_2O_2; therefore higher concentrations of H_2O_2 were used to demonstrate the cellular response. *Chlamydomonas reinhardtii* cells were treated within the range from 200 µM to 4 mM H_2O_2 concentrations while *Dunaliella tertiolecta* cells were treated within the range of 2 mM to 8 mM H_2O_2 concentrations. As shown in Figure 9, we found that in response to H_2O_2 in both species ROS accumulation and lipid contents increased, while cell survival

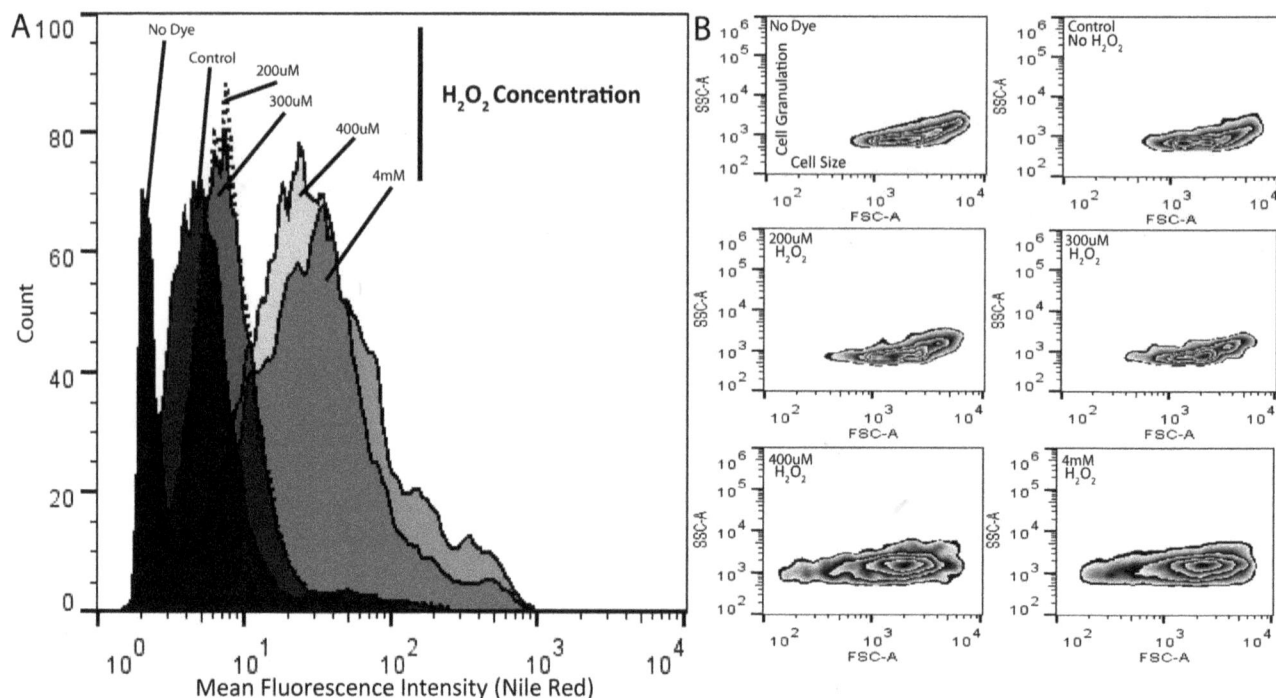

Figure 10. Flow cytometric Nile red, cellular granulation and biovolume analysis of *Chlamydomonas reinhardtii* **cells cultivated under different H₂O₂ concentrations. A)** A representative histogram of flow-cytometric analysis of lipid contents under different H₂O₂ concentrations. **B)** Zebra-plot of SSC (Side Scatter) and FSC (Forward Scatter) expressing cellular granulation and cellular size, respectively, under different H₂O₂ concentrations. Data represent the mean values of triplicates ±1 SE.

dramatically decreased at high H₂O₂ concentrations. In addition, by using flow cytometric analysis, we observed that lipid increase in response to oxidative stress was accompanied with increased granulation and biovolume in *Chlamydomonas* cells (Figure 10). Neither *Dunaliella tertiolecta* nor *Dunaliella salina* strain *Tuz_KS_01* cells showed biovolume increase upon treatment of H₂O₂ at any experimented concentration.

Conclusions

Nitrogen depletion has been shown to be an effective inducer of lipid production of various unicellular green algae by previous studies. But few studies have examined the physiological and biochemical mechanisms underlying this response. In this study, we isolated a new halophilic green alga species. We presented evidence supporting previous observations that nitrogen depletion causes oxidative stress and lipid accumulation. In addition, we

showed that oxidative stress by itself can cause lipid accumulation, suggesting that lipid accumulation under nitrogen depletion is mediated by oxidative stress. These observations are helpful for utilization of green alga for biodiesel production.

Acknowledgments

Authors would like to thank UTEX, The Culture Collection of Algae, for providing kindly of algae species, *Chlamydomonas reinhardtii* and *Dunaliella tertiolecta*. MC was supported by Turkish Academy of Sciences Gebip Programme.

Author Contributions

Conceived and designed the experiments: KY MC SC. Performed the experiments: KY IP. Analyzed the data: KY MC. Contributed reagents/materials/analysis tools: SC MC BE. Wrote the paper: KY MC IP SC.

References

1. Camacho FG, Rodriguez JG, Miron AS, Garcia MC, Belarbi EH, et al. (2007) Biotechnological significance of toxic marine dinoflagellates. Biotechnol Adv 25: 176–194.
2. Li Y, Horsman M, Wang B, Wu N, Lan CQ (2008) Effects of nitrogen sources on cell growth and lipid accumulation of green alga Neochloris oleoabundans. Appl Microbiol Biotechnol 81: 629–636.
3. Csavina JL, Stuart BJ, Riefler RG, Vis ML (2011) Growth optimization of algae for biodiesel production. J Appl Microbiol 111: 312–318.
4. Spolaore P, Joannis-Cassan C, Duran E, Isambert A (2006) Commercial applications of microalgae. J Biosci Bioeng 101: 87–96.
5. Alscher RG, Donahue JL, Cramer CL (1997) Reactive oxygen species and antioxidants: Relationships in green cells. Physiologia Plantarum 100: 224–233.
6. Mallick N, Mohn FH (2000) Reactive oxygen species: response of algal cells. Journal of Plant Physiology 157: 183–193.
7. Liu WH, Huang ZW, Li P, Xia JF, Chen B (2012) Formation of triacylglycerol in Nitzschia closterium f. minutissima under nitrogen limitation and possible

physiological and biochemical mechanisms. Journal of Experimental Marine Biology and Ecology 418: 24–29.
8. Li X, Hu HY, Zhang YP (2011) Growth and lipid accumulation properties of a freshwater microalga Scenedesmus sp. under different cultivation temperature. Bioresour Technol 102: 3098–3102.
9. Hong Y, Hu HY, Li FM (2008) Physiological and biochemical effects of allelochemical ethyl 2-methyl acetoacetate (EMA) on cyanobacterium Microcystis aeruginosa. Ecotoxicol Environ Saf 71: 527–534.
10. Gordillo FL, Goutx M, Figueroa F, Niell FX (1998) Effects of light intensity, CO2 and nitrogen supply on lipid class composition of Dunaliella viridis. Journal of Applied Phycology 10: 135–144.
11. Olmos J, Paniagua J, Contreras R (2000) Molecular identification of Dunaliella sp utilizing the 18S rDNA gene. Letters in Applied Microbiology 30: 80–84.
12. Fredslund J (2006) PHY center dot FI: fast and easy online creation and manipulation of phylogeny color figures. Bmc Bioinformatics 7.

13. Levasseur M, Thompson PA, Harrison PJ (1993) Physiological acclimation of marine-phytoplankton to different nitrogen-sources. Journal of Phycology 29: 587–595.

14. Greenspan P, Mayer EP, Fowler SD (1985) Nile red: a selective fluorescent stain for intracellular lipid droplets. J Cell Biol 100: 965–973.

15. Govender T, Ramanna L, Rawat I, Bux F (2012) BODIPY staining, an alternative to the Nile Red fluorescence method for the evaluation of intracellular lipids in microalgae. Bioresour Technol 114: 507–511.

16. Chen W, Sommerfeld M, Hu Q (2011) Microwave-assisted nile red method for in vivo quantification of neutral lipids in microalgae. Bioresour Technol 102: 135–141.

17. Chen W, Zhang C, Song L, Sommerfeld M, Hu Q (2009) A high throughput Nile red method for quantitative measurement of neutral lipids in microalgae. J Microbiol Methods 77: 41–47.

18. Cole TA, Fok AK, Ueno MS, Allen RD (1990) Use of nile red as a rapid measure of lipid content in ciliates. Eur J Protistol 25: 361–368.

19. Kimura K, Yamaoka M, Kamisaka Y (2004) Rapid estimation of lipids in oleaginous fungi and yeasts using Nile red fluorescence. J Microbiol Methods 56: 331–338.

20. Kalyanaraman B, Darley-Usmar V, Davies KJA, Dennery PA, Forman HJ, et al. (2012) Measuring reactive oxygen and nitrogen species with fluorescent probes: challenges and limitations. Free Radical Biology and Medicine 52: 1–6.

21. Liang ZJ, Ge F, Zeng H, Xu Y, Peng F, et al. (2013) Influence of cetyltrimethyl ammonium bromide on nutrient uptake and cell responses of Chlorella vulgaris. Aquatic Toxicology 138: 81–87.

22. Schonswetter P, Tribsch A, Barfuss M, Niklfeld H (2002) Several Pleistocene refugia detected in the high alpine plant Phyteuma globulariifolium sternb & hoppe (Campanulaceae) in the European Alps. Mol Ecol 11: 2637–2647.

23. Barbarino E, Lourenco SO (2005) An evaluation of methods for extraction and quantification of protein from marine macro- and microalgae. Journal of Applied Phycology 17: 447–460.

24. Bradford MM (1976) A rapid and sensitive method for the quantitation of microgram quantities of protein utilizing the principle of protein-dye binding. Anal Biochem 72: 248–254.

25. Wellburn AR (1994) The spectral determination of chlorophyll-a and chlorophhyll-b, as well as total carotenoids, using various solvents with spectrophotometers of different resolution. Journal of Plant Physiology 144: 307–313.

26. Sabatini SE, Juarez AB, Eppis MR, Bianchi L, Luquet CM, et al. (2009) Oxidative stress and antioxidant defenses in two green microalgae exposed to copper. Ecotoxicology and Environmental Safety 72: 1200–1206.

27. Dhindsa RS, Plumbdhindsa P, Thorpe TA (1981) Leaf senescence - correlated with increased levels of membrane-permeability and lipid-peroxidation, and decreased levels of superoxide-dismutase and catalase. Journal of Experimental Botany 32: 93–101.

28. Aebi H (1984) Catalase invitro. Methods in Enzymology 105: 121–126.

29. Nakano Y, Asada K (1981) Hydrogen-peroxide is scavenged by ascorbate-specific peroxidase in spinach-chloroplasts. Plant and Cell Physiology 22: 867–880.

30. Tang HY, Abunasser N, Garcia MED, Chen M, Ng KYS, et al. (2011) Potential of microalgae oil from Dunaliella tertiolecta as a feedstock for biodiesel. Applied Energy 88: 3324–3330.

31. Ho SH, Chen WM, Chang JS (2010) Scenedesmus obliquus CNW-N as a potential candidate for CO2 mitigation and biodiesel production. Bioresource Technology 101: 8725–8730.

32. Griffiths MJ, van Hille RP, Harrison STL (2012) Lipid productivity, settling potential and fatty acid profile of 11 microalgal species grown under nitrogen replete and limited conditions. Journal of Applied Phycology 24: 989–1001.

33. Chen M, Tang HY, Ma HZ, Holland TC, Ng KYS, et al. (2011) Effect of nutrients on growth and lipid accumulation in the green algae Dunaliella tertiolecta. Bioresource Technology 102: 1649–1655.

34. Sharma KK, Schuhmann H, Schenk PM (2012) High Lipid Induction in Microalgae for Biodiesel Production. Energies 5: 1532–1553.

35. Lee SJ, Yoon BD, Oh HM (1998) Rapid method for the determination of lipid from the green alga Botryococcus braunii. Biotechnology Techniques 12: 553–556.

36. Lin QA, Lin JD (2011) Effects of nitrogen source and concentration on biomass and oil production of a Scenedesmus rubescens like microalga. Bioresource Technology 102: 1615–1621.

37. Bhaduri AM, Fulekar MH (2012) Antioxidant enzyme responses of plants to heavy metal stress. Reviews in Environmental Science and Bio-Technology 11: 55–69.

38. Alscher RG, Erturk N, Heath LS (2002) Role of superoxide dismutases (SODs) in controlling oxidative stress in plants. Journal of Experimental Botany 53: 1331–1341.

39. del Rio LA, Sandalio LM, Corpas FJ, Palma JM, Barroso JB (2006) Reactive oxygen species and reactive nitrogen species in peroxisomes. Production, scavenging, and role in cell signaling. Plant Physiology 141: 330–335.

40. Miller SR, Martin M, Touchton J, Castenholz RW (2002) Effects of nitrogen availability on pigmentation and carbon assimilation in the cyanobacterium Synechococcus sp strain SH-94-5. Archives of Microbiology 177: 392–400.

41. Singh SP, Hader DP, Sinha RP (2010) Cyanobacteria and ultraviolet radiation (UVR) stress: Mitigation strategies. Ageing Research Reviews 9: 79–90.

42. Courchesne NMD, Parisien A, Wang B, Lan CQ (2009) Enhancement of lipid production using biochemical, genetic and transcription factor engineering approaches. Journal of Biotechnology 141: 31–41.

43. Msanne J, Xu D, Konda AR, Casas-Mollano JA, Awada T, et al. (2012) Metabolic and gene expression changes triggered by nitrogen deprivation in the photoautotrophically grown microalgae Chlamydomonas reinhardtii and Coccomyxa sp C-169. Phytochemistry 75: 50–59.

44. Tsukagoshi H (2012) Defective root growth triggered by oxidative stress is controlled through the expression of cell cycle-related genes. Plant Science 197: 30–39.

45. Cash TP, Pan Y, Simon MC (2007) Reactive oxygen species and cellular oxygen sensing. Free Radical Biology and Medicine 43: 1219–1225.

46. Converti A, Casazza AA, Ortiz EY, Perego P, Del Borghi M (2009) Effect of temperature and nitrogen concentration on the growth and lipid content of Nannochloropsis oculata and Chlorella vulgaris for biodiesel production. Chemical Engineering and Processing 48: 1146–1151.

47. Hu Q, Sommerfeld M, Jarvis E, Ghirardi M, Posewitz M, et al. (2008) Microalgal triacylglycerols as feedstocks for biofuel production: perspectives and advances. Plant Journal 54: 621–639.

48. Kobayashi M (2003) Astaxanthin biosynthesis enhanced by reactive oxygen species in the green alga Haematococcus pluvialis. Biotechnology and Bioprocess Engineering 8: 322–330.

49. Pinto E, Sigaud-Kutner TCS, Leitao MAS, Okamoto OK, Morse D, et al. (2003) Heavy metal-induced oxidative stress in algae. Journal of Phycology 39: 1008–1018.

50. Zhang YM, Chen H, He CL, Wang Q (2013) Nitrogen starvation induced oxidative stress in an oil-producing green alga Chlorella sorokiniana C3. PLoS One 8: e69225.

51. Zhang ZD, Shrager J, Jain M, Chang CW, Vallon O, et al. (2004) Insights into the survival of Chlamydomonas reinhardtii during sulfur starvation based on microarray analysis of gene expression. Eukaryotic Cell 3: 1331–1348.

52. Nishiyama Y, Yamamoto H, Allakhverdiev SI, Inaba M, Yokota A, et al. (2001) Oxidative stress inhibits the repair of photodamage to the photosynthetic machinery. EMBO J 20: 5587–5594.

53. Garcia-Ferris C, Moreno J (1993) Redox regulation of enzymatic activity and proteolytic susceptibility of ribulose-1,5-bisphosphate carboxylase/oxygenase fromEuglena gracilis. Photosynth Res 35: 55–66.

54. Cakmak T, Angun P, Ozkan AD, Cakmak Z, Olmez TT, et al. (2012) Nitrogen and sulfur deprivation differentiate lipid accumulation targets of Chlamydomonas reinhardtii. Bioengineered 3: 343–346.

21

Effects of *Escherichia coli* on Mixotrophic Growth of *Chlorella minutissima* and Production of Biofuel Precursors

Brendan T. Higgins, Jean S. VanderGheynst*

Biological and Agricultural Engineering, University of California Davis, Davis, California, United States of America

Abstract

Chlorella minutissima was co-cultured with *Escherichia coli* in airlift reactors under mixotrophic conditions (glucose, glycerol, and acetate substrates) to determine possible effects of bacterial contamination on algal biofuel production. It was hypothesized that *E. coli* would compete with *C. minutissima* for nutrients, displacing algal biomass. However, *C. minutissima* grew more rapidly and to higher densities in the presence of *E. coli*, suggesting a symbiotic relationship between the organisms. At an initial 1% substrate concentration, the co-culture produced 200-587% more algal biomass than the axenic *C. minutissima* cultures. Co-cultures grown on 1% substrate consumed 23–737% more of the available carbon substrate than the sum of substrate consumed by *E. coli* and *C. minutissima* alone. At 1% substrate, total lipid and starch productivity were elevated in co-cultures compared to axenic cultures indicating that bacterial contamination was not detrimental to the production of biofuel precursors in this specific case. Bio-fouling of the reactors observed in co-cultures and acid formation in all mixotrophic cultures, however, could present challenges for scale-up.

Editor: Dwayne Elias, Oak Ridge National Laboratory, United States of America

Funding: This research was supported by NSF grants DGE-0948021 and MCB-1139644. The funders had no role in study design, data collection and analysis, decision to publish, or preparation of the manuscript.

Competing Interests: The authors have declared that no competing interests exist.

* E-mail: jsvander@ucdavis.edu

Introduction

Biofuel production from microalgae has a number of advantages over biofuel from food crops. Microalgae grow rapidly, have lower land-use impacts than food crops, and can utilize saltwater and wastewater resources [1]. Furthermore, certain genera of algae, such as *Chlorella*, have the potential to accumulate lipids and starch, which can be converted to biofuel using existing technology [2]. In addition, many *Chlorella* species have the ability to utilize organic carbon sources, allowing microalgae to act as both a fuel production and conversion platform.

Studies have shown that increasing lipid productivity is the fastest way to achieve cost-effective liquid fuels from microalgae [3,4]. Mixotrophic growth of microalgae (where both organic and inorganic carbon are utilized) has recently gained attention due to the high lipid productivities achieved using this cultivation platform. Numerous researchers have studied algae growth on glucose, glycerol, and acetate and concluded that mixotrophic growth can enhance lipid production by an order of magnitude [5–7].

Still, mixotrophic growth requires supplementation with organic carbon, which can have negative impacts on costs, the food supply, and the environment. In light of these challenges, studies of algae cultivated on organic-rich wastewaters have proliferated in the literature. Theoretically, algae can simultaneously treat the wastewater (removing nutrients) and produce biofuel precursors, thereby reducing cost and environmental impacts [8].

A major challenge with this concept is that other organisms may out-compete algae in nutrient-rich waters. Even if photobioreactors are utilized, contamination is still a concern and can lead to costly system shut-downs if not properly managed. To date, the effect of microbial contamination in mixotrophic cultures is poorly understood. Multiple studies of mixotrophic growth have been conducted using pre-sterilized wastewater (by autoclaving or sterile filtration) prior to inoculating with an algae monoculture [9–11]. These studies reported that organic-rich wastewater increased algae growth rates, and in some cases, lipid content as well. However, these results do not provide insight into how such a system would work in the presence of heterotrophic organisms. Given the presence of organic carbon, heterotrophic organisms, such as bacteria are expected to proliferate in mixotrophic cultures, particularly at high organic carbon concentrations. Such an event could drastically change the projections reported in the literature thus far on mixotrophic algae growth.

Limited research has been conducted on non-axenic algae cultures grown under mixotrophic conditions. Woertz et al. [12] cultivated a mixed algae culture on unsterilized municipal wastewater and diluted anaerobic digester effluent. Their research goal was primarily to study nitrogen and phosphorous removal so it was unclear if these cultures experienced significant mixotrophy. Their results indicated that simultaneous wastewater treatment and biofuel production is feasible, however, no results were presented on the role played by non-algae species in the cultures.

de-Bashan et al. [13] studied the effect of the growth-promoting bacterium, *Azospirillum brasilense*, on three algal species for the purpose of advanced wastewater treatment. They found that *Azospirillum* increased algae growth, lipid content, and fatty acid diversity, potential benefits for biofuel production. *A. brasilense's* growth promotion appeared to stem from the release of the plant hormone indole-3-acetic acid (IAA) [14]. These researchers established co-cultures within alginate beads and used synthetic wastewater whose organic carbon content was unspecified. In addition to *A. brasilense*, Lebsky et al. [15] co-cultured the bacterium *Phyllobacterium myrsinacearum* with *Chlorella vulgaris* in alginate beads in order to elucidate interactions between these species. They found that *P. myrsinacearum* showed no growth enhancing properties compared to *A. brasilense*.

Given limited prior research on mixotrophic algae growth, several lab-scale experiments were conducted in order to address the following question: What effect does bacterial contamination have on mixotrophic algal biomass, lipid, and starch production? Substrate utilization efficiency was also determined on the basis of substrate energy sequestered in lipids and starch. No studies have been identified that address the issue of substrate utilization efficiency, a critical parameter if non-waste carbon sources are used. *Chlorella minutissima* (UTEX 2341) was chosen due to its ability to accumulate lipids [16,17] and utilize a variety of organic compounds. Defined media and controlled "contamination" of the algae culture with a single strain of *E. coli* were used in order to elucidate interactions that may emerge. *E. coli* was chosen because it grows rapidly under the culture conditions tested, competes with *Chlorella* for carbon substrates but is not known to be pathogenic toward algae. It can also be enumerated through plating, and is found in many wastewaters [18]. The carbon sources tested included glucose, glycerol and acetate, all of which enhanced growth in *Chlorella vulgaris* [5]. Glucose and glycerol enter metabolism through glycolysis whereas acetate enters via the TCA cycle, providing an opportunity to observe contrasts in biomass composition. The knowledge gained from these experiments can then be applied toward the study of microbial communities and wastewater of increasing complexity.

Methods and Materials

Algae cultivation

Chlorella minutissima (UTEX 2341) pre-cultures were initiated from selected colonies on ATCC No. 5 agar medium [19]. Pre-cultures were grown in N8-NH$_4$ medium (supplemental material), adjusted to pH 7.2, until a cell density of 10^7 cells per ml was reached (about 7–8 days). Cells were counted with a hemocytometer. The pre-cultures were grown in 1 L bottles filled to 800 ml and aerated with 2% (v/v) CO$_2$ in air at 400 ml per minute under 10,000 lux illumination (16:8 light-dark cycle). Cultures were mixed by magnetic stir bar (~300 rpm) and maintained at ~28°C.

Algae cells in the pre-culture were concentrated by overnight settling before inoculation into autoclaved N8-NH$_4$ media in 250 ml hybridization tubes (filled to 200 ml) to achieve 10^7 cells/ml. Glucose, glycerol, or sodium acetate (pH adjusted to 7.2 with HCl) stock solutions were sterile filtered (0.2 μm) and added to achieve the desired substrate concentration (0, 0.5, 2, and 10 g/L). Each substrate type was tested in sequential batches along with autotrophic control cultures. All three substrates were tested with axenic *C. minutissima*, axenic *E. coli*, and co-culture conditions for a total of nine batch runs, each of which ran for five days.

Stock *Escherichia coli* (ATCC 25922) were cultivated on liquid LB medium with 1% glucose for ~24 hours and then inoculated into the reactors to achieve ~10^4 CFU/ml. This value was chosen because it falls in the range of coliform bacteria concentration found in treated municipal wastewaters (without disinfection). Zhang and Farahbakhsh [20] observed coliform concentrations of 10^3–10^5 CFU/ml in secondary wastewater. Guardabassi et al. [21] measured coliform concentrations of 10^3–10^5 CFU/ml in tertiary wastewater and total bacteria concentrations in the range of 10^5–10^6 CFU/ml.

Axenic algae cultures were supplied with an initial dose of kanamycin at 20 mg/L in order to reduce the risk of contamination in the hybrid tubes. Initial testing showed that kanamycin had no detectable impact on the growth of *C. minutissima*. The hybrid tube reactors were placed in a water bath maintained at 28°C under 10,000 lux illumination (16:8 light-dark cycle). Stir bars were used to sweep the bottom of the reactors (~100 rpm) with aeration tubes providing 125 ml per minute of ambient air. CO$_2$ supplementation was not provided to the hybrid tubes because research has suggested that excess carbon dioxide can inhibit organic carbon uptake in mixotrophic cultures [22]. Optical density (OD$_{550}$) was measured each day in a microplate reader (Spectramax M2, Molecular Devices, Sunnyvale, CA) and correlated to total dry weight at the end of each five-day culture period. Unique correlations were developed for each substrate and culture condition.

Biomass composition

Biomass was harvested at the end of the five-day culture period by centrifugation at 5000 g for 5 minutes (IEC MultiRF, Thermo Electron Corp., Waltham, MA) and the pellet was washed three times with dH$_2$O to remove media salts. The pellet was lyophilized at −45°C (Freezone4.5, Labconco, Kansas City, MO) and weighed. Dry biomass was suspended in Folch solvent (2:1 chloroform and methanol) and cells were lysed with 0.5 mm zirconia-silica beads with a bead beater (FastPrep FP120, Savant Instruments, Holbrook, NY). Beating was performed for six 20-second periods at maximum speed with samples cooled on ice between periods. Beads were removed with a mesh filter and washed with Folch solvent. NaCl solution (0.9%) was added to the disrupted cell suspension to achieve a phase separation between the polar and non-polar fractions. The non-polar (chloroform) fraction was removed and stored at −20°C for later assay with the sulfa-vanillin-phosphoric acid (SPV) assay. The SPV assay has been shown to correlate well with gravimetric lipid found in *Chlorella* [23]. Corn oil was used as the assay standard. Select crude lipid extracts were dried down and weighed. The assay was found to slightly underestimate gravimetric crude lipid.

The polar fraction was washed three times with acetone and then three times with water by centrifugation at 12,000 g for 5 minutes followed by supernatant removal. The resulting pellet was freeze-dried, ground with a micropestle, and the starch was gelatinized by heating the pellet in dH$_2$O for 30 minutes at 80°C. The gelatinized starch suspension was hydrolyzed with α-amylase (7.5 U/ml) and α-amyloglucosidase (3 U/ml) in acetate buffer (pH 5) overnight at 37°C. The suspension was centrifuged and the supernatant removed for analysis. Dinitrosalicylic acid (DNS) was used to detect reducing sugars formed in the enzyme-substrate reaction with glucose standards [24]. Starch mass was determined by multiplying the sample glucose content by 0.9 to correct for water added during starch hydrolysis.

HPLC analysis

Daily culture samples were collected over the five-day growth period, centrifuged at 15,000 g to remove cells, and filtered at 0.2 μm (Titan2 PTFE, SUN-SRi, Rockwood, TN). High performance liquid chromatography (HPLC) was used to quantify organic compounds in the media (Prominence Liquid Chromatograph, Shimadzu Corp., Kyoto, Japan). An Aminex 87H column (BioRad, Hercules, CA) was used to separate glucose, glycerol, acetate, and secreted organic compounds in the culture media. The column oven was set to 60°C and 5 mM sulfuric acid in Milli-Q water was used as the mobile phase under isocratic conditions (0.6 ml/min). Refractive index and UV detectors were used to quantify substrate concentrations.

E. coli enumeration

Viable *E. coli* were determined by plating on LB agar medium with 1% glucose. Plating was performed at 0, 24, 48, and 96 hours for cultures containing *E. coli*. Quantitative PCR of 16S rDNA was used to determine total bacterial biomass in the final freeze-dried co-culture samples. A similar approach has been employed in the quantitation of bacterial populations in a variety of environments [25,26]. Freeze-dried material was re-suspended in dH$_2$O and DNA was extracted using a FastDNA Spin Kit (MP Biomedicals, Solon, OH) according to the manufacturer's instructions. A 110 base portion of *E. coli* 16S rDNA was amplified using the forward primer 5′-CAAGACCAAAGAGGGGGACC-3′ and reverse primer 5′-TCAGACCAGCTAGGGATCGT-3′ (Invitrogen Life Technologies, Grand Island, NY). SYBR Green PCR Master Mix (Applied Biosystems, Warrington, UK) was used in a StepOnePlus qPCR instrument (Applied Biosystems, Foster City, CA). The program ran for thirty cycles: 95°C for 15 sec, 60°C for 15 sec, 72°C for 30 sec. Gel electrophoresis was used to confirm correct amplification. Previously amplified PCR product from *E. coli* was used as a DNA standard. Samples of pure *E. coli* cultured under similar conditions to the co-cultures were used to correlate *E. coli* dry weight to 16S rDNA concentration. Double stranded DNA concentrations were measured using a Qubit assay kit (Invitrogen, Eugene, OR).

Substrate conversion efficiency

Efficiency of substrate utilization was determined on an energy basis (Equation 1):

$$eff = ((lipid)(\Delta H_{c,lipid}) + (starch)(\Delta H_{c,starch}))/ \\ ((substrate_{used})(\Delta H_{c,substrate}))$$ (Equation 1)

ΔH_c = standard enthalpy of combustion. Enthalpy of combustion for lipids and starch were assumed to be 37 and 15.7 kJ/g, respectively [27]. Glucose, glycerol, and acetate were assumed to contain 15.6, 15.6, and 14.6 kJ/g, respectively [27,28]. The rationale for this method is that the substrate could be used for direct combustion or it could be converted to biofuel precursors by algae, thereby losing useable energy in the conversion process.

Data analysis

JMP v.9 software (SAS Institute, Cary, NC) was used to conduct ANOVA and Tukey tests on sample means. These statistical tests require homogeneity of variance between data groups, however, certain measurements exhibited variances that increased with the sample mean. Power transformation of data prior to statistical analysis is the recommended course of action in order to ensure that basic assumptions such as data normality and homogeneity of variance are not violated [29,30]. To correct heteroscedasticity, data were transformed prior to statistical analyses using power transforms based on procedures developed by Box and Cox [31,32]. The form of the Box-Cox transform is expressed in Equation 2 where lambda was optimized using JMP to minimize heteroscedasticity in the transformed data (y′).

$$y' = (y^\lambda - 1)/\lambda$$ (Equation 2)

In the majority of cases, lambda was found to be close to zero, in which case the data were transformed by taking the natural logarithm. Prior to further statistical analyses, all data were visually inspected for heteroscedasticity and checked with Levene's test to ensure non-significance. Significance for all tests was assessed at the 0.05 level.

Results

Biomass productivity

Substrate concentration had a large impact on total biomass and lipid productivity in axenic algae cultures (Table 1). Increasing substrate concentration from 0.5 g/L to 2 g/L consistently yielded a 1.6 fold increase in biomass productivity. Productivity in glucose and glycerol cultures exhibited saturation behavior with increasing substrate concentration: cultures supplied with 10 g/L of substrate did not produce proportionally more biomass than 2 g/L substrate. In contrast, the axenic acetate cultures exhibited apparent substrate inhibition where the 10 g/L cultures produced significantly less biomass than the 2 g/L cultures (p = 0.002, Tukey test). The initial growth rate (first 24 hours) in the 10 g/L acetate culture was about half that observed in the 2 g/L culture (Figure 1). Growth curves for glucose and glycerol can be found in Figures S1 and S2.

In light of saturation behavior, attempts were made to fit biomass productivity to a Monod model based on substrate concentration, however, results only fit for the first 24 hours of growth. Thereafter, growth rates were lower than the model predicted, suggesting that other factors besides substrate concentration had an effect on productivity. Mean comparisons among substrate types were not performed due to the batch effect apparent in autotrophic control cultures. With the exception of 10 g/L acetate, productivity differences among the three substrates (holding concentration constant) were typically within 20% of each other and are unlikely to be of practical significance.

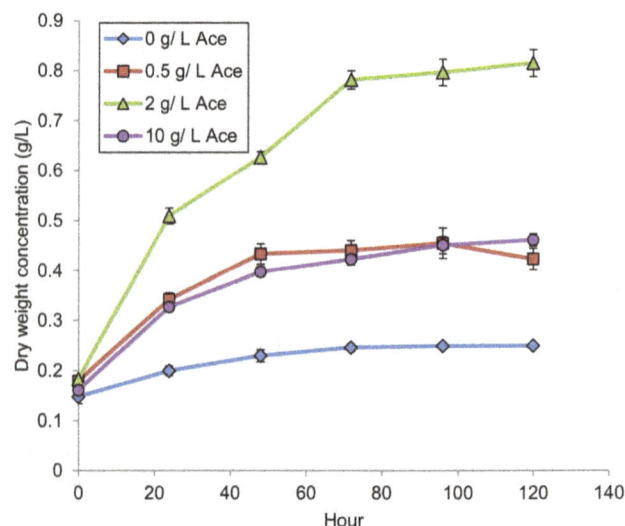

Figure 1. Growth curve for axenic acetate cultures. Error bars are standard deviations based on three biological replicates.

Table 1. Productivity levels and lipid and starch contents for axenic *C. minutissima* (top) and co-cultures of *C. minutissima* and *E. coli* (bottom).

Culture conditions	Biomass productivity (mg/L/d)	Lipid productivity (mg/L/d)	Starch productivity (mg/L/d)	Final lipid content (%)	Final starch content (%)
Axenic *C. minutissima*					
Glucose					
0 g/L	8.4 (1.1) c	2.0 (0.2) b	0.09 (0.01) c	23.5 (2.4) a	1.04 (0.06) bc
0.5 g/L	38.0 (2.4) b	5.1 (1.4) b	0.30 (0.04) c	13.4 (2.8) b	0.78 (0.06) c
2 g/L	98.8 (4.9) a	14.5 (1.3) a	1.86 (0.68) b	14.7 (0.6) b	1.90 (0.68) b
10 g/L	98.7 (6.0) a	13.5 (1.8) a	3.16 (0.34) a	13.7 (1.1) b	3.19 (0.34) a
Glycerol					
0 g/L	13.5 (0.8) d	2.8 (0.1) d	0.27 (0.01) c	20.8 (0.5) a	1.98 (0.15) c
0.5 g/L	42.3 (1.5) c	7.5 (0.4) c	0.66 (0.04) c	17.8 (1.5) a	1.56 (0.05) c
2 g/L	111.0 (0.8) b	19.6 (1.5) b	4.04 (1.01) b	17.7 (1.4) a	3.65 (0.93) b
10 g/L	123.4 (0.9) a	23.3 (1.8) a	7.75 (0.78) a	18.9 (1.6) a	6.28 (0.67) a
Acetate					
0 g/L	19.9 (3.4) d	3.3 (0.6) d	0.20 (0.03) d	16.6 (0.3) a	1.03 (0.06) b
0.5 g/L	48.3 (2.3) c	6.7 (0.6) c	0.42 (0.02) c	13.8 (0.6) b	0.87 (0.02) c
2 g/L	125.8 (2.9) a	18.5 (0.7) a	0.84 (0.04) b	14.7 (0.2) b	0.67 (0.02) d
10 g/L	59.6 (3.9) b	8.8 (0.8) b	0.98 (0.03) a	14.8 (1.3) b	1.65 (0.06) a
C. minutissima, E. coli					
Glucose					
0 g/L	10.5 (1.8) d	2.1 (0.3) d	0.16 (0.02) d	20.2 (0.4) b	1.51 (0.04) bc
0.5 g/L	85.0 (3.5) c	14.5 (1.1) c	1.12 (0.05) c	17.0 (0.9) c	1.31 (0.02) c
2 g/L	190.8 (14.8) b	25.9 (3.5) b	3.48 (0.54) b	13.6 (1.0) d	1.67 (0.04) b
10 g/L	342.8 (65.6) a	83.9 (15.8) a	59.28 (9.70) a	24.5 (1.4) a	17.43 (2.11) a
Glycerol					
0 g/L	12.9 (4.9) d	1.5 (0.5) c	1.25 (0.50) c	12.0 (1.0) b	9.71 (0.58) b
0.5 g/L	74.7 (5.3) c	9.5 (1.8) b	3.79 (0.40) b	12.6 (1.6) b	5.07 (0.32) c
2 g/L	172.4 (15.5) b	22.0 (2.5) b	5.78 (2.50) b	12.2 (0.5) b	4.00 (0.11) c
10 g/L	419.8 (29.6) a	72.2 (8.4) a	76.97 (8.13) a	17.2 (0.8) a	18.31 (0.65) a
Acetate					
0 g/L	34.6 (2.5) d	6.0 (0.1) d	0.97 (0.06) b	17.2 (0.8) a	2.83 (0.36) b
0.5 g/L	87.5 (4.3) c	13.4 (1.3) c	1.06 (0.03) b	15.3 (0.8) b	1.22 (0.06) c
2 g/L	207.7 (1.4) b	26.6 (1.1) b	0.76 (0.11) c	12.9 (0.5) b	0.39 (0.05) d
10 g/L	428.9 (6.3) a	72.5 (4.7) a	26.52 (3.44) a	16.9 (0.8) a	6.18 (0.78) a

Productivities are five-day averages and lipid and starch contents are a percent of dry weight. Three biological replicates were used in all cases except for co-cultures grown on 10 g/L glycerol where only two replicates were used. Mean value is followed by standard deviation in parenthesis. Within substrate batches (e.g. axenic *C. minutissima* w/glucose), values followed by the same letter are not significantly different at the 0.05 level.

Co-cultures, which contained *C. minutissima* and *E. coli*, had higher biomass, lipid, and starch productivity than axenic *C. minutissima* cultures (Table 1). At 10 g/L substrate, total biomass productivities were 240% to 619% higher for co-cultures compared to axenic cultures which was significantly higher than the changes observed in autotrophic control cultures (p<0.0001). Likewise, lipid production was enhanced by 210–722% in these same cultures and the change was also significantly greater than that observed in autotrophic cultures (p<0.0001). Greater growth variability was observed in the glucose and glycerol co-cultures than in the axenic cultures. One possible explanation is the bio-fouling that occurred in all co-cultures under high substrate concentrations (Figure 2). Biomass was scraped from the walls of the reactors and re-suspended on a daily basis prior to sampling.

Significant bio-fouling did not occur until day three of cultivation (late log phase) and appeared to result from media froth caused by *E. coli*. The rising bubbles in the reactor did not break at the liquid interface, allowing them to lift cells out of suspension. In addition, the algae appeared to coagulate and stick to the wall of the reactor. Antifoam was not used to overcome this challenge since the use of such products in large-scale cultivation is unlikely due to cost.

pH fluctuations in mixotrophic cultures

The autotrophic (no carbon substrate) cultures had no detectable pH change over the five-day growth period and did not require pH adjustment. In cultures supplied with glucose and glycerol, pH decreased and required adjustment with 3 M NaOH on a daily basis to restore neutral conditions (Figure 3). Additional

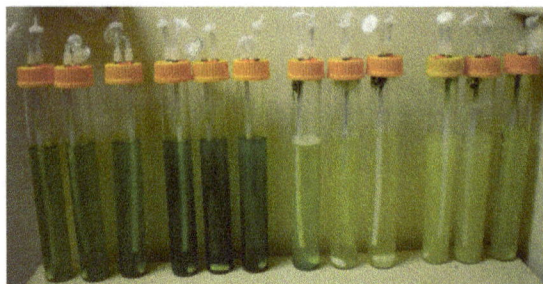

Figure 2. Bio-fouling in glucose co-cultures. From left to right in triplicate: 0 g/L glucose, 0.5 g/L glucose, 2 g/L glucose, 10 g/L glucose. Similar fouling occurred in glycerol and acetate co-cultures. A film of algae can be seen on the upper wall of reactors under high glucose concentrations.

pH graphs can be found in Figures S3 and S4. The decline in pH appeared to coincide with substrate consumption as measured by HPLC (Figure 3). Once a substrate was depleted, the decline in pH ceased and no further pH adjustments were required. Similar pH declines were observed in cultures of axenic *C. minutissima*, axenic *E. coli*, and in co-cultures. Cultures grown on acetate experienced an increase in pH and required daily adjustment to pH 7 using 3 M HCl suggesting that cultures consumed acetate in its protonated form. Secretion of detectable formic and acetic acid was observed via HPLC in axenic *E. coli* cultures grown on glucose and glycerol. At 0.5, 2, and 10 g/L glucose, final formic acid concentrations were 0.014, 0.069, and 0.036 g/L, respectively. For the same cultures, final acetic acid levels were 0, 0.092, and 0.141 g/L, respectively. Neither compound was detected in co-cultures. Identification of compounds secreted by *C. minutissima* under mixotrophic growth is the subject of ongoing investigation but the present analysis showed that *C. minutissima* did not secrete typical anaerobic products such as lactate, formate, or acetate. Some work on secretions of *C. minutissima* grown on solid media has already been published, indicating that this algae secretes a complex mixture of lipids among other metabolites [33].

Bacteria and algae levels in co-cultures

Colony counts of viable *E. coli* concentration in co-cultures increased rapidly on days one and two, then plateaued over the remainder of the culture period (Figures S5, S6, and S7). Order of magnitude differences in CFU concentration were observed in co-cultures for different initial substrate concentrations. Zero, 0.5, 2, and 10 g/L initial glucose concentrations resulted in CFU concentrations of 3.9×10^6, 7.7×10^7, 1.5×10^9, and 1.5×10^9 CFU/ml, respectively, after 96 hours (Table 2). This result indicates saturation of viable *E. coli* with respect to substrate concentration.

Because both non-viable and viable bacteria contributed to co-culture biomass, PCR amplification of 16S rDNA was used to quantify total *E. coli* biomass in the co-cultures (58 g *E. coli* dry weight/µg 16S rDNA). DNA extraction using a bacterial cell-lysing buffer appeared to extract DNA almost exclusively from bacteria (*Chlorella* extract contained 6.4 µg/ml DNA extract compared to 154 µg/ml extract for *E. coli*). The ratio of 16S rDNA to total extracted DNA was also roughly constant in co-cultures (6.8×10^{-6}, 6.6×10^{-6}, 6.6×10^{-6} µg 16S rDNA/µg extracted DNA for glucose, glycerol, and acetate batches respectively). Furthermore, the presence of algae in the samples appeared to have almost no impact on either DNA extraction or the PCR reaction. This was demonstrated using pre-mixed algae

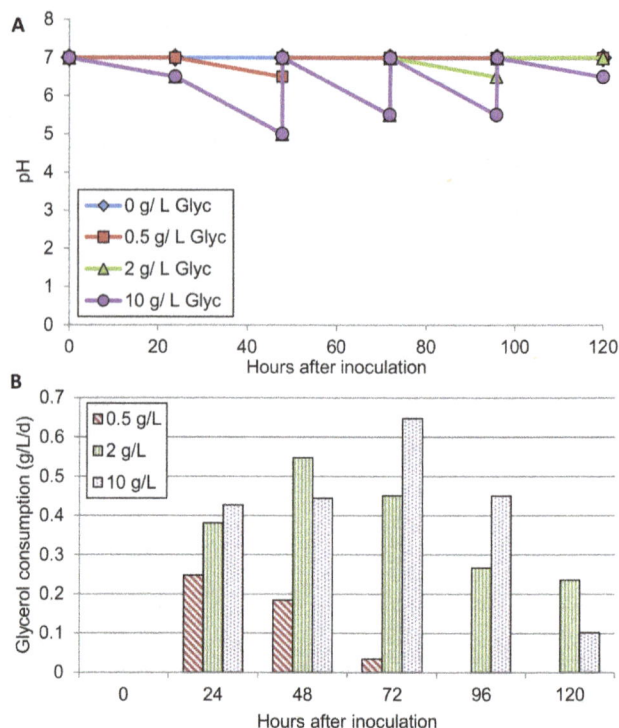

Figure 3. pH change and glycerol consumption in axenic *C. minutissima* cultures. A: The line graph shows a decline in pH. Similar pH declines were observed in glucose cultures. pH was re-adjusted to neutral conditions every 24 hours. B: The bar graph shows the rate of glycerol consumption by the cultures during each 24-hour period.

and *E. coli* cultures that were analyzed as a validation step. The mixtures contained 0%, 25%, 50%, 75%, and 100% *E. coli* by mass and PCR analysis resulted in calculated *E. coli* fractions of 0%, 24%, 46%, 69%, and 100%, respectively.

The qPCR results mirrored the CFU counts except at 10 g/L substrate. The qPCR results indicated that *E. coli* concentrations (g *E. coli* dry weight/L) were lower at 10 g/L substrate than at 2 g/L although this difference was only statistically significant for glucose. However, the percent of biomass as *E. coli* was significantly lower at 10 g/L substrate than at 2 g/L for all substrates (Table 3). This result was surprising given the hypothesis that *E. coli* were expected to occupy a larger fraction of culture biomass with increasing carbon substrate.

One possible explanation for lower apparent *E. coli* levels at higher substrate concentration is that DNA extraction was not representative of *E. coli* biomass in co-cultures. Presumably this would be the result of DNA leakage into the media due to cell lysis. To test this hypothesis, culture samples were tested for their DNA content after high-speed centrifugation (15,000g) and 0.2 µm filtration to remove cells (Titan2 PTFE, SUN-SRi, Rockwood, TN). Supernatants from cultures supplied with 0, 2, and 10 g/L glucose were found to contain 0.018, 2.47, and 0.296 µg/ml DNA extract, respectively. This suggests that leakage cannot explain the lower extracted DNA from 10 g/L cultures.

Based on the PCR result, algal biomass constituted 88.5–95.3% of total biomass in the co-cultures grown on 10 g/L substrate, indicating little displacement of algae by bacteria. More significantly, algal productivity in the co-cultures was 200–587% higher than algal productivity in axenic cultures at 10 g/L substrate suggesting a potential symbiotic relationship between the algae

Table 2. Viable cell densities (CFU/mL) of *E. coli* at 96 hours for co-cultures (top) and axenic *E. coli* cultures (bottom).

Culture condition	Initial substrate concentration			
	0 g/L	0.5 g/L	2 g/L	10 g/L
C. minutissima, E. coli				
Glucose	3.9E+06	7.7E+07	1.5E+09	1.5E+09
Glycerol	7.4E+06	6.4E+07	3.3E+09	4.8E+09
Acetate	9.0E+06	1.5E+08	6.1E+08	1.5E+09
E. coli				
Glucose	7.0E+05	3.1E+07	5.0E+07	3.0E+07
Glycerol	-	6.3E+06	2.3E+07	3.0E+06
Acetate	-	1.4E+07	3.3E+07	2.4E+05

and bacteria. For all three substrates, these increases in algae productivity were significantly greater than differences observed in the autotrophic control cultures (p < 0.001 for all cases) indicating that symbiosis may have depended on the presence of organic substrates. Interestingly, the high bacterial fraction in the 2 g/L co-cultures coincided with only a minimal increase in algal productivity.

Axenic E. coli cultures

To better understand *E. coli* response to substrate treatments, axenic *E. coli* cultures were grown on glucose, glycerol, and acetate under identical conditions to the co-cultures. *E. coli* productivity in axenic cultures was similar to *E. coli* productivity in co-cultures, lending credibility to the qPCR results (Table 4). Moreover, axenic *E. coli* cultures supplied with 10 g/L of glucose and acetate had significantly lower productivity (Table 4) than those supplied with

2 g/L glucose (p = 0.028) and 0.5 g/L acetate (p<0.001), respectively. In fact, axenic *E. coli* exhibited no growth at 10 g/L acetate compared to ~22 mg/L/day at 0.5 g/L. Substrate inhibition is the most likely explanation for this observation since exponential growth was also delayed at 2 g/L acetate (began at 96 hours) compared to 0.5 g/L acetate which began at 24 hours (Figure S8). *E. coli* cultures grown on glucose and glycerol exhibited comparable initial growth rates across all substrate concentrations indicating minimal substrate inhibition. Elevated *E. coli* densities at 2 g/L glucose could be associated with the more favorable (less acidic) pH conditions that prevailed in the culture toward the end of the culture period (Figures S9 and S10 for growth and pH curves). However, a similar increase in growth did not take place in cultures grown on 2 g/L glycerol (Figure S11).

Viable cell counts at nearly all time points for co-cultures were one to three orders of magnitude higher than those in axenic *E. coli*

Table 3. Percent of dry biomass as *E. coli* in co-cultures based on qPCR and the percent increase of algae productivity in the co-culture versus the equivalent axenic culture (Table 1).

Substrate	Final *E. coli* content (%)	*E. coli* productivity (mg/L/d)	Algae productivity (mg/L/d)	Increase in algae productivity vs. axenic (%)
Glucose				
0 g/L	1.3 (0.4) c	0.13 (0.02) c	10.4 (1.8) c	25 (32) c
0.5 g/L	12.0 (2.4) b	10.25 (2.39) b	74.7 (2.6) b	97 (12) b
2 g/L	47.3 (9.6) a	91.23 (25.86) a	99.6 (11.3) b	1 (10) c
10 g/L	6.2 (2.5) b	22.06 (10.61) b	320.7 (57.0) a	224 (40) a
Glycerol				
0 g/L	1.0 (1.0) c	0.16 (0.21) c	12.8 (4.7) d	−6 (34) b
0.5 g/L	18.3 (1.6) ab	13.62 (0.42) b	61.0 (5.4) c	45 (18) b
2 g/L	30.0 (8.2) a	52.56 (18.98) a	119.8 (3.5) b	8 (3) b
10 g/L	11.5 (0.5) bc	48.36 (5.70) a	371.4 (23.9) a	200 (20) a
Acetate				
0 g/L	0.4 (0.2) c	0.13 (0.07) c	34.5 (2.5) d	75 (17) b
0.5 g/L	2.7 (0.3) b	2.35 (0.37) b	85.1 (4.0) c	76 (13) b
2 g/L	18.7 (3.2) a	38.71 (6.46) a	169.0 (7.9) b	34 (3) c
10 g/L	4.7 (0.9) b	20.06 (3.57) a	408.9 (7.9) a	587 (33) a

Productivities are five-day averages. Three biological replicates were used in all cases except for co-cultures grown on 10 g/L glycerol where only two replicates were used. Mean value is followed by standard deviation in parenthesis. Within substrate batches values followed by the same letter are not significantly different at the 0.05 level.

Table 4. Axenic *E. coli* growth measured gravimetrically.

Substrate	*E. coli* productivity (mg/L/d)	Final *E. coli* density (mg/L)
Glucose		
0 g/L	0.05 (0.91) c	1.5 (4.4) c
0.5 g/L	38.01 (0.53) b	194.5 (2.5) b
2 g/L	87.99 (17.91) a	439.1 (90.1) a
10 g/L	56.7 (0.80) b	288.3 (4.0) b
Glycerol		
0 g/L	-	-
0.5 g/L	32.90 (1.00) b	171.8 (4.5) b
2 g/L	62.44 (3.16) a	320.7 (14.7) a
10 g/L	63.09 (1.94) a	325.2 (11.9) a
Acetate		
0 g/L	-	-
0.5 g/L	22.12 (0.41) a	115.3 (1.9) a
2 g/L	40.04 (34.03)*	202.9 (170.3)*
10 g/L	1.27 (0.25) b	11.0 (1.6) b

Productivities are five-day averages. Mean values are followed by standard deviations in parenthesis. Within a batch, values connected by the same letter are not significantly different at the 0.05 level.
*Variability was too large to include this treatment in the multiple comparison.

cultures (Table 2). As an example, viable cell counts at 96 hours at 10 g/L substrate were 51, 1600, and 6400 times higher in co-cultures than in axenic *E. coli* cultures when grown on glucose, glycerol, and acetate, respectively. This suggests that viable *E. coli* constituted a greater fraction of total *E. coli* biomass in co-cultures than in axenic cultures indicating *E. coli* may have benefitted from the presence of algae in co-cultures.

Lipid and starch content

Among axenic *C. minutissima* cultures, the highest lipid contents were observed under autotrophic conditions (Table 1). For co-cultures, the treatments with 10 g/L glucose yielded the highest lipid contents with autotrophic cultures close behind. The high lipid content at 10 g/L substrate is further evidence that algae were the predominant species in these cultures. As expected, axenic *E. coli* cultures had low lipid content (3.4–5% when grown on glucose).

Total starch exhibited greater differences among culture conditions than lipid content (Table 1). The *E. coli* cultures grown on glucose yielded low glycogen levels (<1% of total biomass) that were barely detectable using the DNS assay. Therefore, it is unlikely that *E. coli* contributed significant glycogen to the elevated levels of α-glucan polymers observed in co-cultures. Axenic cultures grown on 10 g/L acetate produced 48 and 74% less starch than cultures grown on 10 g/L glucose and glycerol, respectively. These differences were significantly greater than those observed in the autotrophic control cultures from these respective batches (p<0.001, t-test). In all cases, 10 g/L of substrate resulted in the highest level of starch with very large increases observed in glucose and glycerol co-cultures (>15% starch content).

Effect of kanamycin on biomass composition

Early testing showed that kanamycin had no measureable effect on growth rates of *C. minutissima* compared to no-antibiotic controls whereas tetracycline resulted in substantial growth inhibition (data not shown). A full factorial experiment was later conducted in which axenic *C. minutissima* was grown on 2 and 10 g/L glucose with and without kanamycin in order to determine possible effects on biomass composition. Kanamycin decreased biomass productivity by 9% (p = 0.56) and 13% (p = 0.054) in cultures supplied with 2 and 10 g/L glucose, respectively (Table 5).

Lipid content, as measured by the SPV assay, was nearly identical among treatments. In the 10 g/L glucose cultures, however, kanamycin appeared to significantly suppress starch levels. Two other possible explanations (culture contamination and assay interference) were tested and eliminated: all cultures were tested for bacterial contamination using 16S rDNA qPCR and had a negative result. Likewise, the glucose concentration from hydrolyzed starch was quantified using HPLC in addition to the DNS assay. Both methods showed that kanamycin cultures had about half the starch content as the no-kanamycin control. This difference was less pronounced for 2 g/L glucose.

Substrate consumption

Substrate consumption levels (Table 6) appeared to mirror biomass productivity. Co-cultures consumed more substrate (and consumed it more rapidly) than axenic cultures when provided with 10 g/L substrate. When provided with 0.5 and 2 g/L substrate, axenic algae and co-cultures consumed all substrate within 5 days. Thereafter, growth slowed but did not cease suggesting continued autotrophic growth.

Substrate uptake in the co-cultures was greater than the sum of substrate uptake by axenic *C. minutissima* and *E. coli* cultures. Specifically, the co-cultures consumed 23% more glucose, 59% more glycerol, and 737% more acetate than the sum of the axenic cultures. Axenic *E. coli* cultures consumed acetate at 0.5 and 2 g/L but did not consume any acetate when provided with 10 g/L, further indication of substrate inhibition. Likewise, axenic *C. minutissima* consumed only 0.96 g/L when grown on 10 g/L acetate yet the co-culture consumed 7.89 g/L, indicative of a

Table 5. Productivity levels and starch and lipid contents for axenic *C. minutissima* cultures grown with and without kanamycin.

Culture conditions		Biomass productivity (mg/L/d)	Lipid productivity (mg/L/d)	Starch productivity (mg/L/d)	Final lipid content (%)	Final starch content (%)
2 g/L glucose	No kanamycin	117.2 (2.7) b	16.3 (1.3) ab	4.22 (1.93) c	13.9 (0.8) a	3.58 (1.57) c
	Kanamycin	106.0 (4.8) b	15.3 (2.6) b	2.56 (0.79) c	14.4 (2.2) a	2.40 (0.65) c
10 g/L glucose	No kanamycin	172.4 (2.6) a	23.8 (0.9) a	26.95 (3.07) a	13.8 (0.4) a	15.64 (1.77) a
	Kanamycin	148.4 (15.9) a	20.3 (4.6) ab	10.59 (0.80) b	13.6 (1.8) a	7.16 (0.30) b

Productivities are five-day averages. Mean values are followed by standard deviations in parenthesis based on three biological replicates. Within a column, values connected by the same letter are not significantly different at the 0.05 level.

symbiotic relationship that is linked to substrate uptake and utilization.

Substrate conversion efficiency

Lower substrate concentrations generally led to greater substrate conversion efficiencies (Figure 4) with 0.5 g/L substrate supporting the highest efficiency. This is likely due to the significant contribution of photosynthesis to biomass growth in these cultures. Differences between 2 and 10 g/L substrate were generally not significant. Variability in substrate utilization efficiency was a result of compounding variability across multiple measurement systems. Variability associated with biomass growth, lipid and starch measurement, and substrate utilization measurements all contributed to variability in efficiency.

Discussion

The effects of bacterial contamination

It was initially hypothesized that the presence of a competitive organism such as *E. coli* would slow algae productivity due to resource competition, particularly for the carbon substrate. High substrate concentrations were expected to lead to cultures whose biomass was dominated by *E. coli* despite the significant initial cell density advantage afforded to *C. minutissima*.

The results suggest that neither hypothesis was correct: the co-culture had up to 592% greater algal biomass productivity than the axenic algal culture under 10 g/L substrate conditions. It also had greater lipid and starch production and could consume more of the available substrate than the axenic cultures at the 10 g/L substrate concentration. It appears that the two organisms exhibit something closer to a mutualistic relationship than a competitive one. Furthermore, substrate uptake by the co-culture appeared to be more than the sum of substrate uptake by each organism cultured alone. *E. coli* may also have benefited from the relationship as shown by an increased viable cell fraction in co-cultures compared to axenic cultures.

A number of hypotheses exist for algae-bacteria symbiosis. One hypothesis is that bacteria secrete hormones or co-factors that enhance algal growth. This was observed by de-Bashan et al. in which *Azospirillium* secreted IAA, thereby promoting growth in *Chlorella vulgaris* [14]. Likewise, Bajguz et al. showed that other phytohormones such as indole-3 propionic and indole-3 butyric acid had similar growth-promoting effects on autotrophic *C. vulgaris* cultures, particularly when supplied with brassinosteroids [34]. In addition to hormones, bacteria can synthesize vitamins that may confer growth advantages to algae. Croft et al. demonstrated that many algae species exhibit vitamin B_{12} auxotrophy, due largely to B_{12} dependence of methionine synthase [35]. They demonstrated that bacteria can supply algae with

Table 6. Organic carbon substrate consumption (g/L) over a five-day growth period.

Initial substrate conditions	C. minutissima	C. minutissima & E. coli	E. coli
Glucose			
0.5 g/L	0.40 (0.01) b	0.47 (0.02) c	0.56 (0.01) c
2 g/L	1.70 (0.14) a	1.94 (0.02) b	2.09 (0.10) b
10 g/L	1.26 (0.29) a	6.09 (0.51) a	3.69 (0.56) a
Glycerol			
0.5 g/L	0.47 (0.00) b	0.50 (0.00) c	0.45 (0.02) b
2 g/L	1.88 (0.06) a	2.05 (0.04) b	1.94 (0.02) a
10 g/L	2.07 (0.07) a	6.26 (0.03) a	2.04 (0.26) a
Acetate			
0.5 g/L	0.46 (0.00) c	0.45 (0.00) c	0.46 (0.01) a
2 g/L	1.86 (0.04) a	1.91 (0.00) b	0.67 (0.58)*
10 g/L	0.96 (0.16) b	7.89 (0.20) a	0.00 (0.00) b

Mean values are followed by standard deviations in parenthesis based on three biological replicates (except for 10 g/L glycerol co-culture where n = 2). Within a batch, values connected by the same letter are not significantly different at the 0.05 level.
*Variability was too large to include this treatment in the multiple comparison.

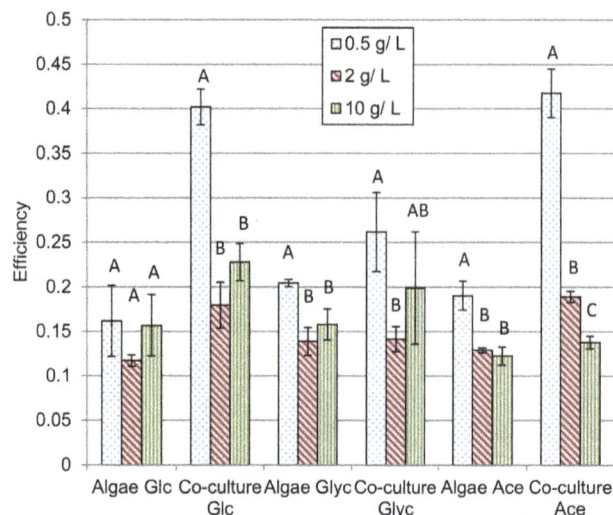

Figure 4. Efficiency of substrate conversion to biofuel precursors calculated using Equation 1. Glc = glucose, Glyc = glycerol, Ace = acetate. Energy obtained through photosynthesis is considered to be "free" so efficiency was not calculated for cultures with no substrate. Error bars are standard deviations based on 3 biological replicates with the exception of co-cultures grown on 10 g/L glycerol where n = 2. Within batches (same substrate and culture type), bars with the same letter are not significantly different.

vitamin B_{12} in exchange for photosynthetic products. Moreover, they showed that *Chlamydomonas reinhardtii* contains two methionine synthase genes – one with vitamin B_{12} dependence and one without. The B_{12}-dependent form has higher catalytic efficiency and is preferentially used when exogenous B_{12} is present indicating that symbiosis can occur even when algae are capable of fully autotrophic growth.

It is possible that the symbiotic relationship between *C. minutissima* and *E. coli* is the result of phytohormone or co-factor secretion. Vitamin B_{12} is not known to be produced by wild-type *E. coli* [36] which must use active transport to obtain it from their environment [37]. Likewise, *E. coli* are not known to produce IAA which is typically synthesized by soil bacteria (such as *Azospirillum*) [38]. Although the specific mechanisms discussed by de-Bashan et al. and Croft et al. are unlikely explanations for symbiosis between *C. minutissima* and *E. coli*, co-factors that assist enzymes specifically involved with organic carbon uptake and central metabolism could play a role in the observed symbiosis. Co-factors such as thiamine, riboflavin, pantothenic acid, biotin, and folic acid are synthesized by *Chlorella* and play a role in central metabolism [39,40]. However, the effect of supplementation of these co-factors has not been well studied in mixotrophic algae cultures.

Another possible symbiotic mechanism is complimentary nutrient uptake and product secretion by the two organisms. For example, carbon dioxide evolution by *E. coli* in exchange for oxygen from algae could enhance overall growth rates in co-cultures. Such a relationship has been widely discussed as it pertains to wastewater treatment [41,42]. This scenario is most plausible in a mixotrophic environment (as opposed to autotrophic conditions) since sufficient organic material must be present for significant bacterial CO_2 production. Likewise, other products of primary metabolism could play a role in the observed symbiosis including metabolites secreted into the media by *C. minutissima* and *E. coli*. The nature of the symbiotic relationship is the subject of ongoing investigation.

Effects of substrate on biomass growth

Productivities achieved in axenic *C. minutissima* cultures were comparable to those achieved in studies using other *Chlorella* strains. Liang et al. [5] achieved 254 mg/L/day, 102 mg/L/day, and 87 mg/L/day growing *Chlorella vulgaris* on 1% glucose, glycerol, and acetate respectively compared to 99, 123, and 60 mg/L/day in this study. Wan et al. [6] achieved 56 mg/L/day with *Chlorella sorokiniana* using 10 g/L glucose. Axenic cultures' substrate uptake appeared to be inhibited after 1–2 grams of substrate were utilized. This in turn, is the likely explanation for the observed saturation growth kinetics. Graphs presented by Liang et al. also indicated saturation behavior in mixotrophic cultures but no indication was made as to the cause.

The fraction of biomass as *C. minutissima* in co-cultures was significantly higher at 10 g/L substrate than it was at 2 g/L substrate, contrary to our hypothesis. *E. coli* biomass density, whether grown on its own or in co-cultures with *C. minutissima*, appeared to not exceed ~0.5 g/L just as viable cell counts appeared to plateau at ~10^9 CFU/ml even though the carbon substrate was not exhausted. Monod identified three factors that can limit bacterial growth: 1) exhaustion of nutrients, 2) accumulation of metabolic products, and 3) changes in pH or ion equilibrium [43]. The plateau in *E. coli* growth could be related to oxygen limitation in spite of the presence of photosynthetic algae in co-cultures, particularly during the dark period. Lee et al. found that dissolved oxygen levels in mixotrophic *Chlorella* cultures could drop as low as 50% of saturation, even in the presence of light [44]. Products of *E. coli* anaerobic metabolism were not detected and were unlikely to play an inhibiting role, however, significant changes in pH over the culture period (despite adjustment) likely inhibited *E. coli* growth. Glass et al. showed that *E. coli* growth was inhibited below pH 6 and above pH 9, and complete inactivation was achieved at pH 4 [45]. In contrast, *C. minutissima* did not appear to have this limitation when grown in co-culture with *E. coli*. Once the maximum *E. coli* level was reached, additional culture growth appeared to come exclusively from *C. minutissima*. The majority of co-cultures provided with 10 g/L substrate were still growing at the end of the five day period suggesting that they still had not reached a biomass density limit. This process was likely truncated due to substrate limitation in the 2 g/L cultures and may explain why algae constituted a greater fraction of biomass in 10 g/L cultures (94% with glucose) than 2 g/L cultures (53% with glucose).

The relatively low efficiency of substrate utilization observed in this study (13–40%) indicates that there is room for improvement in terms of optimizing mixotrophic cultures. For reference, anaerobic fermentation of glucose to ethanol regularly yields a conversion efficiency, based on enthalpy of combustion, approaching 70% (1.43 mol ethanol per mol glucose) [46]. Mixotrophic algae cultures ideally operate as aerobic systems in which intracellular biofuel precursors are produced as opposed to a secreted anaerobic product. Acid formation by *C. minutissima* appeared to correlate with uptake of glucose and glycerol but common anaerobic metabolites were not identified in the media. Future work is required to identify products secreted by *C. minutissima* that are specifically linked to substrate consumption. Axenic *E. coli* cultures secreted detectable formic and acetic acids into the media which are common bi-products of *E. coli* fermentation [47], however, these products were not detected in the co-cultures in spite of the pH decline observed. One explanation is that the two organisms were exchanging metabolites that were secreted into the media (e.g. *C. minutissima* consumed acetate secreted by *E. coli* when grown on glucose) but this explanation requires further investigation.

Effect of substrate and kanamycin on biomass composition

Cultures grown on glucose and glycerol had higher starch levels than those grown on acetate. Glucose and glycerol enter the glycolytic pathway upon phosphorylation by hexokinase and glycerol kinase respectively whereas acetate enters central metabolism as acetyl-CoA [27]. Synthesis of glycolytic intermediates from acetyl-CoA can be accomplished through OAA production via the glyoxylate cycle followed by gluconeogenesis [27]. The gluconeogenic pathway consumes net ATP, which could explain the low starch content in acetate cultures. For axenic *C. minutissima*, autotrophic cultures were found to have higher total lipid content than mixotrophic cultures. This is consistent with the findings of Chen and Johns who observed decreased crude lipid in *Chlorella* when glucose concentration was increased but the cause was unclear [48]. Interestingly, they also observed increased fatty acid content under higher glucose concentrations.

In co-cultures, the substrate concentration appeared to affect the lipid content modestly, and starch content significantly. Specifically, the 10 g/L substrate condition led to large and statistically significant increases in starch content. The likely explanation is that *C. minutissima* was primarily storing the excess substrate as starch rather than lipids. This effect was suppressed in axenic *C. minutissima* cultures but the effect was likely due to kanamycin rather than any effect of *E. coli*. Axenic *C. minutissima* grown on glucose in the absence of kanamycin had starch levels comparable to those found in the co-cultures.

Kanamycin is an inhibitor of the 30 S ribosomal unit found in prokaryotes. Any effect it has on eukaryotic algae is likely to be limited to protein synthesis in the chloroplast. Little research has been conducted on protein synthesis in chloroplasts of *C. minutissima* but synthesis of chloroplast proteins are often shared between the nucleus and chloroplast nucleoids in plants and algae [49]. Theoretically, genes associated with starch synthesis and degradation could be housed in the nucleoids whereas those associated with lipids could be housed in the nucleus. Commercial algae production is unlikely to employ antibiotics unless very high value products are produced. For this reason, the effect observed here has little commercial relevance.

Implications for large-scale cultivation

Until recently, algae have been considered primarily as a tertiary wastewater treatment process in which the majority of organic carbon is consumed in upstream processes [11,50,51]. However, recent success with cultivating algae monocultures on primary wastewaters has led some researchers to conclude that algae can be used for secondary wastewater treatment [52]. *Chlorella* in particular has been identified as a promising candidate for organic carbon removal due to its mixotrophic capabilities [52]. However, these studies overlook the potentially significant impact that heterotrophs would play in such systems. Thus, improved understanding of algal-bacterial interaction in mixotrophic systems is required.

Although the present findings are not representative of the complexity found in a wastewater treatment process, they do shed light on important organism interactions that can take place in organic-rich waters. These results demonstrate that *Chlorella* can compete (and thrive) with a heterotrophic organism, even in nutrient-rich waters. Moreover, simultaneous cultivation of multiple organisms can lead to more productive outcomes (e.g. greater biofuel production and organic carbon removal from water as in the present study) than cultivation of algae monocultures, which have comprised the majority of academic research to date [5,9–11]. Other researchers have shown the benefits of co-

culturing multiple photosynthetic organisms for improved removal of nitrogen, phosphorus, [53] and organic acids [54]. The results of these two studies highlighted the importance of ecological niche in mixed cultures where organisms preferentially consumed different substrates [54] and occupied different culture zones within the reactors [53]. Additional research into co-cultures on various algal and bacterial species is required in order to broaden understanding of microbial interactions in algal cultures.

Despite the productivity benefits of algal-bacterial co-cultures, challenges such as flocculation and bio-fouling remain. Because the fouling in the present study did not begin until culture growth had reached the late-log phase, it is unlikely that it played a significant role in the elevated productivity of co-cultures in the first few days of growth. However, it may have affected cell growth and composition during the final days of the culture period. Discussion of bacterial flocculation in the literature suggests that bacteria can cause algae to flocculate [55,56], indicating that this phenomenon may not be limited to the present study. Raszka et al. discuss the role of polymers secreted by certain bacteria in the flocculation process [55]. Although bio-fouling due to coagulation of co-cultures proved challenging in air-lift reactors, coagulation may provide benefits in large-scale systems. There is great interest in using biological flocculation as an inexpensive harvesting system for microalgal cultures [57]. Salim et al. point out that biological flocculating systems can eliminate the need for costly separation processes required to recover chemical flocculants [57]. The challenge, however, is managing the process such that both high productivity and high harvesting efficiency can be achieved.

Conclusions

The results of the present study indicate that co-culture of the green algae *C. minutissima* with *E. coli* under mixotrophic conditions can enhance algal productivity. Although the co-culture represents a simplification of a "real" algae cultivation system, the results provide insight into how algae can respond to a heterotrophic organism. Future work is necessary to further elucidate the complex interactions between *E. coli* and *C. minutissima*, however. Further research on co-cultures of different algal and bacterial species can provide insight into the types of species interactions that could take place in scaled-up systems. These results, in turn, should lead to improved management practices for mixotrophic algae cultivation.

Supporting Information

Figure S1 Growth curves of axenic *C. minutissima* on glucose. Bars are standard deviations.

Figure S2 Growth curves of axenic *C. minutissima* on glycerol. Bars are standard deviations.

Figure S3 pH of media over time in cultures of axenic *C. minutissima* grown on glucose.

Figure S4 pH of media over time in cultures of axenic *C. minutissima* grown on acetate.

Figure S5 Viable *E. coli* in co-culture grown on glucose.

Figure S6 Viable *E. coli* in co-culture grown on glycerol.

Figure S7   Viable *E. coli* in co-culture grown on acetate.

Figure S8   Growth curves of axenic *E. coli* on acetate. Bars are standard deviations. Large variation was observed at 2 g/L acetate at 120 hours since each culture appeared to enter exponential growth at a different time.

Figure S9   Growth curves of axenic *E. coli* on glucose. Bars are standard deviations. At 96 hours, 0.2 g/L glucose remained in cultures originally supplied with 2 g/L glucose.

Figure S10   pH of media over time in cultures of axenic *E. coli* grown on glucose. pH was adjusted every 24 hours.

Figure S11   Growth curves of axenic *E. coli* on glycerol. Bars are standard deviations.

Acknowledgments

The authors wish to thank Dr. Larry Joh and Lauren Jabusch for their review and comments on the manuscript. The authors also thank Professor David Slaughter for his insight on the statistical analysis methods.

Author Contributions

Conceived and designed the experiments: BH JV. Performed the experiments: BH. Analyzed the data: BH. Contributed reagents/materials/analysis tools: JV. Wrote the paper: BH JV.

References

1. Sheehan J, Dunahay T, Benemann J, Roessler P (1998) A Look Back at the U.S. Department of Energy's Aquatic Species Program—Biodiesel from Algae. Golden, CO: National Renewable Energy Laboratory.
2. Tanadul OU, Vandergheynst JS, Beckles DM, Powell AL, Labavitch JM (2014) The impact of elevated CO concentration on the quality of algal starch as a potential biofuel feedstock. Biotechnol Bioeng: In Press.
3. Davis R, Aden A, Pienkos PT (2011) Techno-economic analysis of autotrophic microalgae for fuel production. Applied Energy 88: 3524–3531.
4. Benemann J, Oswald WJ (1996) Systems and Economic Analysis of Microalgae Ponds for Conversion of CO_2 to Biomass. Berkeley, CA: University of California Berkeley.
5. Liang Y, Sarkany N, Cui Y (2009) Biomass and lipid productivities of *Chlorella vulgaris* under autotrophic, heterotrophic, and mixotrophic growth conditions. Biotechnology Letters 31: 1043–1049.
6. Wan M, Liu P, Xia J, Rosenberg JN, Oyler GA, et al. (2011) The effect of mixotrophy on microalgal growth, lipid content, and expression levels of three pathway genes in *Chlorella sorokiniana*. Applied Microbiology and Biotechnology 91: 835–844.
7. Yan R, Zhu D, Zhang Z, Zeng Q, Chu J (2012) Carbon metabolism and energy conversion of *Synechococcus sp.* PCC 7942 under mixotrophic conditions: comparison with photoautotrophic condition. Journal of Applied Phycology 24: 657–668.
8. Clarens AF, Resurreccion EP, White MA, Colosi LM (2010) Environmental Life Cycle Comparison of Algae to Other Bioenergy Feedstocks. Environmental Science and Technology 44: 1813–1819.
9. Wang H, Xiong H, Hui Z, Zeng X (2012) Mixotrophic cultivation of *Chlorella pyrenoidosa* with diluted primary piggery wastewater to produce lipids. Bioresource Technology 104: 215–230.
10. Zhou W, Min M, Li Y, Hu B, Ma X, et al. (2012) A hetero-photoautotrophic two-stage cultivation process to improve wastewater nutrient removal and enhance algal lipid accumulation. Bioresource Technology 110: 448–455.
11. Sawayama S, Minowa T, Dote Y, Yokoyama S (1992) Growth of the hydrocarbon-rich microalga *Botryococcus braunii* in secondarily treated sewage. Applied Microbiology and Biotechnology 38: 135–138.
12. Woertz I, Feffer A, Lundquist T, Nelson Y (2009) Algae grown on dairy and municipal wastewater for simultaneous nutrient removal and lipid production for biofuel feedstock. Journal of Environmental Engineering 135: 1115–1122.
13. de-Bashan LE, Bashan Y, Moreno M, Lebsky VK, Bustillos JJ (2002) Increased pigment and lipid content, lipid variety, and cell and population size of the microalgae *Chlorella spp.* when co-immobilized in alginate beads with the microalgae-growth-promoting bacterium *Azospirillum brasilense*. Canadian Journal of Microbiology 48: 514–521.
14. de-Bashan LE (2008) Involvement of the indole-3-acetic acid produced by the growth-promoting bacterium *Azospirillum spp.* in promoting growth in *Chlorella vulgaris*. Journal of Phycology 44: 938–947.
15. Lebsky VK, Gonzalez-Bashan LE, Bashan Y (2001) Ultrastructure of interaction in alginate beads between the microalga *Chlorella vulgaris* with its natural associative bacterium *Phyllobacterium myrsinacearum* and with the plant growth promoting bacterium *Azospirillum brasilense*. Canadian Journal of Microbiology 47: 1–8.
16. Gladu PK, Patterson GW, Wikfors GH, Smith BC (1995) Sterol, fatty acid, and pigment characteristics of UTEX 2341, a marine eustigmatophyte identified previously as *Chlorella minutissima* (Chlorophyceae). Journal of Phycology 31: 774–777.
17. Vazhappilly R, Chen F (1998) Eicosapentaenoic acid and docosahexaenoic acid production potential of microalgae and their heterotrophic growth. Journal of the American Oil Chemists' Society 75: 393–397.
18. van der Drift C, van Seggelen E, Stumm C, Hol W, Tuinte J (1977) Removal of *Escherichia coli* in wastewater by activated sludge. Appl Environ Microbiol 34: 315–319.
19. ATCC (2013) ATCC Medium 5: Sporulation Agar.
20. Zhang K, Farahbakhsh K (2007) Removal of native coliphages and coliform bacteria from municipal wastewater by various wastewater treatment processes: Implications to water reuse. Water Research 41: 2816–2824.
21. Guardabassi L, Wong DMALF, Dalsgaard A (2002) The effects of tertiary wastewater treatment on the prevalance of antimicrobial resistant bacteria. Water Research 36: 1955–1964.
22. Sforza E, Cipriani R, Morosinotto T, Bertucco A, Giacometti GM (2012) Excess CO_2 supply inhibits mixotrophic growth of *Chlorella protothecoides* and *Nannochloropsis salina*. Bioresource Technology 104: 523–529.
23. Cheng Y-S, Zheng Y, VanderGheynst J (2011) Rapid quantitative analysis of lipid using a colorimetric method in microplate format. Lipids 46: 95–103.
24. Miller GL (1959) Use of dinitrosalicylic acid reagent for determination of reducing sugar. Anal Chem 31: 426–428.
25. Oppliger A, Charriere N, Droz PO, Rinsoz T (2008) Exposure to bioaerosols in poultry houses at different stages of fattening; use of real-time PCR for airborne bacterial quantification. Ann Occup Hyg 52: 405–412.
26. Rinttila T, Kassinen A, Malinen E, Krogius L, Palva A (2004) Development of an extensive set of 16S rDNA-targeted primers for quantification of pathogenic and indigenous bacteria in faecal samples by real-time PCR. J Appl Microbiol 97: 1166–1177.
27. Garrett RH, Grisham CM (2012) Biochemistry, 5th Edition. Mason, Ohio: Cengage Learning. pp. 578–730.
28. Atkins P, Paula Jd (2002) Atkins' Physical Chemistry. New York, New York: Oxford University Press. pp. 1077.
29. Myers JL, Well AD, Jr RFL (2010) Research Design and Statistical Analysis. 3rd ed. New York: Routledge, Taylor & Francis Group. pp. 191–193.
30. Devore JL (2008) Probability and Statistics for Engineering and the Sciences, Seventh Edition. Belmont, CA: Thomson Higher Education. pp. 391–392.
31. Box GEP, Cox DR (1964) An Analysis of Transformations. Journal of the Royal Statistical Society Series B 26: 211–252.
32. Scott D (2014) Box-Cox Transformations. In: Lane D, editor. Introduction to Statistics. Online: Rice University and University of Houston.
33. Kind T, Meissen JK, Yang D, Nocito F, Vaniya A, et al. (2012) Qualitative analysis of algal secretions with multiple mass spectrometric platforms. J Chromatogr A 1244: 139–147.
34. Bajguz A, Piotrowska-Niczyporuk A (2013) Synergistic effect of auxins and brassinosteroids on the growth and regulation of metabolite content in the green alga *Chlorella vulgaris* (Trebouxiophyceae). Plant Physiol Biochem 71: 290–297.
35. Croft MT, Lawrence AD, Raux-Deery E, Warren MJ, Smith AG (2005) Algae acquire vitamin B12 through a symbiotic relationship with bacteria. Nature 438: 90–93.
36. Roessner CA, Spencer JB, Ozaki S, Min CH, Atshaves BP, et al. (1995) Overexpression in *Escherichia coli* of 12 Vitamin B12 Biosynthetic Enzymes. Protein Expression and Purification 6: 155–163.
37. Bradbeer C, Woodrow ML (1976) Transport of vitamin B12 in *Escherichia coli*: energy dependence. Journal of Bacteriology 128: 99–104.
38. Bianco C, Imperlini E, Calogero R, Senatore B, Amoresano A, et al. (2006) Indole-3-acetic acid improves *Escherichia coli's* defences to stress. Arch Microbiol 185: 373–382.
39. Pratt R, Johnson E (1965) Production of thiamine, riboflavin, folic acid, and biotin by *Chlorella vulgaris* and *Chlorella pyrenoidosa*. J Pharm Sci 54: 871–874.
40. Pratt R, Johnson E (1966) Production of pantothenic acid and inositol by *Chlorella vulgaris* and *C. pyrenoidosa*. J Pharm Sci 55: 799–802.
41. Oswald WJ, Gotaas HB, Ludwig HF, Lynch V (1953) Algae Symbiosis in Oxidation Ponds: III Photosynthetic Oxygenation. Sewage and Industrial Wastes 25: 692–705.
42. Humenik FJ, Hanna Jr GP (1971) Algal-bacterial symbiosis for removal and conservation of wastewater nutrients. Journal (Water Pollution Control Federation): 580–594.

43. Monod J (1949) The growth of bacterial cultures. Annual Review of Microbiology 3: 371–394.

44. Lee Y-K, Ding S-Y, Hoe C-H, Low C-S (1996) Mixotrophic growth of *Chlorella sorokiniana* in outdoor enclosed photobioreactor. Journal of Applied Phycology 8: 163–169.

45. Glass KA, Loeffelholz JM, Ford JP, Doyle MP (1992) Fate of *Escherichia coli* O157:H7 as affected by pH or sodium chloride and in fermented, dry sausage. Appl Environ Microbiol 58: 2513–2516.

46. Albers E, Larsson C, Liden G, Niklasson C, Gustafsson L (1996) Influence of the Nitrogen Source on *Saccharomyces cerevisiae* Anaerobic Growth and Product Formation. Applied and Environmental Microbiology 62: 3187–3195.

47. Clark DP (1989) The fermentation pathways of *Escherichia coli*. FEMS Microbiology Reviews 63: 223–234.

48. Chen F, Johns M (1991) Effect of C/N ratio and aeration on the fatty acid composition of heterotrophic *Chlorella sorokiniana*. Journal of Applied Phycology 3: 203–209.

49. Harris EH, Boynton JE, Gillham NW (1994) Chloroplast ribosomes and protein synthesis. Microbiological Reviews 58: 700–754.

50. Green FB, Bernstone LS, Lundquist TJ, Oswald WJ (1996) Advanced Integrated Wastewater Pond System for Nitrogen Removal. Water Science and Technology 33: 207–217.

51. Craggs RJ, Adey WH, Jessup BK, Oswald WJ (1996) A controlled stream mesocosm for tertiary treatment of sewage. Ecological Engineering 6: 149–169.

52. Wang L, Min M, Li Y, Chen P, Chen Y, et al. (2010) Cultivation of green algae *Chlorella sp.* in different wastewaters from municipal wastewater treatment plant. Appl Biochem Biotechnol 162: 1174–1186.

53. Silva-Benavides A, Torzillo G (2012) Nitrogen and phosphorus removal through laboratory batch cultures of microalga *Chlorella vulgaris* and cyanobacterium *Planktothrix isothrix* grown as monoalgal and as co-cultures. Journal of Applied Phycology 24: 267–276.

54. Ogbonna J, Yoshizawa H, Tanaka H (2000) Treatment of high strength organic wastewater by a mixed culture of photosynthetic microorganisms. Journal of Applied Phycology 12: 277–284.

55. Raszka A, Chorvatova M, Wanner J (2006) The role and significance of extracellular polymers in activated sludge. Part I: Literature review. Acta hydrochimica et hydrobiologica 34: 411–424.

56. Lee A, Lewis D, Ashman P (2009) Microbial flocculation, a potentially low-cost harvesting technique for marine microalgae for the production of biodiesel. Journal of Applied Phycology 21: 559–567.

57. Salim S, Bosma R, Vermuë M, Wijffels R (2011) Harvesting of microalgae by bio-flocculation. Journal of Applied Phycology 23: 849–855.

PERMISSIONS

The contributors of this book come from diverse backgrounds, making this book a truly international effort. This book will bring forth new frontiers with its revolutionizing research information and detailed analysis of the nascent developments around the world.

We would like to thank all the contributing authors for lending their expertise to make the book truly unique. They have played a crucial role in the development of this book. Without their invaluable contributions this book wouldn't have been possible. They have made vital efforts to compile up to date information on the varied aspects of this subject to make this book a valuable addition to the collection of many professionals and students.

This book was conceptualized with the vision of imparting up-to-date information and advanced data in this field. To ensure the same, a matchless editorial board was set up. Every individual on the board went through rigorous rounds of assessment to prove their worth. After which they invested a large part of their time researching and compiling the most relevant data for our readers.

The editorial board has been involved in producing this book since its inception. They have spent rigorous hours researching and exploring the diverse topics which have resulted in the successful publishing of this book. They have passed on their knowledge of decades through this book. To expedite this challenging task, the publisher supported the team at every step. A small team of assistant editors was also appointed to further simplify the editing procedure and attain best results for the readers.

Apart from the editorial board, the designing team has also invested a significant amount of their time in understanding the subject and creating the most relevant covers. They scrutinized every image to scout for the most suitable representation of the subject and create an appropriate cover for the book.

The publishing team has been an ardent support to the editorial, designing and production team. Their endless efforts to recruit the best for this project, has resulted in the accomplishment of this book. They are a veteran in the field of academics and their pool of knowledge is as vast as their experience in printing. Their expertise and guidance has proved useful at every step. Their uncompromising quality standards have made this book an exceptional effort. Their encouragement from time to time has been an inspiration for everyone.

The publisher and the editorial board hope that this book will prove to be a valuable piece of knowledge for researchers, students, practitioners and scholars across the globe.

LIST OF CONTRIBUTORS

David Mouillot
Laboratoire Ecologie des Systèmes Marins Côtiers UMR 5119, Université Montpellier 2, Montpellier, France

Sébastien Villéger
Laboratoire Evolution et Diversité Biologique UMR 5174
Université Paul Sabatier, Toulouse, France

Michael Scherer-Lorenzen
Faculty of Biology - Geobotany, University of Freiburg, Freiburg, Germany

Norman W. H. Mason
Landcare Research, Hamilton, New Zealand

Pedro H. de Paula Silva, Nicholas A. Paul, Rocky de Nys and Leonardo Mata
School of Marine and Tropical Biology & Centre for Sustainable Tropical Fisheries and Aquaculture, James Cook University, Townsville, Australia

Mitchell M. Sewell and Tongming Yin
Bioscience Division, Oak Ridge National Laboratory, Oak Ridge, Tennessee, United States of America

Wellington Muchero, Priya Ranjan, Lee E. Gunter, Timothy J. Tschaplinski and Gerald A. Tuskan
BioEnergy Science Center, Oak Ridge National Laboratory, Oak Ridge, Tennessee, United States of America

Nicolas von Alvensleben, Katherine Stookey and Marie Magnusson
School of Marine and Tropical Biology, James Cook University, Townsville, Queensland, Australia
Centre for Sustainable Fisheries and Aquaculture, James Cook University, Townsville, Queensland, Australia

Kirsten Heimann
School of Marine and Tropical Biology, James Cook University, Townsville, Queensland, Australia
Centre for Sustainable Fisheries and Aquaculture, James Cook University, Townsville, Queensland, Australia
Comparative Genomics Centre, James Cook University, Townsville, Queensland, Australia
Centre for Biodiscovery and Molecular Development of Therapeutics, James Cook University, Townsville, Queensland, Australia

Danielle M. Warfe and Michael M. Douglas
Research Institute for the Environment and Livelihoods, Charles Darwin University, Darwin, Northern Territory, Australia

Timothy D. Jardine
Toxicology Centre, University of Saskatchewan, Saskatoon, Saskatchewan, Canada
Australian Rivers Institute, Griffith University, Nathan, Queensland, Australia

Stuart E. Bunn
Australian Rivers Institute, Griffith University, Nathan, Queensland, Australia

Bradley J. Pusey
Australian Rivers Institute, Griffith University, Nathan, Queensland, Australia
Centre of Excellence in Natural
Centre of Excellence in Natural Resource Management, The University of Western Australia, Albany, Western Australia, Australia

Neil E. Pettit and Peter M. Davies
Centre of Excellence in Natural Resource Management, The University of Western Australia, Albany, Western Australia, Australia

Stephen K. Hamilton
Kellogg Biological Station and Department of Zoology, Michigan State
University, Hickory Corners, Michigan, United States of America

Li Jiang, Tingcheng Zhu, Jixun Guo and Wei Sun
Key Laboratory for Vegetation Ecology, Ministry of Education, Institute of Grassland Science, Northeast Normal University, Changchun, Jilin Province, P. R. China

Rui Guo
Institute of Environment and Sustainable Development in Agriculture, Chinese Academy of Agricultural Sciences, Key Laboratory of Dryland Agriculture, Ministry of Agriculture, Beijing, P. R. China

Xuedun Niu
Key Laboratory of Molecular Enzymology and Engineering of Ministry of Education, College of Life Sciences, Jilin University, Changchun, Jilin Province, P. R. China

Rumiko Kajihara, Seiichiro Shibanuma and Shigeru Montani
Graduate School of Environmental Science, Hokkaido University, Sapporo, Japan

Tomohiro Komorita
Graduate School of Environmental Science, Hokkaido University, Sapporo, Japan
Faculty of Environmental and Symbiotic Science, Prefectural University of Kumamoto

Hiroaki Tsutsumi
Faculty of Environmental and Symbiotic Science, Prefectural University of Kumamoto

Toshiro Yamada
Nishimuragumi Co. Ltd., Hokkaido, Japan

Rebecca J. Lawton, Rocky de Nys and Nicholas A. Paul
School of Marine and Tropical Biology, James Cook University, Townsville, Queensland, Australia

KathiJo Jankowski, Daniel E. Schindler and M. Claire Horner-Devine
University of Washington, School of Aquatic and Fisheries Sciences, Seattle, Washington, United States of America

Jocelyn L. Aycrigg and Anne Davidson
National Gap Analysis Program, Department of Fish and Wildlife Sciences, University of Idaho, Moscow, Idaho, United States of America

Leona K. Svancara
Idaho Department of Fish and Game, Moscow, Idaho, United States of America

Kevin J. Gergely
United States Geological Survey Gap Analysis Program, Boise, Idaho, United States of America

Alexa McKerrow
United States Geological Survey Gap Analysis Program, Raleigh, North Carolina, United States of America

J. Michael Scott
Department of Fish and Wildlife Sciences, University of Idaho, Moscow, Idaho, United States of America

Leigh W. Tait and David R. Schiel
Marine Ecology Research Group, School of Biological Sciences, University of Canterbury, Christchurch, New Zealand

Rebecca J. Lawton, Rocky de Nys and Nicholas A. Paul
School of Marine and Tropical Biology, James Cook University, Townsville, Queensland, Australia

Stephen Skinner
Molonglo Catchment Group, Fyshwick, Australia

Meifang Zhao, Wenhua Xiang, Dalun Tian, Xiangwen Deng and Zhihong Huang
Faculty of Life Science and Technology, Central South University of Forestry and Technology, Changsha, Hunan, People's Republic of China

Xiaolu Zhou
Institute of Environment Sciences, Department of Biological Sciences, University of Quebec at Montreal, Montreal, Quebec, Canada

Changhui Peng
Institute of Environment Sciences, Department of Biological Sciences, University of Quebec at Montreal, Montreal, Quebec, Canada
Faculty of Life Science and Technology, Central South University of Forestry and Technology, Changsha, Hunan, People's Republic of China

Enqing Hou
Key Laboratory of Vegetation Restoration and Management of Degraded Ecosystems, South China Botanical Garden, Chinese Academy of Sciences, Guangzhou, China
Environmental Futures Centre, Griffith School of Environment, Griffith University, Nathan, Queensland, Australia
University of Chinese Academy of Sciences, Beijing, China

Dazhi Wen
Key Laboratory of Vegetation Restoration and Management of Degraded Ecosystems, South China Botanical Garden, Chinese Academy of Sciences, Guangzhou, China
University of Chinese Academy of Sciences, Beijing, China

Chengrong Chen
Environmental Futures Centre, Griffith School of Environment, Griffith University, Nathan, Queensland, Australia

Megan E. McGroddy
Department of Environmental Sciences, NASA/ University of Virginia, Charlottesville, Virginia, United States of America

Elisabeth Marquard
UFZ – Helmholtz Centre for Environmental Research,
Department of Conservation Biology, Leipzig, Germany

Bernhard Schmid and Enrica De Luca
Institute of Evolutionary Biology and Environmental
Studies and Zurich-Basel Plant Science Centre,
University of Zurich, Zurich, Switzerland

Christiane Roscher
UFZ – Helmholtz Centre for Environmental Research,
Department of Community Ecology, Halle, Germany

Karin Nadrowski
Institute of Biology, University of Leipzig, Leipzig,
Germany

Alexandra Weigelt
Institute of Biology, University of Leipzig, Leipzig,
Germany
German Centre for Integrative Biodiversity Research
(iDiv) Halle-Jena-Leipzig, Leipzig, Germany

Wolfgang W. Weisser
Terrestrial Ecology/Department of Ecology and
Ecosystem Management,
Technische Universität München, Freising-
Weihenstephan, Germany

Gunnar Brandt
Systems Ecology, Leibniz Center for Tropical Marine
Ecology, Bremen, Germany

Agostino Merico
Systems Ecology, Leibniz Center for Tropical Marine
Ecology, Bremen, Germany
School of Engineering and Science, Jacobs University,
Bremen, Germany

Björn Vollan
Institutional & Behavioural Economics, Leibniz Center
for Tropical Marine Ecology, Bremen, Germany
Institute of Public Finance, University of Innsbruck,
Innsbruck

Achim Schlüter
Institutional & Behavioural Economics, Leibniz Center
for Tropical Marine Ecology, Bremen, Germany
School of Humanities and Social Sciences, Jacobs
University, Bremen, Germany

Brigitta I. van Tussenbroek
Instituto de Ciencias del Mar y Limnologı´a,
Universidad Nacional Auto´noma de Me´xico, Cancu´
n, Mexico

**Jorge Corte´s, Ana C. Fonseca and Jimena Samper-
Villarreal**
Centro de Investigacio´n en Ciencias del Mar y
Limnologı´a (CIMAR), Universidad de Costa Rica, San
Pedro, Costa Rica

**Rachel Collin, Hector M. Guzma´n and Gabriel E.
Ja´come**
Smithsonian Tropical Research Institute, Panama,
Republic of Panama

Peter M. H. Gayle
Discovery Bay Marine Laboratory, Discovery Bay,
Jamaica,

Rahanna Juman
Institute of Marine Affairs, Trinidad, Trinidad and
Tobago

Karen H. Koltes
Office of Insular Affairs, Department of the Interior,
Washington DC, United States of America

Hazel A. Oxenford
CERMES, University of the West Indies, Barbados,
West Indies

Alberto Rodrı´guez-Ramirez
Instituto de Investigaciones Marinas y Costeras
(INVEMAR), Santa Marta, Colombia

Struan R. Smith
Bermuda Biological Station for Research, St. George,
Bermuda

John J. Tschirky
Garrett Park, Maryland, United States of America

Ernesto Weil
Department of Marine Sciences, University of Puerto
Rico, Mayaguez, Puerto Rico, United States of America

**Fei Peng, Quangang You, Manhou Xu, Jian Guo, Tao
Wang and Xian Xue**
Key Laboratory of Desert and Desertification, Chinese
Academy of Sciences, Cold and Arid Regions
Environmental and Engineering Research Institute,
Chinese Academy of Sciences, Lanzhou, China

Amy S. Groesbeck and Anne K. Salomon
School of Resource and Environmental Management,
Simon Fraser University, Burnaby, British Columbia,
Canada

Kirsten Rowell
Department of Biology, University of Washington, Seattle, Washington, United States of America

Dana Lepofsky
Department of Archaeology, Simon Fraser University, Burnaby, British Columbia, Canada

Kaan Yilancioglu, Inanc Pastirmaci, Batu Erman and Selim Cetiner
Faculty of Engineering and Natural Sciences, Sabanci University, Orhanlı, Istanbul, Turkey

Murat Cokol
Faculty of Engineering and Natural Sciences, Sabanci University, Orhanlı, Istanbul, Turkey
Sabanci University Nanotechnology Research and Application Center,
Orhanlı, Istanbul, Turkey

Brendan T. Higgins, Jean S. VanderGheynst
Biological and Agricultural Engineering, University of California Davis, Davis, California, United States of America

Index